Hans-Georg Elias

An Introduction to
Polymer Science

Distribution:

VCH, P. O. Box 10 11 61, D-69469 Weinheim, Federal Republic of Germany

Switzerland: VCH P.O. Box, CH-4020 Basel, Switzerland

United Kingdom and Ireland: VCH, 8 Wellington Court, Cambridge CB1 1HZ United Kingdom

USA and Canada: VCH, 333 7th Avenue, New York, NY 10001, USA

Japan: VCH, Eikow Building, 10-9 Hongo 1-chome, Bunkyo-ku, Tokyo 113, Japan

ISBN 3-527-28790-6

Hans-Georg Elias

An Introduction to Polymer Science

VCH Weinheim · New York · Basel · Cambridge · Tokyo

Prof. Dr. Hans-Georg Elias
Michigan Molecular Institute
1910 West St. Andrews Road
Midland, MI 48640
USA

This book was carefully produced. Nevertheless, author and publishers do not warrant the information contained therein to be free of errors. Readers are advised to keep in mind that statements, data, illustrations, procedural details or other items may inadvertently be inaccurate.

1st edition 1997

Published jointly by
VCH Verlagsgesellschaft mbH, Weinheim (Federal Republic of Germany)
VCH Publishers, Inc., New York, NY (USA)

Editorial Director: Dr. Barbara Böck
Production Manager: Dipl.-Wirt.-Ing. (FH) Bernd Riedel

Library of Congress Card No. applied for.
A CIP catalogue record for this book is available from the British Library

Die Deutsche Bibliothek — CIP Einheitsaufnahme

Elias, Hans-Georg:
An introduction to polymer science / Hans-Georg Elias. —
1. ed. — Weinheim ; New York ; Basel ; Cambridge ; Tokyo :
VCH, 1997
  ISBN 3-527-28790-6

Printing: Strauss Offsetdruck GmbH, D-69509 Mörlenbach
Bookbinding: J. Schäffer & Co. KG., D-67261 Grünstadt

Printed in the Federal Republic of Germany

Preface

Polymer science began with the discovery and synthesis of macromolecules. Physical methods were then employed to characterize these molecules and their applications as plastics, fibers, elastomers, coatings, adhesives, etc. Most introductory textbooks reflect this historical approach: they emphasize the chemistry of polymers and the characterization of polymers by physical methods.

My colleagues and I have long felt that this traditional approach does not do justice to the interdisciplinary nature of polymer science which is firmly rooted not only in the chemistry of macromolecules but also in their physical chemistry and physics. A modern book should furthermore provide the student with some information on the three most important uses of synthetic polymers as elastomers, fibers, and plastics but should not neglect biopolymers either.

This book is therefore divided into three approximately equal sized parts covering the chemistry, physical chemistry and physics of polymers, followed by a part on technology. The parts are preceded by an Introduction (Chapter 1) which defines the most important terms and reflects on the discovery of macromolecules.

The **chemistry** sections first introduce the constitution, configuration and microconformation of macromolecules (Chapter 2). Syntheses are then presented in Chapter 3 covering chain-growth polymerizations and Chapter 4 dealing with step-growth polymerizations and polymer reactions. This division seems natural since chain polymerizations are most expediently discussed via kinetics whereas non-chain polymerizations such as polycondensation and polyaddition are better treated according to their individual chemistry. Chapter 4 also contains information on biological macromolecules with special emphasis on industrially important ones.

The next three chapters are concerned with the **physical chemistry** of single macromolecules: size and shape (Chapter 5), thermodynamics of polymer solutions (Chapter 6), and polymer hydrodynamics (Chapter 7). Experimental methods are mentioned in the context of the corresponding physical phenomena; they are not listed in the usual handbook style according to the information they deliver. Prior knowledge is assumed for instrumental methods used in organic chemistry (NMR, IR, UV, etc.). Methods specific to polymer science are described but not in detail due to the prescribed size of the book. The reader should also bear in mind that experimental methods cannot be learned from books but need hands-on experience.

The chapters on single molecules are followed by three chapters on **polymer physics**. Chapter 8 discusses the structure of polymer assemblies, i.e., melts, amorphous solids, crystalline solids, liquid-crystalline polymers, domain formation, etc. Chapter 9 describes not only thermal transitions and relaxations but also diffusion and permeation phenomena. Mechanical, electrical and optical properties of solid polymers are treated in Chapter 10.

Chapters 1-10 are rounded off by four chapters on selected aspects of **polymer technology**. These chapters on the three most important applications of polymers intend to show that elastomers (Chapter 12), fibers (Chapter 13) and plastics (Chapter 14) are not simply polymers but polymer systems comprising additives (Chapter 11).

Chapter 14 also discusses reinforced and rubber-modified plastics. Chapters 11-14 have many cross-references to Chapters 1-10 which show the complexity of such polymer systems and the effect of processing on properties.

The book employs recommended IUPAC nomenclature and symbols and SI units. Since American industry continues to use Anglo-Saxon units, Chapter 15 therefore lists SI units and conversions of Anglo-Saxon units, as well as abbreviations and acronyms of polymer names and some important trade names.

Whenever possible, physical equations are not just presented but derived step by step from theories. Exceptions are those few equations that result from sophisticated theories requiring an elaborate mathematical apparatus. In many cases, numerical examples are given so that the student gets some feel for numerical values.

The book was envisioned for students with a basic knowledge of chemistry, physical chemistry and physics. Since it intends not only to acquaint the student with the basic terms, facts, methods, and theories but also to provide a bridge to an understanding of the primary literature, it uses a more formal, compact style than that commonly adopted by other introductory textbooks. Important terms are emphasized by bold type. Since the language of science is part of our heritage, explanations of the Greek, Latin, French, German, etc., linguistic roots of scientific terms are explained throughout the book.

The book is based on a short German-language text published in 1996 but it is not a cover-to-cover translation: sections have been rearranged and new paragraphs, tables and graphs have been added. The German language edition has been thoroughly worked over, keeping in mind the differences in background, knowledge, and cultural heritage of students in the United States and in German-speaking countries based on my experience with graduate students and postdoctoral fellows from 15 countries.

I am indebted to my former colleagues at Michigan Molecular Institute who read and checked the "final" drafts of several chapters: Professors Petar R. Dvornic (all Chapters), Steven A. Keinath (Chapters 1-7), Dale J. Meier (parts of Chapter 10), Robert L. Miller (Chapters 1, 2, 8, 9), and Karel Solc (Chapters 1, 2, 5, and 6 plus all chapters of the German version),

Midland, Michigan, Summer 1996   Hans-Georg Elias

List of Symbols for Physical Quantities

Abbreviations for languages:
- E: English (American spelling)   F: French
- G: Greek   L: Latin

Conventions for symbols of chemical entities (IUPAC) unless noted otherwise:
- R: monovalent ligand, e.g. CH_3-, C_6H_5-
- Z: bifunctional atom or group, e.g., $-CH_2-$, $-(p-C_6H_4)-$
- Y: trifunctional atom or group, e.g., $-N<$
- X: tetrafunctional atom or group

Other conventions in this book:
A, B: monomers leading to monomeric units -a- and -b- respectively, *or* leaving parts of functional groups (e.g., -OH from -COOH)

L = AB: symbol for a leaving molecule, e.g., H_2O from -OH + HOOC-;

$p-C_6H_4$, pPh: in para position (1,4-position) substituted benzene ring (= *para*-phenylene)

Exponents

α exponent of the relationship between intrinsic viscosity and molar mass

ν exponent of the relationship between radius of gyration and molar mass (Flory exponent)

Indices:

1	solvent
2	solute (usually polymer)
a	atom
c	chain
crit	critical (occasionally as c)
cryst	crystalline
end	end group
f	free or functionality
G	glass transformation
i	*i*th component
i	initiation
M	melting
m	molar quantity
mix	mixing, mixture
mol	molecule
mon	monomer
n	number-related quantity
p	polymer, polymerization, or propagation
p	quantity at constant pressure

r   related to end-to-end distance
rel   relative
s   related to radius of gyration
T   quantity at constant temperature
tr   transfer
u   monomeric unit
V   quantity at constant volume
w   mass-related quantity
η   viscosity-related quantity

Physical quantities and units: Usually following IUPAC amd ISO recommendations, see I.Mills, T.Cvitas, K.Homann, N.Kallay, K.Kuchitsu, eds., (International Union of Pure and Applied Chemistry, Division of Physical Chemistry), "Quantities, Units and Symbols in Physical Chemistry", Blackwell Scientific Publications, Oxford 1988.

A area; A_c = cross-sectional area of a chain
A_2 second virial coefficient; A_3 = third virial coefficient
[A]   molar concentration of chemical compound A
a persistence length
a_T shift factor in the WLF equation

B bulk compliance
b bond length; b_{eff} = effective bond length

C electric capacitance
C heat capacity; C_p (isobaric); C_V (isochoric)
C number concentration (number of entities per total volume)
[C]   amount-of-substance concentration = amount of substance per total volume = molar concentration
C_j chain transfer constant (always with index, e.g., j = s (regulator))
C_N characteristic ratio; C_∞ = characteristic ratio at infinite molar mass

c concentration = mass concentration (= mass of substance per total volume)
c_p specific heat capacity at constant pressure

D diffusion coefficient
D tensile compliance
d diameter; d_{blob} = diameter of a blob, d_{sph} = diameter of a sphere

E energy, $E^{\ddagger}$ = activation energy
E tensile modulus (Young's modulus)

F force
f fraction (unless specified)

f functionality of a group or molecule

G Gibbs energy; e.g., $\Delta G_{mix,m}$ = molar Gibbs energy of mixing
G statistical weight fraction ($G_i \equiv g_i/\Sigma_i\, g_i$)
G shear modulus
G electrical conductance
g statistical weight
g branching parameter

H enthalpy; ΔH_{mix} = enthalpy of mixing, $\Delta H_{mix,m}$ = molar enthalpy of mixing
h Planck constant ($h \approx 6.626 \cdot 10^{-34}$ J s)

I intensity
I electric current

J shear compliance

K constant; K_n = equilibrium constant; K_ϑ = optical constant (scattering)
K bulk modulus
k rate constant (always with index); k_p = propagation rate constant, k_t = rate constant of termination, k_{tr} = rate constant of transfer
k_B Boltzmann constant ($k_B \approx 1.3805 \cdot 10^{-23}$ J K^{-1})

L length (always geometric); L_K = Kuhn length

M molar mass (mass per amount of substance); $\overline{M}_n$ = number-average molar mass, $\overline{M}_w$ = mass-average molar mass, M_{end} = molar mass of end group
M_r relative molecular mass (molecular weight; dimensionless); $\overline{M}_{r,n}$ = number-average molecular weight, $\overline{M}_{r,w}$ = weight-average molecular weight
m mass; m_{mol} = mass of a molecule

N number of entities (molecules, segments, groups, atoms, etc.)); N_{end} = number of end groups; N_g = number of lattice sites
N_A Avogadro constant ($N_A \approx 6.023 \cdot 10^{23}$ mol^{-1})
n amount of substance (in mol)
n refractive index

P permeability coefficient
P power
$P(\vartheta)$ scattering function
p extent of reaction; p_A = extent of reaction of group A
p conditional probability
p pressure

Q intermediate variable or constant, usually a ratio
Q electric charge (quantity of electricity)
$Q(\vartheta)$ parameter in the scattering function
q order of a moment
q intermediate variable or constant

R molar gas constant ($R \approx 8.314$ J K^{-1} mol^{-1})
R radius; R_d = Stokes radius, R_v = Einstein radius
R rate of reaction
R electrical surface resistance
R_0 Rayleigh constant at angle zero; R_ϑ = Rayleigh constant at angle ϑ
r spatial end-to-end distance of a chain, usually as $\langle r^2 \rangle^{1/2}$ with various indices;
         r_{cont} = physical contour length of a chain
r_o initial ratio of amounts of substances
r_A copolymerization parameter of chemical compound A

S entropy; ΔS_{mix} = entropy of mixing, $\Delta S_{mix,m}$ = molar entropy of mixing
S solubility coefficient
s radius of gyration (IUPAC)
s sedimentation coefficient
s Staverman coefficient

T thermodynamic temperature (in K); T_c = ceiling temperature, T_G = glass
         temperature, T_M = melting temperature
t time

U electric potential
u excluded volume

V volume; V_m = molar volume, $*V_m$ = partial molar volume
v specific volume; $*v$ = partial specific volume

w mass fraction (weight fraction)

X degree of polymerization; X_u = degree of polymerization of a unit (e.g.,
         monomeric unit, repeating unit, etc.); $\overline{X}_n$ = number-average degree of
         polymerization; $\overline{X}_w$ = mass-average degree of polymerization
x mole fraction (amount-of-substance fraction); x_u = mole fractions of units,
         x_i = mole fraction of isotactic diads, x_{ii} = mole fraction of isotactic triads, etc.

y yield of substance

z number of immediate neighbors
z dissymmetry (light scattering)

α	polarizability
α	linear expansion coefficient of materials or coils (α_s if radius of gyration, α_r if end-to-end distance, α_h if hydrodynamic dimensions)
β	cubic expansion coefficient
γ	shear deformation; $\dot{\gamma}$ = shear rate
γ	surface tension
Γ_s	degree of solvation
δ	solubility parameter
ε	elongation
ε	cohesion energy
ε_r	relative permittivity (formerly: dielectric constant)
η	viscosity; η_o = Newtonian viscosity, η_e = extensional viscosity, η_r = relative viscosity, η_{sp} = specific viscosity, $[\eta]$ = intrinsic viscosity
ϑ	angle, especially scattering angle
θ	torsional angle (conformational angle)
θ	temperature in °C (T if in K)
Θ	theta temperature
κ	isothermal (cubic) compressibility
λ	wavelength (λ_o = wavelength of incident light)
λ	heat conductivity
λ	draw ratio
Λ	aspect ratio
μ	moment of a distribution
μ	chemical potential
μ	Poisson ratio
ξ	frictional coefficient
Ξ	zip length
Π	osmotic pressure
ρ	density
ρ	volume resistivity
σ	tensile stress (= σ_{11})
σ	steric factor
σ	electrical conductance
σ_n	standard deviation with respect to numbers or amounts
ς	degree of coupling of chains
τ	bond angle, valence angle
τ	shear stress (= σ_{21})
Y	cohesion energy density
ϕ	volume fraction; ϕ_f = free volume fraction
Φ	Flory parameter; Φ_Θ = Flory constant
χ	Flory–Huggins interaction parameter
ψ	parameter in an entropy quantity
ω	angular velocity
Ω	thermodynamic probability

Contents

Chemistry

Physical Chemistry

Physics

Technology

1. Introduction

1.1. Fundamental Terms

Polymer science is concerned with the chemistry, physical chemistry and physics of **polymers,** a group of chemical substances composed of macromolecules. It provides the fundamental knowledge for the industrial application of polymers as plastics, fibers, elastomers, adhesives, coatings, etc. Polymer science is also one of the central sciences for molecular biology, biochemistry and biophysics.

A **macromolecule** (G: *makros* = large; L: *molecula* = small mass, diminutive of *moles* = mass) is a molecule composed of a great number of atoms; it thus has a high relative molecular mass or molecular weight. Soluble synthetic macromolecules possess molecular weights between several hundreds and several millions; some biological macromolecules even range into the milliards (American billions). No sharp dividing line exists between macromolecules and molecules of low molecular weight.

IUPAC (International Union of Pure and Applied Chemistry) treats the terms **macromolecule** and **polymer molecule** as synonyms (G: *polys* = many; *meros* = part). In a provisional recommendation (1994), a macromolecule is defined as:

"a molecule of high relative molecular mass, the structure of which essentially comprises the multiple repetition of units derived, actually or conceptually, from molecules of low relative molecular mass."

"Macromolecule" and "polymer molecule" are, however, neither literally nor traditionally synonyms. Linguistically, "macromolecule" contradicts itself since it denotes a small mass (*molecula*) that is large (*makros*). No matter how we define a "molecule", "macromolecule" simply indicates that it is a molecule of high relative molecular mass. Contrary to the IUPAC definition, the term "macromolecule" implies nothing about the structure. Enzyme molecules are, for example, macromolecules that are composed of hundreds or thousands of α-amino acid units –NH–CHR–CO– with up to 20 different substituents R. The α-amino acid units are, however, arranged in a non-repetitive sequence such as ala–gly–lys–leu–glu–trp...gly–gly–leu.

The term "polymer molecule", on the other hand, does imply a multiple repetition. Although "polymer molecule" literally denotes only a molecule that consists of many (*poly*) parts (*meros*), it has come to mean that (a) these parts are units derived from molecules of low relative molecular mass (so-called **mers**), that (b) there is only one species of mer (or only a few species of mers), and that (c) these mers by necessity must be present in multiple repetition. An example of a polymer molecule with equal mers is poly(glycine) $H{+}NH–CH_2–CO{]}_N OH$ where N denotes the number of mers (i.e., glycine units –NH–CH_2–CO–) per molecule. Examples of polymer molecules with few (here: three) species of mers are the polymers which result from the joint reaction (a so-called **copolymerization**) of ethene, propene and a diene.

"Macromolecule" and "polymer molecule" are therefore not synonyms. A "polymer molecule" is always a "macromolecule" but the reverse is not always the case.

"Polymer molecule" is also not synonymous with "polymer". A **polymer** is a *substance* composed of many polymer molecules; it is a **polymeric substance**. A **macromolecular substance** similarly consists of many macromolecules. Different symbols are used in this book to depict structural formulas of molecules and substances. For example, a poly(styrene) *molecule* is characterized by $-[CH_2\text{-}CH(C_6H_5)]_N-$ and the *substance* poly(styrene) by $-[CH_2\text{-}CH(C_6H_5)]_n-$ (endgroups omitted) since N represents numbers and n amounts of substances ("moles", see Chapter 15).

This distinction between *molecules* and *substances* is very important. In the solid state, for example, most properties are not properties of molecules *per se* but properties of substances, i.e., assemblies of molecules that interact physically.

Polymer molecules are (actually or conceptually) generated from **monomers** (G: *monos* = single, sole, alone; *meros* = part). For example, styrene $CH_2=CH(C_6H_5)$ is the monomer for poly(styrene) $-[CH_2\text{–}CH(C_6H_5)]_N-$ with the **monomer(ic)** unit or **mer** $-CH_2\text{–}CH(C_6H_5)-$. The number N of monomeric units per polymer molecule is called the **degree of polymerization. Macromonomers** are large monomers (but smaller than polymers); an example is $CH_2=CH\{COO[CH_2CH(C_6H_5)]_{25}H\}$.

The simplest macromolecules have **linear chains**. Examples are

$-S-S-S-S-$	$-[S]_N-$	polymeric sulfur
$-Si(CH_3)_2-Si(CH_3)_2-Si(CH_3)_2-$	$-[Si(CH_3)_2]_N-$	poly(dimethylsilane)
$-CH_2-CH_2-CH_2-CH_2-$	$-[CH_2]_N-$	poly(methylene)
$-CH_2-CH(CH_3)-CH_2-CH(CH_3)-$	$-[CH_2-CH(CH_3)]_N-$	poly(propene)
$-O-CH_2-CO-O-CH_2-CO-$	$-[O-CH_2-CO]_N-$	poly(glycolic acid)

poly(1,4-phenylene)

Linear chains are comprised of atoms or atomic groups that are linked to two other units, usually by covalent bonds. Examples of such **chain units** are $-S-$, $-O-$. $-CH_2-$, $-CH(CH_3)-$, and $-(C_6H_4)-$. Chain units of the same chain may be identical as in poly(methylene) or different as in poly(propylene) and poly(glycolic acid).

Chain units consist of chain atoms and their substituents. **Chain atoms** of the same chain may be all identical as in polymeric sulfur or poly(propene) or different as in poly(glycolic acid). They may be substituted or unsubstituted, either in the inorganic sense (polymeric sulfur) or in the organic sense (poly(methylene)).

Chains contain **constitutional units** composed of one chain unit or several chain units. "Constitutional unit" refers to the structural groups of a macromolecule and not to the structure of monomers or monomeric units. The constitutional units of polyamide 6 are thus not only $-NH(CH_2)CO-$ but also $-NH-$, $-CH_2-$, $-CO-$, $-NH-CH_2-$, $-CH_2-CH_2-$, $-CH_2-CO-$, $-CO-NH-$, $-NH-CH_2-CH_2-$, etc. (see also Table 1-1).

The chemical structure of poly(ε-caprolactam) is described by its **constitutional repeating unit** (CRU) $-NH(CH_2)_5CO-$. The CRU of this polymer contains 6 carbon atoms; hence, the polymer is also called polyamide 6, PA 6 or nylon 6.

Table 1-1 Monomers, monomeric units, constitutional units and constitutional repeating units of poly(ethylene) PE, poly(methylene) PM, and polyamides PA 6 and 66. Monomers are examples; PA 6 is synthesized industrially from ε-caprolactam and not from ε-aminocaproic acid.

Term	PE	PM	PA 6	PA 66
Process-based terms				
Monomers	$CH_2{=}CH_2$	CH_2N_2	$H_2N(CH_2)_5COOH$	$H_2N(CH_2)_6NH_2$ and $HOOC(CH_2)_4COOH$
Monomeric units	$CH_2{-}CH_2$	CH_2	$NH(CH_2)_5CO$	$NH(CH_2)_6NH$ and $CO(CH_2)_4CO$
Structure-based terms				
Chain atoms	C	C	N, C	N, C
Chain units	CH_2	CH_2	NH, CH_2, CO	NH, CH_2, CO
Constitutional units	CH_2	CH_2	NH, CH_2, CO, NHCO, $NHCH_2$, CH_2CO, etc.	NH, CH_2, CO, NHCO, $NHCH_2$, CH_2CO, etc.
CRU	CH_2	CH_2	$NH(CH_2)_5CO$	$NH(CH_2)_6NHCO(CH_2)_4CO$

The CRU is the smallest constitutional unit which, on repetition and with the same directional connection, generates a **regular macromolecule**. The CRU of poly(hexamethylene adipamide) is thus $-NH(CH_2)_6NH-CO(CH_2)_4CO-$. Since this constitutional repeating unit has 6 carbon atoms from the diamine unit and 6 from the adipic acid unit, it is also called polyamide 66, PA 66 or nylon 66.

PA 66 molecules possess two types of monomeric units if stoichiometric amounts of hexamethylene diamine $H_2N(CH_2)_6NH_2$ and adipic acid $HOOC(CH_2)_4COOH$ were used as starting materials but only one type of monomeric unit if PA 66 was generated from $H_2N(CH_2)_6NHCO(CH_2)_4COOH$ as the monomer.

Since the term *constitutional repeating unit* refers to the structure of the polymer molecule (structure-based term) whereas the term *monomeric unit* points to the units generated by the polymerization process (process-based term), a convention has been made with respect to the meaning of the term **degree of polymerization** (conventional symbol: X). This term is process-based; it thus refers to the number of monomeric units per polymer molecule and not to the number of repeating units per molecule.

For example, the molecule $H[NH(CH_2)_6NHCO(CH_2)_4CO]_N OH$ has a degree of polymerization of $X = N$ if it originated from N $H_2N(CH_2)_6NHCO(CH_2)_4COOH$ as monomers. The number of monomeric units is here identical with the number of repeating units. The degree of polymerization is however $X = 2 N$ if it was synthesized from N $H_2N(CH_2)_6NH_2$ molecules and N $HOOC(CH_2)_4COOH$ molecules since there are now two monomeric units per constitutional repeating unit.

Polymer nomenclature and polymer terms may be either **source-based, process-based**, or **structure-based** (see also Sections 2.1.3. and 2.1.4.). "Source-based" refers to the starting materials (most commonly monomers), "process-based" to the process by which the starting materials are converted into polymers, and "structure-based" to the chemical structure of the resulting products. A distinction between these terms is necessary because "chemical experience" based on low molecular weight chemistry is not always a good guideline in polymer science.

Linear chains possess two **end groups**. The meaning of this term always has to be defined. A polymer molecule $H\text{-}[\text{-}NH(CH_2)_6NH\text{-}CO(CH_2)_4CO\text{-}]_N\text{-}OH$ possesses the end groups H- and -OH if one considers the structural formula but the end groups $H_2N\text{-}$ and -COOH if one reflects on the chemical reactivity. In the latter case, the structure of this molecule must be written as

$$H_2N(CH_2)_6NH[CO(CH_2)_4CONH(CH_2)_6NH]_{N\text{-}2}CO(CH_2)_4COOH$$

End groups are often unknown. The structural formula of this particular molecule is then simply given as $\text{-}[\text{-}NH(CH_2)_6NH\text{--}CO(CH_2)_4CO\text{-}]_N$.

An **oligomer** (G: *oligos* = few) is correspondingly a substance composed of **oligomer molecules**, i.e., molecules with "few" constitutional units, usually less than twenty in synthetic polymers and less than several hundred in nucleic acids. A macromonomer is thus an oligomer with polymerizable groups.

Pre-polymers are polymers or oligomers whose molecules can react further to substances with higher molecular weights. Prepolymers are important as raw materials for so-called thermosetting resins (see Chapter 14).

1.2. History

Life cannot exist without biological macromolecules. Such naturally occurring macromolecules serve as carriers of information (nucleic acids), catalysts (enzymes), structural elements (fiber-forming proteins, cellulose), food reserves (glycogen, starch, polyesters), or transporting molecules (hemoglobin).

Since prehistoric times, man has used huge amounts of biopolymers as materials (see Table 1-2): Wood as a construction material, wool for clothing, and starch as a glue or thickener. Some biopolymers were chemically converted into other polymers in order to make them more resistant against the elements, for example, the collagen of hides into leather by tanning. Biological catastrophes and increasing demand advanced the development of semisynthetic polymers. Examples include the use of wood as a raw material for paper caused by the shortage of rags, and the development of artificial silk (rayon) in response to pest infestation in silkworms. Paper-making involves a physical transformation, rayon-spinning a chemical synthesis.

The development of synthetic polymers was accelerated by shortages of raw materials, embargos, wars, and autarky movements. Early work on substitutes for natural materials was empirical because the macromolecular nature of these materials was unknown. Why should there be something common to such phenomenologically different materials as rigid wood, elastic rubber, tough leather, soft wool, stiff bast fibers, shiny silk, powdery starch, sticky bone glue, or hard phenolic resins?

Synthetic polymers were already known as "organic colloids" in the nineteenth century. Several authors considered these colloids to be polymers, others even postulated chain structures. But neither the polymeric nature nor the chain structure could

Table 1-2   Consumption of important organic polymeric materials in 1990 (without foodstuffs; starch including foodstuffs: $1 \cdot 10^8$ t/a). Population: 5 320 000 000 (world), 247 000 000 (USA). Plastics, rubbers, adhesives, etc. are usually "formulated" by addition of other materials.

Material	Main polymer components	Annual consumption in		
		tons world	kilograms/capita world	USA
Fuel				
Firewood, charcoal	cellulose, lignins	600 000 000	113	250
Working materials, information carriers, etc.				
Lumber	cellulose, lignins	900 000 000	169	474
Other wood materials	cellulose, lignins	400 000 000	75	371
Paper, cardboard	cellulose	400 000 000	75	560
Plastics (as raw materials)	various synthetic	54 500 000	12.0	113
Adhesives, sealants	various synthetic	4 500 000	0.85	?
Resins for graphic purposes	various synthetic	940 000	0.18	?
Rubbers, elastomers				
Synthetic rubbers	various	9 950 000	1.87	?
Natural rubber (raw material)	*cis*-1.4-poly(isoprene)	4 900 000	0.92	?
Thermoplastic elastomers	various synthetic	650 000	0.12	1.04
Fibers				
Synthetic fibers	various	15 900 000	3.0	16
Cotton	cellulose	14 300 000	2.7	5.6
Bast and hard fibers	various polysaccharides	4 000 000	0.75	0.50
Rayon, acetate fibers	cellulose	3 200 000	0.60	2.2
Wool	proteins	2 000 000	0.38	?
Silk	proteins	50 000	0.001	?
Thickeners				
Starch	amylose, amylopectin	6 000 000		1.1 ?
Water-soluble synthetic polymers	various	4 600 000	0.86	?
Gums	various polysaccharides	500 000	0.094	0.14
Oil-soluble synthetic polymers	various	500 000	0.094	0.22

be unequivocally proven due to the lack of experimental methods and theoretical insights. Application of Raoult's law to vapor pressure measurements and van't Hoff's equation to osmotic pressure determinations showed that these colloids must have high molecular weights between 10 000 and 100 000. Since organic colloids behaved similar to inorganic colloids, which were known to be assemblies of small molecules, such high molecular weights were erroneously interpreted as those of "physical molecules", i.e., associations of small organic molecules. These physical molecules were assumed to be held together by intermolecular "partial valences", for example, attractive forces between double bonds of different molecules. Such forces should change upon modification of the chemical structure of the constituent small molecules.

Bromination of natural rubber $-\!\!\left[\text{CH}_2\text{--C(CH}_3)\text{=CH--CH}_2\right]_{\overline{n}}$ by Samuel S. Pickles in 1910 and hydrogenation of poly(styrene) $-\!\!\left[\text{CH}_2\text{--CH(C}_6\text{H}_5)\right]_{\overline{n}}$ by Hermann Staudinger in 1924 (Nobel prize 1953) did not change the degree of polymerization of the alleged physical molecules, however. Staudinger concluded that these organic colloids must be true chemical molecules, i.e., true organic macromolecules.

His opponents countered that well-behaved true organic molecules should crystallize but that such "macromolecules" as poly(styrene) $+CH_2-CH(C_6H_5)\}_N$ did not. These "macromolecules" also exhibited different molecular weights for the same relative composition. Thus, they violated the purity criteria of organic chemistry. It was reasoned that they could not be true covalently bound macromolecules but must be micelle colloids of small molecules held together by secondary forces.

The apparent lack of end groups in these physical molecules was not thought to be contrary to the assumption of micelle colloids because the alleged constituent small molecules could be ring molecules. It was furthermore (wrongly) assumed that reactivities of functional groups decrease with increasing molecular weights.

The inability of organic colloids to crystallize was one of the main supports for the micelle theory. However, James B.Sumner crystallized the enzyme urease in 1926 (Nobel prize, 1946). The crystallized enzyme could be redissolved to furnish again a colloidal solution. This fact ruled out the notion that organic colloids can be crystallized only with the loss of their colloidal character. Combined with the facts that enzymes have high molar masses as shown by ultracentrifugation (The(odor) Svedberg, Nobel prize, 1926) and that proteins always have the same charge per mass according to electrophoresis (Arne Tiselius, Nobel prize, 1948), it seemed clear that naturally occurring organic colloids such as proteins were indeed true macromolecules.

The importance of these findings was apparently not recognized by physical and organic chemists. X-ray measurements of crystallizable organic colloids such as cellulose and natural rubber ruled out ring structures and pointed towards linear chains (Kurt H.Meyer and Hermann F.Mark, 1928). Assuming (wrongly) that the lengths of molecules could not be larger than the dimensions of crystallites, molecular weights of ca. $(5-10)\cdot10^3$ were obtained from these measurements. However, osmotic pressures indicated however much higher molecular weights up to $400\cdot10^3$, which were first explained as molecular weights of solvated chains (wrong) and later as physical molecular weights of associated molecules with chemical molecular weights of $5\cdot10^3$ to $10\cdot10^3$ (possible, but not true for cellulose and natural rubber).

Another problem was the mysterious way in which synthetic polymers were formed from monomers. The reaction of phenol with formaldehyde produced an intractable material that was insoluble in all solvents. Methyl rubber was obtained by the action of sodium and carbon dioxide on 2,3-dimethyl butadiene, and polymers of methyl methacrylate, styrene and vinyl chloride by reaction with free radical initiators. Poly(vinyl chloride) was spun into the first synthetic fiber (Table 1-3).

Wallace H.Carothers decided therefore in 1931 to build synthetic polymers "stepwise" using known condensation reactions of low molecular weight organic chemistry. His work led to the first mass-produced synthetic fiber, nylon 66. A few years later, other researchers established the mechanism of free radical polymerization of vinyl monomers $CH_2=CHR$.

High pressures and temperatures were required for the free radical polymerization of ethylene $CH_2=CH_2$ in 1939. In 1954, Phillips Petroleum and Standard Oil disclosed that ethylene could be polymerized by supported metal oxide catalysts (CrO_3, reduced MoO_3, etc.) to significantly less branched polymers.

Table 1-3 Some early industrial synthetic polymers. † No longer produced.

Polymer	Discovery	Production	Main applications
Phenolic resins	1907	1910	thermosets (electrical insulators)
Methyl rubbers	1912	1915	elastomers †
Alkyd resins	1847	1926	thermosets (coatings)
Amino resins	1915	1928	thermosets
Poly(methyl methacrylate)	1880	1928	plastics (organic glass)
Poly(butadiene)	1911	1929	elastomers (number Bunas)
Poly(vinyl acetate)	1912	1930	adhesive, poly(vinyl alcohol)
Poly(styrene)	1839	1930	thermoplastics, foams
Poly(vinyl chloride)	1838	1931	thermoplastics (synthetic fiber †)
Poly(ethylene oxide)		1931	thickeners, sizes
Poly(chloroprene)	1925	1932	elastomers
Poly(vinyl ethers)	1928	1936	adhesives, plasticizers
Unsaturated polyesters	1930	1936	thermosets
Poly(isobutylene)		1937	elastomers
Styrene–butadiene rubbers	1926	1937	elastomers (letter Bunas)
Poly(hexamethylene adipamide)	1934	1938	fibers, thermoplastics
Poly(vinylidene chloride)	1838	1939	thermoplastics (packaging films)
Poly(N-vinyl pyrrolidone)		1939	blood plasma expander, binders, etc.
Poly(ethylene), low density	1933	1939	thermoplastics
Poly(ε-caprolactam)	1938	1939	fibers, thermoplastics
Polyurethanes	1937	1940	fibers, plastics, elastomers, foams
Poly(acrylonitrile)	1940	1941	fibers
Silicone	1901	1942	fluids, resins, elastomers
Epoxy resins	1938	1946	adhesives
Poly(tetrafluoroethylene)	1939	1950	plastics, fibers
Poly(ethylene terephthalate)	1941	1953	fibers, bottles
Bisphenol A polycarbonate	1898	1953	thermoplastics
Poly(ethylene), high density	1953	1955	thermoplastics, foams
Poly(butadiene), *cis*-1,4-		1956	elastomers
Poly(propylene), isotactic	1954	1957	thermoplastics, fibers
Poly(formaldehyde)	1839	1959	thermoplastics
Aromatic polyamides		1961	high modulus fibers
Styrene–butadiene–styrene block copolymers		1965	thermoplastic elastomers

In the same year, Karl Ziegler (Nobel prize, 1963) obtained a patent for high density poly(ethylene)s which were obtained at room temperature and slightly elevated pressure by using catalysts from aluminum alkyls and titanium tetrachloride. The same catalysts allowed Guilio Natta (Nobel prize, 1963) to prepare configurationally regular poly(propylene)s $+CH_2$-$CH(CH_3)+_{\overline{n}}$ from propylene $CH_2=CH(CH_3)$.

In these and other linear macromolecules, chain atoms can rotate more or less freely around the chain bonds, allowing the chain to assume many spatial positions at any given time. Werner Kuhn recognized in 1930 that the shape of such thread-like macromolecules can be elegantly analyzed by statistical methods. Such methods are now an indispensable tool for the description of many polymer structures and properties, for example, conformations of macromolecules, thermodynamics of polymer solutions, rubber elasticity, and many more (Paul J.Flory, Nobel prize, 1974).

Table 1-4   Production of plastics, fibers and rubbers.

Type	World production in million tons per year					
	1940	1950	1960	1970	1980	1990
Plastics (polymers + additives)	0.36	1.6	6.7	31	59	100
Fibers, synthetic	0.005	0.069	0.70	5.0	11.5	15.9
rayon (regenerated cellulose)	1.1	1.6	2.6	3.4	3.3	3.2
natural (cotton, wool, silk)	8.7	8.0	12.8	14.0	17.7	21
Rubbers, synthetic	0.043	0.54	1.94	5.9	8.7	9.9
natural	1.44	1.89	2.02	3.1	3.9	5.2

Although natural, semisynthetic and even synthetic polymers were utilized long before their macromolecular nature was known, it was the scientific exploration of their synthesis, structure and properties that paved the way for their ever increasing use as plastics, elastomers, fibers, adhesives, thickeners, etc. Their economic success benefitted from petroleum as an inexpensive source of energy and raw materials. In the last 50 years, the annual world production of elastomers increased ca. 200-fold, that of plastics 300-fold, and that of synthetic fibers more than 3000-fold (Table 1-4). The per capita consumption rose correspondingly, for example, that of plastics in highly industrialized countries to 110 kg per year. This in turn generated the demand for recycling and other disposals of plastics waste.

Literature

FUNDAMENTAL TERMS
International Union of Pure and Applied Chemistry, Macromolecular Division, Commission on Macromolecular Nomenclature, Compendium of Macromolecular Nomenclature, Blackwell Scientific, Oxford 1991 (Purple Book); Pure Appl. Chem. **66** (1994) 2483
G.Allen, J.C.Bevington, eds., Comprehensive Polymer Science, Pergamon, Oxford, 7 vols. (1989), First Supplement (1992)
H.-G.Elias, Macromolecules, Plenum, New York, 2nd ed. 1984 (2 vols.); Makromoleküle, Hüthig and Wepf, Basle, 5th ed., vol. I (1990), II (1992)

HISTORY
R.Olby, The Macromolecular Concept and the Origin of Molecular Biology, J.Chem.Educ. **4** (1970) 168
J.K.Craver, R.W.Tess, eds., Applied Polymer Science, Am.Chem.Soc., Washington, DC 1975
R.B.Seymour, ed., History of Polymer Science and Technology, Dekker, New York 1982
Y.Furukawa, "Staudinger, Carothers, and the Emergence of Macromolecular Chemistry", PhD Thesis, University of Oklahoma 1983; University Microfilms 83-24888, Ann Arbor 1984
H.Morawetz, Polymers: The Origin and Growth of a Science, Wiley-Interscience, New York 1985
R.B.Seymour, G.A.Stahl, eds., Genesis of Polymer Science, Am.Chem.Soc., Washington, DC 1985
J.Alper, G.L.Nelson, eds., Polymeric Materials: Chemistry for the Future, Amer.Chem.Soc., Washington DC 1989
R.B.Seymour, ed., Pioneers in Polymer Science, Kluwer, Boston 1989

BIOGRAPHIES AND AUTOBIOGRAPHIES
H.Staudinger, Arbeitserinnerungen, Hüthig, Heidelberg 1961; English translation: From Organic Chemistry to Macromolecules: A Scientific Autobiography Based on My Original Papers, Wiley-Interscience, New York 1970
J.D.Watson, The Double Helix. A Personal Account of the Discovery of the Structure of DNA, Atheneum, New York 1968
E.Trommsdorf: Dr. Otto Röhm, Chemiker und Unternehmer, Econ, Düsseldorf 1976
E.Guth, Birth and Rise of Polymer Science - Myth and Truth, J.Appl.Polym.Sci.-Appl.Polym. Symp. **35** (1979) 1
S.Carra, F.Parisi, I.Pasquon, P.Pino, eds., Giulio Natta: Present Significance of His Scientific Contribution, Edizione Chimia, Milan 1982
H.Kuhn, Leben und Werk von Werner Kuhn, 1899-1963, Chimia **38** (1984) 191
H.-F.Eicke, Peter J.W.Debye's Beiträge zur Makromolekularen Wissenschaft - ein Beispiel zukunftsweisender Forschung, Chimia **38** (1984) 347
H.F.Mark, From Small Organic Molecules to Large: A Century of Progress, American Chemical Society, Washington DC 1993

COLLECTIVE WORKS
H.Mark, G.S.Whitby, eds., Collected Papers of Wallace Hume Carothers on High Polymeric Substances, Interscience, New York 1940
G.Natta, F.Danusso, Stereoregular Polymers and Stereospecific Polymerisation, Pergamon Press, New York 1967, 2 vols.
H.Staudinger, Das wissenschaftliche Werk, Hüthig und Wepf, Heidelberg 1969-1976, 7 vols. in several parts
L.Mandelkern, J.E.Mark, U.W.Suter, D.Y.Yoon, eds., Selected Works of Paul J.Flory, Stanford University Press, Stanford, CA 1985

DICTIONARIES, ENCYCLOPEDIAS, SOURCE BOOKS
H.Mark, C.Overberger, G.Menges, N.M.Bikales, eds., Encyclopdia of Polymer Science and Engineering, Wiley, New York, 2nd ed. (1985 ff.), 18 vols. and 2 supplements
Houben-Weyl, Methoden der Organischen Chemie, G.Thieme, Stuttgart, 5th ed., vol. XX (3 parts), Makromolekulare Stoffe (1986)
M.S.M.Alger, Polymer Science Dictionary, Elsevier Applied Science, Barking, UK 1989
T.Whelan, Polymer Technology Dictionary, Chapman and Hall, London 1994
J.Brandrup, E.H.Immergut, eds., Polymer Handbook, Wiley, New York, 3rd ed. 1989
J.C.Salamone, ed., Polymeric Materials Encyclopedia, CRC Press, Boca Raton 1996

BIBLIOGRAPHIES
J.T.Lee, Literature of Polymers, Encycl.Polym.Sci.Engng., 2nd ed., **9** (1987) 62
R.T.Adkins, ed., Information Sources in Polymers and Plastics, K.G.Saur, New York 1989

2.   Chemical Structure

The chemical structure of a molecule is described by its constitution (Section 2.1), configuration (Section 2.3) and conformation (Section 2.4). Molecular weights are part of the constitution but will be discussed separately (Section 2.2) because of their importance for polymer science. The term "physical structure" refers to the shape of a single macromolecule (Chapter 5) as well as to the arrangement of molecules in molecular assemblies, e.g., melts and solid states (Chapter 8).

2.1.   Constitution

2.1.1. Nomenclature

The **constitution** of a molecule is defined as the sequence of atoms and bonds without regard to spatial arrangements. "Constitution" is thus a subterm of "chemical structure", not a synonym. The constitution of macromolecules and macromolecular substances is described by constitutive names, poly(monomer) names, polygroup names, trivial names, and/or trade names.

Naturally occuring macromolecules often have complex constitutions. They usually carry **trivial names**, which refer to the source (e.g., cellulose), the source and chemical property (e.g., nucleic acid), the chemical action (e.g., catalase), or its main use (e.g., starch).

Synthetic macromolecules are most often given **poly(monomer) names**. These names are *source-based* and consist of the name of the monomer and the prefix "poly". Polymers of ethylene (ethene) are thus called "poly(ethylene)s". Actually, such names usually indicate *groups* of compounds that have the same major monomeric unit; they do not imply the ideal constitution (here $-\!\!-\text{CH}_2\text{CH}_2\!\!-\!\!\frac{1}{n}$).

It is expedient to write the name of the monomer in parentheses in order to avoid confusion with the "poly" compounds of organic chemistry. The prefix "poly" describes in organic chemistry chemical compounds with *two or more intact substituents* whereas in macromolecular chemistry it denotes chemical compounds with *many polymerized monomeric units*. Examples:

| an isocyanate | a polyisocyanate
organic chemistry | a poly(isocyanate)
macromolecular chemistry |

The spelling of poly(monomer) names is not systematic in common English. Simple polymer names are usually written as one word (examples: polyethylene, polystyrene), more complex names as two words (examples: polyvinyl chloride, polyethylene terephthalate), and highly complex names with the name of the monomer(s) in parentheses (example: poly(2,6-dimethyl-1,4-phenylene ether)). Many exceptions exist to this "rule": Polyethyleneimine, for example, is written as one word. This book will use parentheses around monomer names in poly(monomer name)s.

Poly(monomer) names with monomer names in parentheses are not to be confused with polygroup names with group names not in parentheses. **Polygroup names** are *structure-based names* that refer to characteristic groups as chain units; they do not denote intact substituents as in organic chemistry. Examples are "polyamides" with amide groups -NH-CO- in the main chain, polyurethanes with urethane groups -NH-CO-O- and polyesters with ester groups -CO-O-. Polygroup names obviously refer to *classes* of polymers, not to individual polymers.

Individual polymers are unambiguously characterized by **systematic names (generic names, IUPAC names)**. These names consist of the name of a preferred constitutional repeating unit (CRU) prefixed by "poly". The name of the CRU is always written in parentheses. For regular single-strand polymers, the unit is usually a bivalent group; an example is the methylene group $-CH_2-$ in poly(methylene) $+CH_2+_{\overline{n}}$. Systematic names are also always written without spaces between words (see below for some examples).

In each CRU, chain units are arranged in sequential order according to their seniority which in turn is arbitrarily defined by IUPAC. The implicated chain direction in IUPAC names is neither identical with the "natural" sequence of monomeric units nor with the direction of chain growth. An example is the IUPAC structure and IUPAC name for poly(hexamethylene adipamide) $+NH(CH_2)NH-CO(CH_2)_4CO+_{\overline{n}}$:

$+NHCO(CH_2)_4CONH(CH_2)_6+_{\overline{n}}$ poly(imino(1,6-dioxohexamethylene)iminohexamethylene)
or
poly(iminoadipoyliminohexamethylene)

Names of known end groups are written to the left of the poly(CRU). They are prefixed by the italicized Greek letter α for the end group that is attached to the left side of the CRU; the other end group is prefixed by the italicized Greek letter ω (α and ω being the first and last letters of the Greek alphabet). For example, if a polyester $H+O(CH_2)_2OCO(p\text{-}C_6H_4)CO+_{\overline{n}}OH$ from ethylene glycol $HO(CH_2)_2OH$ and terephthalic acid $HOOC(p\text{-}C_6H_4)COOH$ is esterified at both ends by methyl acetate $CH_3CO-OCH_3$, then the structural formula and generic name of the resulting poly(ethylene terephthalate) would be

$CH_3CO+OCH_2CH_2OCO(1,4\text{-}C_6H_4)CO+_{\overline{n}}OCH_3$ α-acetyl-ω-oxymethyl-
poly(oxyethyleneoxyterephthaloyl)

Systematic names are complicated. They are thus mostly used for archival purposes, e.g., in Chemical Abstracts. The names furthermore refer to idealized constitutions. Like most organic reactions, many polymerizations proceed with side reactions. While the byproducts of these reactions can be removed from the main product in organic chemistry, they become part of the structure in polymers. The polymerization of ethylene $CH_2=CH_2$ by free radicals delivers not only monomeric units $-CH_2CH_2-$ and chain units $-CH_2-$ but also constitutional units $-CH_2CH\{(CH_2)_iCH_3\}-$ with branches $-\{(CH_2)_iCH_3\}$, where $i = 1, 3, 4, 5$ or $i \gg 5$ (Section 3.7.7.). These branched polymers are also called poly(ethylene)s.

In industry, polymer names often refer to the name of the main monomeric unit only and not to minor components. The polymer from 92 % ethylene and 8 % 1-octene is, for example, usually called a linear low density poly(ethylene) and neither a branched polymer nor a copolymer of ethylene and 1-octene (Chapter 14). The same trade name may also designate polymers with very different chemical compositions which may be then differentiated by letters, numbers, or combinations thereof.

2.1.2. Monomers for Polymers

Biopolymers are naturally occuring polymers. Some of these are used directly, others after chemical transformation into other polymers (Section 4.3.). Most industrially used polymers are however synthesized from monomers by polymerization. Such monomers must be at least bifunctional (Section 3.1.3.). They usually belong to one of the following three groups:

1. Monomers with two (or more) functional end groups, such as hydroxycarboxylic acids HO–Z–COOH, amino acids H_2N–Z–COOH, diamines H_2N–Z–NH_2, triamines $Y(NH_2)_3$, tetracarboxylic acids $X(COOH)_2$, dichlorides Cl–X–Cl, disodium sulfide Na_2S, etc., where Z denotes a bifunctional unit, Y a trifunctional unit and X a tetrafunctional one. The symbol R is reserved for monofunctional substituents.

2. Monomers with multiple bonds, mostly with carbon-carbon double bonds $>C=C<$, less often with carbon-carbon triple bonds $-C\equiv C-$, carbon-nitrogen triple bonds $-C\equiv N$, or carbon-oxygen double bonds $>C=O$. Examples are propene (propylene) $CH_2=CHCH_3$, 1,3-butadiene $CH_2=CH-CH=CH_2$, formaldehyde $H_2C=O$, and acetylene $HC\equiv CH$.

3. Cyclic monomers with heteroatoms in the ring. Examples are:

ethylene oxide   tetrahydrofuran   hexamethylcyclotrisiloxane   ε-caprolactam

2.1.3. Process-Based Terms

Homopolymers are derived from one species of monomer; some examples are shown in Table 2-1. Syntheses of **copolymers** require two or more species of monomers. The term "copolymer" is sometimes used as a synonym for **bipolymer**, i.e., a

Table 2-1 Examples of homopolymers and their IUPAC formulae and names

Monomer Common name	formula	Polymer IUPAC formula	IUPAC name
Vinyl fluoride	$CH_2=CHF$	$+CHFCH_2\}_{\overline{n}}$	poly(1-fluoroethylene)
Isoprene	$CH_2=C(CH_3)-CH=CH_2$	$+C(CH_3)=CHCH_2CH_2\}_{\overline{n}}$	poly(1-methyl-1-butenylene)
Ethylene oxide	see above	$+OCH_2CH_2\}_{\overline{n}}$	poly(oxyethylene)
ε-Caprolactam	see above	$+NHCO(CH_2)_5\}_{\overline{n}}$	poly[imino(1-oxohexa-methylene)]

copolymer derived from two species of monomers. Copolymers from three, four, five ... species of monomers are accordingly called **terpolymers, quaterpolymers, quin-terpolymers** ... . An older term for "copolymer" is "interpolymer".

The term "copolymer" is historically reserved for polymers from polymerizations of such monomers that can also homopolymerize *in principle*. It is for this reason that the polycondensation of terephthalic acid $HOOC(p-C_6H_4)COOH$ with ethylene glycol $HOCH_2CH_2OH$ to poly-(ethylene terephthalate) $H+O(CH_2)_2OOC(p-C_6H_4)CO\}_{\overline{n}}OH$ is not called a *co*polycondensation because neither terephthalic acid nor ethylene glycol can homopolycondense under polyesterification conditions. The term "copolymerization" is not used systematically, however: the free radical polymerization of stilbene and maleic anhydride is considered a copolymerization although neither of the two monomers will homopolymerize free-radically.

Copolymers are distinguished according to the sequence of the various monomeric units. They are subdivided into statistical, random, alternating, periodic, block, segmented, gradient, and graft copolymers (Table 2-2).

Table 2-2 Sequence of monomeric units a, b and c in various copolymers.

Name	Schematic structure	IUPAC designation
Bipolymer without specific sequence	...a/b...	poly(A-*co*-B)
Statistical bipolymer with statistical sequence	...aabaaaabbabbbabaaababbba...	poly(A-*stat*-B)
Random bipolymer (Bernoullian sequence statistics)	...aaabaabbabbbabaaababbaab...	poly(A-*ran*-B)
Alternating bipolymer	...ababababababababababab...	poly(A-*alt*-B)
Periodic bipolymer	...abbabbabbabbabbabbabb..	poly(A-*per*-B-*per*-B)
Periodic terpolymer	...abcabcabcabcabcabcabc...	poly(A-*per*-B-*per*-C)
Diblock copolymer	a..............ab................b	poly(A)-*block*-poly(B)
Triblock copolymer	a.........ab.........bc.........c	poly(A)-*block*-poly(B)-*block*-poly(C)
Segmented bipolymer (small *n*, *m*, etc.)	$(a)_n-(b)_m-(a)_p-(b)_q(a)_r-(b)_s-(a)_t$	-
Gradient bipolymer	$(a)_n$baaaaabbaaabbaabbbba$(b)_m$	-

Graft bipolymer	...a-a-a.........a-a-a.........aa...
	     $\|$ $\|$
	    b_m b_n
		poly(A)-*graft*-poly(B)

The monomeric units "a" (from monomer A) and "b" (from monomer B) alternate in an **alternating copolymer** which is in turn a special case of a **periodic copolymer**. In periodic copolymers, monomeric units are arranged in periodic sequences. Examples are ...abbabbabb... or $+abb+_n$, ...aabbaabb... or $+aabb+_n$, ...abcabcabc... or $+abc+_n$, etc. In such formulas, this book will distinguish symbols of monomers (in capital letters) from those of monomeric units (small letters).

The sequence of monomeric units in **statistical copolymers** is given by the statistics of the polymerization reactions (see Section 3.1.5.), for example, Markov statistics of first, second ... order. Statistical copolymers with Bernoullian sequence statistics (zeroth order Markov statistics) are called **random copolymers**.

Gradient copolymers exhibit a compositional gradient along the chain: one end of a bipolymer is enriched in a-units and the other end in b-units. Gradient copolymers are also called **tapered copolymers**. **Block copolymers** are limiting cases of gradient copolymers; they consist of blocks of homosequences that are joined via the ends. Multiblock copolymers with short block lengths are known as **segmented copolymers.** If the b-blocks are connected to the a-block as side-chains instead of the ends, **graft copolymers** result. Graft copolymers are a special case of comb polymers (Section 2.1.5.).

All these copolymerizations deliver polymers whose macromolecules differ in the composition and sequence of monomeric units as well as in the degree of polymerization and the molecular weight. Thus, they are **non-uniform** (or "**polydisperse**") with respect to constitution and molecular weight.

2.1.4. Structure-Based Terms

The terms "homopolymer" and "copolymer" are *process-based*. They indicate the monomers used for the synthesis but do not describe the chemical structure of the resulting polymers. Chemists often know by experience which chemical groups react under certain conditions and can thus deduce the probable constitution of the macromolecules from the structure of the monomers. However, this approach is not foolproof; the constitution of polymers must always be proven by standard methods (elemental analysis, nuclear magnetic resonance, infrared spectroscopy, ultraviolet spectroscopy, etc.).

An example is acrolein (I) which can polymerize via the vinyl group (II) or the aldehyde group (III) or both (IV, V). Intermolecular reactions of both the vinyl and the aldehyde group of this monomer result in cross-linked polymers (IV) and intramolecular rections in ring formation (V):

$$CH_2=CH \qquad -CH_2-CH- \qquad CH_2=CH \qquad -CH_2-CH- $$
$$| \qquad\qquad\qquad | \qquad\qquad\qquad | \qquad\qquad\qquad\quad |$$
$$CHO \qquad\qquad\quad CHO \qquad\quad -CH-O- \qquad -CH-O-$$

$$\quad I \qquad\qquad\qquad II \qquad\qquad\qquad III \qquad\qquad\qquad IV \qquad\qquad\qquad V$$

Polymers with only one type of chain atom are called **homochain polymers**. Examples are C-chain polymers such as poly(ethene) $+CH_2CH_2\}_{\overline{n}}$, S-chain polymers such as polymeric sulfur $+S\}_{\overline{n}}$ and Si-chain polymers such as polysilanes $+SiR_2\}_{\overline{n}}$. The main chain of **heterochain polymers** is constructed of more than one type of chain atom. Examples are poly(oxymethylene) $+OCH_2\}_{\overline{n}}$, polyesters $+OXCO\}_{\overline{n}}$ and polyurethanes $+NH\text{-}CO\text{-}O\text{-}X'\}_{\overline{n}}$. In **inorganic polymers**, 50 % or more of the chain atoms are not carbon; an example is poly(dimethylsiloxane) $+OSi(CH_3)_2\}_{\overline{n}}$.

Polymers either occur in nature or are synthesized by man. **Semisynthetic polymers** are naturally occurring polymers that have been transformed by man into other polymers. The term **biopolymer** is used in an ambiguous manner: it may denote either a naturally occurring polymer or a synthetic polymer with a constitution equal or similar to that of a naturally occurring polymer.

Naturally occurring polymers are usually synthesized at matrices; thereby allowing a control of the type, number, linkage and sequence of monomeric units (see Section 4.2.). The macromolecules of these polymers are thus identical with respect to constitution and molecular weight: the polymers are **uniform** ("monodisperse"). All known synthetic polymerizations are statistically controlled, however. The degree of polymerization of synthetic homopolymers varies from macromolecule to macromolecule: these polymers are **non-uniform with respect to molecular weight** and possess a molecular weight distribution (Section 2.2.). Synthetic copolymers exhibit distributions with respect to monomeric units *and* molecular weight (Section 3.8.).

IUPAC recommends the adjective "uniform" if the molecules of a polymer are identical with respect to constitution and molecular weight and the adjective "non-uniform" if they are not. The polymer literature wrongly uses "monodisperse" and "polydisperse" instead. "Disperse" (L: *dispergere* = to scatter on all sides) relates in chemistry to *multiphase* systems (Section 3.7.9.). "Monodisperse" is an oxymoron since something cannot be "mono" and "disperse" at the same time.

Identical monomeric units do not necessarily imply identical constitutions. Macromolecules may exhibit **regioisomerism** by "false" incorporations of monomeric units. Monomeric units -CH_2-CHR- are mostly combined in **head-to-tail** positions (◄-◄) but some may be also added in **head-to-head** (►-◄) or **tail-to-tail** positions (◄-►):

-CH_2-CHR-CH_2-CHR-	-CHR-CH_2-CH_2-CHR-	-CH_2-CHR-CHR-CH_2-
head-to-tail	head-to-head	tail-to-tail

The constitution of a polymer can thus not always be deduced by chemical experience from the monomer constitution. Deviations from the "ideal structure" may range from a few percent of "constitutional mistakes" to structures that drastically alter the macroscopic appearance of a polymer.

For example, monomers with two vinyl groups usually react tetrafunctionally; an example is 1,4-divinyl benzene $CH_2=CH\text{-}(p\text{-}C_6H_4)\text{-}CH=CH_2$. If both vinyl groups have the same reactivity, polymerization will lead to cross-linked and insoluble polymers as, for example, in polymerizations initiated and propagated by free radicals. In polymerizations initiated and propagated by anions, the reactivity of the second vinyl group of 1,4-divinyl benzene is however drastically changed after the polymerization of the first vinyl group. Only the first group polymerizes (except at very high conversions of monomer) and the resulting polymer is soluble and not cross-linked.

The constitution of monomeric units may also vary with the polymerization temperature. 4,4-Dimethyl-1-pentene CH_2=CH-CH_2-C(CH_3)$_3$ polymerizes cationically at −130°C exclusively via the double bond to $+CH_2$-CH{CH_2-C(CH_3)$_3$}$+_n$. At temperatures greater than 0°C, however, monomeric units -CH_2-CH_2-CH{C(CH_3)$_3$}- are predominantly formed. Polymers with monomeric units resulting from non-existing monomers are usually called **phantom polymers** (the polymers are quite real and not phantoms, though; they are derived from phantom monomers!).

A special class of phantom polymers is obtained by the **cyclopolymerization** of multifunctional monomers to intramonomeric ring structures. The two vinyl groups of the tetrafunctional acrylic anhydride CH_2=CH-CO-O-CO-CH=CH_2 (I) are in such proximity to each other that they will form five-membered (II) and six-membered rings (III) by intramolecular cyclization instead of linear (IV) and cross-linked structures (V). The resulting polymers are soluble.

I   II   III   IV   V

2.1.5. One-, Two- and Three-Dimensional Macromolecules

The simplest polymers consist of **regular macromolecules** "with only one species of constitutional repeating units in a single sequential arrangement" (IUPAC). Examples are poly(methylene) $+CH_2+_n$, poly(styrene) $+CH_2$-CH(C_6H_5)$+_n$, and poly-(ε-caprolactam) $+NH$-(CH_2)$_5$-CO$+_n$.

The constitutional units of these macromolecules form **linear chains**. The term "linear" no longer refers to the spatial structure as it did historically. Linear chains do not exist as zigzag chains in dilute solution (Chapter 5); even in the crystalline state, extended zigzag chains are rare (Chapter 8). These chains are now called "linear" because of the one-dimensional connectivity of their constitutional repeating units. Examples of linear chains are polymeric sulfur and poly(α-olefin)s:

polymeric sulfur   a poly(α-olefin)   a spiro polymer

Cyclic polymer molecules result if the two ends of linear macromolecules are intramolecularly connected. They are not called "macrocycles" because this term is used traditionally in organic chemistry for "large rings" of ca. 15-20 chain atoms.

Cyclic deoxyribonucleic acids with many thousands to millions of chain atoms exist in nature. Small amounts of cyclic molecules with rather low degrees of polymerization are often formed as by-products during the synthesis of linear chains.

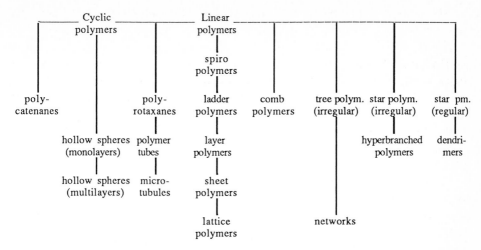

Fig. 2-1   Schematic representation of molecular architectures of polymers. The dimensionality increases from top to bottom.

Special strategies allow industrial syntheses of cyclic macromolecules in high yields (Section 4.1.7.): low concentrations of bifunctional monomers promote intramolecular reactions relative to intermolecular ones (**Ruggli–Ziegler dilution principle**). The low concentrations do not need to be global; they can be local (Section 4.1.7.).

Combinations of linear molecules, of linear molecules with cyclic molecules, and of cyclic molecules themselves lead to a great variety of **molecular architectures** (Fig. 2-1). Cyclic molecules may be linked together like locked rings in **polycatenanes** without the formation of covalent bonds between the rings (Fig. 2-2). Polycatenanes may result as by-products during the synthesis of cyclic polymers.

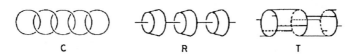

Fig. 2-2   Schematic representation of polycatenanes, polyrotaxanes and polymer tubes.

The threading of small cyclic molecules onto linear chains generates **polyrotaxanes**. Cyclodextrins (cyclic molecules with 6-12 glucose residues per ring) are, for example, attracted to the ether groups in poly(oxyethylene)s $H\text{-}[O\text{-}CH_2\text{-}CH_2]_{\overline{n}}OH$ or imine groups in $H\text{-}[NH\text{-}(CH_2)_i]_{\overline{n}}R$ and slip onto the chains like pearls on a necklace. The chain ends are then sealed with bulky groups in order to prevent the unthreading of the rings from the chains.

These polyrotaxanes may be converted into **polymer tubes**. On reaction with epichlorohydrin + NaOH, three intermolecular $-CH_2CH(OH)CH_2-$ bridges are formed between each pair of adjacent cyclodextrin molecules. Stronger bases remove the bulky end groups and the linear polymer molecules. The resulting polymer tubes have inner diameters of ca. 0.5 nm and outer diameters of ca. 1.5 nm.

Other polymer tubes can be synthesized by **supramolecular chemistry**, the joining of small chemical molecules via non-covalent bonds to larger physical molecules that behave like chemical molecules. On acidification of their alkaline solutions, cyclic octapeptides $cycl\text{-}[\text{-D-(NH-CHR-CO)}\text{-}alt\text{-L-(NH-CHR-CO)}]_{\overline{n}}$ ($n = 4$) cluster together via intermolecular hydrogen bonds to form supramolecular tubes of ca. 0.8 nm inner diameter and lengths of 100 nm to 1000 nm. These tubes behave like true chemical molecules because the combined strength of their eight hydrogen bonds per octapeptide molecule is similar to the strength of one covalent bond.

Polymer tubes also exist in nature. Several silicates are polymer tubes. Carbon black contains not only small proportions of the spherical, hollow buckminster-fullerenes ("buckyballs" C_{60}, C_{72}, etc.) but also smaller proportions of "buckytubes" of 0.34 nm diameter and ca. 1000 nm length. Polymer tubes should not be confused with **microtubules** which can be generated by the controlled precipitation of certain polymers. Microtubules have much larger wall thicknesses of ca. 100 nm; they are not chemical molecules but ordered physical aggregates.

Not only "buckyballs" but several polymer molecules also exist as **hollow spheres**. The protein apoferritin forms a hollow sphere with an inner diameter of 7.3 nm and an outer diameter of 12.2 nm. The inner space is filled with iron(III)hydroxide oxide phosphate in the enzyme ferritin. Even larger are the bag-like protein hulls of certain bacteria.

Aqueous dispersions of amphiphilic monomers with a hydrophilic head group and two polymerizable hydrophobic tails aggregate to **vesicles** upon ultrasonification. These vesicles are hollow spheres with outer diameters of (20-100) nm that can be stabilized by polymerization. The wall is several atom layers thick; it surrounds an aqueous interior which can transport effectors.

Ladder polymers consist of two parallel chains that are interconnected in regular intervals by covalent bonds. They can be synthesized by intramolecular polymerizations of adjacent functional groups of polymer molecules, for example, by the polymerization of the vinyl groups of 1,2-poly(butadiene) $+\text{CH}_2\text{-CH(CH=CH}_2)]_{\overline{n}}$ to polymer I or of the nitrile groups of poly(acrylonitrile) $+\text{CH}_2\text{-CH(C}\equiv\text{N)}]_{\overline{n}}$ to polymer II. Ladder structures are usually not very long because the ladder is interrupted if some functional groups do not polymerize or polymerize intermolecularly and form cross-links. Ladder oligomers are also obtained by multi-step syntheses.

I   II

Repetition of ladders in the plane of the rings leads to **layer** or **parquet polymers**. Graphite is a well-known example. Inorganic layer polymers are known in great numbers, mainly as insoluble solids. Examples are layer silicates such as mica, montmorillonite and bentonite. A soluble layer polymer is poly(silicon monochloride) which is obtained if calcium silicide is treated with iodine chloride ICl.

graphite

silicon monochloride
each Si atom is substituted by a Cl atom (not shown)

Inorganic layer polymers are usually only a few atomic layers thick. Thicker organic layer polymers are obtained by the polymerization of center groups (e.g., vinyl groups) of amphiphilic monomers (Fig. 2-3). These monomers form double layers on surfaces. The polymerized layers are not very stable, however, because of only one connecting bond per monomeric unit.

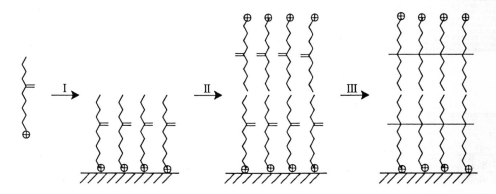

Fig. 2-3 Synthesis of layer polymers from Langmuir-Blodgett double layers. Amphiphilic monomers with a polar head group ⊕ and (e.g.) a vinyl group ══ form an ordered monolayer on a polar surface in a Langmuir trough (I). A second layer is generated via hydrophobic tails (II). The center vinyl groups are polymerized (III).

Sheet polymers are composed of several layer polymers; they are much more stable. Macromonomers with an acrylic end group $CH_2=CH-COO-$ and a nitrile group $N\equiv C-$ at the center (Fig. 2-4, top) dimerize spontaneously (I) to double layers which in turn assemble to smectic liquid crystals (Chapter 8). The acrylic groups in the center of the double layers polymerize to an extent of 90 % (II); the conversion of nitrile groups to -N=C< groups is about 30-50 %. This triple stitching per dimer generates sheet polymers with molecular weights between 10^7 and 10^9 and thicknesses of ca. 10 nm. The incomplete polymerizations lead to many constitutional mistakes and thus to soluble and meltable products.

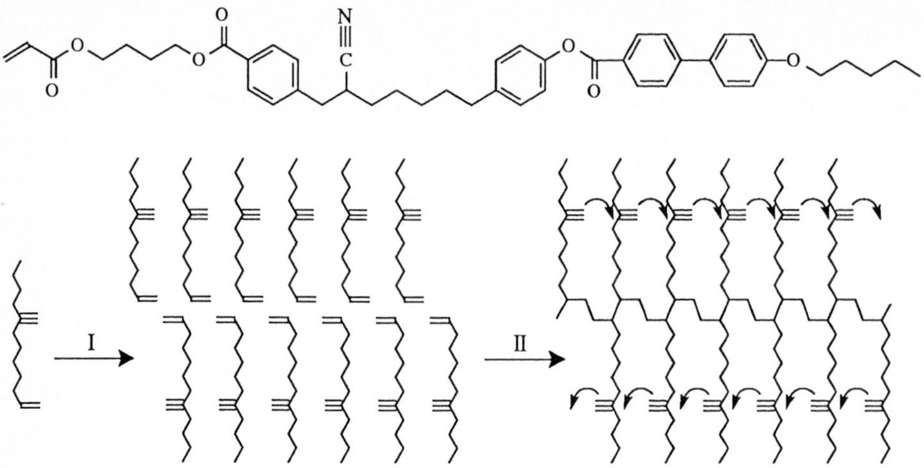

Fig. 2-4 Sheet polymers by polymerization of smectic liquid crystals. Top: Actual chemical structure of monomers. Bottom: schematic representation of bilayer formation and polymerization.

Linear chains may also be arranged at short intervals along a single main chain via trifunctional branch points (Fig. 2-5). These **comb polymers** can be synthesized by polymerization of macromonomers (delivers regular intervals) or by grafting of monomers to or from a main chain (delivers irregular intervals) (see Section 4.3.2.). Comb polymers are a special case of **brushes** which present particular problems with respect to the spatial arrangements of the side chains. The crowded environment for these tethered chains may range from stretched macroconformations in comb and star polymers (Section 5.5.3.) and liquid-crystalline orders of side-chains (Section 8.4.2.) to side-chain crystallizations in the solid state (Section 9.4.4.).

2.1.6. Branched Macromolecules

Branched polymers contain branch points (**junctions**) that connect three or four **subchains** which may be side-chains or a part of the (often fictitious) main chain

Polymers are **statistically branched** if side-chains of different lengths are irregularly distributed along the (fictitious) main chain and along the side-chains. These polymers resemble trees and are thus also called **tree polymers**. Oligomeric offshoots from a main chain are known as **short-chain branches** and polymeric offshoots as **long-chain branches**. These branches result from side reactions, for example, during the free-radical polymerization of ethylene:

$$\sim CH_2{-}\underset{\displaystyle CH_2CH_3}{\underset{|}{CH}}{-}(CH_2)_x{-}\underset{\displaystyle (CH_2)_4H}{\underset{|}{CH}}{-}(CH_2)_y\sim$$

short-chain branches
$(x, y >> 1)$

$$\sim CH_2{-}\underset{\displaystyle (CH_2)_zH}{\underset{|}{CH}}{-}(CH_2)_x{-}\underset{\displaystyle (CH_2)_qH}{\underset{|}{CH}}{-}(CH_2)_y\sim$$

long-chain branches
$(x, y, z, q >> 1)$

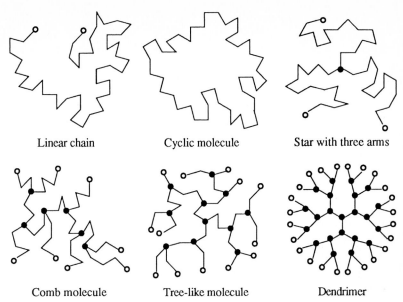

Fig. 2-5 Schematic representation of two-dimensional projections of various polymer molecules with 45 monomer units each. ● Trifunctional branch point, ○ end group.

Polymers from monomers with chain-like substituents are *not* called branched polymers in macromolecular chemistry. Examples are poly(acryl ester)s and so-called linear low density poly(ethylene)s, which are copolymers of ethylene $CH_2=CH_2$ and, e.g., ca. 8 % 1-octene $CH_2=CH(CH_2)_6H$:

$$\sim\!\!\!CH_2\!\!-\!\!CH\!\!-\!\!CH_2\!\!-\!\!CH\!\!-\!\!CH_2\!\!-\!\!CH\sim$$
$$\underset{COOR}{|}\qquad\underset{COOR}{|}\qquad\underset{COOR}{|}$$

a poly(acrylic ester)
an unbranched homopolymer

$$\sim\!\!\!CH_2\!\!-\!\!CH\!\!-\!\!(CH_2)_x\!\!-\!\!CH\!\!-\!\!(CH_2)\sim$$
$$\underset{(CH_2)_6H}{|}\qquad\underset{(CH_2)_6H}{|}$$

linear low density poly(ethylene)
an unbranched copolymer

In **star polymers**, three or more branches (arms) sprout from a common core; an example is the zeroth generation dendrimer (see below). Star polymers with up to 128 arms per core have been synthesized from different monomers. They are called regular if all arms are equally long.

Star polymers with multifunctional ends on the arms can add additional mono-mers. The resulting polymers can be considered either as tree polymers with regular sequences of branches or as star polymers with subsequent secondary branches. They are called **dendrimers** (G: *dendron* = tree) if all branch points have the same func-tionality (with the possible exception of the core) and if all segments between branch points are of equal lengths. An example for a divergent (inside-out) strategy is given:

Ammonia NH_3 can serve as a trifunctional core for the Michael addition of methyl acrylate $CH_2=CHCOOCH_3$ to an intermediate $N(CH_2CHCOOCH_3)_3$. A large excess of $H_2NCH_2CH_2NH_2$ converts this into $N(CH_2CHCONHCH_2CH_2NH_2)_3$ which can be pictured as an extended core cell (0th generation). Further Michael addition of methyl acrylate and subsequent reaction with an excess of $H_2NCH_2CH_2NH_2$ delivers a first generation amidoamine dendrimer with six amino end groups

$$N-CH_2CH_2CONHCH_2CH_2N \begin{cases} CH_2CH_2CONHCH_2CH_2NH_2 \\ CH_2CH_2CONHCH_2CH_2NH_2 \end{cases}$$

<table>
<tr><td></td><td>CH_2CH_2CONHCH_2CH_2N <</td><td>CH_2CH_2CONHCH_2CH_2NH_2
CH_2CH_2CONHCH_2CH_2NH_2</td></tr>
</table>

CH_2CH_2CONHCH_2CH_2N ⟨ CH_2CH_2CONHCH_2CH_2NH_2 / CH_2CH_2CONHCH_2CH_2NH_2

CH_2CH_2CONHCH_2CH_2N ⟨ CH_2CH_2CONHCH_2CH_2NH_2 / CH_2CH_2CONHCH_2CH_2NH_2

0th generation   1st generation

Amidoamine dendrimers with up to 9 generations have been synthesized by this divergent (inside-out) strategy. In principle, they can also be generated by a convergent strategy from the outside-in.

A 3-star core (1 branch point) with trifunctional arms can thus add in each new generation two additional arms per existing arm. After three generations of growth, such a dendrimer will exhibit $(1 + 3 + 6 + 12) = 22$ branching points, $(3 + 6 + 12 + 24) = 45$ segments between branching points or branching points and chain ends, and 24 end groups (Fig. 2-6). These dendrimers have molecular weights up to the millions. With careful experimentation, the distribution of molecular weights is extremely narrow. Since they are caused in theory only by the distribution of isotopes per molecule, such dendrimers are molecularly uniform with respect to degrees of polymerization per molecule (see Section 2.2.2.).

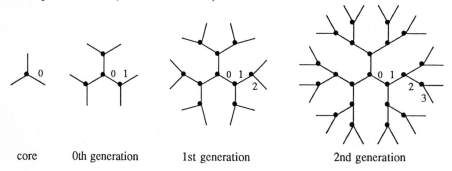

core   0th generation   1st generation   2nd generation

Fig. 2-6   Two generations of a dendrimer with trifunctional branching units. The actual spatial shapes of the molecules differ from the beautiful (schematic!) two-dimensional representations.

In **cross-linked polymers**, all molecules of a specimen are interconnected by many bonds to a **network** consisting of one "infinitely large" molecule (see Fig. 10-7). "Infinite" is used in this context as a figure of speech and not in the mathematical sense. One mole of molecules contains ca. $6.02 \cdot 10^{23}$ molecules according to Avogadro. If 1000 g of a substance are combined to form a single molecule, the molar mass of this molecule would be $M \approx (1000 \text{ g}) \cdot (6.02 \cdot 10^{23} \text{ mol}^{-1}) = 6.02 \cdot 10^{26} \text{ g mol}^{-1}$. Such a molar mass is extremely large but by no means infinite.

Networks can not only be generated by intramolecular covalent bonds, but also by intermolecular physical junctions such as ion clusters, crystallites, microphases, or even mechanical entanglements of chains. Physical (non-covalently bonded) networks can *in principle* be dissolved by solvents to their constituent chemical molecules whereas chemical (covalently bonded) networks are insoluble in *all* solvents.

Chemical properties of chemical and physical network polymers depend strongly on the chemical structure of network chains and the type of junctions. Mechanical properties, on the other hand, are practically determined not by chemical structures but by the cross-link density and the mobility of chain segments. Networks may thus be soft, elastic, brittle or hard (Table 2-3).

Table 2-3   Important cross-linking reactions of low molecular weight (L), oligomeric (O) and polymeric (P) chemical compounds to networks with chemical (c) or physical (p) cross-links.

Process	Reaction			Conversion of material			Cross-link
Drying	L	$\rightarrow$	P	drying oil + air	$\rightarrow$	surface coating	c
Curing	L	$\rightarrow$	P	cement + water	$\rightarrow$	cement stone	c
Hardening	O	$\rightarrow$	P	resin (+ agent)	$\rightarrow$	thermoset	c
Vulcanization	P	$\rightarrow$	P	rubber + agent	$\rightarrow$	elastomer	c
Tanning	P	$\rightarrow$	P	hide (+ agent)	$\rightarrow$	leather	p (or c)
Gelation	P	$\rightarrow$	P	pectin	$\rightarrow$	gel	p
Clotting	P	$\rightarrow$	P	blood	$\rightarrow$	coagulated blood	p

Two independent networks may interpenetrate each other to form an **interpenetrating network**. A material composed of a network in a "non-cross-linked" matrix of another polymer is called a **semi-interpenetrating network**.

Diamond C_n, quartz $(SiO_2)_n$, black phosphorus P_n and a number of other inorganic substances form three-dimensionally ordered networks; they are **lattice polymers**. Such polymers are insoluble in all solvents.

2.2. Molar Mass and Molar Mass Distribution

2.2.1. Degree of Polymerization

The **degree of polymerization** X (DP) of a macromolecule denotes the number of monomeric units in a macromolecule. A polymer consists of many macromolecules which may or may not have the same degree of polymerization. If the polymer is nonuniform with respect to the degree of polymerization of its molecules, then it has a **distribution of degrees of polymerization** which can be described by a **distribution function** (Section 2.2.2.). Types and widths of distribution functions result from synthesis conditions and/or the manipulation of the polymer.

Nonuniform polymers do not possess a single degree of polymerization but various **average degrees of polymerization.** The simplest average is the **number-average degree of polymerization,** $\overline{X}_n$, where the bar over the X symbolizes the averaging process. The number-average degree of polymerization averages over molecules with various degrees of polymerization X_i according to their number-based statistical weights which may be the number N_i, the amount-of-substance $n_i \equiv N_i/N_A$ or the number fraction $x_i = N_i/\Sigma_i N_i = n_i/\Sigma_i n_i$ = number fraction (mole fraction) where $N_A \approx 6.023 \cdot 10^{23}$ mol^{-1}. The number-average degree of polymerization is defined as

$$(2\text{-}1) \qquad \overline{X}_n \equiv \frac{\sum_i N_i X_i}{\sum_i N_i} = \frac{\sum_i n_i X_i}{\sum_i n_i} = \sum_i x_i X_i$$

where $x_i \equiv n_i/\Sigma_i n_i = n_i/n$ is the mole fraction of species i. The **mass-average degree of polymerization** uses mass-based statistical weights, i.e., masses m_i, mass fractions $w_i \equiv m_i/\Sigma_i m_i$ or mass concentrations $c_i = m_i/V$ of species i:

$$(2\text{-}2) \qquad \overline{X}_w \equiv \frac{\sum_i m_i X_i}{\sum_i m_i} = \frac{\sum_i m_i X_i}{m} = \sum_i w_i X_i = \frac{\sum_i c_i X_i}{\sum_i c_i} = \frac{\sum_i c_i X_i}{c}$$

The mass concentration is usually simply called "concentration". It should not be called "weight concentration" because "weight" is the product of mass and acceleration (due to gravity) with the SI unit m kg s^{-2} (the physical unit of a force) whereas "mass" has the SI unit kg. One can however use "weight fraction" instead of "mass fraction" since a "fraction" is a relative quantity (physical unit of unity) (see Chapter 15).

Degrees of polymerization are important theoretical quantities. They cannot be measured directly, however, and have to be calculated from experimentally determined molar masses M or molecular weights M_r (= relative molar masses). Molar masses are given by the degrees of polymerizations X_u and molar masses M_u of the monomeric units and the number N_{end} of end groups per molecule as well as the average molar mass M_{end} of the end groups. The number-average molar mass is thus

$$(2\text{-}3) \qquad \overline{M}_n = \sum_u X_u M_u + \sum_{end} N_{end} M_{end}$$

2.2.2. Molecular Mass, Molecular Weight, Molar Mass

Experimental determinations usually deliver relative molecular masses, reduced molecular masses, or molar masses but not molecular masses themselves. The **molecular mass** m_{mol} is the mass of a molecule of a substance; its physical base unit is the kilogram. For atoms, the molecular mass is the atomic mass m_a of an atom. An example is the mass of an atom of the carbon isotope ^{12}C, $m_a(^{12}$C$)$.

The **relative molecular mass** $M_r \equiv m_{mol}/m_{am}$ is defined as the ratio of the mass m_{mol} of a *molecule* of a substance to the atomic mass constant $m_{am} = m_a(^{12}$C$)/12$. The relative molecular mass is thus a pure number (it is "dimensionless"). It is also called

molecular weight. Nonuniform polymers possess **number-average relative molecular masses** $\overline{M}_{r,n}$ (= number-average molecular weights) and **mass-average relative molecular masses** $\overline{M}_{r,w}$ (= weight average molecular weights) that are defined similarly to the corresponding average degrees of polymerization (Eqns. (2-1) and (2-2)). Number-average molecular weights are usually determined by chemical end group determinations (osmotic measurements do not give *molecular weights* but *molar masses*, see below).

The atomic mass constant m_{am} is equal to the unified atomic mass unit with IUPAC symbol u (and common symbol amu), i.e., $m_{am} = 1$ u. In biochemistry, the unit u is called the dalton, with the symbol Da. Note that the "dalton" has the physical unit of a mass (e.g., gram). Neither the name "dalton" nor the symbol "Da" has been approved by the *Conférence Générale des Poids et Mesures*. If relative quantities are considered, one can use "weight" provided it is done in a consistent manner. One can therefore use either *mass*-average relative molecular *mass* or *weight*-average molecular *weight* but not *weight*-average relative molecular *mass*.

Mass spectroscopy determines the ratio m_{mol}/z of the molecular mass m_{mol} to the electric charge z of a molecule or a molecule fragment. This ratio represents a **reduced molecular mass**.

Most physical methods (osmotic pressure, static light scattering, ultracentrifugation, etc.) do not determine relative molecular masses and thus not molecular weights but molar masses or molar mass averages. The **molar mass**

$$(2\text{-}4) \quad M_i \equiv m_i \,/\, n_i$$

is defined as the ratio of the mass m_i of the molecules i to their amount-of-substances n_i. It is convenient to measure the mass in grams and the amount in moles so that molar masses and relative molecular masses become numerically identical.

Note that molar masses have the physical unit mass/amount-of-substance whereas relative molecular masses (molecular weights) have a physical unit of unity (they are "dimensionless"). The "dalton" has the physical unit of mass and neither unity nor mass/(amount-of-substance). The physical methods mentioned above do *not* give molar masses or molecular weights in daltons.

Physical methods applied to systems in thermodynamic equilibrium (osmotic pressure, static light scattering, sedimentation equilibrium) give simple averages of molar masses because they are based on a series of simple, interrelated statistical weights. Such a statistical weight g may be the amount-of-substance n, the mass m, the z parameter, the $|z+1|$ parameter, etc. For *uniform* polymers, they are defined as

$$n_i \quad m_i \equiv n_i M_i \quad z_i \equiv m_i M_i \equiv n_i M_i^2 \quad |z+1|_i \equiv z_i M_i \equiv m_i M_i^2 \equiv n_i M_i^3, \text{ etc.}$$

Many other statistical weights exist besides n (or N), m (or w or c), and z. An example is the interaction of polymeric materials with fillers (spheres, spheroids, platelets, rods, etc.) where one has to consider surface area statistical weights in addition to number and mass statistical weights.

For *nonuniform* fractions i, M_i has to be replaced by those averages that correspond to the multiplying statistical weight. The statistical weight m_i, if expressed in amounts-of-substance, is thus given by $m_i \equiv n_i(\overline{M}_n)_i$, etc. For molecules with negligible proportions of end groups (see Eqn.(2-3)), the number-average, mass-average and z-average molar masses are thus defined as

(2-5) $\overline{M}_{\mathrm{n}} \equiv \dfrac{\Sigma_i n_i (\overline{M}_{\mathrm{n}})_i}{\Sigma_i n_i}$

(2-6) $\overline{M}_{\mathrm{w}} \equiv \dfrac{\Sigma_i m_i (\overline{M}_{\mathrm{w}})_i}{\Sigma_i m_i} = \dfrac{\Sigma_i n_i (\overline{M}_{\mathrm{n}})_i (\overline{M}_{\mathrm{w}})_i}{\Sigma_i n_i (\overline{M}_{\mathrm{n}})_i}$

(2-7) $\overline{M}_{\mathrm{z}} \equiv \dfrac{\Sigma_i z_i (\overline{M}_{\mathrm{z}})_i}{\Sigma_i z_i} = \dfrac{\Sigma_i m_i (\overline{M}_{\mathrm{w}})_i (\overline{M}_{\mathrm{z}})_i}{\Sigma_i m_i (\overline{M}_{\mathrm{w}})_i} = \dfrac{\Sigma_i n_i (\overline{M}_{\mathrm{n}})_i (\overline{M}_{\mathrm{w}})_i (\overline{M}_{\mathrm{z}})_i}{\Sigma_i n_i (\overline{M}_{\mathrm{n}})_i (\overline{M}_{\mathrm{w}})_i}$

For *molecularly uniform* fractions i (and only those!), Eqns.(2-6) and (2-7) reduce to $\overline{M}_{\mathrm{w}} = (\Sigma_i n_i M_i^2)/(\Sigma_i n_i M_i)$ and $\overline{M}_{\mathrm{z}} = (\Sigma_i m_i M_i^2)/(\Sigma_i m_i M_i) = (\Sigma_i n_i M_i^3)/(\Sigma_i n_i M_i^2)$, respectively.

Certain evaluation techniques deliver z-average molar masses from sedimentation equilibrium experiments (Section 7.3.3.), hence the symbol z for the statistical weight fraction (from German: *Zentrifuge* = centrifuge). Scattering experiments furnish z-averages of molecule dimensions (Chapter 5). All these averages are heavily weighted towards larger molecules since the z-parameter has the physical unit $g^2 \; mol^{-1}$.

Since $M_i > 1$, these equations always imply $\overline{M}_{\mathrm{z}} \geq \overline{M}_{\mathrm{w}} \geq \overline{M}_{\mathrm{n}}$. The ratio of any of these averages (or higher ones) can serve as a measure of non-uniformity with respect to molar masses, i.e., the **polymolecularity** ("polydispersity") of the polymer. The polymolecularity is usually expressed by the **polymolecularity index** $\overline{M}_{\mathrm{w}}/\overline{M}_{\mathrm{n}}$. This index is unity for uniform polymers, ca. 1.04 for "narrowly distributed" polymers from so-called living polymerizations (Section 3.4.4.) and 2.0 for linear polymers from equilibrium polycondensations (Section 4.1.3.). It may reach values of 20-40 for some industrial polymers.

Hydrodynamic methods (diffusion, sedimentation rate, viscosity, etc.) do not deliver simple averages $\overline{M}_{\mathrm{n}}$, $\overline{M}_{\mathrm{w}}$, $\overline{M}_{\mathrm{z}}$, etc., but more complex ones. These averages may be written as combinations of moments of distribution functions. The qth moment $\mu_g^{(\mathrm{q})}$ of the g-distribution of molar masses about the origin (reference value of zero) is defined as

(2-8) $\mu_{\mathrm{g}}^{(\mathrm{q})} \equiv \dfrac{\Sigma_i g_i M_i^{\mathrm{q}}}{\Sigma_i g_i} = \Sigma_i G_i M_i^{\mathrm{q}} \quad ; \quad G_i \equiv g_i / \Sigma_i g_i$

The mass-average molar mass of a polymer with uniform fractions i (but *not* for non-uniform ones!) is therefore the ratio of the second moment of the number distribution of molar masses to the corresponding first moment:

(2-9) $\overline{M}_{\mathrm{w}} \equiv \dfrac{\mu_{\mathrm{n}}^{(2)}(M)}{\mu_{\mathrm{n}}^{(1)}(M)} = \dfrac{\Sigma_i n_i M_i^2}{\Sigma_i n_i M_i}$

The order q of the moment may have any value: positive or negative, rational or irrational, integer or fraction, etc. A moment of the molar mass distribution thus often has a different physical unit than the molar mass.

The existence of molar mass averages instead of a single molar mass seems to be peculiar to polymers but is in fact a general phenomenon for all chemical compounds and many elements. In elements and low-molar mass compounds, it is caused by the presence of isotopes; in polymers in addition by differences in the number of constitutional repeating units per molecule. Furthermore, number-average and mass-average molar masses of atoms and small molecules often differ so little that they are equal within limits of experimental error. Hydrogen contains 99.985 % ^{1}H ($A_r =$ 1.007 825 035) and 0.015 % ^{2}H ($A_r =$ 2.014 101 779); its relative number-average atomic mass (by chemical methods) is $\overline{A}_{n,r} = 1.007\ 975\ 976$ and its relative mass-average atomic mass (by physical methods) is $\overline{A}_{w,r} = 1.008\ 126\ 64$. The polymolecularity index of hydrogen is 1.000 15.

Expressions for averages of molar masses M, relative molecular masses M_r and degrees of polymerization X cannot be schematically transferred to other properties P (end-to-end distances, radii, diffusion coefficients, etc.). The mass-average of a property P of a substance with molecularly uniform molecules is given by:

$$(2\text{-}10) \quad \overline{P}_w \equiv \frac{\sum_i m_i P_i}{\sum_i m_i} = \frac{\sum_i n_i M_i P_i}{\sum_i n_i M_i} \quad \text{and not by} \quad \overline{P}_w = \frac{\sum_i n_i P_i^2}{\sum_i n_i P_i}$$

2.2.3. Distribution Functions

Normalized distribution functions of properties are mathematical functions that describe the proportion of a polymeric substance that has a specific value (or range of values) of the property. Distribution functions must be always specified by their continuity, summation, and statistical weights; they can often by described by certain mathematical functions. These mathematical functions can sometimes be derived by polymerization equilibria or kinetics; sometimes they serve only for curve-fitting.

Continuity: Distribution functions of molar masses and degrees of polymerization are intrinsically **discrete (discontinuous)** because degrees of polymerization can only assume positive integers (Fig. 2-7). Successive degrees of polymerization of

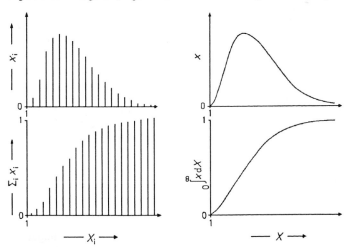

Fig. 2-7 Schematic representation of mole fractions x of degrees of polymerization X (or other properties P). Left: discontinuous; right: continuous. Top: differential; bottom: integral.

molecules differ only by 1, however, so that discontinuous distributions of high molar mass samples can be replaced by **continuous** distribution functions.

Summation: Continuous and discontinuous distribution functions are either integral or differential. **Integral (cumulative)** distribution functions give the proportion of the population for which the property does not exceed a certain value. **Differential** distribution functions describe the proportion of the population to which a certain property value (or a range of values) applies.

Statistical weight: Reaction mechanisms predict theoretical **number-distribution functions** which express the proportions of population by mole fractions. However, most physical methods deliver **mass-distribution functions** that can be converted into number-distribution functions with the help of Eqn.(2-4). Number-distribution and mass-distribution functions differ considerably (Fig. 2-8).

Mathematical functions are usually named after their inventors or discoverers. The **Gaussian distribution** corresponds to the error law in repeated measurements; it is thus called a **normal distribution** in statistics. Gaussian distributions are symmetric about the **median** X_{median} which equals the number-average (Fig. 2-8).

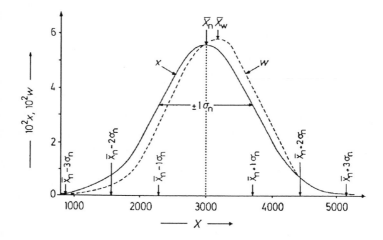

Fig. 2-8   Solid line: "Continuous" differential Gaussian number distribution of degrees of polymerization, $x = f(X)$, for a polymer with $\overline{X}_n = 3000$ and $\overline{X}_w = 3170$, i.e., a number standard deviation of $\sigma_n = 714$ (Eqn.(2-12)). The median is the number-average degree of polymerization. Broken line: The same Gaussian number distribution plotted as $w = f(X)$. Note that the symmetric Gaussian *number* distribution becomes asymmetric if n is replaced by w in the plot but not in the function itself. Conversely, a Gaussian *mass* distribution with $\overline{X}_w = 3170$ would be symmetric around $\overline{X}_w$ in a plot $w = f(X)$ but not in a plot $x = f(X)$.

The differential number-distribution function of a **Gaussian distribution** of property X reads:

$$(2\text{-}11) \quad x(X) = \frac{1}{(2\pi)^{1/2}\,\sigma_n} \exp\left[\frac{-(X - X_{median})^2}{2\,\sigma_n^2}\right]$$

The **number standard deviation** σ_n of the number distribution

$$(2\text{-}12) \quad \sigma_n = (\overline{X}_w \overline{X}_n - \overline{X}_n^2)^{1/2}$$

is an *absolute* measure of the width of a *Gaussian* distribution since a value of ($\overline{X}_n \pm$ 1 σ_n) always corresponds to a mole fraction of $x = 68.26$ %, a value ($\overline{X}_n \pm 2\,\sigma_n$) always to $x = 95.44$ %, and a value of ($\overline{X}_n \pm 3\,\sigma_n$) always to $x = 99.73$ %, etc. The standard deviation, as defined by Eqn.(2-12), can also be used as a *relative* measure of the width of other than Gaussian distributions but the above Gaussian ranges of mole fractions are meaningless for non-Gaussian distributions.

Gaussian distributions are important for many polymer properties. Since they allow negative values of properties, and negative degrees of polymerization do not exist, they should not, strictly speaking, be applied to molar mass distributions. It does not matter, however, for high degrees of polymerization and narrow distributions.

Logarithmic normal distributions (LN distributions) are Gaussian distributions with the natural logarithm of the property as variable instead of the property itself:

$$(2\text{-}13) \quad x(X) = \frac{1}{(2\pi)^{1/2} \sigma_n^*} \exp\left[\frac{-(\log_e X - \log_e X_{median}^*)^2}{2(\sigma_n^*)^2} \right]$$

LN distributions are skewed and not symmetric about the median if X instead of $\log_e X$ is chosen as the abscissa. In contrast to some other types of distribution, they possess a maximum in the differential number distribution $x = f(X)$ (Fig. 2-9); this maximum does not appear at the position $X = \overline{X}_n$, however. The ratio of two consecutive simple averages is constant and independent of the value of the number-average of the degree of polymerization in LN distributions; it is always $\overline{X}_w / \overline{X}_n = \overline{X}_z / \overline{X}_w$.

Many polymerizations deliver **Schulz–Zimm (SZ) distributions** of degrees of polymerization. The differential number and mass distributions are given here by:

$$(2\text{-}14) \quad x = \frac{(\varsigma/\overline{X}_n)^{\varsigma+1} X^{\varsigma-1} \overline{X}_n \exp(-\varsigma X/\overline{X}_n)}{\Gamma(\varsigma+1)} \;\; ; \quad w = \frac{(\varsigma/\overline{X}_n)^\varsigma X^\varsigma \exp(-\varsigma X/\overline{X}_n)}{\Gamma(\varsigma+1)}$$

where $\Gamma(\varsigma + 1)$ is the gamma-function of ($\varsigma+1$). The degree of coupling, ς, denotes the number of independently grown chains that have been coupled to a dead chain.

In SZ distributions, the simple arithmetic averages are interconnected via:

$$(2\text{-}15) \quad \overline{X}_n/\varsigma = \overline{X}_w/(\varsigma+1) = \overline{X}_z/(\varsigma+2)$$

Schulz–Zimm distributions convert to the **Schulz–Flory distribution** ("**Flory distribution**") for $\varsigma = 1$ and $X \to \infty$ (Fig. 2-9). The Schulz–Flory distribution is also called the **most probable distribution** because it is delivered by many processes involving *high* molar mass polymers such as bifunctional equilibrium polycondensations (Section 4.1.3.), free-radical polymerizations with disproportionation (Chapter 3, Appendix A 3-4), and random scissions of polymer chains (Section 4.3.4.).

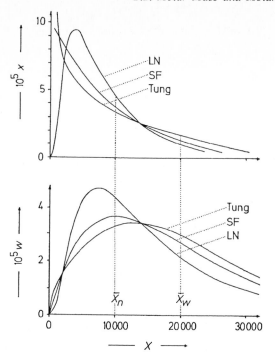

Fig. 2-9 Continuous differential number distributions of degrees of polymerization X for a logarithmic normal distribution (LN), a Schulz–Flory distribution (SF) and a Tung distribution (Tung) shown as number distributions (top) and mass distributions (bottom) for a polymer with a mass-average degree of polymerization of $\overline{X}_w = 20\,000$ and a number-average degree of polymerization of $\overline{X}_n = 10\,000$. Poisson distributions are so narrow for $\overline{X}_n = 10\,000$ that they can be represented only as a vertical line at $X = 10\,000$ since $\overline{X}_w/\overline{X}_n = 1.00010001$.

Poisson distributions exhibit the differential number and mass distributions:

$$(2\text{-}16) \quad x = \frac{(\overline{X}_n - 1)^{X-1}\exp(1 - \overline{X}_n)}{(X-1)!\,(X)} \quad ; \quad w = \frac{X(\overline{X}_n - 1)^{X-1}\exp(1 - \overline{X}_n)}{(X-1)!\,\overline{X}_n}$$

where $(X\text{-}1)!$ is the factorial of $X{-}1$. Poisson distributions are very narrow distributions that are approximately observed for so-called living polymerizations (Section 3.4.4.). For $X \to \infty$, the polymolecularity index approaches $\overline{X}_w/\overline{X}_n = 1$ since:

$$(2\text{-}17) \quad \overline{X}_w/\overline{X}_n = 1 + (1/\overline{X}_n) - (1/\overline{X}_n)^2$$

The **Kubin–Verteilung** is an empirical **generalized exponential distribution (GEX distribution)** with the adjustable parameters β, λ and γ. It includes the regular exponential distribution ($\gamma = 1$), the Schulz–Flory distribution ($\gamma = 1$; $\lambda = \varsigma - 1$) and the Tung distribution ($\lambda = \gamma - 2$). Its mass distribution is given by:

$$(2\text{-}18) \quad w = \gamma\beta^{(\lambda + 2)/\gamma}[\Gamma\{(\lambda + 2)/\gamma\}]^{-1}X^{\lambda+1}\exp(\beta X^{\gamma})$$

2.2.4. Experimental Methods

Experimental methods for the determination of relative molecular weights, reduced molecular masses, and molar masses are subdivided into absolute, equivalent, and relative methods. They deliver various averages and cover different ranges of molar masses (Table 2-4).

Absolute methods deliver experimental data that can be used to calculate molecular weights and molar masses, respectively, without any assumption about the polymer structure. The determination of reduced molecular masses by mass spectroscopy is discussed in this section. Molar mass determinations by absolute methods rest on equilibrium conditions (osmotic pressure, static light scattering, sedimentation equilibrium, etc.); their discussion is more expedient in the context of the various physical phenomena (Table 2-4).

Equivalent methods require information about the chemical structure. **End group determinations** allow the calculation of relative molecular masses if the constitution of end groups and their number per molecule is known. The sensitivity of end group determinations depends strongly on the experimental method. ^{13}C nuclear magnetic resonance delivers molecular weights up to ca. 8000, titration up to ca. 40 000, radioactive groups up to ca. 200 000, and fluorescent groups up to ca. 1 000 000.

Relative methods are influenced by the chemical and physical structure of the polymer as well as the interaction of polymers with solvent molecules. Relative methods must always be calibrated. The most important relative methods are dilute solution viscometry (Section 5.8.3.) and size-exclusion chromatography (this section).

Mass Spectroscopy

Mass spectroscopy is the only method which delivers the mass of the molecule, i.e., as a ratio of the molecule mass m to the number z of electric charges per molecule. The method separates volatile charged compounds in a magnetic field. Since poly-

Table 2-4   Approximate ranges of important methods for molar mass determinations. A = absolute method, R = relative method, E = equivalent method, n = number-average, v = viscosity-average, w = mass-average, z = z-average, (z) ditto, assumptions required. * Upper limit ca. 10^6 g/mol.

Method	Type	Average	Range in g/mol	Section
Light scattering, static	A	w	> 100	5.2.1.
Viscometry of dilute solutions	R	v	> 200	5.8.3.
X-Ray (or neutron) small angle scattering	A	w	> 500	5.2.2.
Combined sedimentation and diffusion	A	various	> 1000	7.3.2.
Size–exclusion chromatography	R	n, w, z	> 1000	2.2.4.
Membrane osmometry*	A	n	> 5 000	6.3.2.
Ebullioscopy, cryoscopy	A	n	< 20 000	6.3.3.
End group determination (by titration)	E	n	< 40 000	2.2.4.
Vapor phase osmometry	A	n	< 50 000	6.3.3.
Mass spectroscopy	A	n, w, z	< 200 000	2.2.4.
Sedimentation equilibrium	A	w, (z)	< 1 000 000	7.3.3.
Light scattering, dynamic	R	(z)	< 10 000 000	7.2.1.

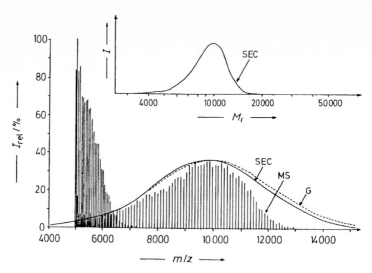

Fig. 2-10   Relative intensities I_{rel} of molecule ions observed by matrix-assisted laser desorption ionization mass spectroscopy (MALDI-MS) of a narrow distribution poly(styrene) as a function of reduced molecular masses m/z. Displacements between signals correspond to the molecular weight $(M_u)_r = 104.15$ of the monomeric unit $-CH_2-CH(C_6H_5)-$. The distribution of singly protonated molecules ranges from $4000 < M_r < 16\,000$; it shows a maximum at $m/z \approx 10\,000$. The distribution curve of doubly charged molecules starts at $m/z \approx 7000$ and extends to lower ratios of m/z; in the figure, it is artificially cut off at $m/z = 5000$. Data of [1].

The distribution from size-exclusion chromatography SEC (solid line) approximates the differential distribution by mass spectroscopy except for high molecular weights. The original SEC plot (insert) shows a logarithmic M_r axis which leads to a non-symmetric SEC curve and an apparent tail at low molecular weights. The true SEC curve is approximately Gaussian.

mers are not volatile and furthermore decompose at elevated temperatures, their molecular mass could not be determined by mass spectroscopy for many years. Newer techniques use a volatile carrier ("matrix") to transport nonvolatile polymers to the magnetic field. These procedures allow the determination of "molecular masses" of uniform or nearly uniform biopolymers up to values of ca. 200 000 g.

The upper limit for the evaluation of "molecular masses" of synthetic polymers by mass spectroscopy is, in general, controlled by the matrix and the width of the molar mass distribution. Because of the latter, conventional synthetic polymers show not only a signal at $m/z \sim M_r + 1$ for a macromolecule charged by a single proton, another signal at $m/z \sim M_r + 2$ for a macromolecule charged by 2 protons, etc., but a spectrum of signals for the singly protonated species, another spectrum for the doubly protonated species, and so on (Fig. 2-10). Mass spectroscopy may fragment the highest molar masses which falsifies the distribution function (see caption to Fig. 2-10).

Size-Exclusion Chromatography

Size-exclusion chromatography SEC is presently the most important method for the determination of molecular weight distributions. It is a special kind of liquid chromatography that separates molecules according to sizes and not according to affinities toward the porous substrate.

Substrates consist of materials with pore sizes between 5 nm and 500 nm. Examples are porous glass or gels such as swollen, cross-linked poly(styrene) for organic solvents, and cellulose, dextrans or poly(acrylamide)s for aqueous solutions. An SEC with a gel as the substrate is also called **gel-permeation chromatography** (GPC) in polymer science and **gel filtration** in the biological sciences.

In SEC and GPC, dilute polymer solutions are placed on the top of a column filled with a solvent-soaked porous substrate. The column is then eluted with the solvent. Very large polymer molecules are too large to enter the pores. Large polymer molecules can move into pores but since there are only a few large pores available to them, they reside, on average, less time in pores than smaller molecules. The large molecules are therefore eluted first. One observes an elution curve, i.e., the concentration of eluted molecules as a function of time or elution volume. Concentrations can be determined via the refractive index of the solutions or spectroscopically. Some instruments also allow the simultaneous determination of molar masses, either directly by light scattering or indirectly by viscometry.

Theoretically, elution curves of uniform polymers should generate sharp signals. In practice, elution curves always have finite widths, never sharp signals, because of diffusion. Elution curves should always be corrected for this diffusion effect.

The maximum of the elution curve is called the **retention volume** V_e. At low molar masses, V_es of homologous polymers approach a molar mass-independent retention volume V_0 (Fig. 2-11) which indicates the total volume that is available for the flow of the eluting solvent. Another molar mass-independent volume V_i is obtained at the high molar mass limit; it is the interstitial volume between substrate particles.

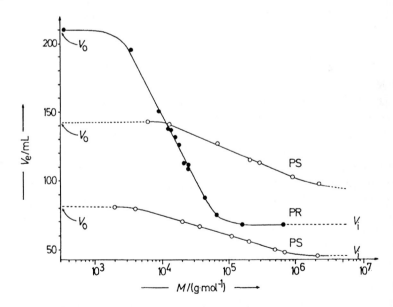

Fig. 2-11 Retention volumes V_e as function of logarithms of molar masses M of spheroidal proteins PR in dilute aqueous buffer solutions on cross-linked dextrans (●) and of poly(styrene)s PS in tetrahydrofuran on two different cross-linked poly(styrene)s (○) as substrate.

In the intermediate molar mass range, retention volumes V_e decrease with increasing molar masses according to $V_e = -K \log_e M$ (Fig. 2-11). The constant K is usually obtained by calibration of the SEC column with narrow distribution poly(styrene)s. Since the calibrating poly(styrene) molecules and the molecules of the test specimen usually exhibit different hydrodynamic volumes at the same molar mass, the resulting molar mass M is not the true molar mass of the specimen. It is rather an apparent molar mass, i.e., the molar mass the specimen molecules would have if they possessed the same hydrodynamic volumes as the poly(styrene) molecules. For example, branched molecules have smaller hydrodynamic volumes than unbranched molecules with the same molar mass and the same monomeric units (see Section 5.5.3.).

Better suited are so-called "universal calibrations" with the product $[\eta]M$ instead of M. Intrinsic viscosities $[\eta]$ are obtained from dilute solutions (Chapter 5.8.). Since they have the physical unit volume/mass and molar masses have the unit mass/amount, the product $[\eta]M$ carries the unit volume/amount (e.g., cm^3/mol), that is, a molar volume. A plot of $V_e = f(\log_{10} [\eta]M)$ often gives a single line for very different polymers in the same solvent and with the same carrier.

2.3. Configuration

2.3.1. Stereoisomers

Isomers are molecules composed of identical numbers of corresponding atoms in different arrangements; isomers thus differ in certain properties. In **constitutional isomers**, atoms of molecules with equal composition are connected to each other in different sequences. **Stereoisomers** are molecules with the same sequence of atoms but different spatial arrangements. An isomer is thus either a constitutional isomer or a stereoisomer; it cannot be both.

Historically, stereoisomers are subdivided into configurational and conformational isomers according to energy barriers or into enantiomers and diastereomers according to symmetry properties:

By definition, two **configurational isomers** are separated by a "high" energy barrier and two **conformational isomers** by a "low" one. Configurational isomers can be isolated as pure materials if their life time is large compared to the time needed for their preparation. The difference between configurational and conformational isomers thus becomes physically meaningless if "fast" methods are used. Conformational isomers with life times of the order of microseconds appear as defined species (i.e., as configurational isomers), for example, on examination by electron spin resonance, a fast method.

Enantiomers relate to each as mirror images (G: *enantios* = opposite, from *en* (in) and *anti*). The two antipodes (enantiomers of the same chemical compound) are always chiral like the left and right hand (G: *cheir* = hand); they rotate the plane of polarized light in opposite directions. A mixture of equal amounts of two antipodes is called a racemic mixture or racemate because it was first found as the optically inactive form of tartaric acid which is called racemic acid (L: *acidum racemicum*; from L: *racemus* = bunch of berries). In racemates, the chirality is compensated externally; a racemate is optically inactive. If a molecule itself contains two centers of opposite chirality, then the overall chirality is compensated internally. Such "meso compounds" are achiral and optically inactive.

Diastereomers do not relate to each other as mirror images. They may be either achiral or chiral.

2.3.2. Ideal Tacticity

In polymer molecules, one has to distinguish between configurational base units, configurational repeating units, and stereorepeating units. Examples are the constitutional repeating units -$CH(CH_3)$- in poly(methylmethylene) PM and -$CH(CH_3)CH_2$- in poly(propylene) PP. In poly(methylmethylene), the two possible constitutional repeating units Ia and Ib are identical with the two possible **configurational repeating units** (here in Fischer projection). In poly(propylene), the two possible constitutional repeating units -$CH(CH_3)CH_2$- (II) and -$CH_2CH(CH_3)$- (III) each have two different configurational repeating units IIa + IIb and IIIa + IIIb, respectively:

The two configurational repeating units IIa and IIb are mirror images of each other; they are **enantiomers**. The units IIIa and IIIb are also enantiomers. The units IIa + IIIa are however **diastereomers** (and likewise the units IIb + IIIb) because they are derived from two different constitutional base units.

Note that the classification of IIa + IIb and IIIa + IIIb as enantiomers and IIa + IIIa and IIb + IIIb as diastereomers refers to the pictured units and *not* to infinitely long chains that result from the repetition of units or groups of units (see below).

Stereorepeating units are configurational repeating units that have defined configurations at all sites of stereoisomerism in the main chain. All units IIa, IIb, IIIa and IIIb have such defined configurations. A sterically undefined configurational unit of poly(propylene) is symbolized by –$CH(CH_3)$–CH_2–.

A **stereoregular polymer** is correspondingly a regular polymer whose molecules consist of only one species of stereorepeating units arranged in only one type of sequence. **Tactic polymers** result if stereorepeating units are replaced by configurational repeating units in regular polymers (G: *taxis* = arrangement). A stereoregular polymer is always tactic but a tactic polymer is not necessarily stereoregular because not *all* centers of stereoisomerism in the main chain need to be defined.

The simplest stereorepeating units of poly(propylene) contain one, two or four monomeric units:

$$
\begin{array}{cccc}
\underset{\mathrm{CH_3}}{\overset{\mathrm{H}}{-\mathrm{C}-\mathrm{CH_2}-}} &
\underset{\mathrm{CH_3}\quad\mathrm{H}}{\overset{\mathrm{H}\quad\quad\mathrm{CH_3}}{-\mathrm{C}-\mathrm{CH_2}-\mathrm{C}-\mathrm{CH_2}-}} &
\underset{\mathrm{CH_3}\quad\mathrm{CH_3}\quad\mathrm{H}\quad\mathrm{H}}{\overset{\mathrm{H}\quad\quad\mathrm{H}\quad\quad\mathrm{CH_3}\quad\mathrm{CH_3}}{-\mathrm{C}-\mathrm{CH_2}-\mathrm{C}-\mathrm{CH_2}-\mathrm{C}-\mathrm{CH_2}-\mathrm{C}-\mathrm{CH_2}-}}
\end{array}
$$

IT   ST   HT

The repetition of these stereorepeating units leads to **isotactic** (IT), **syndiotactic** (ST) or **heterotactic** (HT) polymer chains (G: *isos* = equal, *syn* = together, *dios* = two). An isotactic repeating unit consists of one constitutional repeating unit. A syndiotactic repeating unit is formed by two enantiomeric configurational repeating units. A heterotactic unit is composed of *three* monomeric units (the first three or the last three of HT); the repetition of such a heterotactic triad composed of, for example, the first three units of HT does not lead to a heterotactic polymer $+$IT-*alt*-ST$]_{\overline{n}}$, however, but rather to $+$IT-*per*-ST-*per*-ST$]_{\overline{n}}$.

The classifications "isotactic", "syndiotactic", etc., are based on the **relative configuration** about the centers of stereoisomerism if one proceeds from one end of the main chain to the other (Fig. 2-12). Relative configurations thus differ from the "absolute configurations" of organic chemistry which consider the configuration about *each* center of stereoisomerism relative to the ligand with the lowest seniority.

The relative configurations of isotactic polymers are identical whether one starts with the left or the right end of the chain. In isotactic polymer molecules with -CHR- as constitutional repeating units, ligands H, R and ~~ (chain) always follow each other in the same direction (Figs. 2-12 to 2-14). Poly(α-amino acid)s with constitutional repeating units -NH-*CHR-CO- from D-peptide residues are isotactic and so are poly-

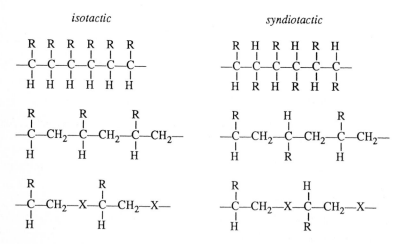

Fig. 2-12   Fischer projections (projections of cis conformations on a plane) of isotactic and syndiotactic polymers. These polymers are stereoregular and tactic if X is, for example, CH$_2$, O or NH. They are tactic but not stereoregular if X is, e.g., CHR'.

(α-amino acid)s with L-peptide units. Poly(α-amino acid)s composed of alternating D- and L-residues are syndiotactic, however. The chain units -*CHR- in these peptide units are chiral because the four immediate ligands NH, H, R, and CO of the chain atom *C are different, and also asymmetric, because they do not lie in a plane.

Poly(propylene) with the constitutional unit -CH_2-$^0CHCH_3$-CH_2- does not possess D- and L-units, however, because two of the four ligands CH_2, H, CH_3 and CH_2 of the chain atom ^0C are identical. With respect to tacticity (relative configuration), prochiral chain units -$^0CHCH_3$- of poly(propylene) behave like chiral ones, however.

Isotactic and syndiotactic poly(propylene)s PP are not optically active. Consider an it-PP with one isopropyl and one isobutyl end group each, i.e., $(CH_3)_2CH[CH_2CH(CH_3)]_nCH_2CH(CH_3)_2$ (Fig. 2-13). These polymer molecules Ia and Ib can begin with either an R or an S configuration which leads indeed to two types of enantiomorphic chains. If however the contribution of the two end groups can be neglected (similar end groups or infinitely long chains), then a mirror plane exists in the center of the chain: one half of the chain has an R configuration and the other half has an S configuration. The chain is meso and not chiral. In st-PP, absolute configurations R and S alternate. An infinitely long chain can therefore not be optically active. Chains of st-PP can thus not be enantiomers. The two chains IIa and IIb are identical; they are not "racemic".

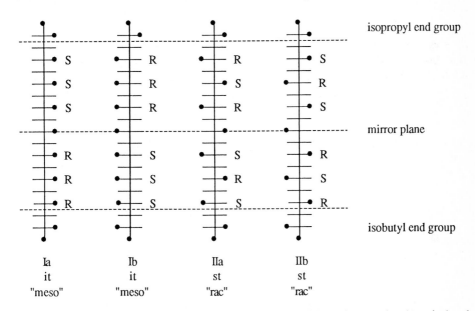

Fig. 2-13 Fischer projections of the heptamers of propylene with one isopropyl and one isobutyl end group each. R and S are defined as absolute configurations. See text.

Ligands appear on the same side or on different sides of a chain in projections of three-dimensional stereo formulas onto a plane if different chain conformations (Section 2.4.) are viewed. Fischer projections are based on cis conformations. Ligands R of $+CH_2-CHR+_n$ chains always appear on the same side if isotactic polymers are presented by Fischer projections and on different sides for syndiotactic polymers (Fig. 2-12). For projections of trans conformations (zigzag chains) this statement applies only if the constitutional repeating units contain even numbers of chain atoms (Fig. 2-14).

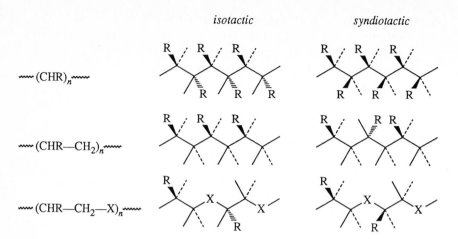

Fig. 2-14 Sections of isotactic and syndiotactic polymers in hypothetical all-trans conformations. The structures correspond for R = CH$_3$ and X = O from top to bottom to poly(methylmethylene), poly(propylene) and poly(propylene oxide).

Hemitactic polymer molecules contain two different, alternating types of central atoms (G: *hemi* = one half). The relative configurations around the first, third, fifth ... central atom are identical but the configurations around the second, fourth, sixth ... are not (Fig. 2-15); only one half of all central atoms are tactic. Hemitactic polymers can be obtained with certain metallocene catalysts (Section 3.6.2.).

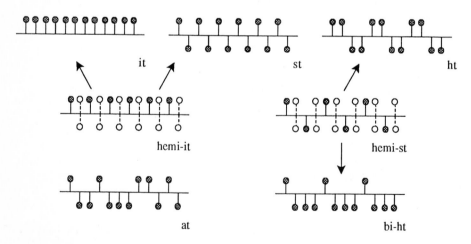

Fig. 2-15   Sections of isotactic (it), syndiotactic (st), heterotactic (ht), hemi-isotactic (hemi-it), hemi-syndiotactic (hemi-st), atactic (at) and bi-heterotactic (bi-ht) chains. Exactly defined configurations are symbolized by ●—, arbitrary configurations by ○- - - -○.

Ditactic polymers possess two centers of stereoisomerism per monomeric unit. Examples are polymers from 1,2-disubstituted ethylenes. Poly(2-pentene) = poly[(1-ethyl)(2-methyl)ethene], $+$CH(C$_2$H$_5$)–CH(CH$_3$)$+_{\overline{n}}$, has four ideal configurations: two different diisotactic ones and two different disyndiotactic ones. Isotacticity can be generated by the relative position of the two central atoms of one monomeric unit or by two central atoms from a pair of adjacent monomer units. The relative positions of the monomeric units may be identical (**erythro** polymers; G: *eruthros* = red) or alternating (**threo** polymers; unknown origin) (Fig. 2-16).

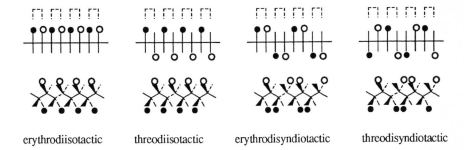

| erythrodiisotactic | threodiisotactic | erythrodisyndiotactic | threodisyndiotactic |

Fig. 2-16 The four possible configurations of ditactic polymers shown for a section composed of four monomeric units. Each monomeric unit contains two different substituents ● and O. Top: Fischer projections (cis conformations); bottom: stereo projections (trans conformations). Groups on wedges are above the paper plane, those on broken lines, below.

Note that the erythro and threo forms of *these* disyndiotactic polymers are indistinguishable. The difference between erythro and threo configurations can be seen for:

erythrodisyndiotactic   threodisyndiotactic

Tacticity is also observed for polymers with **geometric isomerism** of the constitutional repeating units. Examples are 1,4-poly(isoprene)s $+CH_2-C(CH_3)=CH-CH_2+_{\overline{n}}$ with carbon-carbon double bonds in the 2,3-position that are obtained from 2-methyl-1,3-butadiene (isoprene). These polymers may be either cis-tactic (ct) or trans-tactic (tt). The ct-structure corresponds to an E-isomer and the tt-structure to a Z-isomer. In 1,2-poly(isoprene) and 3,4-poly(isoprene), polymer molecules may be either isotactic or syndiotactic:

1,4-cis (E) (ct)   1,4-trans (Z) (tt)   1,2 (it or st)   3,4 (it or st)

Natural rubber (cis) and balata (trans) are examples of natural 1,4-poly(isoprene)s.

2.3.3. Real Tacticity

Most polymer molecules are neither 100 % isotactic nor 100 % syndiotactic. Even protein molecules are not always entirely composed of L-α-amino acid residues; such molecules are not 100 % isotactic.

The tacticity of polymer chains with configurational defects can be described by the proportion and sequence of tactic J-ads (diads, triads, tetrads, etc). Each monomeric unit belongs to two tactic diads (dyads), three tactic triads, four tactic tetrads, etc. A cyclic polymer with 12 mers thus contains 12 tactic diads, 12 tactic triads, 12 tactic tetrads, etc.

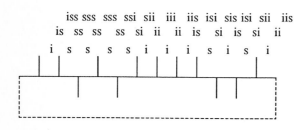

Diads can be either isotactic (i) or syndiotactic (s). The sum of the mole fractions x of the two diad types equals unity ($x_i + x_s \equiv 1$).

Triads are either isotactic (ii), syndiotactic (ss) or heterotactic (is or si); the sum of their mole fractions is also unity ($x_{ii} + x_{ss} + x_{is} + x_{si} \equiv 1$). Many experimental methods deliver only the sum of the mole fractions of the two hetereotactic triads is and si. It is thus convenient to set $x_{is} + x_{si} \equiv x_{|is|}$ so that $x_{ii} + x_{ss} + x_{|is|} \equiv 1$. For tetrads, one obtains $x_{iii} + x_{iis} + x_{isi} + x_{sii} + x_{iss} + x_{sis} + x_{ssi} + x_{sss} \equiv 1$.

General relationships exist between the different types of J-ads because each triad is composed of two diads, each tetrad of two triads, etc. Isotactic diads i show up in isotactic triads ii as well as in heterotactic triads is and si. The interrelationship between the two types of diads, four types of triads, and eight types of tetrads is thus

$$(2\text{-}19) \quad x_i = x_{ii} + (x_{is} + x_{si})/2 = x_{ii} + (1/2)\, x_{|is|} = 1 - x_{ss} - (1/2)\, x_{|is|} \equiv 1 - x_s$$

$$(2\text{-}20) \quad x_{ii} = x_{iii} + (1/2)(x_{iis} + x_{sii}) = 1 - [x_{sss} + (1/2)(x_{ssi} + x_{iss})] - x_{|is|} \equiv 1 - x_{ss} - x_{|is|}$$

A truly **atactic polymer** contains equal amounts of the same class of J-ads (diads: $x_{ii} = x_{is} = x_{si} = x_{ss}$, etc.). The label "atactic" is however generally attached to those polymers that are neither predominantly isotactic nor predominantly syndiotactic.

The cyclic polymer shown above exhibits $x_i = x_s = 1/2$, $x_{ii} = x_{is} = x_{si} = x_{ss} = 1/4$. With respect to diads and triads, it can be considered a truly atactic polymer. However, it is not a true atactic polymer because its tetrads show not only the equalities $x_{iii} = x_{iss} = x_{sis} = x_{ssi}$ and $x_{iis} = x_{isi} = x_{sii} = x_{sss}$ but also the inequality $x_{ssi} \neq x_{iis}$.

Atactic polymers contain isotactic sequences I which are one diad long (i), two diads long (ii), three diads long (iii), and so forth; their number-average degree of polymerization is $\overline{X}_{I,n}$. The same is true for syndiotactic sequences S (= s, ss, sss, ...) with the number-average degree of polymerization $\overline{X}_{S,n}$. An isotactic sequence changes into a syndiotactic sequence via a heterotactic triad [...iiii(is)ssss...] (note that a cyclic polymer with one isotactic block and one syndiotactic block has *two* cross-over sequences!). The number-average degree of polymerization of *all* sequences, $\overline{X}_{seq,n}$, is thus given by the inverse of the sum of the number-average degrees of polymerization of isotactic and syndiotactic sequences:

$$(2\text{-}21) \quad \overline{X}_{seq,n} = 1/(x_{is} + x_{si}) = (1/2)[\overline{X}_{I,n} + \overline{X}_{S,n}]$$

and the number-average degree of polymerization of one type of the sequence by:

$$(2\text{-}22) \quad \overline{X}_{I,n} = 2x_i/(x_{is} + x_{si}) \quad ; \quad \overline{X}_{S,n} = 2x_s/(x_{is} + x_{si})$$

2.3.4. Experimental Methods

Nuclear magnetic resonance spectroscopy (NMR) is the most important method for the determination of tacticity since chemical shifts of signals from ^1H, ^{13}C, ^{19}F, etc. depend on the configuration of the chain atom. In syndiotactic poly(methyl methacrylate) st-PMMA, both H atoms of each CH_2 group of the chain are in chemically equivalent environments since each H is flanked by a CH_3 group and a $COOCH_3$ group. In it-PMMA, the H atoms of the methylene groups are not in chemically equivalent environments, however, because one H is surrounded by two CH_3 groups and the other H by two $COOCH_3$ groups. The two equivalent H atoms of the methylene groups of st-PMMA lead to a single signal but the two non-equivalent H atoms of it-PMMA produce an AB quartet (Fig. 2-17). The conformations of these groups are immaterial since one measures only a time averaged signal.

st—PMMA   it—PMMA

Since the two H atoms of the CH_2 chain units of it-PMMA are NMR-spectroscopically non-equivalent, they have also been called "meso" in the polymer literature. The mole fraction of isotactic diads is then given the symbol m (instead of x_i), the mole fraction of isotactic triads by mm (instead of x_{ii}), etc. In a wrong analogy, H ligands of CH_2 are said to be "racemic" and the mole fraction of st-diads is given the symbol r (instead of x_s) etc. These terms deviate from the accepted definitions of "meso" and "racemic" in organic chemistry; they are misleading and superfluous.

The differences in the signals from the CH_2 groups help to identify the positions of the signals from the protons of the α-CH_3 groups. Since both st-PMMA and it-PMMA exhibit only one signal from α-CH_3 and at-PMMA shows three, these signals must originate from triads. With increasing chemical shift, they are caused by ss, (si + is), and ii triads.

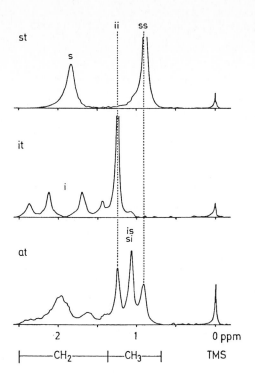

Fig. 2-17   Signals of CH_2 and α-CH_3 groups as shown by the 60 MHz proton magnetic resonance spectra of a syndiotactic (st), an isotactic (it), and an atactic (at) poly(methyl methacrylate).
Signals of protons of methylester groups -$COOCH_3$ are not shown. TMS = Reference signal from
tetramethylsilane $Si(CH_3)_4$.

60 MHz proton magnetic resonance spectra deliver usually only proportions of
the various diads and triads. Tetrads and pentads can be observed at the higher field
strengths of (220-750) MHz. Correlations of signal positions and triad types by proton magnetic resonance spectroscopy can be very difficult, however, and one often
needs X-ray diffraction data of crystallized polymers for calibration. Higher J-ads
can often be determined by ^{13}C spectra. Tacticities may also be obtained from spin–
spin and spin–lattice relaxation times. Two-dimensional NMR allows the unequivocal
correlation of signals and J-ad types. Tacticities can not only be measured in
dissolved polymers but also by ^{13}C NMR spectroscopy of solid polymers.

All other methods for the determination of tacticities are either confined to diad
fractions or to specific chemical structures. Examples are infrared spectroscopy (differences in conformation generated by tactic diads), optical activity (chiral groups
only), X-ray methods (crystallinity required), glass and melting temperatures (qualita-
tive only), or chemical reactions (in special cases).

Industry determines an "isotacticity index" which is the percentage of a poly(propylene) specimen that is insoluble in boiling hexane. Tests based on solubility are, of
course, very ambiguous and can be applied only if the chemical and thermal history
(morphology!) of the specimen is well known.

2.4. Conformation

"Conformation" is the term given by organic chemistry to those spatial arrangements that are produced by rotation of atoms and atomic groups around a single bond. In macromolecular science, such conformations are called **microconformations** or **local conformations**. A polymer molecule possesses a sequence of microconformations along the chain which leads to its **macroconformation** or **molecular conformation** (IUPAC).

2.4.1. Microconformation

Microconformations are mainly determined by constitution and configuration. The population of the various microconformations in a chain depends in addition on the temperature and the interaction of the polymer molecule with the surrounding molecules. Macroconformations may thus differ in solution (Chapter 5), melt (Section 8.2.) and crystalline states (Section 8.3.). Since it is irrelevant for the spatial arrangements of monomeric units whether macroconformations are caused by chemical or physical bonds between chain units, macromolecular physics usually refers to the spatial distribution of chain units as **configurational statistics**.

Three bond lengths $b(A_i,B_j)$, $b(B_j,C_k)$ and $b(C_k,D_l)$, two bond angles $\tau(A_i,B_j,C_k) = \tau(B_i)$ and $\tau(B_j,C_k,D_l) = \tau(C_k)$, and the torsion angle (dihedral angle) θ between the A_i–B_j–C_k plane and the B_j–C_k–D_l plane are required to describe the microconformation of the "chain" of a molecule A–B–C–D (see Fig. 2-18). The torsion angle determines the relative spatial positions of "bonded" ligands to "non-bonded" ones. The H atoms at the $^\alpha C$ of ethane H–$^\alpha CH_2$–$^\beta CH_2$–H are "bonded" to $^\alpha C$, therefore, the H atoms at the $^\beta C$ are "non-bonded" to $^\alpha C$ and *vice versa*.

A rotation of the D_l-atom around the extended B_j–C_k axis creates an infinitely large number of microconformations. Only a few of these microconformations are energetically preferred, however. In methane CH_4, the four angles H–C–H always adopt the dihedral angles of a tetrahedron ($\tau = 109°28'$). The C–C–C angle is,

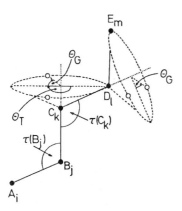

Fig. 2-18   Rotation around chain bonds. The chain atom D_l moves in a circle (broken line) around the extension of the axis B_j–C_k at constant bond length $b(C_k$–$D_l)$ and constant bond angle $\tau(C_k)$. In this example, it resides preferentially in the trans-position (●) and the two gauche positions (O). The chain atom E_m adopts similar positions. All chain atoms ● shown are in the trans conformation with torsional angles $\theta_T = 180°$ (IUPAC and organic chemistry convention); the two gauche conformations (O) are correspondingly at torsional angles of $\theta_G = \pm 60°$. Cis positions are not shown; they are at $\theta_C = 0°$.

however, expanded to 111.5° from 109°28' in poly(methylene) $+CH_2\frac{}{}_n$ because of the different space requirements of C and H atoms and the differing polarities of C-C and C-H bonds. Repulsion between CH_3 groups leads to an energy minimum at the trans conformation T of butane $H_3C-CH_2-CH_2-CH_3$ where the methyl groups are farthest away from each other (Fig. 2-19). Locally shallower energy minima exist for the two gauche conformations G^+ and G^-. Butane has the energy maximum at the cis conformation C and two local maxima at the two anti conformations.

Macromolecular chemistry and organic chemistry use different conventions for the designation of preferred conformations. Microconformations T (trans), A (anti), G (gauche) and C (cis) correspond to conformations ap (antiperiplanar), ac (anticlinal), sc (synclinal) and sp (synperiplanar) in organic chemistry (see top of Fig. 2-19). IUPAC and organic chemistry assign the torsion angle θ = 0° to the cis conformation; most polymer scientists use $\theta = 0°$ for the trans conformation, however. From here on, the polymer convention will be used unless otherwise noted.

The energy difference between the lowest energy microconformation and a higher energy one is called the **conformational energy**, for example, ΔE_{TG} between trans and gauche conformers (Fig. 2-19). The activation energy for the transfer from one type of conformation to the other is the **rotational barrier**; an example is $\Delta E_{TG}^{\ddagger}$ for the transfer from trans to gauche. Such rotational barriers are $\Delta E_{TG}^{\ddagger} \approx 13$ kJ/mol for $CH_3CH_2-CH_2CH_3$ (Fig. 2-18) and $\Delta E_{TG}^{\ddagger} \approx 2.1$ kJ/mol for $\sim CH_2-COOCH_2\sim$. Since such barriers can be overcome by thermal energy ($RT = 2.48$ kJ/mol at 25°C), atoms and groups may undergo (hindered) rotations around chain bonds. At any time, isolated chains can therefore adopt many macroconformations unless they are prevented from doing so by specific intermolecular interactions.

Molecules with low rotational barriers are **dynamically flexible** because they can *rapidly* achieve many different macroconformations. **Statically flexible** molecules possess many accessible potential minima.

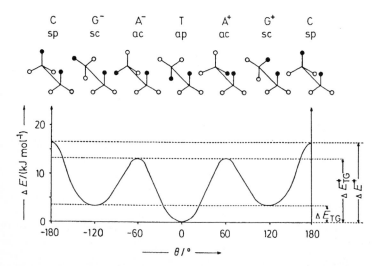

Fig. 2-19   Ideal microconformations and rotational barriers for the center chain bond of butane $CH_3CH_2-CH_2CH_3$ as a function of the torsional angle between methyl groups ● (polymer convention; see Fig. 2-18 for the organic chemistry convention). Nonideal microconformations are given the same names if they do not deviate more than ± 30° from the ideal ones.

In butane $CH_3-CH_2-CH_2-CH_3$, only trans and gauche monads have to be considered. In pentane $CH_3-CH_2-CH_2-CH_2-CH_3$, two conformational diads (C^2-C^3 and C^3-C^4) exist (Fig. 2-20); in hexane, three conformational triads, and so on. The two diads G^+G^- and G^-G^+ are sterically hindered and thus mostly absent in chains with predominant repulsion between adjacent substituents (**pentane effect**).

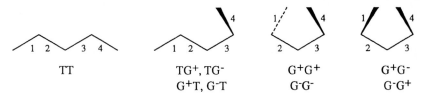

TT TG^+, TG^- G^+G^+ G^+G^-

                    G^+T, G^-T G^-G^- G^-G^+

Fig. 2-20 Conformational diads of pentane $CH_3-CH_2-CH_2-CH_2-CH_3$ (4 chain bonds).

2.4.2. Macroconformation in Crystals

The sequence of microconformations along a chain determines the macroconformation of a polymer molecule, i.e., its shape. In the limiting cases, these sequences can be either completely regular or completely random.

Regular sequences of macroconformations are observed if (1) the polymer molecule itself is regular with respect to its constitution (Section 2.1.5.) and (2) rotations around chain bonds are absent. The last condition is most often fulfilled for polymers in the crystalline state where the close packing of chains in crystal lattices prevents rotations (see Section 8.3. and also Section 9.5.). The type of macroconformation in crystals is determined by the sequence of microconformations and the microconformations are, in turn, controlled by steric and electrostatic effects.

In chains with *predominant repulsion* between ligands at adjacent monomeric units, eclipsed microconformations (cis, anti) are usually absent and one has only to consider trans and gauche microconformations. If all ligands are identical and of small size, the type of microconformation with the lowest energy determines the macroconformation, i.e., most often the trans conformation T (Fig. 2-19). Poly-(ethylene) with its very small H atoms thus crystallizes in the all-trans conformation $[T]_n$ as zigzag chain.

If, however, *steric hindrance* between adjacent groups or each first, third, fifth ... group forces the respective chain bonds to assume gauche conformations and if *gauche* conformations of different signs (+ and −) cannot follow each other, then the polymer chain must adopt a screw-like shape. These chains adopt the macroconformation of a **helix**, a cylindrical spiral (G: *elix* = spiral) (cf. Fig. 2-21).

An example is isotactic poly(propylene) $+CH(CH_3)CH_2\frac{1}{n}$. For this polymer, macroconformations $[TG^+]_n$ and $[TG^-]_n$. are equally likely because the prochiral $-CH(CH_3)-$ group is flanked on both sides by like chain units, i.e., $-CH(CH_3)CH_2-$ and $-CH_2CH(CH_3)-$. it-Poly(propylene) is thus comprised of equal amounts of left-handed and right-handed 3_1 helices. Viewed along its longitudinal axis, a helix is called right-handed if it turns away from the viewer in the clockwise direction.

One screw sense of the helix is preferred over the other if the central group is attached to two unlike chain units. Poly(L-alanine) $\{-NH-CH(CH_3)-CO\}_{\overline{n}}$, for example, forms only right-handed helices because $-CH(CH_3)-$ is attached to $-NH-$ on one side and $-CO-$ on the other (cf. Fig. 5-12).

Electrostatic effects, not repulsive forces, dominate in many polar chains. Such molecules attempt to adopt microconformations with the maximum possible number of gauche interactions between adjacent electron pairs and/or electronegative substituents (**gauche effect**). The two unshared electron pairs of the oxygen atom of poly-(oxymethylene) $\{-O-CH_2\}_{\overline{n}}$ lead to all gauche macroconformations $[G^+]_n$ and $[G^-]_n$ as the lowest energy conformations.

Even anti and cis conformations can be found in polymer chains. The lowest energy form of 1,4-trans-poly(butadiene) $\{-CH_2-CH=CH-CH_2\}_{\overline{n}}$ is $[A^-TA^+T]_n$. Poly-(dimethylsiloxane) $\{-O-Si(CH_3)_2\}_{\overline{n}}$ exists in the macroconformation $[CT]_n$; the all-trans $[T]_n$ conformation of this polymer is more energy rich by ca. 13.3 kJ/mol than the cis–trans conformation $[CT]_n$!

The previous discussion centered around conformations with three-fold rotational potentials. Such potentials are generated by central atoms with four valences (C, Si, etc.). Other chain units lead to two-fold rotational potentials, for example, sulfur atoms in polymeric sulfur $\{-S\}_{\overline{n}}$ or 1,4-substituted phenylene groups.

2.4.3. Macroconformation in Melts and Solutions

All polymer chains with periodic sequences of microconformations ($[T]_n$, $[TG^+]_n$, $[TTG^-]_n$, $[G^+]_n$, etc.), i.e., zigzag chains or helices, have the overall shape of a rod. However, this molecular shape is only preserved if the molecule is externally or internally stabilized against thermal vibrations and/or interactions with solvents.

In crystals, an external stabilization is provided by the close packing of molecules in crystal lattices (Chapter 8). At higher temperatures, vibrations (and sometimes rotations) of segments become stronger. Crystals finally melt (Chapter 9). In melts, chains are no longer forced to adopt a regular periodic sequence of microconformations for the *whole* molecule since thermal energies are sufficiently high to overcome rotational barriers. Periodic sequences of microconformations are occasionally interrupted by "wrong" microconformations. The rod is thus broken several times and eventually adopts the overall conformation of a coil (Fig. 2-21; see also Chapter 5). Depending on polymer constitution and configuration, such a coil may have a random sequence of various types of microconformations. It may also consist of helical segments that are interrupted by random coil segments. Even helices themselves may form random coils if their molar mass is very high. The *overall* shape (macroconformation) is always that of a coil regardless of the local conformations (Chapter 5).

Certain other molecules are internally stabilized by intramolecular hydrogen bonds, for example, most poly(α-amino acid)s. The substituents of these molecules can interact with certain solvents but the solvent is not strong enough to sever the internal hydrogen bonds: the molecule retains its helical structure in solution.

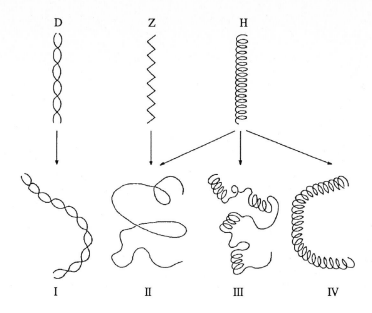

Fig. 2-21 Macroconformations (shapes) of polymer chains. Upper row: in crystalline assemblies; lower row: in dilute solution.

D = Double helix; example: deoxyribonucleic acids.

Z = Zigzag chain; example: poly(ethylene) in extended chain crystals.

H = Single helix; example: crystalline poly(propylene) and poly(γ-benzyl-L-glutamate).

I = double helix forming a worm-like chain at low molar masses M (and a random coil at high M); example: deoxyribonucleic acid in dilute salt solutions at 25°C.

II = random coil; example: poly(ethylene) in xylene at 160°C; poly(γ-benzyl-L-glutamate) in dichloroacetic acid at 25°C.

III = Random coil with helical and random coil segments; example: poly(oxyethylene) in water at 25°C; at-poly(methyl methacrylate) in acetonitrile at 44°C.

IV = Single helix forming a worm-like chain at low M (and a random coil at high M); example: poly(γ-benzyl-L-glutamate) in N,N-dimethylformamide at 25°C.

Literature

H.-G.Elias, Macromolecules, Plenum, New York, 2nd ed. 1984, Vol. I, Chapters 2 (Constitution), 3 (Molar Masses and Molar Mass Distributions), 4 (Configuration), 5 (Conformation).

2.1. CONSTITUTION
W.De Winter, Double Strand Polymers, Revs.Macromol.Sci. **1** (1966) 329

C.G.Overberger, J.A.Moore, Ladder Polymers, Adv.Polym.Sci. **7** (1970) 113

P.A.Small, Long-Chain Branching in Polymers, Adv.Polym.Sci. **18** (1975) 1

Y.Ikeda, Characterization of Graft Copolymers, Adv.Polym.Sci. **29** (1978) 47

S.Bywater, Preparation and Properties of Star-Branched Polymers, Adv.Polym.Sci. **30** (1979) 89

W.W.Graessley, Entangled Linear, Branched and Network Polymer Systems - Molecular Theories, Adv.Polym.Sci. **30** (1979) 89

J.Koenig, Chemical Microstructure of Polymer Chains, Wiley, New York 1980

L.H.Sperling, Interpenetrating Polymer Networks and Related Materials, Plenum, New York 1981
J.M.G.Cowie, Ed., Specialty Polymers, Vol. 1, Alternating Copolymers, Plenum, New York 1984
M.J.Folkes, Ed., Processing, Structure, and Properties of Block Copolymers, Elsevier, New York 1985
J.A.Semlyen, Ed., Cyclic Polymers, Elsevier, New York 1986
N.A.Platé, V.P.Shibaev, Ed., Comb-Shaped Polymers and Liquid Crystals, Plenum, New York 1987
D.A.Tomalia, A.M.Naylor, W.A.Goddard, III, Starburst-Dendrimers, Angew.Chem. **102** (1990) 119-157; Angew.Chem., Int.Ed.Engl. **29** (1990) 138-175
J.E.Mark, H.R.Allcock, R.West, Inorganic Polymers: An Introduction, Prentice-Hall, Englewood Cliffs, NJ, 1992
G.Wenz, B.Keller, Polyrotaxanes, Angew.Chem., Int.Ed.Engl. **31** (1992) 325
D.B.Amabilino, I.W.Parsons, J.F.Stoddart, Polyrotaxanes, Trends in Polym. Sci. **2** (1994) 146
S.I.Stupp, S.Son, H.C.Lin, L.S.Li, Synthesis of Two-Dimensional Polymers, Science **259** (1993) 59-63
Y.Yamashita, Chemistry and Industry of Macromonomers, Hüthig and Wepf, Basel 1993

2.1.1. NOMENCLATURE
International Union of Pure and Applied Chemistry, Macromolecular Division, Compendium of Macromolecular Nomenclature, Blackwell Scientific Publ., Oxford 1991

2.1.1. ANALYSIS
G.M.Kline, Ed., Analytical Chemistry of Polymers, Interscience, New York 1959-1962 (3 parts)
M.Hoffmann, H.Krömer, R.Kuhn, Polymeranalytik, Thieme, Stuttgart 1977 (2 vols.)
J.Urbanski, W.Czerwinski, K.Janicka, F.Majewska, H.Zowall, Handbook of Analysis of Synthetic Polymers and Plastics, Wiley, New York 1977
E.Schröder, G.Müller, K.-F.Arndt, Polymer Characterization, Hanser, Munich 1989
L.S.Bark, N.S.Allen, Ed., Analysis of Polymer Systems, Appl.Sci.Publ., Barking, Essex 1982
A.Krause, A.Lange, M.Ezrin, Plastics Analysis Guide, Hanser, Munich 1984
J.Mitchell, Jr., Ed., Applied Polymer Analysis and Characterization, Hanser, Munich 1987
T.R.Crompton, Analysis of Polymers, Pergamon, Oxford 1989
H.G.Barth, J.W.Mays, Modern Methods of Polymer Characterization, Wiley, New York 1991

2.1.2. INFRARED AND RAMAN SPECTROSCOPY
H.-W.Sisler, K.Holland-Moritz, Infrared and Raman Spectroscopy of Polymers, Dekker, New York 1980
P.C.Painter, M.C.Coleman, J.L.Koenig, The Theory of Vibrational Spectroscopy and its Application to Polymeric Materials, Wiley, New York 1982
W.Klöpffer, Introduction to Polymer Spectroscopy, Springer, Berlin 1984
D.O.Hummel, F.Scholl, Eds., Atlas of Polymer and Plastics Analysis, VCH, Weinheim, 3rd ed. 1991 (several volumes)
J.L.Koenig, Spectroscopy of Polymers, Am.Chem.Soc., Washington, DC, 1992 (FTIR, Raman, IR)

2.1.3. NUCLEAR MAGNETIC RESONANCE SPECTROSCOPY
J.C.Randall, Polymer Sequence Determination - Carbon 13 NMR Method, Academic Press, New York 1977
Q.T.Pham, R.Petiaud, H.Waton, Proton and Carbon NMR Spectroscopy of Polymers, Wiley, New York 1983 (2 vols.)
R.A.Komoroski, Ed., High Resolution NMR Spectroscopy of Synthetic Polymers in Bulk, VCH, Weinheim 1986
A.E.Tonelli, NMR Spectroscopy and Polymer Microstructure, VCH, New York 1989
R.N.Ibbett, Ed., NMR Spectroscopy of Polymers, Chapman and Hall, London 1993

2.2. MOLAR MASS AND MOLAR MASS DISTRIBUTION
L.H.Peebles, Molecular Weight Distribution in Polymers, Interscience, New York 1971
P.E.Slade, Jr., Polymer Molecular Weights, Dekker, New York 1975 (2 vols.)
N.C.Billingham, Molar Mass Measurements in Polymer Science, Wiley, New York 1977

2.2.4.b. MASS SPECTROSCOPY
D.M.Hindenlang, R.D.Sedgwick, Mass Spectrometry, in C.Booth, C.Price, Eds., Polymer
    Characterization, pp. 573-588 (= vol. 1 (1989)), G.Allen, J.C.Bevington, Eds.,
    Comprehensive Polymer Science, Pergamon Press, Oxford

2.2.4.c. SIZE-EXCLUSION CHROMATOGRAPHY
W.W.Yau, I.J.Kirkland, D.D.Bly, Modern Size-Exclusion Liquid Chromatography, Wiley, New
    York 1979
T.Kremer, L.Boross, Gel Chromatography, Wiley, New York 1979
CRC Handbook of Chromatography, Polymers, CRC Press, Boca Raton, FL. C.G.Smith,
    N.E.Skelly, C.D.Chow, R.A.Solomon, Vol. I (1982); C.G.Smith, W.C.Buzanowski,
    J.D.Graham, Z.Iskandarani, Vol. II (1993)
J.Janca, Ed., Steric Exclusion Chromatography of Polymers, Dekker, New York 1984
G.Glöckner, Polymer Characterization by Liquid Chromatography, Elsevier, Amsterdam 1986
G.Glöckner, Gradient HPLC of Copolymers and Chromatographic Cross-Fractionation, Springer,
    Berlin 1991
C.-S.Wu, Ed., Handbook of Size Exclusion Chromatography, Dekker, New York 1995

2.2.4.a. END GROUP ANALYSIS
S.R.Palit, B.M.Mandal, End-Group Studies Using Dye Techniques, J.Macromol.Sci. [Rev.] **C 2**
    (1968) 225
R.G.Garmon, End Group Determinations, Techniques and Methods of Polymer Evaluation
    4/1 (1975) 31

2.3. CONFIGURATION and 2.4. CONFORMATION
A.D.Kelley, Ed., The Stereochemistry of Macromolecules, Dekker, New York 1967 (3 vols.)
F.A.Bovey, Polymer Conformation and Configuration, Academic Press, New York 1969
A.Hopfinger, Conformational Properties of Macromolecules, Academic Press, New York 1973

References
[1] R.C.Beavis, B.T.Chait, Anal.Chem. **62** (1990) 1836-1840, Fig. 2 B (modified)

3. Chain-Growth Polymerizations

3.1. Overview of Polymerizations

3.1.1. Classes of Polymerizations

Macromolecules are synthesized by **polymerization** of monomer molecules or by **polytransformation** of macromolecules into macromolecules of different constitution and/or configuration. Polymerizations and polytransformations require zero or negative Gibbs energies; polymerizations furthermore demand monomers that are at least bifunctional and easy to activate.

Polymerizations are usually subdivided into four classes, depending on whether (1) nonpolymeric byproducts L are generated and whether (2) polymer molecules P_i react in the growth step only with monomer molecules M or also with other polymer molecules P_j. For the purpose of the following sections, it is useful to consider all molecules from the dimer P_2 upwards as polymer molecules (i.e., $i, j \geq 1$). It is also convenient to designate M, P_i, and P_j as **reactants** and L as a **leaving molecule**.

There are thus four classes of polymerizations (Scheme I):

	Step-growth polymerizations	*Chain-growth polymerizations*
	$P_i + P_j \longrightarrow P_{i+j}$	$P_i + M \longrightarrow P_{i+1}$
IUPAC name:	polyaddition	chain polymerization
Historical names:	(rare: adduct formation)	addition polymerization
		chain-growth polymerization
	$P_i + P_j \longrightarrow P_{i+j} + L$	$P_i + M \longrightarrow P_{i+j} + L$
IUPAC name:	polycondensation	condensative chain polymerization
Historical names:	condensation polymerization	no name in use
	step-growth polymerization	polyelimination (this book)

These names of types of polymerizations can be traced back to a classification scheme by the American, Wallace Hume Carothers (1931) who distinguished "addition polymers" and "condensation polymers". "Addition polymers" have a composition of monomeric units that is identical with that of monomer molecules; they are generated by "addition polymerization". "Condensation polymers" have monomeric units that are not identical with monomer molecules; they are produced by "condensation polymerization" with formation of leaving molecules.

"Adduct formations" were discovered by the German, Otto Bayer in 1937. This class of reactions proceeds without formation of leaving molecules (it is thus not a "condensation polymerization"); the structure of the monomeric units is however the same as the structure of the monomer molecules (it is thus similar to an "addition polymer"). In order to distinguish this new type of polymerization from both addition polymerization (German: *Polymerisation*) and condensation polymerization (German: *Polykondensation*), Bayer called it in German a "*Polyaddition*", thus causing monumental confusion. The word "polyaddition" never caught on in English.

It turned out later that the true distinguishing factor was not the composition of monomeric units vs. that of monomer molecules but the growth steps. "Addition polymerizations" proceed as *kinetic chain* reactions whereas "condensation polymerizations" do not. This led to another confusion since the word "chain" denotes both a physical entity (a molecule) as well as a process (kinetic chain reaction). Organic chemists therefore started to refer to "condensation polymerizations" as "step-growth polymerizations" since the reaction products could be easily isolated and reacted again after several "steps" whereas those of the known "addition polymerizations" could not.

This nomenclature was, again, an unlucky choice of words since the basic units of all macromolecules are chains and all polymerizations proceed in steps in the mechanistic sense, i.e., monomer, oligomer, or polymer molecules are added to the growing polymer chain one after the other. The "step" of organic chemistry refers to the easy isolation of reaction products if a reaction is "frozen" by lowering the temperature (polycondensations are often performed at high temperatures; they are very slow at room temperature). The "freezing-in" is much more difficult for "addition polymerizations" which often proceed near room temperature and sometimes have to be frozen-in at the temperature of liquid nitrogen. To distinguish between "step-growth polymerizations" and "chain polymerizations" on the basis of how rapidly a lower temperature can be achieved therefore means to base a physical phenomenon on the skill of the experimentalist which should never be done for scientific definitions.

If "chain polymerization" is the encompassing term, then it is illogical to label one subclass with an adjective ("condensative chain polymerization") but not the other ("chain polymerization"). It is for this reason that the name "polyelimination" is used in this book for IUPAC's "condensative chain polymerization". For all other polymerizations, IUPAC terms are used (polyaddition, polycondensation, chain polymerization).

The various classes of polymerizations differ characteristically in the dependence of the number-average degree of polymerization, $\overline{X}_n$, on the extent of the reaction p (= conversion of functional groups) (Fig. 3-1). Equilibrium polycondensations C

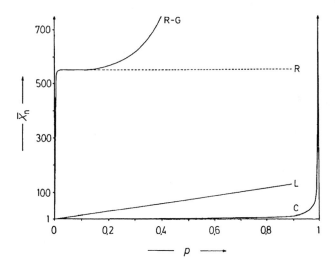

Fig. 3-1 Number-average degree of polymerization, $\overline{X}_n$, as a function of the extent of reaction, p.
C: Bifunctional polycondensations or polyadditions in equilibrium (exact).
L: Living polymerizations. Example: anionic polymerization of styrene S with the monofunctional initiator butyl lithium at an initial molar ratio of $[S]_o/[BuLi]_o = 141:1$.
R: Free-radical polymerizations. Example: polymerization of neat styrene S with *N,N*-azo*bis*-isobutyronitrile AIBN as initiator (initial molar ratio 141:1). Each AIBN molecule forms two radicals but not all initiator radicals start polymer chains.
R-G: Free-radical polymerization with gel effect (schematic; see also Fig. 3-9).

(and also equilibrium polyadditions) of bifunctional monomers require high extents of reaction for high degrees of polymerization, e.g., $p = 99\%$ for $\overline{X}_n = 100$ (see Section 4.1.3.). In so-called living anionic polymerizations L (no chain termination), $\overline{X}_n$ increases linearly with p. Free-radical polymerizations R may generate high degrees of polymerizations at low p; the $\overline{X}_n$ then remains constant with increasing p (although it should decrease according to simple theory) or even grows rapidly due to a "gel effect" (R–G).

Many other classifications of polymerizations are used, for example, according to the chemical structure of monomers (vinyl, ring-opening, etc.), the start reactions (thermal, catalytic, photochemical, etc.), the nature of the active species (anion, cation, radical, etc.), the reaction medium (bulk, solution, emulsion, etc.), and the state of matter (homogeneous, heterogeneous).

3.1.2. Examples of Polymerizations

An example of a **polyaddition** is the reaction of diisocyanates with diols to form polyurethanes:

(3-1) n O=C=N-X-N=C=O + n HO-Y-OH

$\longrightarrow$ HO-Y-O-(CO-NH-X-NH-CO-O-Y-O)$_{n-1}$-CO-NH-X-N=C=O

Monoisocyanates can be polymerized by **anionic chain polymerization** to poly(isocyanate)s using anions $A^{\ominus}$ as initiators:

(3-2) $A^{\ominus}$ + n R-N=C=O $\longrightarrow$ A-(NR-CO)$_n{}^{\ominus}$

The resulting **macroanions** A$\dashv$NR-CO$\}_n^{\ominus}$ remain present ("living") after all n moles of monomer molecules have reacted, provided that terminating impurities are absent (**living polymerization**). In many other chain polymerizations, growing polymer chains are terminated by reaction with other molecules, for example, catalysts, initiators or solvents. In **free-radical chain terminations**, the dominant termination of the kinetic chain is often the combination of two growing macroradicals:

(3-3)   R(CH$_2$-CHR')$_n{}^{\bullet}$ + $^{\bullet}$(CHR'-CH$_2$)$_m$R $\longrightarrow$ R(CH$_2$-CHR')$_n$(CHR'-CH$_2$)$_m$R

The polyesterification of diols by dicarboxylic acids to form polyesters is an example of a **polycondensation**. In this reaction, water molecules leave the reaction mixture if a vacuum is applied. The non-leaving molecules (monomer, oligomer, polymer) will be referred to as **reactants**. Polycondensations are characterized by the indiscriminate reaction of all reactants with each other (see Chapter 4):

(3-4) n HO-X-OH + n HOOC-Y-COOH

$\longrightarrow$ H(O-X-O-OC-Y-CO)$_n$OH + $(2n-1)$ H$_2$O

All biological polymerizations (Section 4.2.) seem to be **polyeliminations**: only monomer molecules are added to the growing chains but in each step, leaving molecules are liberated ("condensative chain polymerizations"). Only a few polyeliminations are known in synthetic polymer chemistry. An example is the polymerization of *N*-carboxy anhydrides (Leuchs anhydrides) of α-amino acids. The growing polypeptide chains react only with the monomer, not with other chains:

(3-5)

$$- CO_2$$

3.1.3. Functionality

The term **"functionality of a molecule"** refers to the number of groups per molecule that can react at a given reaction condition. Molecules can thus adopt different functionalities (see below). An isocyanate group, for example, is monofunctional if it reacts with a hydroxy group (Eqn.(3-1)) but is bifunctional in anionic chain polymerizations (Eqn.(3-2)). Only one isocyanate group per monomer molecule is needed for the formation of linear polymers in anionic chain polymerizations but two isocyanate groups per monomer molecule are required in polycondensations.

3.1.4. Experimental Investigations

Polymerizations require more rigorous experimental conditions than reactions of low molecular weight compounds. Free-radical chain polymerizations are suppressed by oxygen concentrations as low as ca. 10^{-4} % (Section 3.7.7.) and anionic chain polymerizations by traces of water (Sections 3.4.2. and 3.4.5.). In linear polycondensations, monofunctional impurities often reduce the achievable degree of polymerization from "infinite" to only 100-200 (Section 4.1.3.).

The formation of polymers can be followed by the usual experimental methods (gravimetry, spectroscopy, etc.) via the disappearance of monomers, the appearance or isolation of polymers, the emergence of leaving molecules, or the viscosity increase of the reaction mixture. A special method for polymers is **dilatometry,** the continuous measurement of the change of volume with time.

In polymerizations, van der Waals bonds between monomer molecules are replaced by covalent bonds between monomeric units. Bond lengths are approximately (0.3-0.5) nm for van der Waals interactions and ca. (0.14-0.19) nm for covalent bonds. Most polymerizations will thus proceed with a contraction of the polymerization system; only very few systems (containing special ring molecules) will expand. The smaller the monomer molecules (the higher the volume concentration of van der

Waals bonds), the greater the contractions in *chain polymerizations* and *polyadditions*. The volume decreases 66 % for a complete polymerization of ethylene $CH_2=CH_2$ but only 14 % for styrene $CH_2=CH(C_6H_5)$. Volumes shrink 23 % during the ring-opening polymerization of ethylene oxide (three-membered rings) but only 10 % during the polymerization of tetrahydrofuran (five-membered rings).

In *polycondensations* and *polyeliminations*, contractions decrease with increasing size of the leaving molecules. A 22 % contraction is observed if polyamide 66 is formed from hexamethylene diamine and adipic acid (generates H_2O) but a 66 % contraction if dioctyladipate is used instead of adipic acid (generates $C_8H_{17}OH$).

3.1.5. Kinetics and Statistics

Polymerization reactions can be analyzed by two different approaches:

1. The *kinetic method* proposes a plausible reaction scheme. Elementary reactions are described by appropriate differential equations. It is usually assumed that reactivities (i.e., elementary rate constants) are independent of the molecular size (**principle of equal chemical reactivity**). The method is flexible but requires different reaction schemes for different types of polymerization.

2. The *statistical method* evaluates the probability of an event (i.e., a certain elementary reaction) relative to the probabilities of other, competing events. The method delivers generalized equations which may be applied to very different phenomena. Examples are the degrees of polymerization in equilibrium polymerizations, the sequence statistics of monomer units in copolymerizations, and the tacticities of vinyl polymers.

All chain polymerizations have certain mechanistic features in common; they will thus be discussed in this chapter with emphasis on their physical chemistry. Polycondensations, polyadditions, polyeliminations, and polytransformations are better explored with respect to their particular chemistry; they will be treated in a separate chapter (Chapter 4).

3.2.   Thermodynamics of Polymerizations

3.2.1. Polymerization Equilibria

Equilibrium reactions are controlled by thermodynamics. The standard Gibbs energy $\Delta G_p{}^o$ of polymerization is given by the standard polymerization enthalpy $\Delta H_p{}^o$, the temperature T, the standard polymerization entropy $\Delta S_p{}^o$, and the equilibrium constant K_n, respectively:

(3-6) $\Delta G_p{}^o = \Delta H_p{}^o - T\Delta S_p{}^o = -RT \log_e K_n$ (at constant pressure p)

Low molecular weight reaction participants such as monomers M and leaving molecules L are always defined species. Polymers P_i and P_j are however *mixtures* of species with $i, j = 2, 3, ... n$ which are present in **consecutive equilibria**. These equilibria must be carefully defined in order to obtain correct expressions for the interrelationships between equilibrium constant K_n, initial monomer concentration $[M]_o$, equilibrium concentration $[M]$, and number-average degree of polymerization, $\overline{X}_n$, as the following example shows:

R^1–R^2 reacts with monomer M to form molecules R^1MR^2, $R^1M_2R^2$, $R^1M_3R^2$... (Type I). The equilibrium constant $K_1 = [R^1MR^2]/\{[R^1R^2][M]\}$ is assumed to be different from $K_2 = K_3 = ... = K_n$ for $K_n = [R^1M_nR^2]/\{[R^1M_iR^2][M]\}$ where $n \geq 2, i \geq 1$, and $n > i$ (Appendix A 3-1). The number average degree of polymerization includes all species $R^1M_iR^2$, thus also R^1MR^2, but not the monomer M. The number-average degree of polymerization of polymers from such equilibria is $\overline{X}_n = (1 - K_n[M])^{-1}$. The same expression is obtained for equilibrium polymerizations where monomers M are first converted into activated monomers M*, then into activated dimers M_2*, activated trimers M_3*, etc. (Type II). If however the monomer is directly converted into dimers M_2, trimers M_3 ... with equilibrium constants $K_1 = K_2 = ... = K_n$ (as in the self-condensation of HO-R-COOH), then the expression will read $\overline{X}_n = 1 + (1 - K_n[M])^{-1}$ (Type III).

Although Types I and II feature the same dependence of number-average degrees of polymerization on equilibrium constants K_n and equilibrium concentrations $[M]_o$, they do not share the same dependence of degrees of polymerization and equilibrium constants on the initial concentrations $[M]_o$ and $[R^1R^2]_o$. These equations are rather lengthy and will not be reproduced here.

3.2.2. Critical Polymerization Temperatures

Standard Gibbs polymerization energies become zero at $K_n = 1$ and thus also for $T_{crit} = \Delta H_p^{\,o}/\Delta S_p^{\,o}$ (Eqn.(3-6)). At this critical polymerization temperature T_{crit}, polymerizations cannot proceed to *high* molar masses although oligomers may be obtained because all reactants participate in consecutive equilibria.

Assume a polymerization enthalpy of $\Delta H_p^{\,o} = -30$ kJ/mol and a polymerization entropy of $\Delta S_p^{\,o} = -100$ J K^{-1} mol^{-1} for the polymerization of a liquid monomer with $M = 166.67$ g/mol and $[M]_o = 6$ mol/kg. If this monomer is polymerized in equilibrium to an insoluble, crystalline polymer, then the activity coefficients are unity and the monomer is completely converted into a polymer at the critical polymerization temperature $T_{crit} = (-30\,000$ J/mol$)/(-100$ J K^{-1} mol$^{-1}) = 300$ K. If the same monomer polymerizes in bulk to a dissolved polymer according to Type III ($K_1 = K_n$), then an equilibrium concentration of $[M] = 0.667$ mol/kg and an $\overline{X}_n = 4$ are observed for all polymer molecules ($X \geq 2$) at $T_{crit} = 300$ K.

For Type I equilibria, degrees of polymerization at T_{crit} depend strongly on initial initiator concentrations and less strongly on equilibrium constants K_1. The critical polymerization temperature can thus *not* be obtained by extrapolation of the degrees of polymerization as a function of the various temperatures to a degree of polymerization of unity.

The same considerations apply to kinetic interpretations of, e.g., Type I equilibrium polymerizations. Reaction rates of polymerization and depolymerizations equal each other at equilibrium ($R_p = R_{dp}$). For R-M_i + M $\rightleftarrows$ R-M_{i+1}, these two rates are $R_p = k_{p,i}[RM_i][M]$ and $R_{dp} = k_{dp,i+1}[RM_{i+1}]$. Equilibrium constants are given by the ratio of rate constants of forward and backward reactions, i.e., by $K_n = k_{p,i}/k_{dp,i+1} = [RM_{i+1}]/([RM_i][M])$. At the critical polymerization temperature, rate constants may equal each other ($k_{p,i} = k_{dp,i+1}$) but concentrations will not because of consecutive equilibria ($[RM_i] < [RM_{i+1}]$). Insertion of $[RM_{i+1}]/[RM_i] = K_n[M] < 1$ into Eqn.(3-6) leads to $\log_e \{[M]([RM_i]/[RM_{i+1}])\} = -(\Delta S_p{}^o/R) + (\Delta H_p{}^o/R)(1/T)$. The extrapolation of $1/T$ as a function of $\log_e [M]$ to $\log_e[M] = \log_e[M]_o$ thus does not deliver the critical polymerization temperature.

3.2.3. Ceiling and Floor Temperatures

Polymerizations proceed only if the standard Gibbs energy $\Delta G_p{}^o$ is zero or negative. Eqn.(3-6) thus predicts four cases for $\Delta G_p{}^o \leq 0$:

Case 1: Negative enthalpy and negative entropy. The entropy term $-T\Delta S_p{}^o$ becomes more positive with increasing temperature. At a temperature T_{crit}, the entropy term compensates the enthalpy term ($\Delta G_p{}^o = 0$ for $\Delta H_p{}^o = T_{crit}\Delta S_p{}^o$). Polymerizations can thus only be performed *below* a critical **ceiling temperature**. An example is the polymerization of neat liquid α-methylstyrene to a poly(α-methylstyrene) dissolved in its own monomer ($T_{crit} = 60°C$). Case 1 is encountered most often.

Case 2: Negative or zero enthalpy and positive entropy. Polymerizations are possible at *all* temperatures. This case has never been observed.

Case 3: Positive enthalpy and negative entropy. Polymerizations are *impossible* at all temperatures. Case 3 cannot be proven at all because the absence of a thing or phenomenon (here: non-observance of polymerization) cannot be used as proof for the presence of something else (here: positive enthalpy and negative entropy).

Case 4: Positive enthalpy and positive entropy. Since the entropy term $-T\Delta S_p{}^o$ becomes less negative at decreasing temperatures, polymerizations can only be performed *above* a critical **floor temperature**. This case is very rare; an example is the polymerization of liquid cyclooctasulfur to poly(sulfur) which proceeds at temperatures above 159°C.

3.2.4. Polymerization Enthalpies and Entropies

Activity coefficients are usually not unity. Thermodynamic quantities therefore have to be related to the gaseous state (index g; usually for 1 bar = 10^5 Pa $\approx$ 1 atm), the melt (index 1 = liquid), the solid amorphous state (index c = condensed), the crystalline state (index c'), or the solution (index s). The standard polymerization enthalpy for the conversion of a liquid monomer to an amorphous polymer thus has the symbol $\Delta H_{lc}{}^o$.

Polymerization entropies depend on constitutions and thermodynamic states of monomers and polymers (Table 3-1). For polymerizations of gaseous monomers to (hypothetical) gaseous polymers, they are given by the contributions of the translational entropy $\Delta S_{tr}{}^o$, external rotational entropy $\Delta S_{er}{}^o$, internal rotational entropy $\Delta S_{ir}{}^o$, and vibrational entropy $\Delta S_{vb}{}^o$. For olefins and olefin derivatives, the loss of $\Delta S_{er}{}^o$ is just compensated by gains of $\Delta S_{ir}{}^o + \Delta S_{vb}{}^o$; one obtains $\Delta S_{gg}{}^o \approx \Delta S_{tr}{}^o$.

Table 3-1 Polymerization entropies and enthalpies.

Monomers	$\Delta S°/(\text{J K}^{-1}\ \text{mol}^{-1})$				$\Delta H°/(\text{kJ mol}^{-1})$			
	gc'	gc	gg	lc	gc'	gc	gg	lc
Tetrafluoroethylene					-172		-155	-155
Ethylene	-173	-156	-142		-108	-102	-94	-92
Propylene	-198		-167	-115	-104		-87	-84
1-Butene	-219	-190	-166	-113			-87	-84
Styrene			-149	-108			-75	-71
α-Methylstyrene				-110			-35	-35
Formaldehyde	-174	-169	-124		-60			-31
Acetone						-10	$+12$	$+25$
Cyclopropane				-69				-113
Cyclohexane				-11				$+3$
Cyclooctane				-3				-35
Ethylene oxide	-174			-78	-134		-104	-95
Tetrahydrofuran	-177	-140		-48		-43	-21	-12
Oxepane $C_6H_{12}O$ (7-ring)		-92		$+12$		-39		$+2$
1,3-Dioxolane $C_3H_6O_2$ (5-ring)	-205	-154		-61		-50	-26	-21

Vaporization entropies have to be considered if gaseous monomers are polymerized to condensed polymers: $\Delta S_{gc}°$ is always more negative than $\Delta S_{lc}°$. Polymerization entropies $\Delta S_{gc'}°$ are similarly always more negative than $\Delta S_{gc}°$ because of the additional effect of positive melt entropies $\Delta S_M°$.

The polymerization of small cyclic monomers liberates plenty of rotational entropy (negative $\Delta S_{lc}°$). Large cyclic monomers have similar rotations as chains, however. Polymerization entropies thus become less negative with increasing ring size; they may even become positive.

Many polymerizations are exothermic (negative polymerization enthalpy). Their heats of polymerization can be considerable: the adiabatic polymerization of ethylene leads to a temperature increase of 1800 K at complete monomer conversion! Similar to polymerization entropies, one also observes less negative values for polymerization enthalpies in the series gc' - gc - gg - lc.

The polymerization of olefinic monomers (ethylene, α-olefins, vinyl compounds, etc.) proceeds by opening of π-bonds and formation of σ-bonds. Polymerization enthalpies depend strongly on van der Waals radii (and not on atomic radii) of ligands. These radii increase in the series $F < H < CH_2 \approx CH_3 < C_6H_5$. Polymerization enthalpies thus become less negative in the series $CF_2=CF_2 < CH_2=CH_2 < CH_2=CH(CH_3) \approx CH_2=CH(CH_2CH_3) < CH_2=CH(C_6H_5) << CH_2=C(CH_3)(C_6H_5)$.

The polymerization of C=O double bonds to C-O single bonds is likewise controlled by ligands. Formaldehyde $H_2C=O$ has a negative polymerization enthalpy $\Delta H_{lc}°$ but acetone $(CH_3)_2C=O$ has a positive one. Acetaldehyde exists mainly as the aldehyde in its keto–enol equilibrium $CH_3CHO \rightleftarrows CH_2=CHOH$. Both isomers can be polymerized from this equilibrium by suitable catalysts. The polymerization enthalpy $\Delta H_{lc}°$ is 0 kJ/mol for the polymerization to poly(acetaldehyde) $+O\text{-}CH(CH_3)\}_{\overline{n}}$ but -63 kJ/mol for poly(vinyl alcohol) $+CH_2\text{-}CH(OH)\}_{\overline{n}}$.

The polymerization of cyclic monomers proceeds by opening and closing of σ-bonds. Since the bond energies of cyclic monomers and their polymers are approximately equal, polymerization enthalpies are controlled by strain and delocalization energies. The polymerization enthalpies of strongly strained small rings (cyclopropane, ethylene oxide) are very negative whereas those of strain-free rings (cyclohexane) are zero. Larger rings (cyclooctane, oxepane = oxacycloheptane $C_6H_{12}O$) are also not strained. The trans-annular hindrance of hydrogen atoms in rings causes increases in ring strain, however, and polymerization enthapies become negative.

3.2.5. Chain and Ring Formation

Intermolecular chain formation always competes with intramolecular ring formation in bifunctional polymerizations. The ratio of chains to rings can be controlled by either thermodynamics or kinetics. In both cases, it is affected by concentrations and temperature; in addition, it is influenced by the degree of polymerization under thermodynamic control and by the catalyst under kinetic control.

Kinetic control. Chain polymerizations $P_i^* + M \rightarrow P_{i+1}^*$ are bimolecular reactions; the polymerization rate is $R_p = k_p[P_i^*][M]$. Ring formations $P_i^* \rightarrow$ cyclo-P_i are, however, monomolecular reactions with a rate of $R_c = k_c[P_i^*]$. The fraction of cyclics formed is given by $f_c = R_c/(R_c + R_p)$. Thus:

(3-7) $1/f_c = 1 + (k_p/k_c)[M]$

The smaller the monomer concentration [M], the higher is the proportion of cyclic molecules (**Ruggli–Ziegler dilution principle**). The rate constants k_p and k_c are affected differently by various catalysts, solvents and temperatures. Some catalysts deliver 100 % linear chains, others 100 % ring molecules or mixtures of both. The proportion of rings may also vary with time due to equilibration.

Thermodynamic control. Equilibria between chains $\sim M_q \sim$ ($q = i, i - j$) and cyclic molecules c-M_j can be described by:

(3-8) $\sim M_i \sim \rightleftarrows \sim M_{i-j} \sim + $ c-M_j ; $K_c = [\sim M_{i-j} \sim][\text{c-}M_j]/[\sim M_i \sim]$

At very high molar masses, $[\sim M_{i-j} \sim]$ approximately equals $[\sim M_i \sim]$ and one obtains $K_c \approx [\text{c-}M_j]$. The molar concentrations [c-M_j] of cyclic molecules are very low: practically only linear chains are produced in thermodynamic equilibrium.

The equilibrium constant of cyclization is a function of the probability density $W(r_j)$ that the two ends of a chain will meet; the molar probability density is thus $W(r_j)/N_A$. Statistical theory calculates this probability density as $W(r_j) \sim X^{-1/2}$ for the one-dimensional case (Chapter 5) and $W(r_j) \sim X^{-3/2}$ for the three-dimensional case.

The more ring bonds $N_c = N_{sym}X$ can be opened per cyclic molecule with degree of polymerization X, the lower is the concentration of ring molecules (provided that the probability density is equal). N_{sym} is a symmetry number that equals 1 for cyclic polylactams because only one bond at ~NH–CO~ can be severed per monomeric unit –NH–CO–X–. Two possibilities exist, however, for each monomeric unit of cyclic poly(siloxane)s $\left(\text{O–SiR}_2\right)_n$ since these units can be opened at ~O–SiR$_2$~ as well as at ~SiR$_2$–O~. Here, the probability of a ring-opening doubles and $N_{sym} = 2$.

Chain units number N_b for monomeric units and $N = N_b X$ for cyclic molecules with a degree of polymerization X. The equilibrium constant of cyclization thus decreases with the (5/2) power of the number N of chain units per ring (Fig. 3-2):

$$(3\text{-}9) \qquad K_c \approx [c\text{-}M_j] = \frac{W(r_j)}{N_A N_c} = \frac{W(r_j)}{N_A N_{sym} X} = \frac{const \cdot X^{-3/2}}{N_A N_{sym} X} = \frac{const \cdot (N_b)^{5/2}}{N_A N_{sym}} \cdot \frac{1}{N^{5/2}}$$

This dependence is approximately obeyed for high numbers N of chain units. At small N, theoretical assumptions are no longer valid (infinitely long chains, rings without strains). Deviations are especially prevalent in thermodynamically good solvents with strong interactions between macromolecules and solvent molecules.

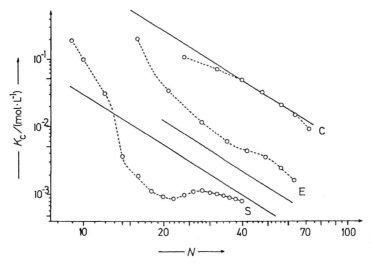

Fig. 3-2 Experimental equilibrium constants K_c of cyclization as a function of the number N of ring atoms for bulk polymerizations of cyclooctene C (25°C) and octamethylcyclotetrasiloxane S (110°C) and polycondensation E of ethylene glycol and terephthalic acid (270°C). Solid lines indicate theoretical predictions with slopes of –2.5.

3.3. Statistics of Chain Polymerizations

In chain polymerizations, a monomer molecule can only be joined with a polymer chain if the chain possesses a so-called **active center**. These active centers are usually introduced into the chain by external agents, the **initiators**. An example is the anion $A^\ominus$ of an initiator $A^\ominus R^\oplus$ which initiates the anionic polymerization of monoisocyanates (Eqn.(3-2)). Another example is the free-radical polymerization of vinyl monomers $CH_2=CHR'$ to growing macroradicals $R(CH_2-CHR')_n{}^\bullet$ (Eqn.(3-3)) which is initiated by free radicals $R^\bullet$ from the thermal decomposition of initiators R-R, for example, dibenzoyl peroxide $C_6H_5COO-OOCC_6H_5$. The initiator fragments become permanently attached to the growing chain: Chain polymerizations are *initiated* and not catalyzed. A catalyst, on the other hand, would move from chain to chain and emerge unchanged at the end of the reaction. Polycondensations and polyadditions are catalyzed polymerizations (Chapter 4).

Since chain polymerizations sometimes do occur even without deliberately added initiators (mostly due to initiating impurities) and since the addition of initiators causes increases of polymerization rates, **initiators** are often called **accelerators** in industry.

An active center may be at the growing end of a chain where it may be an anion (see Eqn.(3-2)), a cation, or a free radical (see Eqn.(3-3)). In these **chain-end-controlled** chain polymerizations, monomers must have electron-withdrawing groups (anionic polymerizations, Section 3.4.), or electron-donating groups (cationic polymerizations, Section 3.5) or be able to be added to free radicals (Section 3.7.). In all these reactions, the newly formed chain end must have the same structure as the disappearing one.

The active center may also be a catalytic site, for example, a transition metal atom [Mt] with some ligands. These catalytic sites are bound to the chain; the monomer is not added to the chain end but inserted between the active center and the polymer chain in these **catalytic-site-controlled** chain polymerizations (Section 3.6.).

$$\text{\raisebox{1ex}{$\frown$}}\;\; \text{\textapprox M—M}^* + M \longrightarrow \text{\textapprox M—M—M}^* \qquad\qquad \text{\textapprox M—M—[Mt] + M} \longrightarrow \text{\textapprox M—M—M—[Mt]}$$

chain-end control   catalytic site control

Each type may proceed by either a single propagation state or by multiple propagation states. Multiple states can be concurrent; two different types of catalytic centers may coexist, for example, at any given time (static multiple sites). Multiple states may also exist consecutively: free macroions may convert reversibly into ion pairs or ion associates (dynamic multiple states).

Chain-end-controlled single propagation states are subdivided into **Markov mechanisms** of zeroth, first, second ... order depending on the number of monomeric units at the active ends that influence the growth step: no effect (0th order), last monomeric unit only (1st order), penultimate unit (2nd order)... A zeroth order Markov mechanism is also called a **Bernoulli mechanism.**

The polymerization of two constitutionally different monomers A and B (or enantiomers D and L) to polymers with monomeric units a and b or of a prochiral monomer A = B to polymers with configurational diads i (= a) and s (= b) leads to the following **conditional probabilities** for the various propagation steps:

Bernoulli:	p_A, p_B
Markov 1st order:	$p_{a/A}, p_{a/B}, p_{b/A}, p_{b/B}$
Markov 2nd order:	$p_{aa/A}, p_{aa/B}, p_{ba/A}, p_{ba/B}, p_{ab/A}, p_{ab/B}, p_{bb/A}, p_{bb/B}$

In Bernoulli mechanisms, probabilities do not depend on the type of chain end. One only has to consider the two conditional probabilities for the addition (or insertion) of A monomers and B monomers. The sum of these probabilities is unity ($p_A + p_B \equiv 1$).

In Markov mechanisms, an a-group can only attach an A monomer or a B monomer. Thus: $p_{a/A} + p_{a/B} \equiv 1$, $p_{aa/A} + p_{aa/B} \equiv 1$, etc. Whether or not a certain type of diad (aa, ab, ba, bb) is formed, depends on both the mole fractions of a-units and b-units that are present at chain ends (or insertion centers) *and* the conditional probabilities. The total probability is the product of the single probabilities if the latter refer to independent events. The mole fractions of diads aa, ab, ba, and bb in polymers generated by a **Markov first order mechanism** are therefore given by:

$$(3\text{-}10) \quad \begin{aligned} x_{aa} &= x_a p_{a/A} = x_a(1 - p_{a/B}) \\ x_{bb} &= x_b p_{b/B} = x_b(1 - p_{b/A}) \\ x_{ab} &= x_a p_{a/B}; \quad x_{ba} = x_b p_{b/A} \end{aligned}$$

The probability of finding an a/b bond is given by the product of the mole fraction of a-units, x_a, and the conditional probability of forming the a/b bond, $p_{a/B}$, i.e., $x_a p_{a/B}$. The probability of finding an a/b bond must also equal the probability of finding a b/a bond if the chains are infinitely long (or cyclic). Thus: $x_a p_{a/B} = x_b p_{b/A}$.

An **asymmetric Bernoulli mechanism** is characterized by $x_a = p_A \neq p_B = x_B$. The mole fractions of diads are thus:

$$(3\text{-}11) \quad x_{aa} = x_a^2; \quad x_{|ab|} = x_a x_b + x_b x_a = 2\, x_a(1 - x_a); \quad x_{bb} = x_b^2 = (1 - x_a)^2$$

Growth steps are equally probable for **symmetric Bernoulli mechanisms**: $x_a = x_b = 1/2$, $x_{aa} = x_{bb} = x_{ab} = x_{ba} = 1/4$, etc. (see also Section 2.3.3.).

The isotactic diads resulting from a *copolymerization of chiral or prochiral monomers* D and L by a Markov first order mechanism may be either dd or ll diads; syndiotactic diads are composed of dl and ld diads. The probability of generating an ll diad is given by the product of the mole fraction of the l-units and the conditional probability $p_{l/L}$. Since $p_{d/D} + p_{d/L} \equiv 1$ and $p_{l/L} + p_{l/D} \equiv 1$, the mole fraction $x_i \equiv 1 - x_s$ of isotactic diads is thus:

$$(3\text{-}12) \quad x_i = x_{ll} + x_{dd} = x_l p_{l/L} + x_d p_{d/D} = 1 - 2\, p_{l/D} p_{d/L}/(p_{l/D} + p_{d/L})$$

In the derivation of the last expression, use has been made of the conditions $x_l + x_d \equiv 1$, $p_{d/D} + p_{d/L} \equiv 1$, $p_{l/L} + p_{l/D} \equiv 1$ and $x_l p_{l/D} = x_d p_{d/L}$ (see above).

If the polymerization of such (pro)chiral monomers follows an asymmetric Bernoulli mechanism, then the addition of a new monomer molecule to the growing chain will be independent of the type of the ultimate unit, i.e., $p_{d/L} \equiv 1 - p_{d/D} = 1 - p_{l/D} \equiv p_{l/L}$. Insertion of this relationship into Eqn.(3-12) results in:

$$(3\text{-}13) \quad x_i = 1 - 2\, p_{l/D}(1 - p_{l/D}) \equiv 1 - x_s$$

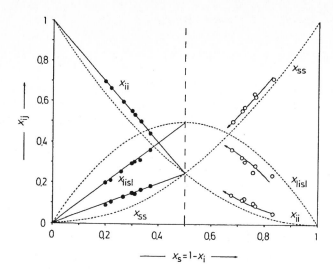

Fig. 3-3   Mole fractions x_{ij} of tactic triads (ij = ii, ss or |is| = is + si) as a function of the mole fraction of syndiotactic diads. Dotted lines indicate asymmetric Bernoulli statistics (Eqn.(3-11)).

● Polymerization of the prochiral methylvinylether $CH_2=CH(OCH_3)$ with $Al_2(SO_4)_3 \cdot H_2SO_4$ in toluene. Solid lines ——— correspond to polymerizations with enantiomorphic catalysts (Eqn.(3-14)). The vertical, broken line — — — indicates the maximum diad syndiotacticity obtainable by enantiomorphic catalysts. Diad tacticities are always $x_s \leq 1/2$.

○ Radical polymerization of 9.1 mol-% methyl methacrylate in acetonitrile at temperatures (in arrow direction) from –5°C to +120°C. The polymerization obeys a Markov first order statistics.

The fractions of tactic *triads* in Markov first order mechanisms are given by the mole fractions of units and *two* conditional probabilities. In asymmetric Bernoulli statistics, only conditional probabilities have to be considered since the addition is independent of the tacticity of the last pair of monomeric units (equations to the right of the semicolon):

(3-14)

$$x_{ss} = x_d p_{d/L} p_{l/D} + x_l p_{l/D} p_{d/L} \qquad \text{\textit{Markov first order trial}}$$
$$x_{ss} = p_{l/D} p_{d/L} = p_{l/D}(1 - p_{l/D}) = (1/2)\, x_s \qquad \text{\textit{Asymmetric Bernoulli trial}}$$

$$x_{|is|} = x_d p_{d/L} p_{l/L} + x_l p_{l/D} p_{d/D} + x_l p_{l/L} p_{l/D} + x_d p_{d/D} p_{d/L} \quad \text{\textit{Markov first order trial}}$$
$$x_{|is|} = 2\, p_{l/D}(1 - p_{l/D}) = x_s = 2\, x_{ss} \qquad \text{\textit{Asymmetric Bernoulli trial}}$$

$$x_{ii} = x_d (p_{d/D})^2 + x_l (p_{l/L})^2 \qquad \text{\textit{Markov first order trial}}$$
$$x_{ii} = 1 - 3\, p_{l/D}(1 - p_{l/D}) = 1 - (3/2)\, x_s = 1 - 3\, x_{ss} \qquad \text{\textit{Asymmetric Bernoulli trial}}$$

Polymerizations of chiral or prochiral monomers according to asymmetric Bernoulli mechanisms are known as **enantiomorphic catalyst polymerizations**. Polymers from such polymerizations are restricted with respect to their tacticities. Their syndiotactic diad fraction can never exceed 1/2: these polymers are predominantly isotactic (Fig. 3-3).

3.4. Anionic Polymerizations

3.4.1. Ion Equilibria

In ionic polymerizations, monomer molecules are added to a growing chain with a polymerization-active ionic chain end, i.e., to a **macroion**. Anionic polymerizations proceed via **macroanions**; cationic polymerizations, via **macrocations**:

$$(3\text{-}15) \quad R(M_i)^\ominus \;\to\; R(M_{i+1})^\ominus \;\to\; R(M_{i+2})^\ominus \quad \text{etc.} \qquad \text{(anionic)}$$
$$(3\text{-}16) \quad R(M_i)^\oplus \;\to\; R(M_{i+1})^\oplus \;\to\; R(M_{i+2})^\oplus \quad \text{etc.} \qquad \text{(cationic)}$$

Ionic polymerizations are not spontaneous. They are started by added initiators $R^\ominus X^\oplus$ (anionic) or $R^\oplus X^\ominus$ (cationic) which provide the initiator ions $R^\ominus$ and $R^\oplus$, respectively. Initiator ions form so-called monomer ions with monomers M: **monomer anions** in anionic polymerizations according to $R^\ominus + M \to RM^\ominus$ and **monomer cations** in cationic polymerizations according to $R^\oplus + M \to RM^\oplus$. These names do not refer to the ions $M^\ominus$ and $M^\oplus$ but rather to the ions $RM^\ominus$ and $RM^\oplus$. Monomer ions then add further monomer molecules in the propagation reactions (see Eqns.(3-15) and (3-16)).

Initiator ions can attack only those prospective monomers whose electrons can be moved in such a way that monomer ions result. Anionically polymerizable monomer molecules must therefore contain electron-accepting groups so that they can be attacked nucleophilically by initiator anions $R^\ominus$, monomer anions $RM^\ominus$ and macroanions $R(M_i)^\ominus$. Cationic polymerizations proceed via electrophilic additions of macrocations to monomers molecules.

Polymerization rates and stereospecificities of ionic polymerizations are strongly influenced by **counterions (gegenions**; from German: *gegen* = counter) of growing macroions. "Free ions" are usually present in only very small proportions. In general, solvent-separated ion pairs as well as contact ion pairs and ion pair associates are present simultaneously, especialy in anionic polymerizations. The generation of these species can be represented as follows:

Covalent compound	Polarized molecule	Contact ion pair	Solvent-separated ion pair	Free ions

$$(3\text{-}17) \quad R\text{---}X \;\rightleftharpoons\; \overset{\delta^-}{R}\text{---}\overset{\delta^+}{X} \;\rightleftharpoons\; R^\ominus\text{---}X^\oplus \;\rightleftharpoons\; R^\ominus /\!/ X^\oplus \;\rightleftharpoons\; R^\ominus + X^\oplus$$

Polarization	Ionization	Solvation	Dissociation

An ion pair is composed of two oppositely charged ions. It is held together by Coulomb interactions without formation of covalent bonds. Both ions are in direct contact in **contact ion pairs**. They are separated by solvent molecules in **solvent-separated ion pairs**. The presence and concentration of ion pairs and ions can be determined by spectroscopy and also by kinetic measurements.

Ion pairs can furthermore form ion associates (dimers, trimers, etc.). Many ionic polymerizations are thus propagated by different types of macroions which are simultananeously present and are interconverted in dynamic equilibria.

Examples of
ion associates

3.4.2. Monomers

Anionic polymerizations can be peformed with monomers that contain electron-withdrawing groups in ligands or rings. These comprise styrene $CH_2=CH(C_6H_5)$ and its derivatives, acrylic monomers $CH_2=CHR$ and $CH_2=CRR'$ (R, R' e.g. CN, COOR), 1,3-dienes $CH_2=CHR-CH=CH_2$, certain aldehydes R-CHO and ketones R-CO-R', and isocyanates R-N=C=O. Oxiranes, thiiranes, N-carboxy anhydrides, glycolide, lactams, and lactones polymerize by ring-opening.

| Ethylene oxide | Methyl thiirane | N-Carboxy anhydride | Glycolide | ε-Capro- lactam | ε-Capro- lactone |

Anionic polymerizations (Table 3-2) are used less frequently in industry than free radical polymerizations because fewer monomers can be subjected to anionic poly-merizations than to radical ones. Anionic copolymerizations are also often difficult

Table 3-2 Industrial anionic homopolymerizations.

Monomer	Polymer repeating unit	Use
Butadiene	$-[CH_2-CH=CH-CH_2]_{\overline{n}}$	Elastomers (1,4-cis)
Isoprene	$-[CH_2-C(CH_3)=CH-CH_2]_{\overline{n}}$	Elastomers (1,4-cis)
Cyanoacrylates	$-[CH_2-C(CN)(COOR)]_{\overline{n}}$	Adhesives
Formaldehyde	$-[O-CH_2]_{\overline{n}}$	Engineering plastics
Ethylene oxide	$-[O-CH_2-CH_2]_{\overline{n}}$	Thickeners
Glycolide (R = H) and lactide (R = CH$_3$)	$-[O-CO-CHR]_{\overline{n}}$	Sutures
ε-Caprolactone	$-[O-CO-(CH_2)_5]_{\overline{n}}$	Polymer plasticizers
ε-Caprolactam	$-[NH-CO-(CH_2)_5]_{\overline{n}}$	Fibers, thermoplastics
Laurolactam	$-[NH-CO-(CH_2)_{11}]_{\overline{n}}$	Fibers, casings
Octamethylcyclotrisiloxane	$-[O-Si(CH_3)_2]_{\overline{n}}$	Elastomers

because macroanions differ strongly in their polarities. The latter feature, however, makes anionically polymerizable monomers excellent candidates for block copolymerizations. Anionic polymerizations also need organic solvents and expensive initiators. Organic solvents have to be recovered; their use is ecologically problematical.

Growing macroanions react rapidly with proton donating impurities (water!) and usually also with oxygen. These reactions terminate polymerizations. Traces of water on reactor surfaces must be rigorously excluded which is more difficult in the laboratory than in industry because laboratory glassware has a higher surface/volume ratio than a big industrial reactor.

3.4.3. Initiators

Anionic polymerizations are initiated by Brønsted or Lewis bases, for example, by alkali metals, alkoxides, amines, phosphines or sodium naphthalate in solvents such as ethers (tetrahydrofuran, ethylene glycol dimethylether) or pyridine. Weak electron-withdrawing monomers require strong bases for initiation. Strong electron-withdrawing monomers such as alkyl cyanoacrylates $CH_2=C(CN)(COOR)$, on the other hand, can be polymerized by the very weak base H_2O (action of super glue).

In suitable solvents, many of these initiators dissociate "spontaneously" and completely into the initiating anions. Examples are amyl potassium $C_5H_{11}K \rightarrow C_5H_{11}^{\ominus}$ + $K^{\oplus}$ and alkoxides $ROK \rightarrow RO^{\ominus} + K^{\oplus}$. These dissociations and the subsequent start reactions with monomers require little or no thermal activation energy. Polymerization rates are thus frequently high, even at temperatures of $-100°C$.

The initiatior anions initiate the polymerization. An example is the initiation of styrene polymerization by the amyl anion:

$$(3\text{-}18) \quad C_5H_{11}^{\ominus} \xrightarrow{+ C_8H_8} C_5H_{11}-CH_2-\underset{\underset{C_6H_5}{|}}{CH}^{\ominus} \xrightarrow{+ C_8H_8} C_5H_{11}-CH_2-\underset{\underset{C_6H_5}{|}}{CH}-CH_2-\underset{\underset{C_6H_5}{|}}{CH}^{\ominus}$$

The primary initiator anion formed is however not always the true initiating species. Strong bases such as $t\text{-}C_4H_9OK$ may react with solvents such as dimethyl sulfoxide; the resulting anion is the true initiator:

$$(3\text{-}19) \quad C_4H_9O^{\ominus}K^{\oplus} + (CH_3)_2SO \longrightarrow C_4H_9OH + CH_3\text{-}SO\text{-}CH_2^{\ominus}K^{\oplus}$$

In laboratories, lithium butyl LiC_4H_9 is probably the most often used anionic initiator. Its actions are complex, however, since it provides not only free anions $C_4H_9^{\ominus}$ and ion pairs but also ion associates (dimers, trimers, hexamers). It is commonly called butyl lithium although, by definition, positive partners should be named first, hence lithium butyl.

Initiator ions may also be generated from monomers by electron transfer. For example, naphthalene reacts in tetrahydrofuran (THF) with sodium metal to form naphthalide radical anions I which dissolve in THF to a green-blue solution. These

radical anions I are stabilized by interaction with THF. The electrons of the naphthalide anions are transferred to added styrene molecules in an equilibrium reaction.
The resulting styryl radical anions II dimerize immediately to distyryl dianions III
because II is less resonance stabilized than III. The distyryl dianions dissolve in THF
with a bright red color; they start the polymerization by adding monomer molecules:

3.4.4. Living Polymerizations

Many macroanions proceed to grow until all initially present monomer molecules
are polymerized. The macroanions remain active; addition of new monomer leads to
further polymerization without newly added initiator. Such polymerizations without
internal or external termination reactions are called **living polymerizations**; examples
are polymerizations shown in Eqns.(3-18) and (3-20). Living polymerizations can be
used to synthesize polymers with narrow molar mass distributions (see below). They
also serve for syntheses of star and block copolymers (Section 4.3.2.).

The start reaction $R^{\ominus} + M \rightarrow RM^{\ominus}$ (e.g., left side of Eqn.(3-18)) consumes only
one monomer molecule per initiator anion whereas the subsequent propagation reac-
tion $RM_i^{\ominus} + M \rightarrow RM_{i+1}^{\ominus}$ ($i \geq 1$; e.g., right side of Eqn.(3-18)) adds many hundreds
and thousands of monomer molecules to the growing chain. The consumption of
monomer by the start reaction can thus be neglected and the total polymerization rate
R_{gross} equals approximately the propagation rate $R_p = -d[M]/dt$ in kinetically controlled living polymerizations with fast initiation.

If both initiator dissociation $RX \rightarrow R^{\ominus} + X^{\oplus}$ and start reaction $R^{\ominus} + M \rightarrow RM^{\ominus}$ are
fast and complete, one has $[I]_o = [R^{\ominus}] = [RM^{\ominus}] = [P^{\ominus}]$ and:

(3-21) $R_{gross} \approx R_p = -d[M]/dt = k_p[RM_i^{\ominus}][M] = k_p[I]_o[M]$

Integration of this equation gives:

(3-22) $\log_e([M]_o/[M]) = k_p[I]_o t = k_p[P^{\ominus}]t$

where $[M]_o$ and $[M]$, respectively, are monomer concentrations at time 0 and time t. The slope of a plot of $\log_e([M]_o/[M]) = f(t)$ thus delivers the product $k_p[P^\ominus]$ where $[P^\ominus]$ is the molar concentration of macroanions (see Fig. 3-4).

The rate constants k_p are *average* constants since they average over the proportions and rate constants of all propagating species (ion pairs, free ions, etc., see Eqn.(3-17)). The k_ps of macroions do not depend on the solvent but those of ion pairs and ion associates do since they are affected by the ion–solvent interaction. Free ions, ion pairs and ion associates are furthermore present in various proportions.

Ion pairs have much lower propagation rate constants than free ions since the accepting electron pair is not free. They contribute, however, considerably more to the overall rate constant k_p than free ions since they are present in much higher proportions due to their low dissociation constants. A plot of $k_p = f(T)$ thus does not follow an Arrhenius-type relationship since dissociations of ion pairs are strongly temperature dependent and the overall propagation constants k_p are averages over the contributions of different proportions of ion pairs at different temperatures.

Depolymerization leads to the back reaction of Eqn.(3-18). The expression on the left side of Eqn.(3-22) is then replaced by $\log_e \{([M]_o - [M]_\infty)/([M] - [M]_\infty)\}$, where $[M]_\infty$ is the monomer concentration at equilibrium (at infinite time). An example is the anionic polymerization of lactams at 280°C.

For ideal living polymerizations with fast initiation and no chain transfer (see below), number average degrees of polymerization, $\overline{X}_n$, can be calculated from initial molar ratios $[M]_o/[I]_o$ of monomer and initiator, extent p of monomer conversion, and functionalities f of growing chains ($f = 1$ for Eqn.(3-18); $f = 2$ for Eqn.(3-20)):

(3-23) $\quad \overline{X}_n = fp[M]_o/[I]_o$

Number average degrees of polymerization thus increase linearly with increasing extent of reaction, p. High molar masses can be obtained only at very low initial initiator concentrations $[I]_o$ which in turn cause slow polymerizations.

In (hypothetical) ideal living polymerizations, all initiator anions and monomer anions, respectively, would be completely present at the beginning of the polymerization ($k_i \gg k_p$). They also would be homogeneously distributed throughout the reaction vessel. All polymer chains would be therefore started simultaneously; each chain would have the same chance to add irreversibly the same number of monomer molecules. All resulting polymer chains would be molecularly uniform; there would be no distribution of degrees of polymerization ($\overline{X}_w/\overline{X}_n = 1$).

The addition of a monomer molecule to a growing chain is however a reversible reaction that leads to equilibria $RM_{i+1}^\ominus \rightleftarrows RM_i^\ominus + M$. These equilibria are established within 5 seconds according to theoretical calculations. As a result, the numbers of monomeric units per chain fluctuate somewhat; the result is a Poisson distribution of degrees of polymerization of polymer molecules (see Section 2.2.3. and Appendix A 3.2.). The non-uniformity with respect to degrees of polymerization is given by $\overline{X}_w/\overline{X}_n = 1 + (1/\overline{X}_n) - (1/\overline{X}_n)^2$ (see Eqn.(2-17)). A living polymer with $\overline{X}_n = 500$ should thus exhibit $\overline{X}_w/\overline{X}_n \approx 1.002$.

Experimental values of $\overline{X}_w/\overline{X}_n$ are usually higher, however, and mainly in the range of $1.03 \leq \overline{X}_w/\overline{X}_n \leq 1.05$, even with careful experiments. Polymers with those non-uniformities are commonly called "practically molecularly uniform". Such polymers are however not as narrowly distributed as the numbers suggest. A polymer with $\overline{X}_n = 500$ and $\overline{X}_w/\overline{X}_n = 1.04$ possesses a standard deviation of $\sigma_n = 100$ (Eqn.(2-12)). For Gaussian distributions, this means that 15.87 % of the molecules possess degrees of polymerization smaller than 400 and 15.87 % higher than 600!

The reason for this broadening of distributions is a diffusion effect. At time zero, monomer and initiator are not homogeneously mixed. If the initiator is added to the monomer solution, then the first polymer molecules will start to grow before the last initiator molecules are added. A better technique is to add the monomer solution to the initiator solution. Far better are **seeding techniques** which circumvent diffusion effects at least in part:

Monomer and initiator are conventionally mixed. The living polymer solution is then heated to a temperature far above the ceiling temperature until all polymer chains are depolymerized to monomer anions. After quenching the solution to the desired polymerization temperature (avoiding temperature gradients!), chains will grow and add monomer molecules simultaneously. If ceiling temperatures are too high to be practical, the initiator may also be reacted with a small proportion of monomer and the resulting oligomers used as "seed".

Living macroanions try to reach thermodynamic equilibrium not just between the growing end and monomers but also between the chains themselves. These consecutive equilibria take years to establish; at equilibrium, a Schulz-Flory distribution will result with $\overline{X}_w/\overline{X}_n = 2$. Ratios $\overline{X}_w/\overline{X}_n$ of living polymers do indeed increase on aging albeit less by the approach to equilibrium than by an increase in transfer reactions (see below).

Anionic polymerizations are used industrially therefore only if: (1) the monomer cannot be polymerized by other means or only with difficulty (formaldehyde to poly(oxymethylene)), (2) very high polymerization rates are desired (ε-caprolactam by reaction injection molding (RIM) to polyamide 6 thermoplastics, see Chapter 14), or (3) a certain stereoregularity is required (butadiene to 1,4-cis-poly(butadiene); thermoplastic elastomers of the styrene–butadiene–styrene type) (see Table 3-2).

3.4.5. Termination and Chain Transfer

Ideal living polymerizations exhibit neither termination nor transfer reactions (see below). In real anionic polymerizations, spontaneous termination reactions are quite rare. Many anionic polymerizations thus remain living until they are deliberately killed by added agents, for example, by water ($RM_i^\ominus + H_2O \rightarrow RM_iH + HO^\ominus$) or alcohols. This **termination reaction** seals the chain ends and protects them against depolymerization and other unwanted after-reactions. A reaction is defined as a termination reaction if it delivers a new anion (here: $HO^\ominus$) that is unable to initiate polymerizations.

Chain transfer reactions also kill macroanions but the new anions produced by chain transfer may initiate new polymer chains. An example is the reaction between macroanions $RM_i^\ominus$ and chlorohydrocarbons Cl-R':

$$(3\text{-}24) \quad RM_iCl + {}^\ominus R' \xleftarrow{\hspace{1cm}} RM_i^\ominus + Cl\text{-}R' \xrightarrow{\hspace{1cm}} RM_i\text{-}R' + Cl^\ominus$$

$$\text{transfer} \qquad\qquad\qquad \text{termination}$$

Note that *matter* is transferred in both termination and transfer reactions. The distinction between these two types of reactions is a *kinetic* one: in a chain transfer reaction, the kinetic chain is preserved whereas in a termination reaction, it is terminated.

Transfer reactions may also occur with monomers according to $M_i^\ominus + M \rightarrow M_i +$ $M^\ominus$ ($i \geq 1$). A new monomer anion $M^\ominus$ is formed for each disappearing macroanion $M_i^\ominus$. Monomer anions are consumed in the start reaction $M^\ominus + M \rightarrow M_2^\ominus$. The rate $R_{st} = k_{st}[M^\ominus][M]$ of the start reaction must thus equal the rate $R_{tr} = k_{tr}[M_i^\ominus][M]$ of the transfer reaction. It follows that $[M_i^\ominus]/[M^\ominus] = k_{st}/k_{tr} \equiv C$. Since all initiator molecules will start a chain ($[I]_o = [M^\ominus] + [M_i^\ominus]$), the polymerization rate is given by:

$$(3\text{-}25) \quad R_p = k_p[M_i^\ominus][M] = k_pC(1 + C)^{-1}[I]_o[M] = const\cdot[I]_o[M] ; \quad const < k_p$$

The transfer to the monomer lowers polymerization rates but the proportionality factor is still a constant (albeit smaller than k_p). A transfer to monomer thus cannot be detected from $R_p = f([M])$ (see Fig. 3-4, II and Eqn.(3-21)). However, the number-average degree of polymerization is no longer proportional to the extent of reaction.

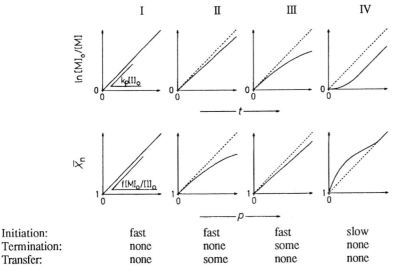

	I	II	III	IV
Initiation:	fast	fast	fast	slow
Termination:	none	none	some	none
Transfer:	none	some	none	none

Fig. 3-4 Living and pseudoliving polymerizations. Top: Change of natural logarithm of monomer concentration with time. Bottom: Change of number average degree of polymerization with monomer conversion p = ([M]$_o$ – [M])/[M]$_o$. I: Ideal living polymerization without termination and chain transfer; II: pseudoideal living polymerization with some chain transfer; III: polymerization with termination; IV: living polymerization with slow initiation.

The occurence of termination causes a nonlinear increase of $\log_e([M]_o/[M])$ with time if chain transfer reactions are absent (Fig. 3-4, III). Slow start reactions generate an **induction period** (Fig. 3-4, IV). Since only few initiating anions are initially present in such reactions, degrees of polymerization will be relatively too high during the early polymerization stages.

Termination reactions allow one to introduce desired end groups (**functionalization**). For example, macroanions $RM_i^{\ominus}$ are converted into polymers $RM_iCOO^{\ominus}$ with carboxyl end groups on reaction with carbon dioxide. Ethylene oxide converts $RM_i^{\ominus}$ into $RM_iCH_2CH_2O^{\ominus}$ and, on further reaction with water, $RM_iCH_2CH_2OH$.

3.4.6. Stereocontrol

Active centers of growing macroanions of acrylates, methacrylates, vinyl arylenes, and dienes have planar delocalized configurations. They exist each in one Z and one E form because the rotation barrier is high around the $\beta-\gamma$ bond. An example is the anionic polymerization of acrylate esters with metal cations $Mt^{\oplus}$ as counter ions:

(3-26)

These polymerizations proceed according to a 2-way mechanism; they do not follow simple Bernoulli and Markov statistics (Section 3.3.). The type and extent of stereocontrol are determined by the polarities of solvents and monomers, the type and concentration of counter ions, and temperature (Table 3-3). Lithium compounds control steric structures especially well because the bond between lithium and carbon has a highly covalent character. In apolar solvents, fast (kinetically controlled) diene

Table 3-3 Stereocontrol in the anionic polymerization of dienes $CH_2=CHR-CH=CH_2$.

R	Initiator	Solvent	Temperature in °C	Proportion (in %) of mers as			
				1,4-cis	1,4-trans	1,2	3,4
H	Li	Heptane	25	35	52	13	–
	Na	Heptane	25	10	25	65	–
	Na	Tetrahydrofuran	0	6	14	80	–
	Li	Tetrahydrofuran	0	6	6	88	–
CH₃	Li	Heptane	25	94	0	0	6
	LiC₄H₉	Heptane	40	70	22	0	7
	Na	Heptane	25	0	43	6	51
	Na	Diethyl ether	25	--------19-------		6	75
C₄H₉	LiC₄H₉	Heptane	40	62	35	3	0
	Et₃Al-TiCl₄	Hexane	10	60	33	7	0

polymerizations prefer the E-form (cis center) whereas slow (thermodynamically controlled) diene polymerizations favor the Z-form (trans center). In polar solvents, it is just the opposite: the cis form is thermodynamically more stable and the trans form is kinetically preferred.

Magnesium ions are strongly associated in non-polar solvents; they polymerize 2-vinyl pyridine to highly isotactic polymers ($x_{ii} \approx 90$ %). Ion associations are concentration dependent and so are the resulting tacticities: the polymerization of α-methyl styrene in tetrahydrofuran by lithium butyl delivers syndiotactic triad fractions of x_{ss} = 40 % if [LiBu] = $2 \cdot 10^{-5}$ mol/L but x_{ss} = 70 % if [LiBu] = 0.5 mol/L.

In general, stereocontrol of anionic polymerizations is low for strongly electron-withdrawing monomers in polar solvents but high for weakly electron-withdrawing monomers in apolar solvents. The latter thus resemble insertion polymerizations (Section 3.6.).

3.5. Cationic Polymerizations

3.5.1. Monomers

Cationic polymerizations are started by the reaction of electrophilic initiator cations with electron-donating monomer molecules. The resulting monomer cations add additional monomer molecules if the nucleophilic groups of monomers reside in a part of the monomer molecule that can participate directly in the polymerization. Cationic polymerizations can thus be performed with (1) olefins $CH_2=CHR$ with electron-rich substitutents R, (2) compounds $R_2C=Z$ with hetero atoms or hetero groups Z, and (3) cyclic molecules with hetero atoms as part of the ring structure.

Electron-rich olefin derivatives include π-donors such as olefins $CH_2=CRR'$, dienes $CH_2=CR–CH=CH_2$, and vinyl aromatics $CH_2=CHAr$ as wells as $(\pi+n)$-donors such as N-substituted vinyl amines $CH_2=CH(NRR')$ and vinyl ethers $CH_2=CH(OR)$. The initiator cation always attacks the most nucleophilic group of a monomer molecule; this group must be able to propagate if polymers are to result.

For example, the nitrile group is the most nucleophilic group of acrylonitrile $CH_2=CH–C\equiv N$. Addition of an initiator cation $R^\oplus$ thus generates $[CH_2=CH–CN–R]^\oplus$. This monomer cation cannot start the cationic polymerization of acrylonitrile, however, because of the resonance stabilization by an equilibrium $CH_2=CH–C\equiv N^\oplus–R \rightleftarrows CH_2=CH–C^\oplus=N–R$.

Double bonds within rings can also be polymerized; examples are benzofuran (coumarone), indene, cyclopentadiene, and α-pinene. However, α-pinene does not polymerize via the endo double bond because it isomerizes first to the true monomer, D,L-limonene $CH_3–(p$-$(1,2$-$C_6H_8))–C(CH_3)=CH_2$. β-Pinene polymerizes cationically without added initiator to a phantom polymer $+CH_2–(p$-$(1,2$-$C_6H_8))–C(CH_3)_2\}_{\overline{n}}$, where $-1,2$-C_6H_8- is the cyclohexene residue.

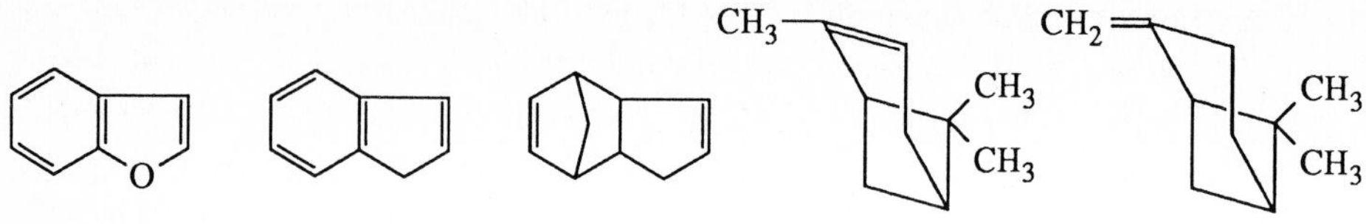

Benzofuran   Indene   Dicyclopentadiene   α-Pinene   β-Pinene
(Coumarone)

Monomers with *heteronuclear multiple bonds* can also be polymerized cationically. Examples are aldehydes $RHC=O$, ketones $RR'C=O$ (but not acetone because of a
ceiling temperature of $-273°C$), thioketones $RR'C=S$ and diazoalkanes $RR'CN_2$.

Cationically polymerizable *ring compounds* comprise cyclic ethers, imines, acetals,
sulfides, esters (lactones) and amides (lactams), e.g.:

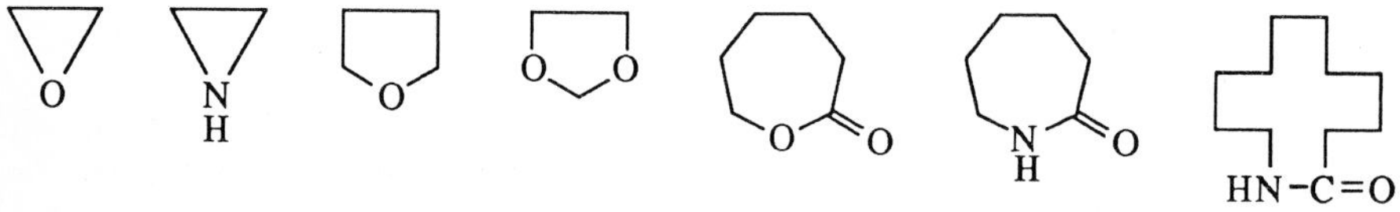

Ethylene Ethylene Tetrahydro- 1,3-Dioxo- ε-Capro- ε-Capro- Lauro-
oxide   imine   furan   lane   lactone   lactam   lactam

Relatively few cationic polymerizations are performed industrially although there
are many more cationically polymerizable monomers than anionically polymerizable
ones (Table 3-4). The main reason is the inherent instability (high reactivity) of many
macrocations which undergo termination and transfer reactions much more frequently than macroanions; these reactions result in oligomers.

3.5.2. Initiators

Cationic initiators ideally add to monomers to give monomer cations which add
more monomer molecules and form macrocations. Both monomer cations and
macrocations are, however, fairly reactive. They often try to add their counter ions instead of monomer molecules which terminates the polymerization (macrocations) or
does not get it started (monomer cations). Counter ions should thus not be too nucleophilic. Since the nucleophilicity of counter anions is changed by solvation, only
few solvents are suitable for cationic polymerizations. Such solvents include benzene,
nitrobenzene, and methylene chloride.

In rare cases, cationic polymerizations are initiated homolytically by one-electron
mechanisms, i.e., by charge transfer or by direct oxidation of radicals. Most initia-
tions take place heterolytically by two-electron mechanisms, though. The initiators
for these mechanisms can be subdivided into three classes: carbenium salts, Brønsted
acids, and Lewis acids.

Table 3-4   Industrial cationic polymerizations. * With 4 % isoprene as comonomer; c = cyclic.

Monomer	Structure	Initiator	Application
Isobutene*	$CH_2=C(CH_3)_2$	$BF_3 + H_2O$	Elastomers, adhesives, viscosity improvers
Alkyl vinyl ether	$CH_2=CHOR$	$BF_3 + H_2O$	Adhesives, plasticizers, textile additives
Coumarone+indene	see above		Sealants, coatings
Formaldehyde	HCHO		Engineering plastics
Ethylene imine	$c\text{-}(NHCH_2CH_2)$	Proton acids	Paper additives, flocculants
Tetrahydrofuran	$c\text{-}(O(CH_2)_4)$		Soft segments of elastomers

Carbenium salts dissociate to some extent into carbenium ions and counter anions in solvents used for cationic polymerizations. An example is the dissociation of "trityl chloride" $(C_6H_5)_3CCl$ into the triphenylcarbenium ion $(C_6H_5)_3C^\oplus$ and $Cl^\ominus$. The dissociation is promoted by complexation of the counter anion. For example, antimony pentachloride $SbCl_5$ adds a chloride anion $Cl^\oplus$ to form the hexachloroantimonate ion $[SbCl_6]^\ominus$. Other complex anions are $[AsF_6]^\ominus$, $[PF_6]^\ominus$ and $[BF_4]^\ominus$.

Complex anions can be easily fragmented; $[SbCl_6]^\ominus$ is, for example, stable only up to ca. 30°C, $[AsF_6]^\ominus$ however up to 80°C. The dissociation products can undergo many side reactions. Non-complex anions such as $[ClO_4]^\ominus$, $[CF_3SO_3]^\ominus$ or $[FSO_3]^\ominus$ do not dissociate. However, they may terminate chains by bonding to macrocations.

Brønsted acids such as perchloric acid $HClO_4$ dissociate into protons $H^\oplus$ and counter anions $[ClO_4]^\ominus$ according to the classic scheme. Perchloric acid is, however, a covalent compound in oxygen-free solvents since it does not conduct an electric current. It does dissociate in the presence of Brønsted bases such as isobutene according to $HClO_4 + CH_2=C(CH_3)_2 \rightleftarrows H\text{-}CH_2\text{-}^\oplus C(CH_3)_2 + [ClO_4]^\ominus$. This acid–base reaction explains why not all proton acids are able to start cationic polymerizations.

Monomer cations and macrocations should also not combine irreversibly with the counter ion. Addition of trifluoroacetic acid to styrene only results in a small yield of low molar mass poly(styrene). If styrene is poured into trifluoroacetic acid, however, high molar mass poly(styrene) is obtained in high yield. In the latter case, $CF_3COO^\ominus$ ions are stabilized by excess acid but not in the former case.

Lewis acids such as $AlCl_3$, BF_3, I_2, $RAlCl_2$, etc., can be subdivided into two classes. Some Lewis acids form electrically conducting solutions, obviously by self-ionization. Some examples: $2\ AlCl_3 \rightleftarrows [AlCl_2]^\oplus + [AlCl_4]^\ominus$; $2\ I_2 \rightleftarrows I^\oplus [I_3]^\ominus$; $2\ TiCl_4 \rightleftarrows [TiCl_3]^\oplus [TiCl_5]^\ominus$; $2\ RAlCl_2 \rightleftarrows [RAlCl]^\oplus [RAlCl_3]^\ominus$; $2\ PF_5 \rightleftarrows [PF_4]^\oplus [PF_6]^\ominus$.

Other Lewis acids cannot undergo self-ionization; they need "co-catalysts" such as water, trichloroacetic acid, alkylhalogenides, ethers, or the monomer itself. With these co-catalysts, Lewis acids form dissociating compounds, for example, $BF_3 + H_2O \rightleftarrows H^\oplus [BF_3OH]^\ominus$ or $R_2AlCl + C_2H_5Cl \rightleftarrows C_2H_5^\oplus [R_2AlCl_2]^\ominus$. Cations resulting from these reactions add monomers and start polymerizations. "Co-catalysts" are thus incorrectly named: they do not act catalytically and by no means *co*-catalytically. They rather provide the initiating species and are thus often called "initiators". Again, a wrong name: the true initiating compound is the salt from a Lewis acid and a "co-catalyst".

3.5.3. Propagation

In the simplest case, cationic polymerizations are propagated by repeated addition of monomer molecules to enium or onium ions. Examples:

$$\text{~~CH}_2\text{—}\underset{\underset{\text{CH}_3}{|}}{\overset{\overset{\text{CH}_3}{|}}{\text{C}}}^{\oplus}$$

| Carbenium | Dioxolenium | Silicenium | Carbonium | Oxonium |

Enium ions are cations with electron-deficient centers. They comprise carbenium ions (trivalent carbocations), dioxolenium ions and silicenium ions.

Onium ions are formally obtained from enium ions by further addition of ligands. Carbonium ions are four-fold or five-fold coordinated, non-classical carbocations. Carboxonium ions are generated by addition of oxygen compounds to carbenium ions. They are much more stable than carbocations because of the strong nucleophilic character of the heteroatom.

In polymerizations by *carbenium ions*, a center with high density of positive charge attacks the partially electronegative β-atom of an olefin $^{\beta}CH_2{=}^{\alpha}CHR$. The dipole moment of the transition state is largely created by the attacking cation. The transition state must be therefore almost linear; the activation energy is thus low.

Oxonium ions are strongly solvated; their charge density is smaller than that of carbenium ions. The monomer molecules possess strong dipoles; their negative oxygen atoms approach the oxonium ions. As a result, the transition state is not linear and the polymerizations have high activation energies.

For these reasons and because of the presence of very different propagating species (free ions, various ion pairs and ion associates), *average propagation rate constants* of cationic polymerizations cover an extraordinarily large range from $k_p \approx 10^9$ L mol^{-1} s^{-1} for the γ-ray-initiated polymerization of neat cyclopentadiene at $-78°C$ to $k_p \approx 10^{-5}$ L mol^{-1} s^{-1} for the polymerization of 3,3-diethylthietane in CH_2Cl_2 at 20°C by initiation with $[(C_2H_5)_3O]^{\oplus}[BF_4]^{\ominus}$.

Within a given class of monomers, propagation rate constants often do not differ greatly, however. For example, ion equilibria have been observed for olefin monomers but not for cyclic monomers, probably, because the positive charges on such rings (oxonium, dioxolenium, etc.) are much better screened.

The *stereocontrol* of cationic polymerizations depends strongly on the system monomer–initiator–solvent–temperature, i.e., on the complexation of polar chain ends with counter ions, on the presence of ion associates, etc. Sometimes highly iso–tactic polymers are observed and at other times highly syndiotactic ones. The polymerization of methyl methacrylate by C_6H_5MgBr in toluene at 30°C, for example, leads to a poly(methyl methacrylate) with 99 % isotactic triads. If α-methylstyrene is polymerized by BF_3 in CH_2Cl_2 at $-78°C$, triads are found to be 89 % syndiotactic and 11 % heterotactic.

3.5.4. Termination and Chain Transfer

Living cationic polymerizations can be achieved if unstable carbenium ions are nucleophilically stabilized. One possibility is to reduce the cationic charge by addition of certain counter ions B. At –40°C, hydrogen iodide HI in toluene forms adducts with isobutyl vinyl ether, but no polymers. Addition of the electrophilic I_2 converts HI into $H^{\oplus}[I \cdot I_2]^{\ominus}$ which initiates a living polymerization. $ZnCl_2$ reacts similarly with other protonic acids, for example, CCl_3COOH. Change of the solvent or the temperature may convert these living conditions into non-living ones:

$$(3\text{-}27) \quad \begin{array}{c} H \;\; H \\ | \;\;\; | \\ \sim\!\!\!\sim\!\!C - C \overset{\delta^+}{} \cdots B^{\delta^-} \\ | \;\;\; | \\ H \;\; R \end{array} \xleftarrow[-A]{+B} \begin{array}{c} {}^{\beta}H \;\; {}^{\alpha}H \\ | \;\;\; | \\ \sim\!\!\!\sim\!\!C - C \overset{\oplus}{} \cdots A^{\ominus} \\ | \;\;\; | \\ H \;\; R \end{array} \xrightarrow{+Y} \begin{array}{c} H \;\; H \\ | \;\;\; | \\ \sim\!\!\!\sim\!\!C - C \overset{\oplus}{} \cdots Y \cdots A^{\ominus} \\ | \;\;\; | \\ H \;\; R \end{array}$$

Strong Lewis acids $H^{\oplus}A^{\ominus}$ lead to non-living polymerizations. $HCl + SnCl_4$ form nucleophilic counter ions according to $[SnCl_5]^{\ominus} \rightleftarrows Cl^{\ominus} \cdot SnCl_4$ but these counter ions are too weak to stabilize the growing macrocations. The acidity of the β-proton of vinyl monomers can, however, be decreased by the addition of a weak Lewis base Y. An example is the addition of *p*-dioxane to vinyl ethers where macrocation and counter ion create a system $\sim CH_2-CH(OR) \cdots O^{\oplus} \text{-} \cdots Cl^{\ominus} \cdots SnCl_4$. The effectiveness of such additives depends on their pK_a values: as the pK_a increases, one observes a change from polymerizations with termination to living polymerizations, then to poly`mers with broader molar mass distributions and, finally, no polymerization at all.

Most cationic polymerizations exhibit termination and/or chain transfer reactions. Due to the high reactivity of macrocations, **terminations** by dissociation of complex counter ions are frequent, for example, according to $\sim CH_2C^{\oplus}R_2 + [BF_4]^{\ominus} \rightarrow \sim CH_2CR_2F + BF_3$. Non-complex counter ions do not fragment but often form covalent bonds, especially in cationic polymerizations that are initiated by super acids. Such *super acids* have, by definition, higher acid strengths than 100 % sulfuric acid. Esters CF_3SO_2OR of the super acid trifluoromethanesulfonic acid (triflic acid) CF_3SO_3H lead to cationic polymerizations where macrocations I are in equilibrium with macroesters II:

$$(3\text{-}28) \quad CF_3SO_3R + O{\bigcirc} \rightleftarrows R\overset{\oplus}{-}O{\bigcirc} \; [CF_3SO_3]^{\ominus} \rightleftarrows RO(CH_2)_4OSO_2CF_3$$

$$\qquad\qquad\qquad\qquad\qquad\qquad\qquad\quad \text{I} \qquad\qquad\qquad\qquad\qquad \text{II}$$

Macrocations I are predominantly formed in polar solvents and macroesters II mainly in non-polar solvents. The polymerization proceeds in any case almost exclusively via macrocations because macrocations are much more reactive than macroesters. The molar mass distribution is bimodal: high molar masses are generated by macrocations and low molar masses by macroesters. Macrocations react with traces of water whereas macroesters are relatively tolerant to large proportions of H_2O.

A **termination by monomer** is a kind of suicide. An example of this is the cationic polymerization of propylene. A hydride transfer from the monomer to the cation generates a resonance-stabilized allylic monomer structure that cannot add new monomer molecules; the degree of polymerization remains low:

(3-29) $\sim\sim$CH$_2$—$\overset{\oplus}{\text{CH}}$ + CH$_2$=CH—CH$_3$ $\longrightarrow$ $\sim\sim$CH$_2$—CH$_2$ + CH$_2$ $\equiv$ $\overset{\oplus}{\text{CH}}$ $\equiv$ CH$_2$
              | |
              CH$_3$ CH$_3$

Chain transfer reactions occur frequently in cationic polymerizations. An example is the polymerization of isobutene where the transfer to monomer produces monomer cations that initiate further polymerizations:

                 CH$_3$ CH$_3$ CH$_3$ CH$_3$
                 | | | |
(3-30) $\sim\sim$CH$_2$—$\overset{\oplus}{\text{C}}$ + CH$_2$=C $\longrightarrow$ $\sim\sim$CH$_2$—C   + H—CH$_2$—$\overset{\oplus}{\text{C}}$
                 | | ‖ |
                 CH$_3$ CH$_3$ CH$_2$ CH$_3$

Additional chain transfer reactions to polymers, solvents or initiators depend on the system. A special case is that of salts R[BCl$_4$] when RCl is added in excess, for example, as Cl–C(CH$_3$)$_2$–(p-C$_6$H$_4$)–C(CH$_3$)$_2$–Cl. The growing macrocation is terminated here by the counter ion [BCl$_4$]$^\ominus$:

                 CH$_3$ CH$_3$
                 | |
(3-31) $\sim\sim$CH$_2$—$\overset{\oplus}{\text{C}}$ [BCl$_4$]$^\ominus$ $\longrightarrow$ $\sim\sim$CH$_2$—C—Cl + BCl$_3$
                 | |
                 CH$_3$ CH$_3$

The Lewis acid BCl$_3$ complexes the excess RCl and forms the salt R$^\oplus$[BCl$_4$]$^\ominus$, which starts a new polymer chain. Possibly, the reaction does not occur in two steps (termination and susequent complexation) but in one step:

                 CH$_3$ CH$_3$
                 | |
(3-32) $\sim\sim$CH$_2$—$\overset{\oplus}{\text{C}}$ [BCl$_4$]$^\ominus$ + RCl $\longrightarrow$ $\sim\sim$CH$_2$—C—Cl + R[BCl$_4$]
                 | |
                 CH$_3$ CH$_3$

Since this reaction involves a chain trans*fer* to the *ini*tiator, it is called an **inifer** reaction. Such polymerizations are pseudo-living (see Fig. 3-4, II) if chain transfer to monomer is absent.

The inifer technique involves the formation of a carbocation salt *in situ*. The technique works especially well with BCl$_3$ and isobutylene since it leads to growing poly(isobutylene) cations $\sim$CH$_2$–C$^\oplus$(CH$_3$)$_2$ which, in the presence of [BCl$_4$]$^\ominus$ anions, show at temperatures below 0°C little or no transfer to monomer and no irreversible termination reactions.

3.6. Insertion Polymerizations

3.6.1. Introduction

Classical anionic, cationic and free radical polymerizations proceed by *addition* of monomers to active chain ends that are separated from covalently bound initiator fragments by many monomeric units (*chain-end control*). An example is the anionic polymerization of styrene $CH_2=CHPh$ by $Li^{\oplus}C_4H_9^{\ominus}$:

$$(3\text{-}33) \quad C_4H_9(CH_2\text{-}CHPh)_n{}^{\ominus} + CH_2=CHPh \longrightarrow C_4H_9(CH_2\text{-}CHPh)_nCH_2\text{-}CHPh^{\ominus}$$

In insertion polymerizations, on the other hand, monomer molecules are *inserted* between the initiator complex and the next monomeric unit. For example, the initiator complex $TiCl_4$-AlR_3 (symbolized by [Ti]) initiates an isospecific polymerization of propylene which proceeds via an α-insertion of the monomer:

$$(3\text{-}34) \quad \sim(CHR\text{-}CH_2)_n\text{-}[Ti] + CH_2=CHR \longrightarrow \sim(CHR\text{-}CH_2)_n\text{-}CHR\text{-}CH_2\text{-}[Ti]$$

Polyinsertions include Ziegler–Natta polymerizations (Section 3.6.2.), metathesis polymerizations (Section 3.6.3.) and probably also the so-called group transfer polymerizations (Section 3.6.4.).

Polyinsertions are very important industrial polymerizations. They allow the polymerization of ethylene and inexpensive olefins at low pressures and the stereospecific polymerization of α-olefins, dienes and cycloolefins (Table 3-5).

Table 3-5 Industrial polymerizations with transition group metal catalysts. [a] Copolymer of ethylene with (6-10) % 1-butene or 1-octene; [b] terpolymer with small proportions of a non-conjugated diene; [c] after addition of mineral oil; [d] partially reduced; [e] main process (Phillips); [f] Unipol process. In industry, butadiene and isoprene are polymerized by lithium organic compounds.

Polymers	Industrial catalysts	Use
Ziegler–Natta polymerizations		
High-density poly(ethylene)	$TiCl_4$-R_2AlCl-$MgCl_2$, Cr/Ti/Mg [f], Cr_2O_3 [d] on Al_2O_3 [e]	Thermoplastics
Linear low density PE [a]	e.g., metallocene catalyst	Thermoplastics
Ethylene–propylene polymer [b]		Elastomers
it-poly(propylene)	$TiCl_4$-AlR_3-$MgCl_2$	Thermoplastics, fibers
it-poly(1-butene)		Thermoplastics
it-poly(4-methyl-1-pentene)	Thermoplastics	
Metathesis polymerizations		
trans-Poly(cyclopentene)	WCl_6-R_3Al-C_2H_5OH	Elastomers
Poly(dicyclopentadiene)	WCl_6-$WOCl_4$-$(C_2H_5)_2AlCl$	Thermosets
Poly(norbornene)		Thermoplastics, elastomers [c]

In these polyinsertions, insertion steps are preceded by a coordination of the monomer molecule with the polymer–initiator "complex". Such insertion polymerizations are therefore also called **coordinative polymerizations** or **anionic-coordinative polymerizations**. However, these names are neither useful nor correct. A coordination of a monomer molecule with an initiator molecule or complex does not necessarily mean that the resulting coordination compound participates in the propagation steps. An example is the coordination of ethylene with silver nitrate where the propagation is by free radicals and not via insertion. Polyinsertions do involve electrophilic species but no ions are detected by measurements of electrical conductivity: these polymerizations are thus not "anionic".

In insertion polymerizations, initiating compounds are called *catalysts* for historic reasons. Unlike true catalysts, they do not emerge intact from the polymerization; rather, they are consumed and reside as initiator fragments at one end of the polymer chain. The initiator also does not hop from chain to chain as a true catalyst would do (note that this statement is *apparently* violated by chain transfer reactions but these reactions are side reactions with respect to the main reaction, i.e., propagation).

3.6.2. Ziegler–Natta Polymerizations

Ziegler–Natta (ZN) polymerizations are polymerizations that are initiated and propagated by so-called Ziegler catalysts. These "catalysts" were discovered by Karl Ziegler during attempts to synthesize organometallic compounds; he also utilized them for low-pressure polymerizations of ethylene. The same and similar catalysts were later used by Guilio Natta for the stereospecific polymerization of α-olefins and dienes.

Classical **Ziegler catalysts** are a group of organometallic compounds that result from the combination of metal compounds of groups IVB–VIIIB of the periodic table with hydrides or alkyl or aryl compounds of metals from the main groups I–III. A typical Ziegler catalyst is generated from $TiCl_4$ and $(C_2H_5)_3Al$. Millions of possible combinations exist but only a few are industrially useful. Nowadays, Ziegler catalysts with organometallic compounds from main group IV elements are also known.

The components of Ziegler catalysts may undergo exchange reactions. $TiCl_4$ + $(C_2H_5)_3Al$ exchange ligands which results in the formation of β-$TiCl_3$, $(C_2H_5)_2AlCl$ and other compounds. In addition to the brown, amorphous β-$TiCl_3$, three other crystalline $TiCl_3$ modifications are known (α, γ, δ), all with different polymerization activities. Structures of the polymerization-active compounds are thus different from the structures of the initial compounds.

All complexes with unoccupied coordination sites and uneven electron distributions are potential Ziegler catalysts. Such catalysts may be complexes from compounds of two different metals (e.g., Ti and Al), from two species of the same metal with different valences (e.g., Ti(II) and Ti(III)), or from two different compounds of the same metal with the same valence but with different ligands (e.g., $RTiCl_2$ and $TiCl_3$). The resulting active species of Ziegler catalysts are very often mixtures of different complexes which each exhibit two metallic centers. Modern catalysts for insertion polymerizations are compounds with monometallic centers (see below).

Heterogeneous Ziegler Catalysts

Heterogeneous Ziegler catalysts are insoluble in the reaction mixture. Proposed structures of polymerization-active species include the bimetallic complexes I for monometallic mechanisms and II for bimetallic mechanisms (participation of two metal atoms in the propagation step). In these structures, X is an "anion" (e.g. Cl), P_n a polymer chain, and O an unoccupied ligand site. In addition, monometallic complexes such as III are often proposed for monometallic mechanisms:

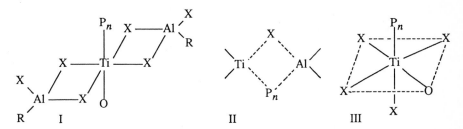

Fig. 3-5 Proposed structures of heterogeneous Ziegler catalysts.

The ligands X must be balanced in such a way with respect to their electron-donation properties that the right degree of destabilization is obtained. An unstable bond $Ti-P_n$ (or metal–R) would fragment and would not be polymerization-active. On the other hand, if too stable a bond is formed after coordination of the monomer with the unoccupied ligand site, monomer molecules will not insert into the chain.

The monometallic mechanism of ethylene polymerization by a monometallic titanium complex I assumes that the π-bond of the monomer molecule approaches the unoccupied ligand site O of the transition metal (Fig. 3-6). The subsequent monomer coordination (II) destabilizes the bond [Ti]–P_n so that it reacts with the double bond of the complexed ethylene (III) which is inserted between [Ti] and P_n (IV):

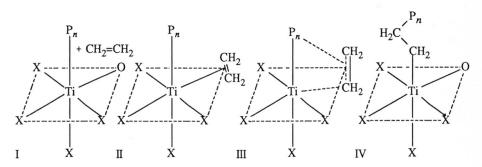

Fig. 3-6 Polymerization of ethylene according to a monometallic mechanism.

Ziegler catalysts polymerize ethylene at low pressures. The resulting polymers are only slightly branched which leads to fairly high crystallinities and thus high densities (high density poly(ethylene)s HDPE or PE-HD). Poly(ethylene)s from free radical polymerizations at high pressure (Section 3.7.7.) are strongly branched which results in lower crystallinities and lower densities (LDPE or PE-LD). So-called "linear low-density poly(ethylene)s" LLDPE are copolymers of ethylene with small proportions of 1-butene, 1-hexene, 1-octene, or 4-methyl-1-pentene (see below).

In analogy to butene, pentene, hexene, etc., ethylene is now called "ethene" and propylene "propene" though not in industry. Curiously, one still speaks of "ethylene oxide' and not "ethene oxide". IUPAC has also decreed that the synonym for 1-butene (formerly butene-1) is "butene' and not "*n*-butene" whereas 2-butene can still be called "*i*-butene" (industry: isobutylene).

α-Olefins and dienes possess nonsymmetric electron distributions and can thus be coordinated only in very specific positions which leads to a stereocontrol of the insertion step. Propylene $CH_2=CH(CH_3)$ is polymerized heterogeneously via α-insertion to isotactic poly(propylene) $+CH(CH_3)-CH_2\}_n$[Ti] by the insoluble catalyst $TiCl_4-R_3Al$ (Eqn.(3-34)). The stereocontrol follows the enantiomorphic catalyst model (Section 3.3.). The soluble system $VCl_4-(C_2H_5)_2AlCl$–anisole polymerizes propylene homogeneously by β-insertion to the syndiotactic polymer $+CH_2-CH(CH_3)\}_n$[V].

The first industrial polymerizations of propylene by $TiCl_3-(C_2H_5)_2AlCl$ as initiating system generated ca. 4 kg PP/(g Ti). The polymers were ca. 90 % isotactic. Atactic fractions were extracted by solvent because they lowered mechanical properties. Remaining catalyst residues were made harmless by reaction with alcohol to TiO_2.

The second generation catalysts used additional electron donors (ethers, esters, amines, etc.). [Ti] complexes became better stereoregulating by steric effects and aluminum alkyls less reducing. The catalyst productivity climbed to (12-20) kg PP/(g Ti) and the isotacticity to 95 %.

The third generation of catalysts used complexes of $TiCl_3$ on crystal surfaces of $MgCl_2$ which has the same lattice constants as δ-$TiCl_3$. The catalyst productivity increased to ca. 300 kg PP/g Ti. The small proportion of catalyst residues in these polymers is negligible and need not be extracted.

The newest, highly reactive catalyst systems employ $TiCl_4$. They deliver spherical granules of highly isotactic poly(propylene)s that are only slightly crystalline in the native state. The polymers thus require little energy for their melting (see Section 9.4.) prior to processing (Section 14.2.2.). Goods from melt-processed poly(propylene) are highly crystalline, however, because the high isotacticity leads to rapid crystallization from the melt.

A great number of industrial "poly(propylene)s" are not homopolymers but copolymers. Statistical bipolymers contain less than 8 wt-% of ethylene units and statistical terpolymers less than 12 wt-% of ethylene and 1-butene units. So-called block copolymers of propene are in reality heterophase blends of it-poly(propylene), poly(propylene-*co*-ethylene) and poly(ethylene) where the copolymers serve as compatibilizers for the two homopolymers. They are generated by a polymerization in cascade reactors where a homopolymerization of propene is followed by a copolymerization of propylene and ethylene.

Catalyst systems similar to the ones delivering isotactic polymers from α-olefins generate diene polymers that are rich in 1,4-cis units. Such polymerizations probably follow bimetallic mechanisms. Industry polymerizes dienes with organolithium compounds, however, not with Ziegler catalysts.

Homogeneous Ziegler Catalysts

Soluble vanadium-based Ziegler catalysts serve for the synthesis of an amorphous terpolymer from ethene, propylene and a small proportion of a non-conjugated diene (usually 5-ethylidene-2-norbornene). These polymers are known as EPDM elastomers. Soluble vanadium compounds also polymerize propene to a predominantly syndiotactic poly(propylene) which is not used industrially, however.

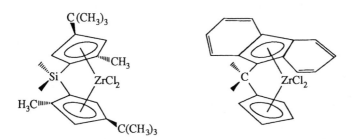

| Cyclopentadiene | Norbornadiene | Dicyclopentadiene | 5-Ethylidene-2-norbornene |

The newly developed "metallocene polymers" are poly(olefin)s that are generated by soluble **metallocene catalysts**. These catalysts consist of only one catalytic active species with a single metal atom (**single-site catalysts**). They deliver high catalyst productivities of (150-700) kg polymer/g metal atom. Depending on the catalyst structure, these monometallic catalysts produce isotactic or syndiotactic homopolymers of propylene or styrene, isotactic stereoblock copolymers of propene with inversion of the chirality, or copolymers of ethylene with α-olefins.

Metallocene catalysts for the stereospecific polymerization of propylene require "open structures", e.g., with zirconium atoms (Fig. 3-7). Since these structures also allow access of other α-olefins to the zirconium atom, various copolymers of ethylene and α-olefins can be obtained. The product range of linear low density poly-(ethylene)s with less than 10 % α-olefin units is thus extended to so-called polyolefin plastomers POP with less than 20 % α-olefin units and polyolefin elastomers POE with 20 % to 40 % α-olefin units. These metallocene catalysts are more expensive than conventional Ziegler catalysts but they lead to polymers with less extractables and excellent clarity.

Fig. 3-7 Metallocenes for the polymerization of propylene to it-PP (left) and st-PP (right).

Other metallocene catalysts possess "closed structures" which allow only the polymerization of propylene (Fig. 3-8). Depending on the catalyst structure, either isotactic stereoblocks (with reversal of the chirality in the chain) or it-poly(propylene)s with the same handedness are obtained.

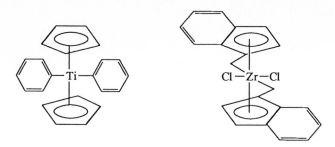

Fig. 3-8 Metallocene catalysts for the polymerization of propylene to isotactic stereoblocks (left) or isotactic homopolymers (right).

Kinetics

Ziegler–Natta polymerizations are living polymerizations if preformed, stable, soluble Ziegler catalysts are used. The polymerization rate is given by the initial concentration $[I]_0$ of the Ziegler catalyst and the actual monomer concentration $[M]$ according to Eqn.(3-21). The degree of polymerization is controlled by the molar ratio monomer/catalyst and the monomer conversion (Eqn.(3-23)).

Active centers are, in general, however, neither preformed nor stable or soluble. Their concentrations are not constant with time because of continuing exchange reactions, blockage of active centers by precipitating polymer, fracture of solid catalyst particles which generates new, catalytically effective surfaces, etc.

The catalyst concentration is therefore not necessarily constant in $R_p = - d[M]/dt = k_p[Cat][M]$. With $TiCl_4$-$(i$-butyl$)_3Al$–$MgCl_2$ as catalyst, the polymerization rate of propylene in heptane at 60°C fast approaches a maximum and then decreases rapidly and later more slowly. In the same system with ethylene as monomer, the rate first increases and then becomes constant.

True termination reactions are rare for Ziegler–Natta polymerizations at room temperature. At higher temperatures, the growth of individual polymer chains is stopped by spontaneous β-elimination of transition metal hydrides H-Mt:

(3-35) $\{CHR\text{-}CH_2\}_n CHR\text{-}CH_2\text{-}Mt \longrightarrow \{CHR\text{-}CH_2\}_n CR=CH_2 + H\text{-}Mt$

The metal hydride may be, however, realkylated by the monomer $CH_2=CHR$:

(3-36) $Mt\text{-}H + CHR=CH_2 \longrightarrow Mt\text{-}CHR\text{-}CH_2\text{-}H$

This reaction is not a termination since the number of active centers remains constant. It is rather a chain transfer: the rate stays constant while the degree of polymerization decreases (see II in Fig. 3-4). The realkylation of metal hydrides by monomers allows industry to use hydrogen as a chain transfer agent for the regulation of molar masses:

(3-37) $Mt\{CH_2\text{-}CHR\}_n + H_2 \longrightarrow Mt\text{-}H + H\{CH_2\text{-}CHR\}_n$

Kinetic **chain transfers** may be to monomers or to alkyl aluminum:

(3-38)   ~CHR'-CH$_2$-Mt + CHR'=CH$_2$ $\longrightarrow$ ~CR'=CH$_2$ + H-CHR'-CH$_2$-Mt

(3-39)   ~CHR'-CH$_2$-Mt + R$_3$Al $\longrightarrow$ ~CHR'-CH$_2$-AlR$_2$ + R-Mt

Ziegler–Natta polymerizations proceed in solution, in slurry, or in gas phase and gas phase/bulk, respectively, by the fluidized bed process. All heterogeneous processes use highly active Ziegler catalysts on solid supports, for example, Al$_2$O$_3$ for ethene and MgCl$_2$ for propylene. Fluidized bed polymerizations require polymerization temperatures above the dew point of the monomer. Linear low density poly(ethene)s can thus be obtained in fluidized beds from ethene and 1-butene but not from ethylene and 1-octene.

3.6.3. Metathesis Polymerizations

Metatheses

Metatheses are exchange and disproportionation reactions of carbon–carbon double bonds in olefins and cycloolefins. The metathesis of simple, acyclic olefins is a transalkylidenation with exchange of parts of the molecule. If WCl$_6$–C$_2$H$_5$AlCl$_2$–C$_2$H$_5$OH is added to 2-pentene, a mixture of 2-butene, 2-pentene and 3-hexene results within minutes:

(3-40)   CH$_3$—CH CH—CH$_3$ CH$_3$—CH=CH—CH$_3$
         ‖   +   ‖ $\rightleftharpoons$ +
         C$_2$H$_5$—CH   CH—C$_2$H$_5$ C$_2$H$_5$—CH=CH—C$_2$H$_5$

The molar ratio of 1:2:1 corresponds to the statistical expectation for an exchange of parts of molecules about the double bond. The enthalpy of this reaction is zero since similar bonds are exchanged. The reaction is thus promoted by an entropy increase (generation of 3 species from 2). The reaction enthalpy is, however, not zero for sterically hindred compounds; the molar ratio then deviates from 1:2:1.

Metatheses between two olefin molecules are not spontaneous but proceed via intermediate metal carbenes CR$_2$=[Mt] where [Mt] is the complexed transition metal atom of the catalyst. Catalyst and olefin form an unstable four-membered intermediate which subsequently disproportionates:

        R—CH   [Mt]   R— CH —[Mt]   R—CH=[Mt]
(3-41) ‖   + ‖ $\rightleftharpoons$ | | $\longrightarrow$ +
        R'—CH   CHR   R'— CH ——CHR   R'—CH=CH—R

The groups [Mt]= and R-CH= become end groups of the resulting polymer chains, for example, in the polymerization of cyclooctene:

(3-42)   R—CH=[Mt]   +   [structure] ⟶ [structure] ⟶ [structure]

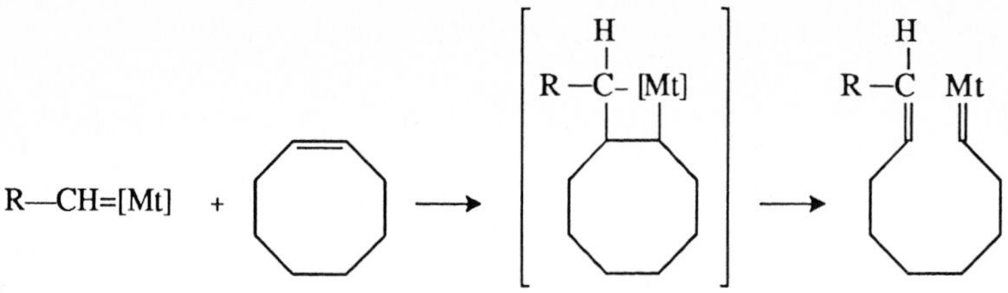

Metathesis catalysts are thus not catalysts but initiators. Propagation proceeds by insertion of additional monomer molecules into the –CH=[Mt] bond. Such polymerizations are frequently living. Terminations may occur by a reversal of the start reaction, Eqn.(3-42).

Metathesis catalysts are mainly tungsten compounds; molybdenum, ruthenium and rhenium compounds are also used. They may be subdivided into three classes:

(a) Preformed metal carbene complexes such as $(C_6H_5)_2$=W(CO) need to first lose a ligand by thermal or photochemical fragmentation or by reaction with another compound before they become active metathesis catalysts.

(b) The most effective metathesis catalysts are usually generated by reaction of a transition metal compound with alkyl- or allyl-group-containing compounds. An example is the reaction of WCl_6 + $(CH_3)_4Sn$ to CH_3WCl_5 which loses HCl and gives the catalyzing carbene CH_2=WCl_4.

(c) A third group generates the carbene by reaction of a transition metal compound with the monomer. An example is Re_2O_7 + Al_2O_3.

Polyelimination of Aliphatic Olefins

Open-chain olefins usually do not polymerize under metathesis conditions but rather undergo disproportionation reactions (see Eq.(3-40)). Exceptions are acyclic olefins where the carbon–carbon double bond is conjugated with a carbonyl group. Mesityl oxide (4-methyl-3-pentene-2-one) releases acetone and forms poly(methylacetylene):

(3-43)   [structure: CH_3\ /CH_3 C=CH—C=O with CH_3 below] ⟶ [CH_3\ /CH_3 C=O] + [~~CH=C~~ with CH_3 below]

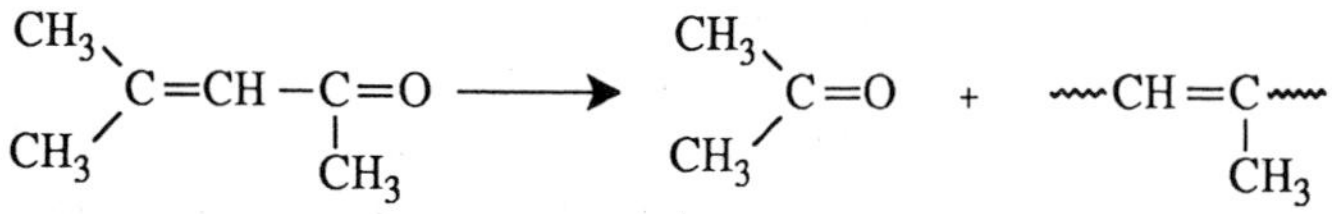

Ring-Opening Polymerization of Cycloolefins

Ring-opening metathesis polymerizations (ROMP) can be performed with cyclopentene, cyclooctene, norbornene, dicyclopentadiene and other cycloolefins above a critical initial monomer concentration. Below this concentration, only cyclic oligomers are obtained. Examples are:

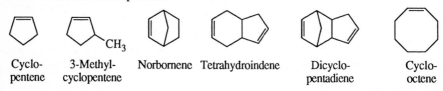

Cyclo-   3-Methyl-   Norbornene Tetrahydroindene   Dicyclo-   Cyclopentene   cyclopentene   pentadiene   octene

Metathesis polymerizations cannot be performed with similar cycloolefins, however:

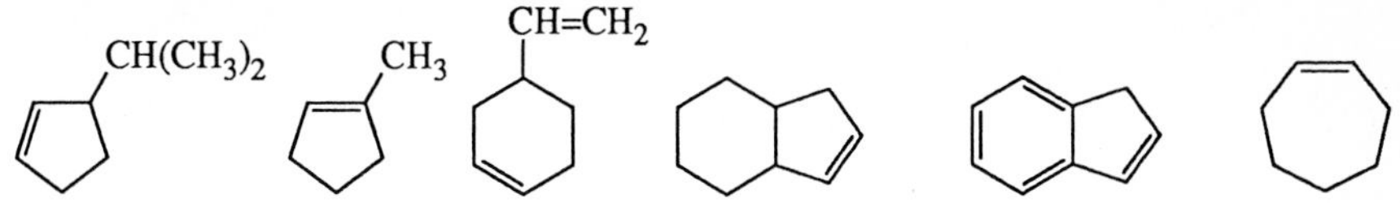

The metathesis ring-opening polymerization of cyclooctene results in the so-called poly(octenamer) $\mathrm{+CH{=}CH{-}(CH_2)_6\}_{\overline{n}}$. Both the cis and the trans isomer of poly(octenamer) are elastomers. Norbornene polymerizes to a thermoplastic:

(3-44)

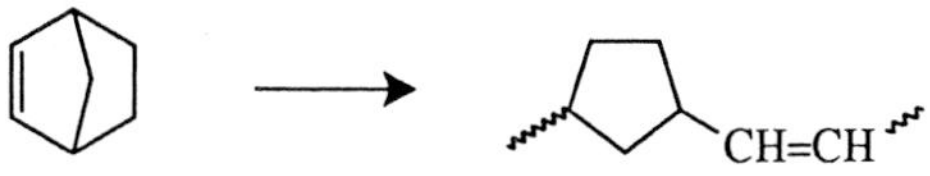

which becomes an elastomer on plasticization by mineral oil. Dicyclopentadiene polymerizes during reaction injection molding (RIM) to a cross-linked plastic:

(3-45)

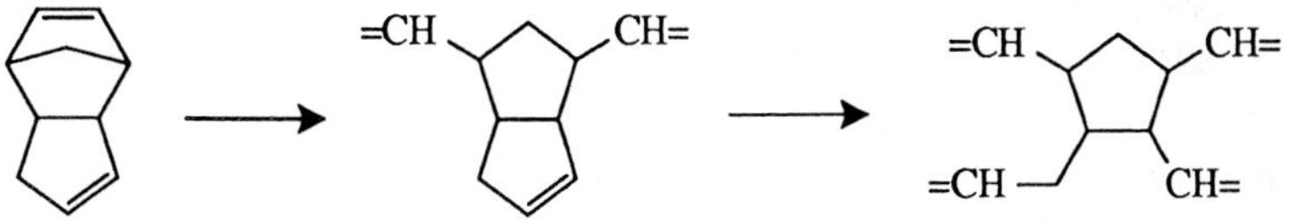

The configuration of polymers from metathesis polymerizations is kinetically, not thermodynamically, controlled. Primarily formed cis-structures are later isomerized: cis-olefins are easily polymerized under metathesis conditions but not trans-olefins.

3.6.4. Group Transfer Polymerizations

Group transfer polymerizations are defined as polymerizations in which an initiator molecule transfers its active group to a monomer molecule (or vice versa) under the action of a catalyst. These polymerizations do not involve a transfer reaction to a monomer molecule, however, but an insertion of the monomer molecule into the bond between the active group and the rest of the chain. Depending on the monomer, either a nucleophilic or an electrophilic catalyst must be used. The mechanisms of group transfer reactions are still not well understood.

The nucleophilic (with, e.g., $[HF_2]^{\ominus}$) catalyzed polymerization of methyl methacrylate with the corresponding silylketene acetals as initiators involves formally a transfer of the silylketal group and an insertion of the monomer into the α-position:

(3-46)

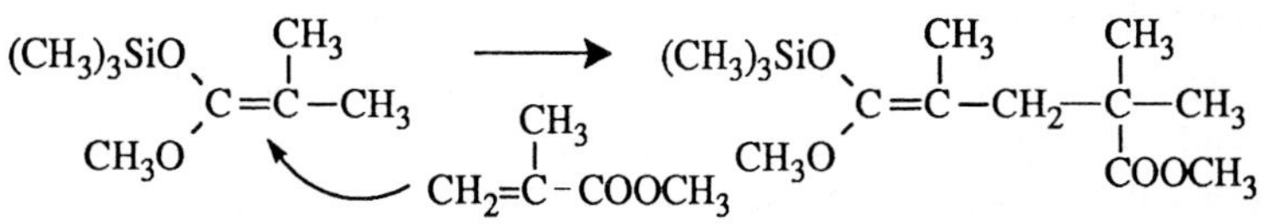

Conversely, aldol group transfer polymerizations consist of a transfer of the silyl group to the initiator and an insertion of the monomer into the β-position:

(3-47) $OHC-$⟨⟩ $\longrightarrow$ $OHC-CH_2-CH-$⟨⟩
                    $CH_2=CH$ $OSi(CH_3)_2(t-C_4H_9)$
                        |
                    $OSi(CH_3)_2(t-C_4H_9)$

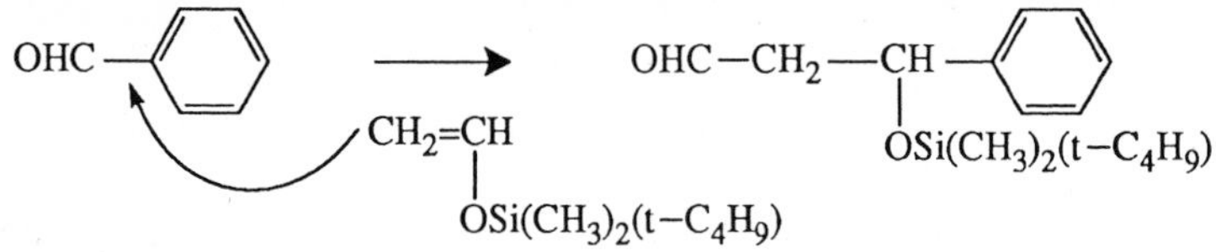

3.7. Free Radical Polymerizations

3.7.1. Introduction

Free radical polymerizations are initiated by radicals and propagated by macroradicals. These radicals exhibit an unpaired electron. They are historically called *free* radicals in order to distinguish them from "bound" radicals (nowadays called substituents or ligands (IUPAC)).

Initiating radicals are rarely formed by monomers themselves but rather thermally, electrochemically or photochemically from deliberately added initiators. In the start reaction, an initiator radical I˙ adds a monomer molecule M and becomes a monomer radical IM˙, which adds further monomer molecules in propagation reactions. In contrast to ionic polymerizations, growing macroradicals $I+M]_n^•$ may react with their own kind or with initiator radicals which terminates both the individual polymer chain and the kinetic chain. These termination reactions limit the total concentration of active centers to ca. 10^{-8} mol/L (vs. ca. 10^{-3} mol/L in ionic polymerizations). Free radicals may further undergo many chain transfer reactions to monomers, polymers, solvents, initiators, etc.

Depending on polymerization conditions, the same monomer may deliver polymers that differ in constitution (degree of branching), configuration (tacticity), molar mass, and molar mass distribution, and, subsequently, in properties. Free radical polymerization is the most important class of industrial polymerizations since many different monomers can be polymerized free-radically and the reaction can be easily controlled by additives, the reaction temperature and the state of matter (Table 3-6).

3.7.2. Formation of Free Radicals

Self-Initiated Polymerizations

In very rare cases, monomers themselves can start free radical polymerizations without added initiators or initiating impurities. Such **self-initiating polymerizations** (so-called **thermal polymerizations**) are to be distinguished from conventional

Table 3-6 Industrial free radical homopolymerizations. Polymerizations in the gas phase G (with subsequent precipitation), in bulk B, in suspension S, in emulsion E, in solution L, or with precipitation P from solution. Applications as adhesives A, coatings C, thermosets D, elastomers E, fibers F, thermoplastics T, and various other V. + Major processes, (+) less often used.

| Monomer | Polymerizations | | | | | | Appli- |
	G	B	S	E	L	P	cation
Living polymerization							
p-Xylene	+						C
Irreversible polymerizations to linear or slightly branched polymers							
Vinyl chloride	+	(+)	(+)	+	(+)		T
Vinyl acetate		(+)	(+)	+	(+)		C,A
Ethylene		+	+	(+)	(+)	(+)	T,F
Styrene		+	+	(+)	(+)		T
Methyl methacrylate		+	+	+	+		T
N-Vinyl pyrrolidone		+			+		V
Vinyl fluoride		+					T,C
p-Methylstyrene			+	+			T
Vinylidene fluoride			+	+			T,C
Trifluorochloroethylene			+				T
Tetrafluoroethylene			+				T,C
Acrylic esters				+			A
Chloroprene				+			E
Acrylamide					+		V
Acrylic acid					+		V
Acrylonitrile						+	F
Cross-linking polymerization							
Diallyl monomers		+					D

"spontaneous" polymerizations that are initiated by light, traces of impurities, etc. These "spontaneous polymerizations" are sometimes also incorrectly called "thermal polymerizations".

Self-initiated polymerizations are known for styrene, some styrene derivatives, methyl methacrylate, 2-vinyl pyridine, 2-vinyl furan, and acenaphthylene. Some of these polymerizations are free-radical polymerizations because they can be stopped by radical inhibitors (see below) but others are probably not radical reactions. These polymerizations are difficult to investigate because they are slow: at 29°C, a methyl methacrylate conversion of 50 % is achieved after 5.3 years!

The self-initiated polymerization of styrene in the dark proceeds with high overall activation energy: a 50 % monomer conversion needs 400 days at 29°C but only 4 hours at 127°C. Since initiators and other additives are absent, this thermal polymerization generates an especially pure, non-crystalline poly(styrene) that is called "crystal poly(styrene)" because of its optical clarity.

The slow thermal polymerization of styrene at 29°C is probably caused by a slow formation of free radicals. Two styrene molecules generate a Diels–Alder product which, on reaction with another styrene molecule, generates initiating radicals:

(3-48)

Start and propagation rates are composed of a thermal part at temperatures greater than 60°C and an athermal part at temperatures lower than 60°C. The athermal part is caused by natural radiation (from ^{40}K in rocks, by radon or by cosmic radiation).

Thermal Initiators

Most free radical polymerizations are started by thermally decomposing initiators (sometimes called "accelerators"). Scientific investigations prefer N,N-azo*bis*isobutyronitrile AIBN which fragments mainly into isobutyronitrile radicals:

(3-49)

$$NC-\underset{\underset{CH_3}{|}}{\overset{\overset{CH_3}{|}}{C}}-N=N-\underset{\underset{CH_3}{|}}{\overset{\overset{CH_3}{|}}{C}}-CN \longrightarrow NC-\underset{\underset{CH_3}{|}}{\overset{\overset{CH_3}{|}}{C}}{}^{\bullet} + N_2 + {}^{\bullet}\underset{\underset{CH_3}{|}}{\overset{\overset{CH_3}{|}}{C}}-CN$$

A portion of these radicals combines to $(CH_3)_2C(CN)-C(CN)(CH_3)_2$ or disproportionates into $(CH_3)_2CH(CN) + CH_2=C(CN)CH_3$. Only a fraction of the primarily formed isobutyronitrile radicals can thus add monomer molecules and start the polymerization. The **radical yield** f from AIBN-initiated polymerizations of neat styrene is $f = 60$ % at a monomer conversion of $u \approx 1$ % but only $f = 30$ % at $u \approx 70$ % (see also Fig. 3-12); it depends only slightly on temperature.

Radical yields can be calculated theoretically from the ratio of formation of initiator radicals to the sum of the rate of all reactions of these radicals. This theoretical radical yield is only ca. 10^{-4} %! The experimental radical yield is larger than the theoretical because initiator radicals reside in a "cage" of monomer molecules and react with nearby monomer molecules rather than with other (distant) initiator radicals.

The driving force for the initiator decomposition is the formation of stable N_2 molecules. AIBN is thus used as a foaming agent. In industrial polymerizations, gas formation is undesirable, however. Instead of AIBN, peroxides, peresters, etc., are used. Dibenzoyl peroxide BPO is often used in laboratory experiments. It decomposes into benzoyloxy radicals $C_6H_5COO^{\bullet}$:

(3-50)

$$C_6H_5-\underset{\underset{O}{\|}}{C}-O-O-\underset{\underset{O}{\|}}{C}-C_6H_5 \longrightarrow 2\,C_6H_5-\underset{\underset{O}{\|}}{C}-O^{\bullet}$$

The resulting benzoyloxy radicals initiate polymerizations. They fragment however into $C_6H_5^{\bullet} + CO_2$ if monomer is absent.

BPO is only infrequently used as an industrial polymerization initiator. For reasons of price, shelf life, half-life of decomposition, etc., one prefers different compounds, depending on monomer and process. Examples are:

	Acetylcyclohexanesulfonyl peroxide	$CH_3COO\text{-}OSO_2C_6H_{11}$
IPP	Di(isopropylperoxy)dicarbonate	$(CH_3)_2CHOCOO)_2$
CHP	Cumene hydroperoxide	$C_6H_5C(CH_3)_2O\text{-}OH$
Dicup	Dicumyl peroxide	$(C_6H_5C(CH_3)_2O)_2$
	Vinylsilane triacetate	$CH_2=CHSi(OOCCH_3)_3$,
	Benzpinacol	$((C_6H_5)_2C(OH))_2$
	Dipotassium persulfate	$K_2S_2O_8$

Dipotassium persulfate decomposes in alkaline to neutral solution into two radical anions according to $(S_2O_8)^{2\ominus} \rightarrow 2\,{}^{\bullet}(SO_4)^{\ominus}$.

Decomposition rates $-d[I]/dt = k_d[I]$ of initiators vary widely; integration gives $[I] = [I]_o\exp(-k_dt)$. They depend on the temperature and usually also on the solvent and the presence of monomer and polymer although not as strongly as in ionic polymerizations (Table 3-7). Solvent effects are relatively large for dibenzoyl peroxide and less pronounced for AIBN (hence the preferred use of the latter in scientific investigations).

The action of a free radical initiator is usually characterized by its activation energy $E_d^{\ddagger}$ from the Arrhenius equation $k_d = A_d\exp(-E^{\ddagger}/RT)$ or by the time $t_{5\%} \approx 0.0513/k_d$ for a 5 % decomposition ($[I] = 0.95\,[I]_o$) or the time $t_{50\%} \approx 0.693/k_d$ for a 50 % decomposition ($[I] = 0.50\,[I]_o$). For scientific investigations, the time $t_{5\%}$ is preferred since the initiator concentration can then be treated as constant in kinetic equations. Industry prefers half-life $t_{1/2}$ since it wants to make good use of expensive initiators. American industry often uses a temperature T_{10h} at which 50 % of the initiator have decomposed after 10 hours.

Table 3-7 Activation energies $E_d^{\ddagger}$, half-lifes $t_{1/2}$, and 10-hour temperatures of the decomposition of some free radical initiators. Note that "solvent" does not refer to the T_{10h} data since these are given by industry without reference to the presence of solvents, monomers, or polymers.

Initiator	Solvent	$E_d^{\ddagger}$ in kJ/mol	$t_{1/2}$/h 40°C	70°C	110°C	T_{10h}/°C
AIBN	Dibutyl phthalate	122.2	303	5.0	0.057	64
	Benzene	125.5	354	6.1	0.076	
	Styrene	127.6	414	5.7	0.054	
BPO	Acetone	111.3	443	10.6	0.180	73
	Styrene	132.8	3 525	29.2	0.231	
	Poly(styrene)	146.9	11 730	84.6	0.392	
IPP	Dibutyl phthalate	115.0	21	0.32	0.0044	45
Dicup	Benzene	170	3 000 000	11 200	27	117
CHP	Benzene	100	4 000 000	60 000	760	155
$K_2S_2O_8$	0.1 mol NaOH/L H_2O	140	1 850	11.9		

Industry chooses thermal initiators that generate reasonable polymerization rates at the usual polymerization temperatures of 50°C to 120°C. These rates should be neither too small (no high yields) nor too high (danger of explosion because the generated heat of reaction cannot be removed fast enough).

Redox Initiators

Redox initiators generate initiating radicals by the reaction of a reducing agent with an oxidizing agent. Such redox reactions require only a little activation energy: redox polymerizations can be started at far lower temperatures than those with thermally decomposing initiators.

The classical redox system consists of dihydrogen peroxide and iron(II) salts. It generates hydroxyl radicals according to:

(3-51) $Fe^{2\oplus} + HO\text{-}OH \rightarrow Fe^{3\oplus} + HO^{\ominus} + {}^{\bullet}OH$

Hydroxyl radicals are small and can thus react easily not only with the β-carbon atom of vinyl compounds ${}^{\beta}CH_2={}^{\alpha}CHR$ but also with the α-carbon atom. Other water-soluble redox systems consist of $K_2S_2O_8$ plus catalytic amounts of iron(II) ions, peroxides plus glucose (oxidation to glucuronic acid), etc.

These redox systems are sensitive to the biradical oxygen. Far less sensitive to oxygen O_2 are systems composed of peroxides and amines, e.g., the water-insoluble system dibenzoyl peroxide–dimethylaniline. The decomposition of BPO is here induced by the solvent; it may result in an explosion:

(3-52) $(CH_3)_2NC_6H_5 + (C_6H_5COO)_2 \rightarrow [(CH_3)_2N^{\bullet}C_6H_5]^{\oplus} \; {}^{\ominus}OOCC_6H_5 + C_6H_5COO^{\bullet}$
$[(CH_3)_2N^{\bullet}C_6H_5]^{\oplus} \; {}^{\ominus}OOCC_6H_5 \rightarrow (CH_3)_2NC_6H_4^{\bullet} + C_6H_5COOH$

Photo Initiators

Initiating radicals may also be generated photochemically or by electron beams. Such radical formations do not need thermal activiation energies; polymerizations proceed with high speed even at very low temperatures. Monomers are usually acryl esters of hydroxyl-group-containing industrial compounds (polyols, polyesters, epoxides, or polyurethanes) that are used in the printing ink industry or for coatings.

3.7.3. Initiation, Propagation and Termination

Initiator radicals $I^{\bullet}$ are formed by initiator decomposition. They add monomer molecules M; the resulting monomer radicals I–M$^{\bullet}$ start the polymerization. In the subsequent **propagation reaction**, more monomer molecules are added and macro-radicals $I\text{-}[M]_{\overline{n}}^{\bullet}$ are formed. The addition proceeds mainly in head-to-tail fashion. **Regiospecificities** are, however, never 100 %. Larger proportions of head-to-head and tail-to-tail structures (see Section 2.1.4.) are obtained from monomers $CH_2=CRR'$ with small substituents R and R' and little resonance stabilization of their macroradicals I$+$M$+_n^{\bullet}$. For example, head-to-head structures amount to (1-2) % in poly(vinyl acetate) $+$CH$_2$–CH(OOCH$_3$)$+_n$, (6-10) % in poly(vinyl fluoride) $+$CH$_2$–CHF$+_n$, and (10-12) % in poly(vinylidene fluoride) $+$CH$_2$–CF$_2+_n$.

In contrast to many polyinsertions and ionic polymerizations, radical polymerizations are not very stereospecific. The conformation of the carbon radicals in highly reactive macroradicals of vinyl and acrylic monomers is either planar or that of fast interconverting pyramids. Propagation steps are therefore controlled by the prochiral side of the threefold substituted carbon radicals, in most compounds by repulsion of the substituents. Most vinyl and acrylic polymers are therefore predominantly syndiotactic. Exceptions are possible for strong effects between adjacent ligands, for example, in the polymerization of vinyl formate CH_2=CH(CHO).

At any given moment, at least initiator radicals I$^{\bullet}$, monomer radicals IM$^{\bullet}$ and macroradicals IM$_n^{\bullet}$ are present in free radical polymerizations. All these radicals can react with each other. They are eliminated by **coupling (combination;** often called **recombination** for unclear reasons), for example, in polymerizations of vinyl and acrylic monomers by:

(3-53)   ~CH$_2$–$^{\bullet}$CRR' + $^{\bullet}$CRR'–CH$_2$~ $\longrightarrow$ ~CH$_2$–CRR'–CRR'–CH$_2$~

or by **disproportionation** with transfer of atoms or groups between two macroradicals:

(3-54)   ~CH$_2$–$^{\bullet}$CHR' + $^{\bullet}$CHR'–CH$_2$~ $\longrightarrow$ ~CH$_2$–CH$_2$R' + CHR'=CH~

or, with allyl monomers, by **termination by the monomer.** This reaction generates a resonance stabilized monomer radical which is too stable to add monomer molecules:

(3-55)   ~CH$_2^{\bullet}$CH(CH$_2$R) + CH$_2$=CH(CH$_2$R) $\rightarrow$ ~CH$_2$CH$_2$CH$_2$R + [CH$_2 \doteq$CH$\doteq$CHR]

The monomer radical thus commits suicide. Only oligomers ($X \approx 20$) are obtained from *uncharged mono*allyl compounds. With diallyl monomers, cross-linking is observed at higher monomer conversions.

At higher initiator concentrations, a **termination by initiator radicals** I$^{\bullet}$ has to be considered:

(3-56)   ~CH$_2$–$^{\bullet}$CHR + I$^{\bullet} \rightarrow$ ~CH$_2$-CHR-I

Free radical polymerizations with macro*mono*radicals can thus never be living. **Pseudoliving free radical polymerizations** are obtained by reversible termination reactions, for example, with nitroxide radicals, and with **iniferter** molecules that *ini*ti-ate, trans*fer* and *ter*minate kinetic chains.

True living free radical polymerizations are however possible (in principle) by **polyrecombination** of macro*di*radicals. Di-*p*-xylene ([2.2]-paracyclophane) dissoci-ates at 600°C into *p*-xylene. On solid surfaces subjected to the vapor, diradicals com-bine to poly(*p*-xylene) (Parylene®) and form coatings:

(3-57)

3.7.4. Steady State

Free radical polymerizations involve the simultaneous generation and disappearance of initiator radicals, monomer radicals, and macroradicals. After a few seconds, the total concentration of *all* radicals becomes constant (but not necessarily that of individual radical types). This **steady state** (or **stationary state**) is obtained at very small monomer conversions. The total free radical concentration becomes stationary: the rate of radical formation equals the rate of radical disappearance.

Initiator radicals I dissociate according to $I \rightarrow 2\ I^{\bullet}$ with a rate $R_d = -d[I]/dt = k_d[I]$. Since two initiator radicals are formed per one initiator molecule, the rate of radical formation $R_r = d[I^{\bullet}]/dt = 2\ k_d[I] = 2\ R_d$ is double the rate of dissociation. Since however the initiator molecule dissociates in a "cage" of surrounding monomer and/or solvent molecules, some of the newly formed radicals recombine immediately. Due to this **cage effect**, only a fraction f of initiator radicals $I^{\bullet}$ reacts with monomer in the start reaction $I^{\bullet} + M \rightarrow IM^{\bullet}$. The rate of radical formation is thus $R_r = 2\ fk_d[I]$.

Monomer radicals $RM^{\bullet}$ are generated with a rate $R_{st} = -d[I^{\bullet}]/dt = k_{st}[I^{\bullet}][M]$. The rate R_{st} of the start reaction is however much greater than the rate R_d of initiator decomposition. Initiator radicals are therefore consumed as fast as they are generated. The rate determining step is thus the initiator decomposition. Initiator radicals are consequently formed with a rate $d[I^{\bullet}]/dt = 2\ fk_d[I] - k_{st}[I^{\bullet}][M] = 0$. The rate of the start reaction is thus:

(3-58) $R_{st} = k_{st}[I^{\bullet}][M] = 2\ fk_d[I]$

At low monomer conversions, only a little initiator is consumed and $[I] \approx [I]_0$.

At sufficiently low initiator concentrations, growing chains are mainly terminated by mutual deactivation of two macroradicals (combination and disproportionation). The rate of termination is thus given by $R_{t(pp)} = k_{t(pp)}[P^{\bullet}]^2$. American scientific literature often writes $R_{t(pp)} = 2\ k_{t(pp)}[P^{\bullet}]^2$ instead because two radicals are eliminated (note the differences in reported values of $k_{t(pp)}$).

The rate constants k_t of the various termination reactions are distinguished by indices: $k_{t(pp)}$ indicates termination by 2 polymers (coupling or disproportionation); $k_{t(pm)}$ termination reaction between a polymer and a monomer (e.g., allyl polymerization); $k_{t(i)}$ termination by initiator.

In the steady state, as many radicals are formed with the rate R_{st} as disappear by the rate $R_{t(pp)}$ (**Bodenstein steady state principle**). This results in $R_{st} = R_{t(pp)}$ and therefore also in $2 fk_d[I]_o = k_{t(pp)}[P^\bullet]^2$, respectively. The solution of this equation for the radical concentration in the steady state ($[P^\bullet] = [P^\bullet]_{stat}$) gives:

$$(3\text{-}59) \quad [P^\bullet]_{stat} = (2 fk_d[I]/k_{t(pp)})^{1/2}$$

Example. The following data have been found for the polymerization of neat styrene at 50°C initiated by $[I] = 5 \cdot 10^{-3}$ mol AIBN/L: $f = 1/2$, $k_d = 2 \cdot 10^{-6}$ s^{-1}, and $k_{t(pp)} = 10^8$ L mol s^{-1}. According to Eqn.(3-59), the radical concentration is 10^{-8} mol/L in the steady state. This steady state concentration is obtained after 3 s and a monomer conversion of $6 \cdot 10^{-4}$ % (see Appendix A 3-3).

3.7.5. Ideal Kinetics

Per radical, only one monomer molecule is consumed in the start reaction, but many hundreds in the subsequent propagation reaction. Monomers are thus practically only consumed by propagation; their consumption by start, termination and transfer to monomer is negligible. For irreversible reactions, the gross rate of polymerization, R_{gross}, thus approximates the rate of propagation, R_p:

$$(3\text{-}60) \quad R_{gross} = - \, d[M]/dt \approx R_p = k_p[P^\bullet][M]$$

The following derivation of **ideal polymerization kinetics** assumes:
(a) only initiator decomposition and start, propagation and termination have to be considered (no kinetic chain transfer; see below);
(b) all reactions are irreversible;
(c) the effective concentration of initiator radicals is steady ($[I^\bullet] = [I^\bullet]_{stat}$);
(d) the concentration of macroradicals is also stationary ($[P^\bullet] = [P^\bullet]_{stat}$);
(e) the principle of equal chemical reactivity applies to propagation and termination (no dependence on molar mass);
(f) termination occurs only by mutual deactivation of two macroradicals (combination or disproportionation).
Combination of Eqn.(3-59) and Eqn.(3-60) delivers with condition (d):

$$(3\text{-}61) \quad R_{gross} \approx R_p = - \, d[M]/dt = k_p(2 fk_d/k_{t(pp)})^{1/2}[M][I]^{1/2}$$

(g) Ideal kinetics is concerned with small monomer conversions which implies a constant initiator concentration and therefore $[I] \approx [I]_o$.

The **rate of polymerization** $R_{gross} \approx R_p$ is thus directly proportional to the monomer concentration in ideal kinetics. Polymerizations in bulk are therefore always faster than in solution, provided that the solvent does not affect the initiator decomposition (Section 3.7.2.). Polymerization rates are predicted to decrease linearly with decreasing monomer concentration (increasing polymer yield) as has been found experimentally for small monomer conversions (Fig. 3-9).

Polymerization rates are also predicted to be proportional to the square root of the initiator concentration: a four-fold increase of the initiator concentration should only double the polymerization rate. Experimentally, increases are somewhat smaller than predicted: 5 % lower at II and 14 % lower at III compared to I (data of Fig. 3-9).

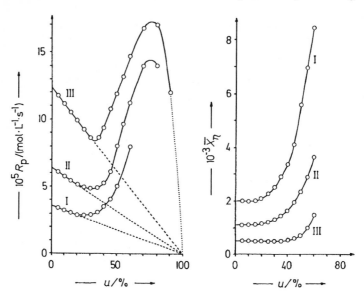

Fig. 3-9 Polymerization rates (left) and viscosimetric degrees of polymerization (right) as function of the monomer conversion u for the polymerization of neat styrene by AIBN at 50°C. Initial initiator concentrations of 0.018 mol/L (I), 0.061 mol/L (II), and 0.28 mol/L (III). Data of [1].

Larger initiator concentrations lead not only to higher polymerization rates but also to lower kinetic chain lengths. The **kinetic chain length** v denotes the number of monomer molecules that are added to an initiator radical before the macroradical is terminated. It is given by the ratio of the rate R_p of propagation to the sum of rates $R_{t,i}$ of all termination reactions, i.e., $v = R_p/\sum_i R_{t,i}$; the type of termination does not matter.

The type of termination does control the **number-average degree of polymerization**, however. A termination by disproportionation generates one dead macromolecule from one growing macroradical (the **degree of coupling** is $\varsigma = 1$). A termination by combination couples two growing chains to one dead macromolecule ($\varsigma = 2$). Conversely, one initiator radical produces more than one macromolecule if chain transfer reactions are present ($\varsigma < 1$; Section 3.7.7.). If R_{st} is given by Eqn.(3-58) and R_p by Eqn.(3-61), termination rates $R_{t(pp)}$ thus have to be divided by the degree of coupling for the calculation of number-average degrees of polymerization:

$$(3\text{-}62) \quad \overline{X}_n = \frac{R_p}{R_{t(pp)}/\varsigma} = \varsigma \frac{R_p}{R_{st}} = \varsigma \frac{k_p}{(2\,f k_d k_{t(pp)})^{1/2}} \cdot \frac{[M]}{[I]_o^{1/2}} \quad ; \text{ if } [I] \approx [I]_o$$

Eqn.(3-62) predicts that the number-average degree of polymerization decreases linearly with decreasing monomer concentration [M] (i.e., increasing monomer conversion $u = ([M]_o - [M])/[M]_o$). However, experiments indicate constant degrees of polymerization at small conversions ($u < 20$ %) and steep increases at higher ones ($u > 20$ %) (Fig. 3-9). Theory deviates from experiment because the former neglects both the chain transfer from initiator radicals to polymer molecules and the decrease of $k_{t(pp)}$ with increasing polymer formation (Section 3.7.8.). The distribution of degrees of polymerization conforms to a Schulz–Zimm distribution (Appendix A 3-4).

In ideal kinetics, the product of polymerization rate and number-average degree of polymerization depends only on the square of the monomer concentration but not on the initiator concentration. The higher the polymerization rate, the lower is the degree of polymerization:

$$(3\text{-}63) \quad R_p \overline{X}_n = (\varsigma k_p^2/k_{t(pp)})[M]^2$$

Newer electron spin resonance spectrometers are often sensitive enough to allow the measurement of small radical concentrations [P$^{\bullet}$]. Propagation rate constants k_p can then be calculated directly by Eqn.(3-60) using [P$^{\bullet}$], [M] and $-d[M]/dt$.

In most cases, [P$^{\bullet}$] is too small to be measured directly. Eqns.(3-61) and (3-62) both eliminate [P$^{\bullet}$] but deliver only combinations of various rate constants. Some of the constants can be obtained by other means. Experiments with added inhibitors (see Section 3.7.7.) deliver the product fk_d since the inhibitor catches initiator radicals according to $-d[\text{inhibitor}]/dt = fk_d[I]$.

The remaining unknown ratio $k_p/k_{t(pp)}^{1/2}$ is obtained from the average life time of a chain, $\tau = [P^{\bullet}]/R_{t(pp)} = (k_p[M])/(k_{t(pp)}R_p)$. The average life time τ can be calculated from the time to reach a steady state (in most cases too small for measurement), from the decrease of polymerization rate if the radical formation is restricted with respect to time or space (rotating sector method), or from the maxima of molar mass distribution curves (using laser flashes).

The **temperature dependence of polymerization rates** is controlled by the temperature dependence of the rate constants k_a of elementary reactions. Each rate constant k_a (i.e., k_p, $k_{t(pp)}$, k_d, etc.) follows the Arrhenius law $k_a = A_a^{\ddagger}\exp(-E_a^{\ddagger}/RT)$. The overall activation energy of the total polymerization rate equals the sum of the properly weighted activation energies of the elementary reactions (see Eqn.(3-61)):

$$(3\text{-}64) \quad E_{\text{gross}}^{\ddagger} = E_p^{\ddagger} + (1/2)\, E_d^{\ddagger} - (1/2)\, E_{t(pp)}^{\ddagger}$$

The overall activation energy for the formation of the number-average degree of polymerization is similarly given by (see Eqn.(3-62)):

$$(3\text{-}65) \quad E_X^{\ddagger} = E_p^{\ddagger} - (1/2)\, E_d^{\ddagger} - (1/2)\, E_{t(pp)}^{\ddagger}$$

Example: Styrene polymerization by AIBN yields $E_p^{\ddagger} = 25$ kJ/mol, $E_d^{\ddagger} = 128$ kJ/mol, and $E_{t(pp)}^{\ddagger} = 2$ kJ/mol and therefore $E_{\text{gross}}^{\ddagger} = 88$ kJ/mol and $E_X^{\ddagger} = -40$ kJ/mol. Polymerization rates thus increase with temperature whereas degrees of polymerization decrease.

3.7.6. Allyl Polymerizations

In free radical polymerization of allylic monomers, macroradicals are generated similar to the free radical polymerization of vinyl monomers. However, terminations of kinetic chains proceed by **degradative chain transfer** to monomer which leads to a **self-inhibition** of the polymerization (Eqn.(3-55)). In the steady state, the rate of formation of macroradicals is $d[P^\bullet]/dt = k_{st}[I^\bullet][M] - k_{t(pm)}[P^\bullet][M] = 0$. Replacement of $k_{st}[I^\bullet][M]$ via Eqn.(3-58) and $[P^\bullet]$ via Eqn.(3-60) shows that the polymerization rate of allyl polymerizations is only a function of the initiator concentration but not the monomer concentration (compare Eqn.(3-61)):

$$(3\text{-}66) \quad R_{gross} = (2\,f k_d k_p/k_{t(pm)})[I]_o \qquad ; \text{if } [I] \approx [I]_o$$

The number-average degree of polymerization is independent of monomer and initiator concentrations:

$$(3\text{-}67) \quad \overline{X}_n = R_p/R_{t(pm)} = (k_p[P^\bullet][M])/(k_{t(pm)}[P^\bullet][M]) = k_p/k_{t(pm)}$$

In both propagation and degradative chain transfer, monomer radicals $P^\bullet$ react with monomers M. Propagation constants k_p are thus not much greater than the transfer constants $k_{t(pm)}$: the degree of polymerization of polymers from monoallyl monomers is small, usually not greater than 10-20. Diallyl monomers polymerize however to cross-linked polymers with "infinitely" high degrees of polymerization.

Eqn.(3-66) delivers the overall activation energy of polymerization rate as $E_{gross}^{\ddagger} = E_d^{\ddagger} + E_p^{\ddagger} - E_{t(pm)}^{\ddagger}$ and Eqn.(3-67) the overall activation energy of degree of polymerization as $E_X^{\ddagger} = E_p^{\ddagger} - E_{t(pm)}^{\ddagger}$. Since AIBN-initiated polymerization of allyl acetate results in $E_d^{\ddagger} = 128$ kJ/mol, $E_p^{\ddagger} = 25$ kJ/mol and $E_{t(pm)}^{\ddagger} = 25$ kJ/mol, values of $E_{gross}^{\ddagger} = 128$ kJ/mol and $E_X^{\ddagger} = 0$ kJ/mol are obtained. Polymerization rates of allyl polymerizations are thus much more strongly dependent on temperature than those of vinyl or acrylic polymerizations. Degrees of polymerization are independent of temperature, however.

3.7.7. Kinetic Chain Transfer

Kinetic chain transfer may occur to all species in the polymerization system: monomers, polymers, initiators, solvents, or certain additives. A **transfer to polymers** produces branched polymers. An *inter*molecular chain transfer to polymer molecules by macroradicals or initiator radicals generates long-chain branches:

(3-68)

$$\text{\textasciitilde CH}_2\text{—CH}_2\text{\textasciitilde} \quad \xrightarrow[-RH]{+R^\bullet} \quad \text{\textasciitilde CH}_2\text{—} \overset{\bullet}{\text{CH}} \text{\textasciitilde} \quad \xrightarrow{+\,CH_2{=}CH_2} \quad \text{\textasciitilde CH}_2\text{—CH} \text{\textasciitilde} \quad \text{etc.}$$
$$\underset{\underset{\text{CH}_2\text{—}\overset{\bullet}{\text{CH}}_2}{|}}{}$$

This transfer does not change the number-average degree of polymerization. Molar mass distributions become broader, however.

An *intra*molecular chain transfer to polymer occurs in the high-pressure polymerization of ethylene. Polyradicals generated by this **back-biting reaction** add ethylene monomer which results in ethyl, butyl and hexyl groups; methyl and propyl groups are not observed:

(3-69)

The number-average degree of polymerization decreases if **transfers to monomers, solvents,** or so-called **regulators** (see below) are present. Similar to the treatment of ideal kinetics, degrees of polymerization can be calculated from the ratio of propagation rate to the sum of all reactions that terminate kinetic chains. The first part of the right side of Eqn.(3-62) therefore has to be amended by the expressions for transfers to monomer M and substance S (solvent, regulator): $\overline{X}_n = R_p/\{(R_{t(pp)}/k) + R_{tr(m)} + R_{tr(s)}\}$. Inversion of this expression and introduction of $v = R_p/R_{t(pp)}$ and the rates $R_p = k_p[P^\bullet][M]$ (propagation), $R_{tr(m)} = k_{tr(m)}[P^\bullet][M]$ (transfer to monomer), and $R_{tr(s)} = k_{tr(s)}[P^\bullet][S]$ (transfer to added substance S) gives:

$$(3\text{-}70) \quad \frac{1}{\overline{X}_n} = \frac{(1/\varsigma)R_{t(pp)} + R_{tr(m)} + R_{tr(s)}}{R_p} = \frac{1}{\varsigma v} + \frac{k_{tr(m)}}{k_p} + \frac{k_{tr(s)}}{k_p} \cdot \frac{[S]}{[M]}$$

where ς = coupling constant (Section 2.2.3.). After introduction of the transfer constants $C_m \equiv k_{tr(m)}/k_p$ and $C_s \equiv k_{tr(s)}/k_p$, one obtains the **Mayo equation:**

$$(3\text{-}71) \quad \frac{1}{\overline{X}_n} - \frac{1}{\varsigma v} = C_m + C_s \frac{[S]}{[M]}$$

A plot of the left side of Eqn.(3-71) against the molar ratio [S]/[M] gives C_m from the intercept and C_s from the slope. A $C_m = 6 \cdot 10^{-5}$ was observed for the neat polymerization of styrene at 60°C: a transfer to monomer occurs after $1/C_m = 16\,700$ propagation steps. Added solvents deliver transfer constants of $C_s = 2 \cdot 10^{-6}$ (benzene), $C_s = 1.2 \cdot 10^{-2}$ (carbon tetrachloride) and $C_s = 15$ (1-dodecanethiol). The degree of coupling, ς, can be obtained from the ratio $\overline{X}_w/\overline{X}_n$ (Section 2.2.3.).

The small radicals generated by these transfer reactions diffuse more easily than macroradicals. Termination reactions thus become more probable which in turn reduces the degree of polymerization. The concentration of macroradicals may also be lowered and therefore the polymerization rate. This **degradative chain transfer** causes the polymerization rate to decrease more strongly then one may expect from the effects of dilution.

In the polymerization of vinyl chloride $CH_2=CHCl$, the rate of transfer to mono-mer, $\sim CH_2\text{-}^{\bullet}CHCl + CH_2=CHCl \rightarrow \sim CH_2\text{-}CHCl_2 + CH_2=^{\bullet}CH$, is much greater than the rate of mutual deactivation of two macroradicals. The degree of polymerization be-comes independent of the initiator concentration; it is controlled industrially by changing the polymerization temperature.

The strong transfer action of chlorine-containing compounds is utilized in the synthesis of oligomers with functional end groups. A typical reaction is $R^{\bullet} + CCl_4 \rightarrow RCl + {}^{\bullet}CCl_3$; $Cl_3C^{\bullet} + n\,M \rightarrow Cl_3C\text{-}M_n{}^{\bullet}$; $Cl_3C\text{-}M_n{}^{\bullet} + CCl_4 \rightarrow Cl_3C\text{-}M_n\text{-}Cl + {}^{\bullet}CCl_3$.

Oligomers with functional end groups from transfer reactions are called **telomers** in preparative macromolecular chemistry, those from other reactions **telechelic poly-mers**, and those with polymerizable end groups (e.g., $CH_2=CH\text{-}$) **macromonomers** (Macromer® is a protected trade mark).

Thiols (mercaptanes) and disulfides are even stronger transfer compounds than chloro compounds. They are used as molar mass **regulators** in free-radical diene polymerizations. In such polymerizations, initiator concentrations cannot be in-creased at will in order to achieve greater polymerization rates; the reactor would ex-plode. Without regulators, polymerizations would lead to very high molar masses at the safe (low) concentrations of initiators; such polymers would exhibit far too high melt viscosities for processing.

At higher monomer conversions, these polymerizations also undergo multiple chain transfer to the same polymer molecule. The resulting polyradicals can combine to form branched polymers (1 combination per primary polymer molecule) and then to cross-linked polymers (2 or more combinations per primary polymer molecule). Cross-linked polymers can only be removed from the reactor by "mining"; they are also very difficult to process. A chain transfer to regulators decreases the degree of polymerization and therefore also the probabilities of transfer to polymers and cross-linking. Since they modify the properties, regulators are also called **modifiers**.

Chain transfer agents that generate less reactive radicals are called **retarding agents**. An example is nitrobenzene (Fig. 3-10). Radicals formed from nitrobenzene act as initiators and as chain terminators; they retard polymerizations (lower rates). One nitrobenzene molecule eliminates two radicals:

(3-72)

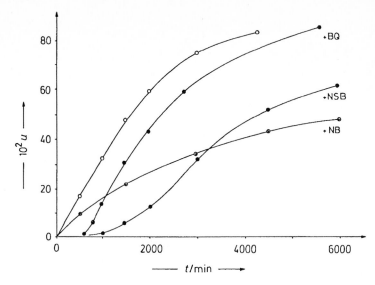

Fig. 3-10   Time dependence of monomer conversion $u = ([M]_o - [M])/[M]_o$ during the polymerization of neat styrene at 100°C (○) or with addition of 0.1 % benzoquinone BQ (⊕), 0.2 % nitrosobenzene NSB (●), or 0.5 % nitrobenzene NB (◉) [2].

Benzoquinone does not change the polymerization rate but causes an inhibition period before the polymerization starts (Fig. 3-10): benzoquinone is an **inhibitor**. It delivers highly resonance stabilized radicals that are unable to start polymerizations:

(3-73)

In industry, stored monomers are often **stabilized** by addition of copper(II) salts or hydroquinone against a premature polymerization by oxygen in air. Oxygen O_2 is a biradical which adds to other radicals $R^\bullet$ and gives peroxy radicals $R\text{-}O\text{-}O^\bullet$. These radicals are not very reactive; at first, O_2 thus acts as an inhibitor. However, $R\text{-}O\text{-}O^\bullet$ can also form hydroperoxides ROOH by chain transfer to RH, peroxides ROOR by addition of $R^\bullet$, and alternating copolymers $R\text{-}[O\text{-}O\text{-}M]_n\text{-}O\text{-}O^\bullet$ by copolymerization with monomers M. All these compounds can decompose into polymerization active radicals $RO^\bullet$, $HO^\bullet$, $ROOMO^\bullet$, etc. Peroxides may also be converted into aldehydes and acids which are strong transfer agents.

Hydroquinone is oxidized by oxygen to benzoquinone. It thus catches the biradicals of O_2 before these can form peroxy radicals. Benzoquinone then acts as a retarding agent; hydroquinone itself is neither an inhibitor nor a retarding agent.

3.7.8. Non-Ideal Kinetics

The ideal polymerization kinetics described in Section 3.7.5. applies only to the limiting cases of vanishingly small monomer conversions and initiator concentrations (see also Fig. 3-9). In general, rates $R_p = const \cdot [M][I]^{1/2}$ have to be replaced by $R_p = const \cdot [M]^\alpha [I]^\beta$ with $\alpha \neq 1$ and $\beta \neq 1/2$. Exponents $\beta > 1$ usually result from induced initiator decompositions (Section 3.7.2.). Exponents α may adopt values greater or smaller than 1 which may be caused by an inconstancy of $k_{t(pp)}$, a termination by primary radicals, or an induced initiator decomposition.

Termination constants $k_{t(pp)}$ decrease with increasing viscosity of the polymerizing liquid, even at very small monomer conversion. They are also lowered with increasing degrees of polymerization of polymers because mutual deactivation of macroradicals can occur only if the radical-carrying segments adopt certain relative positions. In order to move to these positions, segments must diffuse. This diffusion is controlled by the size of the macroradicals and the viscosity of the system. Termination constants are thus averages.

It is also observed that polymerization rates in bulk (and sometimes also in concentrated solutions) often increase strongly at monomer conversions higher than 15% to 25% (Figs. 3-9 and 3-11). This increase in polymerization rate is accompanied by an increase of the degree of polymerization (Fig. 3-9). Such a **self-acceleration** (**auto-acceleration**) of the polymerization is also found (although less pronounced) for isothermal polymerizations; it thus cannot be *caused* by a heat build-up.

This phenomenon is called a **gel effect** (**Trommsdorff–Norrish effect, Norrish–Smith effect**). It is especially pronounced in highly viscous systems. Since termination

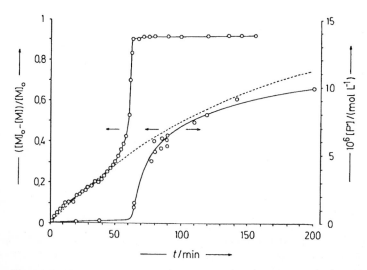

Fig. 3-11   Time dependence of monomer conversion u and concentration [P·] of macroradicals during the free radical polymerization of neat methyl methacrylate at 60°C ([M]$_o$ = 9.25 mol/L) by [AIBN]$_o$ = 0.1 mol/L [AIBN]$_o$. The broken curve indicates the conversion expected for ideal polymerization kinetics (absence of a gel effect). Data of [3].

constants are already diffusion-controlled at very small monomer conversions (Fig. 3-12) *before* a gel effect is observed, such a self-acceleration cannot be *caused* by an onset of diffusion control. It must rather come from a *changed* diffusion control. The diffusion rate changes when the concentration of chain entanglements (see Section 7.5.1.) surpasses a certain critical value (the "onset" of **entanglements**). Isolated polymer chains form random coils with low coil densities (Section 5.3.4.) which start to overlap (Section 6.3.5.) and finally to entangle at higher polymer concentrations, i.e., greater monomer conversions.

Entanglements form a temporary (physical) network which impedes the diffusion of radical-carrying segments and reduces the mutual deactivation of macroradicals. Since new initiator radicals, and thus new macroradicals, are steadily formed, concentrations of macroradicals jump upon the onset of entanglements, i.e., after the gel effect sets in. The probability of termination by mutual deactivation of macroradicals decreases and the degree of polymerization increases strongly (Fig. 3-9) which leads to more entanglements (Section 5.7.2.). The simultaneous formation of new macroradicals results in broadened, bimodal molar mass distributions.

At even higher monomer conversions, the polymerizing system solidifies. This **glass effect** reduces not only the diffusion of macroradicals and the termination reactions but also the diffusion of monomer molecules and thus the chain propagation. The polymerization rate approaches zero (Fig. 3-9, III): the monomer can no longer be polymerized completely.

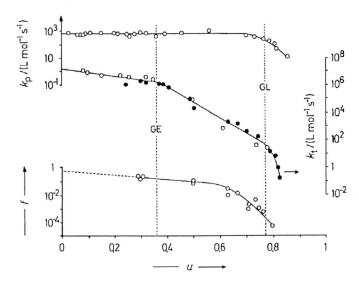

Fig. 3-12   Dependence of rate constants of propagation, k_p, and termination, k_t, on the monomer conversion $u = ([M]_o - [M])/[M]_o$ in the free radical polymerization of methyl methacrylate in bulk at 60°C. Polymerization by UV radiation with $[I]_o = 0.109$ mol/L dimethyl-2,2'-azodiiso-butyrate (k_p), $[I]_o = 0.337$ mol/L dicumyl peroxide at 60°C (O) (k_t) or 0.335 mol/L dibenzoyl per-oxide at 26°C (●) (k_t). The yield f of initiator radicals was determined for dimethyl-2,2'-azodiiso-butyrate. Dotted vertical lines indicate the onset of the gel effect (GE) and the glass effect (GL) according to monomer conversion as function of polymerization time. Data of [6].

Polymerizations also come to an end if the initiator is used up (**dead-end polymerization**). Since this effect is not due to thermodynamics, polymerizations proceed upon addition of fresh initiator.

The kinetics of dead-end polymerizations can be derived as follows. Integration of equation $-d[I]/dt = k_d[I]$ for the initiator decomposition (Section 3.7.4.) gives $[I] = [I]_0(\exp - k_dt)$. Insertion of this equation into Eqn.(3-61), separation of variables, and integration from $[M]_0$ to $[M]$ and from $t = 0$ to $t = t$, respectively, results in:

$$(3\text{-}74) \quad - \log_e ([M]/[M]_0) = 2\, k_p[2\, f/(k_dk_{t(pp)})]^{1/2}[I]_0^{1/2}\{1 - \exp[-\,(1/2)\, k_dt]\}$$

At infinitely long times, $[M] \to [M]_\infty$, and Eqn.(3-74) becomes:

$$(3\text{-}75) \quad - \log_e ([M]_\infty/[M]_0) = 2\, k_p[2\, f/(k_dk_{t(pp)})]^{1/2}[I]_0^{1/2}$$

Eqn.(3-75) shows that not all monomer can be polymerized even at the usual initiator concentrations. The following data are obtained for the polymerization of neat styrene at 60°C by 1 wt-% AIBN: $[M]_0 = 8.5$ mol/L, $[I]_0 = 0.0582$ mol/L, $k_d = 1.35 \cdot 10^{-5}$ s^{-1}, $k_p = 285$ L mol^{-1} s^{-1}, $k_{t(pp)} = 1.2 \cdot 10^8$ L mol^{-1} s^{-1}, and $f = 1/2$. According to Eqn.(3-75), the maximum monomer conversion $u_\infty = ([M]_0 - [M]_\infty)/[M]_0$ is only 96.7 %! In polymers, residual monomer is undesirable for toxicological and other reasons. Residual monomer concentrations usually should not exceed 1 ppm = 10^{-4} %. Industrially, residual monomer is thus either polymerized by renewed addition of initiator, removed by steam, or, if volatile, burned since recovery is often too expensive.

3.7.9. Industrial Polymerizations

Reactors

Industrial polymerizations differ from their low molar mass analogs by large exothermic heats of reaction, high viscosities, low diffusion rates, and small heat conductivities. These properties lead to a strong coupling between chemical reactions and the transport of matter and heat. As a result, not only reaction rates are affected but also constitution, configuration, molar mass, and molar mass distribution of the polymers.

The central property is the viscosity of the system. It not only controls the diffusion rates of reactants but also the heat conductivity which is important for the transfer of heats of polymerization. During polymerization, polymer concentrations increase and often so do molar masses (see Figs. 3-1, 3-4, and 3-9). Increases in polymer concentrations and molar masses increase viscosities of melts and solutions (Fig. 3-13), for example, by 6 decades in the bulk polymerization of styrene for monomer conversions from 0 % to 80 %. Viscosities of emulsions and suspensions are only slightly affected, however (Fig. 3-13).

High viscosities cause heat build-ups that have to be prevented by suitable reactors and/or stirring. Polymerizations are conducted in batch reactors BR (stirred tank reactors STR, charge reactors), continuous-flow stirred tank reactors CSTR, cascades C of STRs, or continuous plug flow reactors CPFR (Fig. 3-14). These reactors generate different time dependencies of monomer concentrations and different residence times which dictate the types of reactors and stirrers for a certain polymerization.

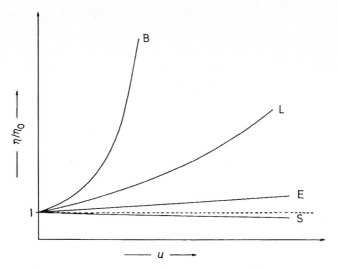

Fig. 3-13   Schematic dependence of relative viscosities, η/η_o, of polymerizing systems on monomer conversion u for polymerizations in bulk (B), solution (L), emulsion (E), and suspension (S).

Polymerization in Bulk

Polymerizations in bulk (polymerization of neat monomer) are polymerizations of monomer melts without added solvent, non-solvent, suspending or emulsifying agents. Many industrial "bulk polymerizations" do use (5-25) % solvent as a polymerization aid, however.

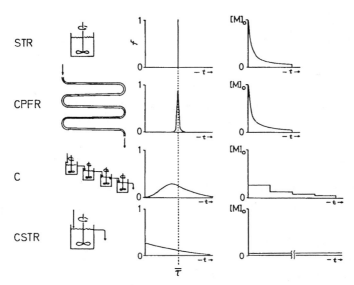

Fig. 3-14   Left: Various reactors: STR = stirred tank reactor, CPFR = continuous plug flow reactor, C = cascade of STRs, CSTR = continuous stirred tank reactor. Center: Time dependence of the differential distribution of fractions f of average residence times. Right: Time dependence of monomer concentrations (2nd order reactions) [4].

Polycondensations and polyadditions are usually performed at such high temperatures that both monomers and polymers are molten at all times. In chain polymerizations, only monomers are usually liquid; the resulting polymers solidify to a block (hence the outdated term "block polymerization" for polymerizations in bulk).

Chain polymerizations in bulk are rarely used for other than free radical polymerizations. Very pure polymers are formed by polymerization of neat monomers since only monomers, polymers, and initiators are present. Bulk polymerizations are usually performed in stirred tank reactors. Heats of polymerizations are removed via jackets for reactor volumes up to ca. 30 m³; cooling coils are required for larger reactors with volumes up to 200 m³.

Polymerization in Solution and by Precipitation

Inert solvents act as diluents. They lower monomer concentrations and thus polymerization rates. Chain transfer to polymer becomes less prominent: the degree of branching is lowered and the molar mass distribution becomes less broad. Diluents also reduce the gel effect and allow a better transport of heat. Disadvantages are the costs of solvents and their removal and recovery as well as toxicological and environmental problems. For these reasons, several countries have mandated that less and less organic solvents have to be used for paints.

Some polymers are insoluble in their own monomers. In these **precipitation polymerizations**, macroradicals become included in the precipitated polymer. An example is the polymerization of neat vinyl chloride. Since the diffusion rate of macroradicals is very low in such particles, termination by mutual deactivation is lessened and molar masses increase. Polymerization rates remain high, however, since monomer molecules can still diffuse easily to macroradicals. A further advantage of precipitation polymerization is the solid (often even powdery) state of polymers which do not need to be pelletized.

Polymerization in Suspension

In **suspension polymerizations**, water-insoluble monomers are dispersed as small droplets by "suspending" agents such as ionic detergents, poly(vinyl alcohol), barium sulfate, etc. Suspension polymerizations can be pictured as "water-cooled" bulk polymerizations "in microreactors" that allow a rapid removal of the heat of polymerization via the water.

The polymerization is started in the monomer droplets by free radical, "oil-soluble" initiators. On polymerization, monomer droplets are converted into beads of (50-400) μm diameter. Suspension polymerizations are thus often called **bead** or **pearl polymerizations**. **Dispersion polymerizations** are "inverse" bead polymerizations of dispersions of aqueous monomer droplets in organic solvents.

The term "**suspension polymerization**" used by polymer chemists does not adhere to the terminology of colloid chemists. In colloid chemistry, all systems composed of a continuous and a discontinuous phase are called **dispersions**. Dispersions thus comprise solids or liquids that are dispersed in gases (**aerosols**), liquids dispersed in other liquids (**emulsions**), and solids dispersed in liquids (**suspensions**). A "suspension polymerization" starts with an *emulsion* of the monomer in water. Depending on the polymerization temperature T and the glass temperature T_G of the polymer, one obtains after polymerization either a suspension ($T_G > T$; for example, poly-(styrene)) or an emulsion ($T_G < T$; for example, styrene–butadiene rubbers), both in water.

Suspension polymerizations are advantageous because the polymerization can be easily controlled and the product is obtained as beads (no additional pelletizing required). Disadvantages are the necessary work-up of the water and the presence of small proportions of suspending agents in polymers that are difficult to remove and that may affect polymer properties (appearance, aging).

Polymerization in Emulsion

Emulsion polymerizations are free radical polymerizations of water-insoluble monomers that are emulsified in water by surface-active compounds and polymerized by water-soluble initiators ($K_2S_2O_8$, redox initiators). N emulsifier molecules E associate in water according to $N\,E \rightleftarrows E_N$ and form micelles E_N with $15 < N < 100$ emulsifier molecules per micelle of (4-10) nm diameter (Fig. 3-15). Hydrophobic monomer molecules can enter micelles and reside in their interior which leads to a slight swelling of the micelles. Large monomer droplets of (1000-10 000) nm diameter are formed in addition; these droplets are stabilized by emulsifier molecules.

Initiator molecules dissociate in water to initiator radicals which add some of the dissolved monomer molecules. The hydrophilic oligomer radicals can enter micelles where they start the polymerization. Each micelle contains either one growing macroradical or none because in such a small compartment, an incoming second radical would immediately terminate the kinetic chain.

Emulsion polymerizations do not start in the emulsion, i.e., the emulsified monomer droplets, but in micelles. A start in a monomer droplet is very unlikely: per liter, a typical system contains ca. 10^{20} monomer-filled micelles (diameter ca. 8 nm) and ca. 10^{12} monomer droplets (diameter ca. 5000 nm). The total surface of micelles ($2000 \cdot 10^{19}$ nm^2/L) is much larger than that of monomer droplets (ca. $8 \cdot 10^{19}$ nm^2/L).

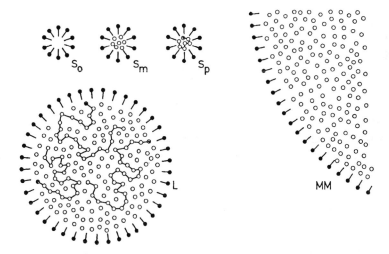

Fig. 3-15 Polymerizing emulsion with empty micelles (S_o), monomer-filled micelles (S_m), micelles with growing polymer chains (S_p), monomer droplets (MM) and latex particles (L) with partially polymerized monomers. ●– Surfactant molecule, ○ monomer molecule.

Three time periods can be distinguished in emulsion polymerizations (Fig. 3-16). Period I is characterized by a constant surface tension: the concentration of micelles must thus be greater than the critical micelle concentration. In this period, oligomer radicals are formed that successively start the polymerization in the micelles: the polymerization rate rises. The polymerization would stop after all monomer molecules have polymerized that were initially present in micelles. Since it does not stop, additional monomer molecules must have diffused from monomer droplets through the aqueous phase into polymerizing micelles. This additional polymerization lets the micelles increase to latex particles of ca. (20-500) nm diameter.

The surface tension increases dramatically at the end of period I, indicating a complete transformation of micelles into latex particles. The polymerization then proceeds in latex particles. This period II is characterized by a constant polymerization rate $R_p = k_p[P^\bullet][M]$ since the monomer concentration in the latex particles is continuously replenished by a monomer diffusion to latex particles from monomer droplets. Diffusion and subsequent polymerization increase the diameter of latex particles.

Each latex particle is generated from an oligoradical-containing micelle. In principle, this radical can polymerize all monomer molecules in a latex particle. Monomers with $M = 100$ g/mol and an interior (neat) concentration of $[M]_o = 8$ mol/L in a latex particle with a diameter of 100 nm would thus lead to a single polymer molecule with $M \approx 2 \cdot 10^{12}$ g/mol! These "theoretical" molar masses are much larger than the experimentally observed molar masses of several millions. Two possibilities exist: (1) another oligoradical enters the particle and terminates the chain and/or (2) growing radicals diffuse out of the polymerizing micelle or latex particle.

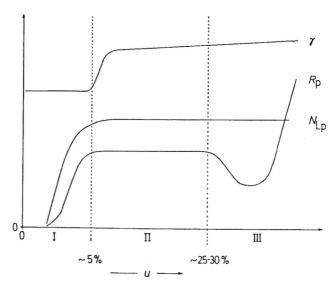

Fig. 3-16 Dependence of surface tension γ, polymerization rate R_p, and number N_{Lp} of latex particles as a function of monomer conversion u at an ab-initio emulsion polymerization of water-insoluble monomers. I = Formation of latex particles, II = polymerization of latex particles and diffusion of monomer from monomer droplets into latex particles, III = polymerization after all monomer droplets have disappeared.

All monomer droplets have vanished in Period III. Only monomer in latex particles can still polymerize. The monomer concentration in latex particles decreases steadily and the polymerization rate $R_p = -d[M]/dt = k_p[P^\bullet][M]$ is first order with respect to the monomer. Finally, no terminating oligoradicals can be formed anymore and the polymerization rate increases strongly because of the gel effect.

Emulsion polymerizations have several advantages: easy control of temperature by water, high polymerization rates due to redox initiators, large degrees of polymerization, simple removal of residual monomer by steam. A disadvantage is the difficult removal of emulsifier residues.

3.8. Copolymerizations

3.8.1. Introduction

According to IUPAC, a **copolymer** is "a polymer derived from more than one species of monomer" (process-based). A **copolymerization** is thus a polymerization in which a copolymer is formed; these polymerizations have also been called interpolymerizations or mixed polymerizations in the older literature. **Pseudocopolymers** are polymers with two or more types of monomeric units that are not generated by a copolymerization of more than one species of monomer; an example is partially saponified poly(vinyl acetate).

Copolymerizations are subdivided into bi-, ter-, quater-, quinter- ... polymerizations according to the number of monomer species involved. "Copolymerization" is however often also used as a synonym for "bipolymerization".

Copolymerizations allow syntheses of many polymers with very different properties from a limited number of monomer species. Most copolymerizations are free radical polymerizations (Table 3-8) because (a) many monomers can be copolymerized free radically, (b) sequence statistics are more easily modified than in other types of polymerization, and (c) monomers are fairly inexpensive. For statistical ionic copolymerizations, monomers often differ too much in their nucleophilicities and electrophilicities, respectively; instead, block copolymers are produced.

3.8.2. Copolymerization Equation

In simple bipolymerizations, both types of active chain ends ~a* and ~b* react irreversibly with both types A and B of monomers:

$$(3\text{-}76) \quad \sim\!a^* + A \rightarrow \sim\!a\text{-}a^* \quad ; \quad R_{aA} = k_{aA}[a^*][A]$$
$$(3\text{-}77) \quad \sim\!a^* + B \rightarrow \sim\!a\text{-}b^* \quad ; \quad R_{aB} = k_{aB}[a^*][B]$$
$$(3\text{-}78) \quad \sim\!b^* + A \rightarrow \sim\!b\text{-}a^* \quad ; \quad R_{bA} = k_{bA}[b^*][A]$$
$$(3\text{-}79) \quad \sim\!b^* + B \rightarrow \sim\!b\text{-}b^* \quad ; \quad R_{bB} = k_{bB}[b^*][B]$$

Table 3-8   Industrial copolymerizations in bulk B, solution (liquid) L, emulsion E, or by precipitation P. [a] In t-butanol; [b] in acetone, 1,4-dioxane or hexane; [c] in water. Copolymerization parameters r_a and r_b indicate relative reactivities of monomers (see below). They are independent of initiator and solvent in free radical copolymerizations but not in ionic and Ziegler–Natta copolymerizations (see Section 3.8.6.). * After sulfonation; ** for polyurethanes.

Monomer A	Comonomer B	r_a	r_b	Type	Application
Free radical copolymerizations to linear or branched copolymers					
Ethylene	Vinyl acetate (10 %)	0.88	1.03	B	Shrink films
Ethylene	Vinyl acetate (10-35 %)	0.88	1.03	B	Thermoplastics
Ethylene	Vinyl acetate (35-40 %)	0.88	1.03	P[a]	Films
Ethylene	Vinyl acetate (> 60 %)	0.88	1.03	E	Elastomers
Ethylene	Methacrylic acid (< 10 %)			B	Coatings
Ethylene	Chlorotrifluoroethylene	0.25	0.0025		Thermoplastics
Butadiene	Styrene	1.44	0.84	E	Elastomers
Butadiene	Acrylonitrile (37 %)	0.36	0.04	E	Elastomers
Vinyl chloride	Vinyl acetate (3-20 %)	1.7	0.23	L[b]	Binders
Vinyl chloride	Propylene (3-10 %)			B	Thermoplastics
Vinylidene chloride	Acrylonitrile	0.32	0.92		Packaging films
Acryl esters	Acrylonitrile (5-15 %)	0.72	1.2		Elastomers
Acrylonitrile	Various (4 %)			P[c]	Fibers
Acrylonitrile	Styrene	0.04	0.41		Thermoplastics
Acrylonitrile	Styrene + butadiene				Thermoplastics
Tetrafluoroethylene	Propylene	1.0	0.06	E	Thermoplastics
Methacrylic acid	Methacrylonitrile	0.60	1.64		Hard foams
Free radical copolymerizations to cross-linked polymers					
Glycol methacrylate	Glycol dimethacrylate (2-4 %)				Contact lenses
Styrene	Unsaturated polyesters			B	Thermosets
Styrene	Divinyl benzenes				Ion exchangers*
Anionic copolymerizations					
Styrene	Butadiene			L	Elastomers
Cationic copolymerizations					
Isobutene	Isoprene (4 %)			L	Butyl rubber
Trioxane	Ethylene oxide			L	Thermoplastics
Ethylene oxide	Propylene oxide			L	Thickeners
Propylene oxide	Non-conjugated dienes			L	Elastomers
Coumarone	Indene				Cements
Ziegler–Natta copolymerizations (without block copolymers)					
Ethylene	Propylene + non-conjugated diene				Elastomers
Ethylene	1-Butene, 1-hexene, 1-octene, 4-methyl-1-pentene				Thermoplastics

This **terminal model** considers four different rates R_{iJ} and four propagation rate constants k_{iJ}; it corresponds to first order Markov statistics. The two ratios r_i (i = a, b) of propagation rate constants of homo and cross reactions:

(3-80) $r_a \equiv k_{aA}/k_{aB}$; $r_b \equiv k_{bB}/k_{bA}$

are called **copolymerization parameters (copolymerization ratios, copolymerization coefficients)**. Five different cases can be distinguished for $r_i = r_a$ and $r_i = r_b$:

$r_i = 0$ Rate constants of homopolymerizations are zero. The active chain end adds only the other type of monomer.

$r_i < 1$ The other monomer is added preferentially.

$r_i = 1$ Both monomers are added with equal probability if $[A] = [B]$.

$r_i > 1$ The same monomer is added preferentially.

$r_i = \infty$ Only homopolymerization, no copolymerization.

Monomers are consumed by the two homopropagation reactions, Eqns. (3-76) and (3-79), and the two cross-propagations, Eqns. (3-77) and (3-78), but not by any other reaction if molar masses are large. The relative monomer consumption is thus:

$$(3-81) \qquad \frac{-d[A]/dt}{-d[B]/dt} = \frac{R_{aA}+R_{bA}}{R_{bB}+R_{aB}} = \left(\frac{k_{bA}+k_{aA}([a^*]/[b^*])}{k_{bB}+k_{aB}([a^*]/[b^*])} \right) \cdot \frac{[A]}{[B]} = \frac{d[A]}{d[B]}$$

In the two cross-propagations, a species ~a* is replaced by a species ~b* or vice versa. In the steady state, the concentrations of these two species are time-independent: the rates R_{aB} and R_{bA} must equal each other. Eqn.(3-81) thus converts to the **Lewis–Mayo equation** for the relative change of monomer concentrations, d[A]/d[B], as a function of the *instantaneous* monomer ratio [A]/[B]:

$$(3-82) \qquad \frac{d[A]}{d[B]} = \frac{1+r_a([A]/[B])}{1+r_b([B]/[A])}$$

The relative change of monomer concentrations equals the molar ratio of monomeric units at small increments of conversion, for example, from 0 % to 2 % or from 49 % to 51 % ($d[A]/d[B] = x_a/x_b$). In this case, the Lewis–Mayo equation describes the instantaneous relative polymer composition x_a/x_b as a function of the instantaneous relative monomer composition [A]/[B].

The Lewis–Mayo equation does not apply to larger intervals of conversion. It has to be integrated before r_a and r_b can be evaluated. Most researchers do not do this but use small initial monomer conversions (e.g., from $u = 0$ % to $u = 4$ %) and evaluate r_a and r_b graphically. Setting $d[A]/d[B] = [a]/[b] = f$ and $[A]/[B] = F$, Eqn.(3-82) turns into the linearized **Fineman–Ross equations**

$(3-83) \quad (f-1)/F = r_a - r_b(f/F^2)$

$(3-84) \quad F(f-1) = -r_a + r_b(F^2/f)$

By plotting $(f-1)/F$ vs. f/F^2 and $F(f-1)$ vs. F^2/f, respectively, copolymerization parameters are obtained from intercepts and slopes. Many other linearizations can be found in the literature.

Initially formed macromolecules from copolymerizations with $r_a \neq r_b$ have different compositions than those formed later. Polymers composed of such macromolecules may demix (Chapter 6) which leads to inferior mechanical properties. Copoly-

mers with conversion-invariant compositions of macromolecules are obtained if the
faster polymerizing monomer is constantly replenished during polymerization. Such
copolymers also result from **azeotropic conditions**, i.e., $d[A]/d[B] = [A]/[B]$ and, with
Eqn.(3-82), $[A]/[B] = (1 - r_b)/(1 - r_a)$. **Alternating copolymers** are obtained for $r_a = r_b = 0$ and **block copolymers** for $r_a \to \infty$ and $r_b \to \infty$.

The type of copolymerization is affected by the type of initiator. The free radical
polymerization of styrene + methyl methacrylate is almost azeotropic (Fig. 3-17).
The same monomer pair forms very long styrene sequences in cationic copolymerizations ($r_S \gg r_{MMA}$) and very long methyl methacrylate sequences in anionic ones
($r_S \ll r_{MMA}$). Ziegler–Natta copolymerizations by $(C_2H_5)_3Al_2Cl_3$ and traces of oxygen lead to almost alternating copolymers.

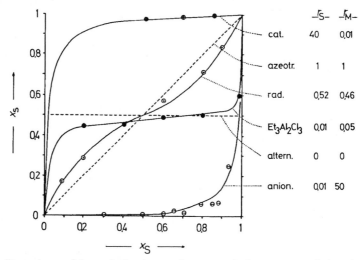

Fig. 3-17   Dependence of the mole fraction x_s of styrene units in styrene-methyl methacrylate copolymers on the mole fraction x_S of styrene monomer (small monomer conversions). - - - Theory
for azeotropic copolymerizations (diagonal line) and alternating copolymerizations (horizontal
line). ------ Lines calculated with indicated copolymerization parameters from experimental data for
cationic ($\oplus$), anionic ($\ominus$), Ziegler–Natta ($\bullet$), and free radical polymerizations ($\circ$).

3.8.3. Sequence Statistics

Copolymerization parameters and instantaneous monomer concentrations determine the lengths and distributions of sequences of monomeric units in copolymers.
The conditional probability $p_{a/A}$ (see Section 3.3.) for the formation of a constitutional diad of two a-units is given by the ratio of the rate R_{aA} of addition of monomer
molecules A to active chain ends a* to the sum of two possible rates of reactions at
a* ends, i.e., R_{aA} and R_{aB}:

$$(3\text{-}85) \quad p_{a/A} = \frac{R_{aA}}{R_{aA} + R_{aB}} = \frac{k_{aA}[A]}{k_{aA}[A] + k_{aB}[B]} = \frac{r_a}{r_a + ([B]/[A])}$$

In order to obtain an a-sequence of i a-units, $(i-1)$ monomer molecules have to be added to a chain end ~b–a*. The probability for this event is $(p_{a/A})^{i-1}$. In order to have a sequence of exactly i a-units, the a-sequence must be terminated by a b-unit; this probability is $p_{a/B} \equiv 1 - p_{a/A}$. The probability $(P_{a\text{-seq}})_i$, of forming an a-sequence with exactly i monomeric units is thus the product of these two probabilities:

$$(3\text{-}86) \quad (P_{a\text{-seq}})_i = (p_{a/A})^{i-1}(1 - p_{a/A}) = (x_{a\text{-seq}})_i$$

This sequence probability $(P_{a\text{-seq}})_i$ equals the mole fraction $(x_{a\text{-seq}})_i$ of sequences of i a-units with respect to all a-sequences of different lengths i. It depends only on $p_{a/A}$ and thus r_A and the instantaneous ratio [A]/[B] (see Eqn.(3-85)). Equimolar initial concentrations of A and B, small monomer conversions, and small copolymerization parameters r_a all lead to a preponderance of isolated a-units (monads). Large copolymerization parameters r_a and excesses of A over B result in longer a-sequences and a flatter distribution of sequence lengths.

At [A]/[B] = 1 and r_a = 1, the following percentages of a-sequences are obtained: 90.91 % monads, 8.26 % diads, 0.75 % triads, 0.07 % tetrads, etc. Quite different sequence distributions are obtained for [A]/[B] = 1 and r_a = 10: 9.09 % monads, 8.26 % diads, 7.51 % triads, etc.

The number-average sequence length of a-sequences, $(\overline{N}_{a\text{-seq}})_n$, is defined by

$$(3\text{-}87) \quad (\overline{N}_{a\text{-seq}})_n \equiv \Sigma_i \, (x_{a\text{-seq}})_i(N_{a\text{-seq}})_i$$

$(N_{a\text{-seq}})_i$ can adopt only the values 1, 2, 3 ... . Mole fractions $(x_{a\text{-seq}})_i$ are given by conditional probabilities (see Eqn.(3-86)). One thus obtains for monads ~b<u>a</u>b~ the expression $(x_{a\text{-seq}})_1 = p_{a/B}$, for diads ~b<u>aa</u>b~ the expression $(x_{a\text{-seq}})_2 = p_{a/A}p_{a/B}$, etc.:

$$(3\text{-}88) \quad (\overline{N}_{a\text{-seq}})_n = p_{a/B} + 2\, p_{a/A}p_{a/B} + 3\, (p_{a/A})^2 p_{a/B} + ..$$
$$= p_{a/B}(1 + 2\, p_{a/A} + 3\, (p_{a/A})^2 + ...) = p_{a/B}/(1 - p_{a/A})^2$$

This series can be transformed into a closed expression because of the condition $p_{a/A} = 1 - p_{a/B} \le 1$. Introduction of Eqn.(3-85) results in:

$$(3\text{-}89) \quad (\overline{N}_{a\text{-seq}})_n = p_{a/B}/(1 - p_{a/A})^2 = 1/(1 - p_{a/A}) = 1 + (r_a[A]/[B])$$

Average sequence lengths of copolymers are usually short. They adopt larger values only for large parameters r_a or r_b and/or great molar excess of A over B. Example: [A]/[B] = 1 results in $(\overline{N}_{a\text{-seq}})_n$ = 1.1 at r_a = 0.1 and $(\overline{N}_{a\text{-seq}})_n$ = 11 at r_a = 10.

3.8.4. Kinetics of Radical Copolymerizations

The Lewis–Mayo model describes the *relative* kinetics of copolymerizations quite well but this is not necessarily true for the *absolute* kinetics. A constancy of the *ratio* of two rate constants (i.e., the copolymerization parameters) for the total composition range does not necessarily imply a constancy of the rate constants themselves.

The copolymerization parameters of styrene–methyl methacrylate are both smaller than unity (r_s = 0.50, r_m = 0.42). The copolymerization rate at all initial monomer compositions ($0 < x_m = 1 - x_s < 1$) should thus be higher than both homopolymerization rates ($x_m = 0$ *and* $x_s = 0$). Experimentally, however, smaller copolymerization rates are found. It is assumed that this effect is caused by enhanced **cross termination** rates between unlike macroradicals. Enhanced terminations lead to smaller polymerization rates if the generation of radicals is unchanged.

Cross terminations are characterized by the geometric mean of the termination rate constants of homo terminations ($k_{t,aa}$, $k_{t,bb}$) and the cross termination ($k_{t,ab}$):

$$(3\text{-}90) \quad \Phi = \frac{k_{t,ab}}{2(k_{t,aa}k_{t,bb})^{1/2}}$$

Far higher Φ values are observed, however, than predicted by statistical probability. Since these values have been calculated by the terminal model of copolymerization, an additional effect of penultimate units must be assumed.

Table 3-9   Copolymerization rates R_p, propagation rate constants k_p, and Φ values of the radical copolymerization of methyl methacrylate M with styrene S at 60°C at different initial mole fractions $x_{m,o}$ of methyl methacrylate in the methyl methacrylate–styrene mixture.

$x_{m,o}$	$10^5 R_p/(\text{mol L}^{-1} \text{ s}^{-1})$	Φ	$k_p/(\text{L mol}^{-1} \text{ s}^{-1})$			
			~m• + M	~m• + S	~s• + M	~s• + S
1.00	26.30	-	734			
0.87	9.58	9				
0.73	7.49	12				
0.52	5.90	17		1740	352	
0.30	5.32	20				
0.15	4.93	23				
0.00	5.45	-				176

3.8.5. *Q,e*-Scheme

Copolymerization parameters vary widely, not only for a certain pair of comonomers with different types of initiators (Fig. 3-17) but also for different monomer pairs with the same type of initiator. The free radical bipolymerization of styrene and methyl methacrylate delivers, for example, r_s = 0.52 and r_m = 0.46 (Table 3-9), but that of styrene and octyl vinyl ether gives r_s = 100 and r_{ove} = 0. Since copolymerization parameters indicate relative reactivities of two monomers, they must be determined separately for each comonomer pair–initiator–temperature.

Propagation rate constants k_{iJ}, and thus copolymerization parameters $r_a = k_{iI}/k_{iJ}$, depend on polarizations, resonance stabilizations, and steric effects of monomers and macroradicals. The first two of these parameters must be contained in the activation energy $E_{aB}^{\ddagger}$ of the Arrhenius equation $k_{aB} = A_{aB} \cdot \exp(- E_{aB}^{\ddagger}/RT)$ for the temperature dependence of the rate constant k_{aB} of the addition of monomer B to a growing chain end ~a*. Similar equations can be written for rate constants k_{aA}, k_{bA} and k_{bB}.

$E_{aB}^{\ddagger}/RT$ can be subdivided into several parts. Monoradicals possess a resonance factor $p_a^{\bullet}$ and an electrostatic interaction $e_a^{\bullet}$, monomer molecules the corresponding factors q_B and e_B:

(3-91) $k_{aB} = A_{aB} \cdot \exp(- E_{aB}^{\ddagger}/RT) = A_{aB} \cdot \exp(- p_a^{\bullet} + q_B + e_a^{\bullet} e_B)$

Each macroradical usually attacks the methylene group in monomers of type $CH_2=CRR'$. Since the pre-exponential factor A_{aB} counts the frequency of successful collisions, it can thus be assumed to be independent of the structure of these monomers. The terms $\exp(p_a^{\bullet})$ and $\exp(q_B)$ are united with the pre-exponential factor to new constants P_a and Q_B. It is further assumed that monomer A and macroradical ~a* exhibit equal charges ($e_a^{\bullet} = e_A$):

(3-92) $k_{aB} = P_a Q_B \cdot \exp(- e_a^{\bullet} e_B) = P_a Q_B \cdot \exp(- e_A e_B)$

Introduction of the corresponding equations for k_{aA}, k_{bA} and k_{bB} and the definitions of the copolymerization parameters, $r_a \equiv k_{aA}/k_{aB}$ and $r_b \equiv k_{bB}/k_{bA}$, results in the **Alfrey–Price Q,e-scheme**:

(3-93) $r_a = (Q_A/Q_B) \exp[- e_A(e_A - e_B)]$; $r_b = (Q_B/Q_A) \exp[- e_B(e_B - e_A)]$
$r_a r_b = \exp[- (e_A - e_B)^2]$

Each monomer A or B is thus assigned a Q value for the resonance and an e value for the polarity. Q values depend on the temperature and the type of polymerization (free radical, ionic type and solvent, etc.).

Eqns.(3-93) contain two unknowns; they can be solved only if one monomer is chosen as a reference. Historically, this monomer was styrene since it can be copolymerized free radically with many monomers. Styrene was assigned a Q value of unity

Table 3-10 Copolymerization parameters by the terminal model for cationic (C), free radical (R) and anionic (A) bipolymerizations of acrylonitrile AN, butadiene BU, isoprene I, methyl methacrylate MMA, and styrene S.

| Comonomers | | T/°C | Initiator + solvent | | r_a | r_a | r_a | r_b | r_b | r_b |
A	B		cationic	anionic	C	R	A	C	R	A
S	BU	50	R_3B	BuLi in C_6H_6	0.01	0.44	0.05	1.30	1.59	20
I	BU	50	R_2AlCl	BuLi in C_6H_{14}	1.0	0.47		1.0	3.4	
I	S	27	$TiCl_4$–$LiAlH_4$	BuLi in Toluol	0.1	1.2	1.0	9.0	0.66	0.80
		27	R_3Al	BuLi in DMF	0.5	1.2	7.0	0.46	0.66	0.14
AN	BU	50	R_3B		0.04	0.05		0.40	0.35	
AN	MMA	– 8	R_3B	–BuLi	0.12	0.10	7.0	1.34	1.30	0.39
S	MMA	35	R_3B	SO_2	0.56	0.5	0.55	0.50	0.46	0.23

because it was assumed to be the monomer with the greatest resonance stabilization. This choice of Q turned out to be incorrect: Q values are presently known to vary from 0.001 for tetrachloroethylene to 16.0 for vinyl chloromethylketone. Styrene was also assigned an e value of –0.800; however, known e values range from –8.53 for vinyl *o*-cresyl ether to +3.76 for carbon monoxide.

The Q,e scheme allows to estimate unknown copolymerization parameters. In general, two rules apply: (1) monomers with very different Q values do not copolymerize; (2) at approximately equal Q values, similar e values lead to azeotropic copolymers and very different e values to alternating ones. Q,e data are not always reliable, especially not for sterically hindred and/or non-homopolymerizable monomers.

3.8.6. Ionic Copolymerizations

Ionic copolymerizations lead to totally different copolymerization parameters than free radical copolymerizations because of the vastly different polarities of the propagating active species. They also depend on the type of initiator and the solvent because of the different ion equilibria and ion associations (Table 3-10).

Copolymerization parameters should thus be able to distinguish between different types of initiation. For example, approximately the same copolymerization parameters are obtained for copolymerizations of acrylonitrile and butadiene by free radicals or by boron alkyls. Copolymerizations by boron alkyls are thus probably proceeding free radically. The method is not always reliable, however.

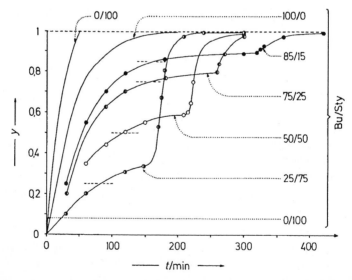

Fig. 3-18 Time dependence of polymer yields *y* for the anionic copolymerization of butadiene Bu and styrene Sty by butyl lithium in cyclohexane at 50°C for different initial monomer ratios of butadiene and styrene. Broken horizontal lines indicate complete polymerization of butadiene for independent and successive polymerizations of the two monomers. With permission by the Rubber Division, American Chemical Society [5].

The kinetics of ionic copolymerizations can be very complicated. Butyl lithium homopolymerizes styrene in cyclohexane much faster than butadiene ($k_{sS} > k_{bB}$; (Fig. 3-18). The bipolymerization of the monomers is, however, much slower than the respective homopolymerizations. Polymerization rates increase steeply some time after the complete consumption of butadiene. This change to higher rates occurs at a greater total monomer consumption than the initial butadiene concentration. Styrene is thus consumed at the smallest monomer conversion. If, however, styrene slows the polymerization despite $k_{sS} > k_{bB}$, then a large cross propagation constant k_{sB} must exist: all styrene chain ends are rapidly converted into butadiene chain ends.

Appendix to Chapter 3

A 3-1: Equilibria

The first compound is formed by $R^1R^2 + M \rightleftarrows R^1MR^2$ with an equilibrium constant $K_1 = [R^1MR^2]/([R^1R^2][M])$. For simplification, successive compounds R^1MR^2, $R^1M_2R^2$, $R^1M_3R^3$... are symbolized by a_1, a_2, a_3.... The total concentration of all molecules is $\Sigma_i [a_i] = [a_1] + [a_2] + [a_3] +...$ and that of all monomeric units $\Sigma_i i \cdot [a_i] = [a_1] + 2[a_2] + 3[a_3] +...$. Introduction of the equilibrium constants $K_2 = [a_2]/[a_1]^2$, $K_3 = [a_3]/([a_2][a_1]) = [a_3]/(K_2[a_1]^3)$, $K_4 = [a_4]/([a_3][a_1]) = [a_4]/(K_3[a_2][a_1]^2) = [a_4]/(K_3K_2[a_1]^4)$, etc., leads to the series

$$\Sigma_i [a_i] = [a_1](1 + K_2[a_1] + K_3K_2[a_1]^2 + K_4K_3K_2[a_1]^3+...)$$
$$\Sigma_i i \cdot [a_i] = [a_1](1 + 2K_2[a_1] + 3 K_3K_2[a_1]^2 + 4K_4K_3K_2[a_1]^3+...).$$

The assumption $K_2 = K_3 = K_4 = ... = K_n$ converts these series into the closed expressions: $\Sigma_i [a_i] = [a_1]/(1 - K_n[a_1])$ and $\Sigma_i i \cdot [a_i] = [a_1]/(1 - K_n[a_1])^2$. The number-average degree of polymerization is given by $\overline{X}_n = \Sigma_i i \cdot [a_i]/\Sigma_i [a_i]$, if one sets $X_i = i$ and replaces the mole fractions x_i by mole concentrations $[a_i]$. The $\overline{X}_n$ of a type I equilibrium is thus given by $\overline{X}_n = \Sigma_i i \cdot [a_i]/\Sigma_i [a_i] = 1/(1 - K_n[M])$.

Type II equilibria between M, M*, M_2*, M_3* ... deliver the same expression for the number-average degree of polymerization because the same species are formally present, i.e., monomers M and "activated" monomers $R^1MR^2 (= M*)$ with one mer, "activated" dimers $R^1M_2R^2 (= M_2*)$ with two mers, etc. Type III equilibria are different, however, because the species are now M, M_2, M_3 ... ; the number average degree of polymerization is here $\overline{X}_n = 1 + 1/(1 - K_n[M])$.

The product $K_n[M]$ must be always smaller than unity because the series in braces would diverge in the following example (Type I with $K_2 = K_3 = ... = K_n$):

$$[M]_o = [M]_{free} + [M]_{bound} = [M] + \Sigma_{i=1}^{i=\infty} i \cdot [R^1M_iR^2]$$
$$[M]_o = [M] + [R^1M_1R^2] + 2 [R^1M_2R^2] + 3 [R^1M_3R^2] + ...$$
$$[M]_o = [M] + K_1[R^1R^2][M] + 2 K_2[R^1M_1R^2][M] + 3 K_3[R^1M_2R^2][M] + ...$$
$$[M]_o = [M] + K_1[R^1R^2][M] + 2 K_2K_1[R^1R^2][M]^2 + 3 K_3K_2K_1[R^1R^2][M]^3 + ...$$
$$[M]_o = [M] + K_1[R^1R^2][M]\{1 + 2 K_n[M] + 3 (K_n[M])^2 + ...\}$$

A 3-2: Poisson Distribution

In living polymerizations, a monomer concentration $[M]_o$ is initially present. A fast initiation reaction $I* + M \rightarrow IM*$ converts initiator molecules $I*$ into monomer ions $IM* \equiv P_1*$ that add further monomer molecules and become macroions P_i* (with $i \geq 2$). The total concentration $[C*] = \Sigma_i [P_i*]$ of growing ions stays constant during the polymerization (no termination); $[C*] = [P_1*]$ must be valid at time $t = 0$.

The kinetic chain length v is the number of monomer molecules added to a monomer ion. The number-average degree of polymerization also includes the monomeric unit in the monomer ion, however; thus $v = \overline{X}_n - 1 = ([M]_o - [M])/[C*]$.

The polymerization rate is $-d[M]/dt = k_p[M][C*]$. The difference between the concentrations of initial monomer and the monomer ion is $\Delta M = [M]_o - [P_1*] - [M] = [M]_o - [C*] - [M] = [C*](\overline{X}_n - 1) = [C*]v$. The polymerization rate is thus $-d[M]/[C*] = k_p[M]dt = dv$. The rate equations for the different propagation steps can thus be written:

(1) $- d[P_1*]/dt = k_p[M][P_1*]$ as $- d[P_1*] = [P_1*]dv$

(2) $d[P_2*]/dt = k_p[P_1*][M] - k_p[P_2*][M]$ as $d[P_2*] = [P_1*]dv - [P_2*]dv$

....

(3) $d[P_i*]/dt = k_p[P_{i-1}*][M] - k_p[P_i*][M]$ as $d[P_i*] = [P_{i-1}*]dv - [P_i*]dv$

Integration of Eqn.(1) gives $\log_e [P_1*] = \text{const} - v$. At time $t = 0$, $v = 0$; since $[P_1*] = [C*]$, one has $\log_e [P_1*] = \log_e [C*] - v$ and $[P_1*] = [C*]\cdot\exp(- v)$.

Introduction of this expression into Eqn.(2) delivers $d[P_2*] = [C*]\cdot\exp(- v)dv - [P_2*]dv$. Integration and multiplication by $\exp(v)$ results in $\exp(v)\cdot d[P_2*] + \exp(v)\cdot[P_2*]\cdot dv = [C*]\cdot dv$. With $d\{\exp(v)\}/dv = \exp(v)$, one obtains the complete differential $\exp(v)\cdot d[P_2*] + [P_2*]\cdot d\{\exp(v)\} = [C*]\cdot dv$. Integration delivers $\exp(v)\cdot[P_2*] = [C*]v + \text{const}'$. Since $[P_2*] = 0$ at $t = 0$ ($v = 0$), one has $\text{const}' = 0$ and $[P_2*] = \exp(- v)\cdot v[C*]$.

By analogy, $d[P_3*] = [P_2*]dv - [P_3*]dv = \exp(- v)[C*]vdv - [P_3*]dv$ delivers $[P_3*] = \exp(- v)\cdot[C*]\cdot v^2/2!$. The general expression for an i-mer is thus $[P_i*] = \exp(- v)\cdot[C*]\cdot v^{i+1})/(i-1)!$.

The mole fraction $x_i \equiv [P_i*]/[C*]$ of i-meric ions is also the mole fraction of i-mers: $x_i = \exp(- v)\cdot v^{i-1}/(i-1)!$ and, with $i \equiv X_i$, also $x_i = \exp(- v)\cdot v^{i-1}/(X_i-1)!$.

The mass fraction is defined as $w_i \equiv x_i i/\overline{X}_n = x_i X_i/\overline{X}_n = x_i X_i/(v + 1)$. Introduction of the expression for x_i delivers $w_i = \{\exp(- v)\cdot v^{i-1}\cdot X_i\}/\{(X_i -1)!(v + 1)\}$.

The mass-average degree of polymerization is defined as $\overline{X}_w \equiv \Sigma_i w_i X_i$ and for Poisson distributions therefore given by $\overline{X}_w = \Sigma_i \{[\exp(-v)\cdot v^{i-1}]/[(X_i-1)!(v+1)]\} = [v^2 + 3v + 1]/[v + 1]$ which converts with $v = \overline{X}_n - 1$ into $\overline{X}_w = \overline{X}_n + 1 - (1/\overline{X}_n)$.

A 3-3: Stationary State

Macroradicals are produced with a rate $d[P^\bullet]/dt = R_{st} - R_{t(pp)} = R_{st} - k_{t(pp)}[P^\bullet]^2$. The integrals are $\int dt = \int (R_{st} - k_{t(pp)}[P^\bullet]^2)^{-1}d[P^\bullet]$ with the limits 0 to t and 0 to $[P^\bullet]_t$, respectively. Integration leads to $[P^\bullet]_t = (R_{st}/k_{t(pp)})^{1/2} \tanh\{(R_{st}k_{t(pp)})^{1/2}t\}$. The radical concentration $[P^\bullet]_t$ becomes the radical concentration $[P^\bullet]_{stat}$ in the stationary state; thus $[P^\bullet]_t = [P^\bullet]_{stat}$. The steady state is practically obtained for a concentration ratio of $[P^\bullet]_t/[P^\bullet]_{stat} = 0.995$; thus $\tanh\{(R_{st}k_{t(pp)})^{1/2}t\} \geq 0.995$. According to the tanh tables, this corresponds to $(R_{st}k_{t(pp)})^{1/2}t \geq 3$. For the example in Section 3.7.4., the steady state is obtained after $t \approx 3$ s.

The monomer conversion $p = ([M]_0 - [M])/[M]_0$ at the steady state is calculated from $\Delta[M]/\Delta t = k_p[P^\bullet][M]$. For an initial monomer concentration of $[M]_0 = 8.5$ mol/L, a stationary radical concentration of $[P^\bullet]_{stat} = 10^{-8}$ mol/L, and an experimental propagation rate constant $k_p = 215$ L mol^{-1} s^{-1}, one obtains a monomer conversion of $p = 6 \cdot 10^{-4}$ % for the onset of the steady state.

A 3-4: Schulz–Flory Distribution

Free radical polymerizations and equilibrium polycondensations differ in both the participation of reactants in the propagation steps (addition of monomer molecules to growing polymer *vs.* indiscriminate mutual reaction of all polymer molecules) as well as the absence and presence of leaving molecules (Section 3.1.1.). Nevertheless, both types of polymerization lead to the same type of distribution (Schulz–Flory distribution) because the addition of monomers in free radical chain polymerization with termination by disproportionation and the reaction of reactants in polycondensation both occur at random.

The following derivation of the Schulz–Flory (SF) distribution is based on the kinetics of elementary reactions. The SF distribution can also be obtained from probabilities similar to derivations used for polycondensation equilibria.

For a derivation of the SF equation, ideal polymerization kinetics (Section 3.7.5.) and a chain termination by disproportionation is assumed: two macroradicals are converted into two dead macromolecules. In the start reaction, light quanta of intensity $I_{h\nu}$ generate monomer radicals $M^\bullet \equiv P_1^\bullet$ with a rate $R_{st} = k_I[M]I_{h\nu}$. The rate of formation of monomer radicals is thus:

(1) $d[P_1^\bullet]/dt = k_I[M]I_{h\nu} - k_p[P_1^\bullet][M] - k_t[P_1^\bullet][P_1^\bullet] - k_t[P_1^\bullet][P_2^\bullet]... - k_t[P_1^\bullet][P_i^\bullet]$
$= k_I[M]I_{h\nu} - k_p[P_1^\bullet][M] - k_t[P_1^\bullet][P^\bullet]$

where $[P_1^\bullet] + [P_2^\bullet]... +[P_i^\bullet] = \Sigma_j [P_j^\bullet] \equiv [P^\bullet]$. In the steady state, $d[P_1^\bullet]/dt = 0$ and thus

(2) $(k_I[M]I_{h\nu})/(k_t[M]) = [P_1^\bullet](1 + \beta)$; $\beta \equiv (k_t[P^\bullet])/(k_p[M]) = (k_Ik_tI_{h\nu})^{1/2}/(k_p[M]^{1/2})$

The last expression results from the steady state condition $k_I[M]I_{h\nu} = k_t[P^\bullet]^2$. The concentrations $[P^\bullet]$ and $[M]$ are constant at small conversion. β is thus a constant, too.

Dimer, trimer, ... polymer radicals ($i \geq 2$) are formed with a rate

(3) $d[P_i^\bullet]/dt = k_p[P_{i-1}^\bullet][M] - k_p[P_i^\bullet][M] - k_t[P_i^\bullet][P_1^\bullet] - k_t[P_i^\bullet][P_2^\bullet]... - k_t[P_i^\bullet][P_i^\bullet]$

Analogous to the formation of monomer radicals, one obtains $[P_{i-1}^{\bullet}] = [P_i^{\bullet}](1 + \beta)$. Successive insertion of the expressions for all $i \geq 1$ and application of Eqns.(2) gives

(4) $[P_i^{\bullet}] = \{(k_I[M]I_{hv})/(k_t[M])\}(1 + \beta)^{-i}$

Dead polymers are generated with a rate $d[P_i]/dt = k_t[P_i^{\bullet}][P^{\bullet}]$. Integration leads to $[P_i] = k_t[P_i^{\bullet}][P^{\bullet}]t$. $[P_i^{\bullet}]$ is expressed by Eqn.(4) and $[P^{\bullet}]$ by the equation for β. The result is $[P_i^{\bullet}] = k_I[M]I_{hv}\beta(1 + \beta)^{-i}t$. During time t, a total of $k_I[M]I_{hv}t$ radicals are produced. Each radical leads to a polymer molecule. The sum of all polymer concentrations is given by $\Sigma_i [P_i] = \Sigma_i k_I[M]I_{hv}\beta(1 + \beta)^{-i}t = k_I[M]I_{hv}t$ since $\Sigma_i \beta(1+\beta)^{-i} = 1$. The mole fraction of polymer molecules i with a degree of polymerization, X_i, is $x_i = [P_i]/\Sigma_i [P_i] = \beta(1 + \beta)^{-i} = \beta[\exp(\beta)]^{-1}$ since $\beta \ll 1$.

The number-average degree of polymerization can be calculated if the sum is replaced by an integral. After integration from zero to infinity, one obtains for $i \equiv X_i$

(5) $$\overline{X}_n = \frac{\Sigma_i x_i X_i}{\Sigma_i x_i} = \frac{\Sigma_i (1+\beta)^{-i} X_i}{\Sigma_i (1+\beta)^{-i}} = \frac{\int X_i[\exp(-\beta X_i)]dX_i}{\int [\exp(-\beta X_i)]dX_i} = \beta^{-2}/\beta^{-1} = 1/\beta$$

Similar calculations deliver the mass average degree of polymerization, $\overline{X}_w = 2/\beta$. The ratio of degrees of polymerization is thus $\overline{X}_w/\overline{X}_n = 2$ for a termination by disproportionation.

Literature

H.-G.Elias, Makromoleküle (in German), Hüthig and Wepf, Basel, 5th ed., Vol. I (1990); Chapters 6 (polymerization), 9 (polymerization equilibrium), 10 (ionic polymerization), 11 (polyinsertion), 12 (free radical polymerization), 14 (copolymerization).
H.-G.Elias, Makromoleküle (in German), Hüthig and Wepf, Basel, 5th ed., Vol II (1992); Chapters 4 (industrial syntheses), 5-11 (carbon chains to inorganic polymers).
G.Allen, J.C.Bevington, Eds., Comprehensive Polymer Science, Pergamon Press, Oxford 1989; Vols. 3-6, G.C.Eastmond, A.Ledwith, S.Russo, P.Sigwalt, Eds., Vol. 3: Chain Polymerization I (anionic, cationic, free radical, copolymerization); Vol. 4: Chain Polymerization II (polyinsertion, heterogeneous radical polymerization); Vol. 5: Step Polymerization (polycondensation); Vol. 6: Polymer Reactions.
F.M.McMillan, The Chain Straighteners: Fruitful Innovation. The Discovery of Linear and Stereoregular Polymers, MacMillan, London 1981

3.1. FUNDAMENTALS
R.W.Lenz, Organic Chemistry of Synthetic High Polymers, Interscience, New York 1967
P.Rempp, E.W.Merrill, Polymer Synthesis, Hüthig und Wepf, Basel, 2nd ed. 1991
H.R.Kricheldorf, Ed., Handbook of Polymer Synthesis, Dekker, New York 1992 (2 vols.)
G.Odian, Principles of Polymerization, Wiley, New York, 3rd ed. 1992
W.J.Mijs, Ed., New Methods for Polymer Synthesis, Plenum, New York 1992
G.B.Butler, Cyclopolymerization and Cyclocopolymerization, Dekker, New York 1992
M.Kucera, Mechanism and Kinetics of Addition Polymerizations, Elsevier, Amsterdam 1992

R.K.Sadhir, R.M.Luck, Eds., Expanding Monomers: Synthesis, Characterization and
Applications, CRC Press, Boca Raton, FL 1992
D.J.Brunelle, Ed., Ring-Opening Polymerization, Hanser, Munich 1993
Y.Yamashita, Ed., Chemistry and Industry of Macromonomers, Hüthig and Wepf, Basle 1993

3.2. THERMODYNAMICS
H.Sawada, Thermodynamics of Polymerization, Dekker, New York 1976

3.4. ANIONIC POLYMERIZATIONS
M.Morton, Anionic Polymerization: Principles and Practice, Academic Press, New York 1983
M.Szwarc, M.Van Beylen, Ionic Polymerization and Living Polymers, Chapman and Hall, New
York 1993
H.L.Hsieh, R.P.Quirk, Anionic Polymerization, Dekker, New York 1996

3.5. CATIONIC POLYMERIZATIONS
J.P.Kennedy, B.Iván, Designed Polymers by Carbocationic Macromolecular Engineering: Theory
and Practice, Hanser, Munich 1992
K.Matyjaszewski, ed., Cationic Polymerizations, Dekker, New York 1996

3.6. INSERTION POLYMERIZATIONS
Y.V.Kissin, Isospecific Polymerization of Olefins, Springer, Berlin 1986
K.J.Ivin, Olefin Metathesis, Academic Press, New York 1983
K.Soga, M.Terano, eds., Catalyst Design for Tailor-Made Polyolefins, Kodansha, Tokyo 1994
G.Fink, R.Mülhaupt, H.H.Brintzinger, Eds., Ziegler Catalysts, Springer, Berlin 1995

3.7. FREE RADICAL POLYMERIZATIONS
C.H.Bamford, C.F.H.Tipper, Eds., Free-Radical Polymerization (Comprehensive Chemical
Kinetics **14 A**), Elsevier, Amsterdam 1976
D.C.Blackley, Emulsion Polymerization: Theory and Practice, Halsted, New York 1975
J.A.Biesenberger, D.A.Sebastian, Principles of Polymerization Engineering, Wiley, New York
1983

3.8. COPOLYMERIZATIONS
G.E.Ham, Ed., Copolymerization, Interscience, New York 1964

References

[1] G. Henrici-Olivé, S. Olivé, Kunststoffe-Plastics **5** (1958) 315, data of Table 1
[2] G.V.Schulz, Ber.Deutsch.Chem.Ges. **80** (1947) 232, Fig. 1
[3] T.G.Carswell, D.J.T.Hill, D.I.Londero, J.H.O'Donnell, P.J.Pomery, C.L.Winzor, Polymer
33 (1992) 137, data of Figs. 2 and 3
[4] H.Gerrens, Chem.-Ing.Techn. **52** (1980) 477, Fig. 1
[5] H.L.Hsieh, W.H.Glaze, Rubber Chem.Technol. **43** (1970) 22, Fig. 21
[6] J.Shen, Y.Tian, G.Wang, M.Yang, Makromol.Chem. **192** (1991) 2669, Figs. 3, 5, 9

4. Step-Growth Polymerizations and Polymer Reactions

4.1. Polycondensations and Polyadditions

4.1.1. Introduction

Chain-growth polymerizations are characterized by successive addition of monomer molecules to growing chains (Sections 3.4.-3.8.). They are always initiated; initiator fragments become end groups of polymer chains.

Polycondensations and polyadditions, however, are polymerizations with random addition of *all* reactants (monomer *and* polymer molecules) to polymer chains (see Section 3.1.). They are *catalyzed*, not initiated: catalyst molecules do not become part of the polymer molecules.

An example is the polycondensation of stoichiometric amounts of hexamethylene diamine $H[NH(CH_2)_6NH]H$ (symbol: H-D-H) and adipic acid $HO[OC(CH_2)_4CO]OH$ (symbol: HO-A-OH). This reaction generates H_2O from H and OH and delivers the monomeric units $D = [NH(CH_2)_6NH]$ and $A = [OC(CH_2)_4CO]$. In the first step, dimers H-D-A-OH are formed with a degree of polymerization $X = 2$. These dimers either add monomer to form two different types of trimer ($X = 3$) or add another dimer to give a tetramer ($X = 4$). Two different tetramers are obtained by reaction of trimers with monomers, etc.:

$$
\begin{array}{llll}
(4\text{-}1) & \text{H-D-H} + \text{HO-A-OH} & \rightleftharpoons \text{H-D-A-OH} + H_2O & X = 2 \\
& \text{H-D-H} + \text{HO-A-D-H} & \rightleftharpoons \text{H-D-A-D-H} + H_2O & X = 3 \\
& \text{HO-A-OH} + \text{H-D-A-OH} & \rightleftharpoons \text{HO-A-D-A-OH} + H_2O & X = 3 \\
& \text{HO-A-OH} + \text{H-D-A-D-H} & \rightleftharpoons \text{HO-A-D-A-D-H} + H_2O & X = 4 \\
& \text{H-D-H} + \text{HO-A-D-A-OH} & \rightleftharpoons \text{H-D-A-D-A-OH} + H_2O & X = 4 \\
& \text{H-D-A-OH} + \text{H-D-A-OH} & \rightleftharpoons \text{H-D-A-D-A-OH} + H_2O & X = 4 \\
\hline
& N \text{ H-D-H} + N \text{ HO-A-OH} & \rightleftharpoons \text{H(D-A)}_N\text{OH} + (2N - 1) H_2O & X = 2N
\end{array}
$$

All reactants H-D-H, HO-A-OH, H-D-A-OH, H-D-A-D-H....H(D-A)$_N$OH are in equilibrium with each other and the leaving molecules H_2O. It is important to note that the degree of polymerization of polymer molecules refers to the number of monomeric units per molecule and not the the number of repeating units. It is also useful to distinguish between the **extent of reaction** p as conversion of functional groups, the **conversion** u of monomer molecules, and the **yield** y of polymers.

Polycondensations and polyadditions share the same relationships between monomer conversions and degrees of polymerizations for equilibrium reactions and also for simple kinetics. Statements that apply to polycondensations are thus also valid for polyadditions; exceptions are noted.

Polycondensations and polyadditions with bifunctional monomers are subdivided into AB reactions and AA/BB reactions. **AB reactions** are "self-polymerizations" between functional groups of monomers of the same type. An example is the polycondensation of hydroxy acids HO-X-COOH to polyesters $H\text{-}[\text{O-X-CO}]_{\overline{n}}OH$. **AA/BB reactions** are polymerizations of functional groups of different types of monomers. An example is the formation of polyamides according to Eqn.(4-1).

Polycondensations result in the formation of **leaving molecules** (water in Eqn.(4-1)) whereas no such low molar mass molecules are produced by polyadditions. This difference is very important for industrial reactions since the removal of leaving molecules requires special techniques. Residual leaving molecules may also worsen properties and performance of polymers.

Polycondensations and polyadditions may be performed in melts, in solutions, at interfaces, in crystals, and even in the gas phase, depending on stabilities and reactivities of monomers and polymers. They may proceed in homogeneous or heterogeneous phases and may be controlled by thermodynamics or kinetics. In each case, different relationships are obtained between extents of reaction and degrees of polymerization and molar mass distributions.

Very few polycondensations and polyadditions are spontaneous; most of these reactions require added catalysts. The chemistry of elementary reactions usually does not differ from those of low molar mass analogs. Only very few reactions of low molar mass chemistry are suitable, however, for the synthesis of linear polymers since most suffer from side reactions that prevent the attainment of high extents of reaction (cf. Fig. 3-1). Side reactions are less important for the synthesis of cross-linked polymers from multifunctional monomers.

In equilibrium, neat bifunctional monomers form almost exclusively linear macromolecules. The proportion of cyclic polymer molecules is promoted by diffusion-controlled reactions. In general, multifunctional monomers deliver branched polymer molecules at small monomer conversions, and cross-linked polymers at higher ones. Exceptions are monomers of the AB_2 type (see Section 4.1.9.).

4.1.2. Chemistry of Bifunctional Polycondensations

AB Polycondensations require bifunctional monomers of the AB type. In rare cases, AB monomers are directly polycondensed after their synthesis. An example is the microbiological synthesis of lactic acid HO-CH(CH$_3$)-COOH. Removal of water from the broth causes a polycondensation to *polylactide* $H\text{-}[\text{O-CH(CH}_3)\text{-CO}]_{\overline{n}}OH$. This polymer can also be obtained by ring-opening polymerization of lactide, the cyclic dimer of lactic acid. It is a biodegradable polymer that can be used for sutures.

Most AB monomers are isolated, stored and later polycondensed at elevated temperatures. 11-Aminoundecanoic acid $H_2N(CH_2)_{10}COOH$ (from castor oil in several steps) undergoes melt polycondensation to $H\text{-}[\text{NH(CH}_2)_{10}\text{CO}]_{\overline{n}}OH$ (*polyamide 11*, PA 11). A new industrial synthesis makes *polyamide 12* by polycondensation of 12-aminolauric acid $H_2N(CH_2)_{11}COOH$; the standard process is the anionic ring-open-

ing polymerization of laurolactam (from 1,5,9-cyclododecatriene → cyclododeca-
none → oxime → lactam). *Poly(p-benzamide)* is obtained from the polycondensation
of the acid chloride of *p*-aminobenzoic acid, $H_2N(p\text{-}C_6H_4)COCl$.

Polyamide 6 is obtained from ε-caprolactam by the so-called **hydrolytic lactam
polymerization.** ε-Caprolactam is partially hydrolyzed to $H_2N(CH_2)_5COOH$; this
amino acid initiates the ring-opening polymerization of the lactam. The amino acid
simultaneously undergoes polycondensation.

Agricultural refuse delivers furfuryl alcohol I which polymerizes on heating with
acids to a soluble *furan resin* with monomeric units II and III. In iron foundries, mixtures of furan resins with phenol-formaldehyde and urea-formaldehyde resins serve
as binders for molding sands.

In polysulfonations, aromatically bound hydrogen is substituted electrophilically
by sulfonylium ions. The resulting *polyethersulfone* is a thermoplastic:

(4-2)

Phospholene oxides (mono-unsaturated five-membered heterocycles with phosphorus atoms in the ring) act as catalysts for the polymerization of diisocyanates to
polycarbodiimides. The leaving CO_2 converts the polymers into rigid foams:

(4-3) $n\ O=C=N\text{-}X\text{-}N=C=O \longrightarrow \text{-}[N=C=N\text{-}X\text{-}]_n + n\,CO_2$

AA/BB polycondensations work with monomers A-X-A and B-Y-B that do not
homopolymerize and thus have excellent shelf lives. In industry, most often used are
the functional groups -COOH, -COOR, -COCl, -SO$_2$Cl, and -Cl on one hand and
-NH$_2$, -OH, -ONa and -OK on the other (Table 4-1). Functional groups differ from
leaving groups since parts of the functional groups remain with the monomeric unit
(e.g., -CO- of -COOH, -NH- of -NH$_2$).

All AA/BB polycondensations are described by the general reaction:

(4-4) $n\ A\text{-}X\text{-}A + n\ B\text{-}Y\text{-}B \longrightarrow A\text{-}[X\text{-}Y\text{-}]_n B + (2n-1)\ AB$

For example, functional groups -COCl and NaO- are converted into polymeric ester groups -CO-O- and leaving molecules NaCl (= AB). If ring molecules are used as
monomers, only (*n*–1) leaving molecules are formed instead of (2*n*–1). An example
is the polycondensation of ethylene glycol and maleic anhydride to an *unsaturated
polyester* (most maleic acid units isomerize to fumaric acid units during this reaction):

Table 4-1 Important industrial linear AA/BB polycondensations to polymers $A \text{-}[X\text{-}Y\text{-}]_n B$ (see Eqn.(4-4)). Ar = Aromatic residue, pPh = *para*-substituted phenylene group, R = organic residue (usually aliphatic). [a] From maleic anhydride.

Name of Polymer	Chemical structure of A X	Y	B	Application
Poly(ethylene terephthalate)	H O(CH₂)₂O H	CO-pPh-CO	OH OCH₃	Fibers, thermoplastics
Unsat. polyester	H O(CH₂)₂O	CO-CH=CH-CO[a]	O₁/₂	Thermosets
Polycarbonate A	H O-pPh-C(CH₃)₂-pPh-O	CO	OC₆H₅	Thermoplastics
	Na O-pPh-C(CH₃)₂-pPh-O	CO	Cl	
Polyarylates	Na O-Ar-O	CO-Ar-CO	Cl	Engineering plastics
Poly(phenylene sulfide)	Na S	pPh	Cl	Engineering plastics
Polysulfide	Na S	R	Cl	Elastomers
Polysulfone	K O-Ar-O	O(Ar-SO₂)₂	Cl	Engineering plastics
Polyamides	H NH-Z-NH	CO-Z'-CO	OH	Fibers, thermoplastics
Aramides	H NH-Ar-NH	CO-Ar'-CO	OH	Fibers
Polyether-etherketone	K O-pPh-CO-pPh-O-	pPh-CO-pPh	F	Engineering plastics

(4-5) HOCH₂CH₂OH +

Poly(ethylene terephthalate) (PET) $\text{-}[OCH_2CH_2OCO(p\text{-}C_6H_4)CO\text{-}]_n$ is synthesized by polycondensation of ethylene glycol $HOCH_2CH_2OH$ with either terephthalic acid $HOOC(p\text{-}C_6H_4)COOH$ or dimethyl terephthalate $CH_3OOC(p\text{-}C_6H_4)COOCH_3$. Industry first used the dimethyl ester because industrial terephthalic acid was not pure enough for a direct polyesterification. Nowadays, terephthalic acid is preferred since one avoids the costly recovery of methanol.

Polycarbonate A (PC) is also synthesized industrially by two different processes: transesterification of bisphenol A $HO(p\text{-}C_6H_4)C(CH_3)_2(p\text{-}C_6H_4)OH$ with dimethyl-carbonate $(CH_3O)_2CO$ or reaction of the sodium salt of bisphenol A with phosgene $COCl_2$. Some industrial polycarbonates are copolymers with other aromatic residues.

Industrial *polyarylates*, *polysulfones*, *aramides*, and *polyetherketones* are manufactured in small amounts but with many different aromatic monomer units; an example is polyetheretherketone $\text{-}[O\text{-}(p\text{-}C_6H_4)\text{-}O\text{-}(p\text{-}C_6H_4)\text{-}CO\text{-}(p\text{-}C_6H_4)\text{-}]_n$ (PEEK).

Polysulfides are synthesized by polyetherification of sulfur-containing diols, for example, $HO(CH_2CH_2SCH_2CH_2O)_nH$ (H_2O as leaving molecule). Note that the characteristic sulfide group is preformed in the monomer molecules and is *not* generated by the polycondensation itself (cf., polycarbonates, polyamides, etc.). Aromatic *poly-(phenylenesulfide)s* $\text{-}[S\text{-}(p\text{-}C_6H_4)\text{-}]_n$ (PPS) owe their sulfide groups to the polycondensation of Na_2S and $Cl(p\text{-}C_6H_4)Cl$, however.

Aliphatic *polyamides* PA are also known as nylons. They are characterized by the numbers i and j of carbon atoms in diamines and dicarboxylic acids, respectively.

Polyamide 66 (PA 66, PA 6.6) is therefore the polyamide from hexamethylene-diamine $H_2N(CH_2)_6NH_2$ and adipic acid $HOOC(CH_2)_4COOH$. Industrial aliphatic polyamides comprise PA 46, 66, 69, 610, 612, and 6.66. The acid components of aromatic polyamides are usually symbolized by capital letters, for example, T for terephthalic acid and I for isophthalic acid.

4.1.3. Equilibria of Bifunctional Polycondensations

Equilibrium Constants

Many (but not all!) polycondensations are equilibrium reactions. The two monomers A-X-A and B-Y-B react to form polymeric -X-Y- groups and leaving molecules AB. Equilibrium constants can be expressed by the molar concentrations of these entities or their mole fractions x, i.e.:

$$(4\text{-}6) \quad K = \frac{[XY][AB]}{[A][B]} = \frac{x_{XY}x_{AB}}{x_A x_B} = \frac{x_{XY}x_{AB}}{(1-x_{XY})^2}$$

At high degrees of polymerization, equilibrium constants K are independent of molar masses. Most equilibrium constants are small. At 280°C, an equilibrium constant of only $K = 0.42$ is observed for the polytransesterification of $HO(CH_2)_2OH +$ $CH_3O-OC(p-C_6H_4)CO-OCH_3$ to $-[OC(p-C_6H_4)CO-OCH_2CH_2O]_{\overline{n}}$ + CH_3OH. The direct polyesterification of $HOOC(p-C_6H_4)COOH$ and $HOCH_2CH_2OH$ delivers $K = 7.0$. The equilibrium constant $K = 300$ of the polycondensation of 11-amino undecanoic acid $HOOC(CH_2)_{10}NH_2$ is exceptionally high. All these polycondensations are mainly controlled by the enthalpy of polycondensation. The entropy contributions are small since the reaction of two molecules generates another two molecules (polymer and leaving molecule).

In low molar mass chemistry, concentrations of XY bonds can be enlarged by increases in the concentrations of A *or* B groups. This strategy is not possible in polymer chemistry. In AB polycondensations, A and B groups are both united in one monomer molecule. In AA/BB polycondensations with excess AA molecules, some A groups would not find a partner. If two moles of A-X-A are reacted with one mole of B-Y-B, then only A-X-Y-X-A + 2 AB would result at 100 % conversion of B groups. The composition A-X-Y-X-A is however an *average composition*. Since the reaction involves consecutive equilibria (see Eqn.(4-1)), larger molecules such as $A(XY)_2XA$, $A(XY)_3XA$, etc., will be formed in addition to $A(XY)_1XA$. This can only happen if some monomer molecules A-X-A remain unreacted. The number-average degree of polymerization is just $\overline{X}_n = 3$ (3 monomeric units X, Y and X in A-X-Y-X-A) but the distribution of mers in polymers ranges from $1 \le X \le \infty$.

Number-Average Degrees of Polymerization

Number-average degrees of polymerization can be calculated as follows from the initial molar ratio $r_0 \equiv n_{A,o}/n_{B,o}$ of amounts n of A and B groups and the extent of reaction of A groups, $p \equiv (n_{A,o} - n_A)/n_{A,o}$. In AA/BB polycondensations, each mono-mer molecule carries either two A groups or two B groups. The amount n_u of *mono-meric units* u in reactants (from monomers to polymers!) thus equals the total initial amount $n_{M,o}$ of both monomers M or one-half of the total initial amount of (A + B) groups: $n_u = n_{M,o} = (n_{A,o} + n_{B,o})/2$.

The instantaneous amount n_R of reactant molecules (monomers and polymers) is given by the initial amount of monomer molecules, $n_{M,o}$, minus the amount $n_{M,el}$ of monomer molecules that have been eliminated by reaction with another monomer or polymer molecule. Each of these reactions removes an A group. The amount $n_{M,el} = n_{A,o} - n_A$ thus equals the difference between the initial amount $n_{A,o}$ of A groups and the amount n_A that remains after an extent of reaction p of A groups (initial excess of B groups, if any). The number-average degree of polymerization of the *reactant molecules* ($1 \leq X \leq \infty$) is given by the ratio of amounts of units (n_U) to reactants (n_R):

$$(4\text{-}7) \qquad \overline{X}_n = \frac{n_U}{n_R} = \frac{n_{M,o}}{n_{M,o} - n_{M,el}} = \frac{(n_{A,o} + n_{B,o})/2}{[(n_{A,o} + n_{B,o})/2] - (n_{A,o} - n_A)} = \frac{1 + r_o}{1 + r_o - 2r_o p_A}$$

Eqn.(4-7) contains two special cases:

(1) $r_o \equiv 1$ for stoichiometric polycondensations; Eqn.(4-7) thus reduces to the **Carothers equation** $\overline{X}_n = 1/(1 - p_A)$. The number-average degree of polymerization is small at low extents of reaction and increases steeply at almost complete reaction (Fig. 3-1): $\overline{X}_n = 2$ at $p_A = 1/2$ (50 % conversion of A groups) but $\overline{X}_n = 100$ at $p_A = 0.99$ and $\overline{X}_n = \infty$ at $p_A = 1.000$. Equilibrium polycondensations require at least $p_A = 0.99\text{-}0.995$ for the syntheses of industrially useful aliphatic polyamides, etc.

(2) $r_o < 1$ for non-stoichiometric polycondensations (excess of B groups). If all A groups are completely reacted ($p_A \equiv 1$), Eqn.(4-7) becomes $\overline{X}_n = (1 + r_o)/(1 - r_o)$. The number-average degree of polymerization of reactants is limited to $\overline{X}_n = 199$ if $r_o = 0.99$ (1 % deviation from stoichiometry; $r_o \equiv 1$). It is lowered to $\overline{X}_n \approx 66.8$ if also $p_A = 0.99$. Since perfect stoichiometry is difficult to achieve by direct mixing of the two monomers, a monomer salt is often prepared first, isolated, and reacted later. An example is the so-called AH salt $[NH_3(CH_2)_6NH_3]^{2\oplus}[OOC(CH_2)_4COO]^{2\ominus}$ from hexamethylene diamine and adipic acid (also called nylon salt).

The desired degree of polymerization is difficult to obtain by stopping the poly-condensation at a certain extent of reaction since small variations in p_A can lead to large differences in $\overline{X}_n$. Unprotected end groups (= functional groups!) of polymers can furthermore continue to react under fabrication conditions (injection molding, extrusion, etc.); this causes melt viscosities to rise and hinders fabrication control. Polycondensations are therefore regulated by addition of monofunctional **regulators** Z (also called **chain terminators** or **stabilizers**). Eqn.(4-7) has then to be modified by adding the amount $n_{Z,o}$ of the regulator to both the numerator and denominator:

$$(4\text{-}8) \qquad \overline{X}_n = \frac{[(n_{A,o} + n_{B,o})/2] + n_{Z,o}}{[(n_{A,o} + n_{B,o})/2] - (n_{A,o} - n_A) + n_{Z,o}} = \frac{1 + (1/r_o) + (2n_{Z,o}/n_{A,o})}{1 + (1/r_o) + (2n_{Z,o}/n_{A,o}) - 2p_A}$$

One mole-percent of a monofunctional regulator ($n_{Z,o}/n_{A,o} = 0.01$) reduces the number-average degree of polymerization to 50.5 from 100 at an otherwise stoichiometric polycondensation at an extent of reaction of 99 % ($p_A = 0.99$).

Distributions of Degrees of Polymerization

Polycondensations result in mixtures of polymer molecules with different degrees of polymerization. The distribution of these molecules can be calculated by simple statistical methods. In *stoichiometric* polycondensations and polyadditions, two functional groups A and B are reacted with a probability $p_A = p_B = p$. Probabilities must be multiplied: the probability for 3 connections of 4 molecules is p^3 and that for the formation of a molecule with a degree of polymerization of $X_i = i$ equals p^{i-1}.

The probability P_i for the appearance of *one* polymer molecule with the degree i of polymerization is given by the product of the probabilities p^{i-1} for the $i-1$ bonds and the probability $1 - p$ for the appearance of unreacted end groups; it is thus $P_i = (1 - p)p^{i-1}$. The amount n_i of *all* molecules with the degree of polymerization i is calculated from the probability P_i for the appearance of one molecule and the amount n_R of all reactants. One obtains $n_i = P_i n_R = n_R(1 - p)p^{i-1}$. The mole fraction x_i of molecules with a degree of polymerization, X_i, in the reactant is thus:

$$(4\text{-}9) \qquad x_i \equiv n_i/n_R = P_i n_R/n_R = (1 - p)p^{i-1}$$

The mass fraction w_i of molecules with a degree of polymerization, X_i, is given by the ratio of the mass m_i of all i molecules to the mass m_R of all reactant molecules. Masses are products of amounts n and molar masses M ($m = nM$). Molar masses are calculated from the X_i monomeric units with a molar mass M_u each and the sum of the molar masses M_e of the two leavings groups, i.e., the molar mass M_L of the leaving molecules. With these definitions, the Carothers equation $\overline{X}_n = 1/(1 - p)$, and Eqn.(4-9), one obtains:

$$(4\text{-}10) \qquad w_i = \frac{m_i}{m_R} = \frac{n_i M_i}{n_R(\overline{M}_R)_n} = \frac{n_i(M_L + M_u X_i)}{n_R(M_L + M_u(\overline{X}_R)_n)} = \frac{M_L + M_u X_i}{M_u + M_L(1 - p)}(1 - p)^2 p^{i-1}$$

For negligibly small molar masses of leaving molecules ($M_L \to 0$) (and *not* for infinitely high molar masses of polymers, $M_u X_i \to \infty$!), Eqn.(4-10) is reduced to:

$$(4\text{-}11) \qquad w_i \approx X_i(1 - p)^2 p^{i-1}$$

Eqn.(4-11) is exact for polyadditions. For polycondensations, Eqn.(4-10) and the conventional Eqn.(4-11) may differ considerably. For example, the mass fractions of residual phenyl-p-hydroxybenzoate at $p = 0.99$ are calculated as $w_i = 1.78 \cdot 10^{-4}$ (Eqn.(4-10)) and $w_i = 1.00 \cdot 10^{-4}$ (Eqn.(4-11)), respectively.

The function $x_i = f(X_i)$ is universal and not specific for a particular monomer (Fig. 4-1). Mole fractions x_i of species i decrease with increasing degree of polymerization at constant p. The function $w_i = f(X_i)$ is universal for polyadditions but not for poly-

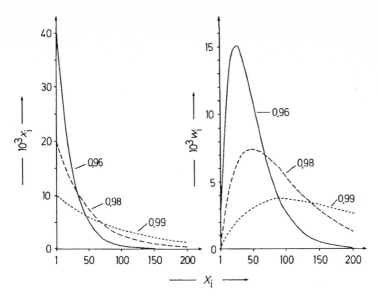

Fig. 4-1 Continuous differential distributions of mole fractions x_i and mass fractions w_i of degrees of polymerization $X_i = i$ at $p = 0.96$, 0.98 and 0.99, respectively. Molar distributions apply to both polycondensations and polyadditions; the mass distributions of Fig. 4-1 were calculated for the polycondensation of $HO-C_6H_4-COOC_6H_5$ ($M_u = 120.10$ g/mol; $M_L = 95.11$ g/mol).

condensations where it depends on the chemical structures of monomer and leaving molecules. Unlike number distributions, mass distributions show maxima. These functions are described by Schulz–Flory distributions (Section 2.2.3.).

Number-average degrees of polymerization are defined as $\overline{X}_n = \Sigma_i \, x_i X_i$. Introduction of Eqn.(4-9) delivers $\overline{X}_n = (1-p) \, \Sigma_i \, p^{i-1} X_i$. The sum is then developed into a series $\Sigma_i \, p^{i-1} X_i = 1 + 2\,p + 3\,p^2 + \ldots = 1/(1-p)^2$ which lets one recover the Carothers equation $\overline{X}_n = 1/(1-p)$ for both polycondensations and polyadditions.

Mass-average degrees of polymerization $\overline{X}_w \equiv \Sigma_i \, w_i X_i = (1-p)^2 \, \Sigma_i \, p^{i-1} X_i^2$ are obtained from Eqn.(4-11). The sum can be developed into a series $\Sigma_i \, p^{i-1} X_i^2 = 1 + 2^2 \, p + 3^2 \, p^2 + \ldots = (1+p)/(1-p)^3$ for $p < 1$. At any p, the mass-average degree of polymerization is thus $\overline{X}_w = (1+p)/(1-p) = 2\overline{X}_n - 1$ and the molar mass ratio $\overline{X}_w/\overline{X}_n = 2 - (1/\overline{X}_n)$. These equations are universal for polyadditions and also for polycondensations with $M_L \to 0$. Somewhat more complicated expressions are obtained if one cannot neglect the molar mass of leaving molecules. All average degrees of polymerization include the monomer, contrary to chain-growth polymerizations.

4.1.4. Kinetics of Bifunctional Reactions

Many polycondensations and polyadditions are not controlled by thermodynamics but by kinetics. Such irreversible reactions of the AB and AA/BB type are either catalyzed by deliberately added catalysts ("external catalysis") or by the functional groups of the monomers themselves ("internal catalysis", "self-catalysis"). The poly-esterification of dicarboxylic acids HOOC-X-COOH and diols HO-Y-OH is, for example, self-catalyzed by COOH groups if no external catalyst is added.

A "self-catalysis" seems to contradict the definition of a catalyst which should be unchanged at the completion of a reaction. COOH groups are consumed, however. A close inspection of such a self-catalyzed polyesterification shows that the reaction of a COOH group of a molecule I is catalyzed by a second COOH group from the same molecule I or from another molecule II. An intermediate is formed consisting of two COOH groups and one OH group. If one COOH group and the OH group are united to form an ester group -CO-O-, the other COOH group is released intact. It does not matter that this group may be reacted later to an ester group: it leaves unchanged that reaction in which it acted as a catalyst. Such reactions are thus not uncatalyzed.

In the simplest case, a reaction is of 3rd order with respect to all participants. For stoichiometric reactions ($[A]_o = [B]_o$ and thus $[A] = [B]$), the decrease of A groups with time is given by:

$$(4\text{-}12) \quad -d[A]/dt = k_3[A][B][C] = k_3[A]^2[C]$$

In **self-catalyzed polyadditions**, A-groups act as catalysts C and $[C]$ becomes $[A]$. Eqn.(4-12) thus becomes $-d[A]/dt = k_3[A]^3$ and, after integration, $1/[A]^2 = (1/[A]_o^2) + 2\,k_3 t$. After insertion of the extent of reaction, $p_A = ([A]_o - [A])/[A]_o$, and the Carothers equation, $\overline{X}_n = (1 - p_A)^{-1}$, one obtains for internally catalyzed reactions:

$$(4\text{-}13) \quad \overline{X}_n^2 = 1/(1 - p_A)^2 = 1 + 2\,k_3[A]_o^2 t$$

In externally catalyzed polyadditions, the concentration of the catalyst remains unchanged ($[C] = [C]_o$). Uniting $[C]_o$ with k_3, Eqn(4-12) thus becomes:

$$(4\text{-}14) \quad \overline{X}_n = 1/(1 - p_A) = 1 + k_3[C]_o[A]_o t = 1 + k_2[A]_o t \; ; \quad k_2 \equiv k_3[C]_o$$

Eqns.(4-13) and (4-14) are exact for polyadditions but only approximations for **polycondensations** (except for $M_L \to 0$) as one can see from the following considerations. The molar concentration $[A] = n_A/V$ is defined as the ratio of amount of substance A, n_A, to the total volume V of the reacting mixture. This volume is affected by the presence of A-groups since one obtains $V = V_o - [(n_{A,o} - n_A)M_L/\rho_L]$ for volume-invariant mixtures (ρ_L = density of leaving substance). The molar concentration $[A]$ thus contains the variable n_A in both the numerator and the denominator. Since variables must be separated *before* integration, one cannot proceed as in the derivation of Eqns.(4-13) and (4-14). One rather obtains more complicated equations which, for example, lead to an apparent induction period and a wrong value for $k_2[A]_o$ if the experimental data are plotted according to Eqn.(4-14). Direct plots of $\overline{X}_n = f(t)$ according to Eqn.(4-14) or $\overline{X}_n^2 = f(t)$ according to Eqn.(4-13) are often ambiguous since various reaction intervals are weighted differently. An increase of the number-average degree of polymerization from $\overline{X}_n = 1$ to $\overline{X}_n = 2$ amounts to a change from $p = 0$ to $p = 0.5$, but one from $\overline{X}_n = 100$ to $\overline{X}_n = 101$ corresponds to a change from $p = 0.99$ to $p = 0.990099...$. Logarithmic plots are thus more expedient, e.g., for Eqn.(4-13):

(4-15) $\log_{10} [\bar{X}_n^2 - 1] = \log_{10} [2 \, k_3[A]_o^2] + \log_{10} t$

The self-catalyzed polycondensation of succinic acid $HOOC(CH_2)_2COOH$ and ethylene glycol $HO(CH_2)_2OH$ is a 3rd order reaction for the total range of reaction $(0.09 \leq p \leq 0.82)$ according to Fig. 4-2. The self-catalyzed polycondensation of adipic acid $HOOC(CH_2)_4COOH$ and diethylene glycol $HO(CH_2)_2O(CH_2)_2OH$ shows 3rd order regions for small times (low p) and large times (high p) but not for intermediate p (behavior caused by complexation of the ether groups?).

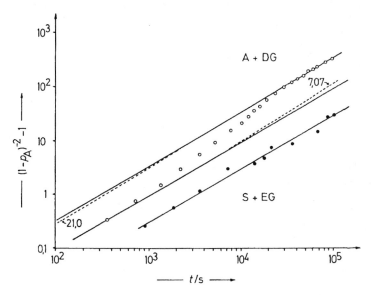

Fig. 4-2 Self-catalyzed, stoichiometric polycondensations of (O) adipic acid A and diethylene glycol DG at 166°C [1] and (●) succinic acid S and ethylene glycol EG at 123°C [2]. Solid lines: theoretical slopes of unity for self-catalyzed 3rd order polyadditions; broken lines: calculated with $k_3/(g^2 \, mol^{-2} \, s^{-1}) = 21.0$ and 7.07, respectively, for self-catalyzed 3rd order polycondensations.

4.1.5. Diffusion-Controlled Polycondensations

Rates of polycondensations are strongly affected by the chemical structure of functional groups. The syntheses of aliphatic polyamides by direct amidation of carboxylic groups (Eqn.(4-1)) are very slow, even at 250°C, because of the resonance stabilization of COOH groups. Polyamide formation from diamines and diesters is much faster and that of diamines and dicarboxylic acid dichlorides:

(4-16) $H_2N–Z–NH_2 + ClOC–Z'–COCl \longrightarrow \{NH–Z–NH–CO–Z'–CO\}_n + 2 \, HCl$

is so fast that it proceeds at room temperature in minutes.

This *Schotten–Baumann* reaction can also be performed in a heterogeneous phase reaction as **interfacial polyconcensation**. The two monomers are dissolved in two immiscible solvents, for example, diamines in water and dicarboxylic acid dichlorides in chloroform. On layering the aqueous solution above the more dense chloroform solution, a polymer film (or powder, depending on polymer constitution) is formed at the interace. The locus of the reaction is at the side of the film facing the organic solution. The polycondensation stops if the newly formed polymer is not continuously removed.

Interfacial polycondensations are diffusion controlled since the rate of diamine diffusion through the film is slower than the rate of polycondensation. In this kinetically controlled reaction, the ratio of functional groups thus does not need to be stoichiometric as in equilibrium polycondensations. The maximum molar mass is obtained at a ratio [diamine]$_o$/[dicarboxylic acid dichloride]$_o$ < 1, probably because some acid chloride is hydrolyzed. This optimal ratio is controlled by the distribution coefficient of the diamine between aqueous and organic phases.

Schotten–Baumann reactions are used industrially for (1) the synthesis of bisphenol A polycarbonate from $NaO(p\text{-}C_6H_4\text{-}C(CH_3)_2\text{-}(p\text{-}C_6H_4)ONa$ and $COCl_2$, (2) the formation of aromatic polyamides (aramides), and (3) the anti-felt treatment of wool. The latter reaction consists of an interfacial polycondensation of sebacic acid dichloride and hexamethylene diamine (to PA 610) on the surface of the wool.

Aramides are defined as long-chain synthetic polyamides where at least 85 % of the amide groups are bound to aromatic rings. Interfacial polycondensations are industrially used for the syntheses of (a) poly(p-phenylene terephthalamide) from $ClOC(p\text{-}C_6H_4)COCl$ and $H_2N(p\text{-}C_6H_4)_2NH_2\cdot 2$ HCl in hexamethylphosphoric triamide or N-methylpyrrolidone-CaCl$_2$ (Kevlar®) and (b) poly(m-phenylene isophthalamide) in N,N-dimethylacetamide with trimethylamine hydrochloride as catalyst and NaOH as HCl acceptor (Nomex®).

4.1.6. Activated Polycondensations

AB polycondensations cannot employ highly reactive monomers due to their low shelf life. Sluggishly reactive AB monomers can be subjected to **activated polycondensations**. In these reactions, activators and monomers form highly reactive intermediates; activators are consumed, however.

Neat p-aminobenzoic acid $H_2N(p\text{-}C_6H_4)COOH$ polycondenses only in the melt albeit with degradation because of high reaction temperatures. However, it reacts in solution at room temperature if thionyl chloride is added. $H_2N(p\text{-}C_6H_4)COOH$ and $SOCl_2$ form an intermediate, $OSN(p\text{-}C_6H_4)COCl$, which is converted into p-aminobenzoylchloride hydrochloride. The latter undergoes polycondensation:

(4-17) $H_2N(p\text{-}C_6H_4)COOH + 2\ SOCl_2 \longrightarrow OSN(p\text{-}C_6H_4)COCl + SO_2 + 3\ HCl$

$OSN(p\text{-}C_6H_4)COCl\ +\ 3\ HCl \longrightarrow [H_3N(p\text{-}C_6H_4)COCl]^+Cl^-\ +\ SOCl_2$

$[H_3N(p\text{-}C_6H_4)COCl]^+Cl^- \longrightarrow \sim NH(p\text{-}C_6H_4)CO\sim\ +\ 2\ HCl$

Polymers obtained by activated or diffusion-controlled polycondensations and by ring-opening polymerization exhibit molar mass distributions that differ from those obtained by equilibrium polycondensations. These molar mass distributions try to equilibrate by **exchange reactions** between chain segments. An example is the exchange reaction of poly(dimethyl siloxane)s from the ring-opening polymerization of hexamethylcyclotrisiloxane or octamethylcyclotetrasiloxane:

$$(4\text{-}18) \quad \sim\!\!\sim[SiR_2-O]_i \quad + \quad [SiR_2-O]_m\!\!\sim\!\!\sim \quad \rightleftharpoons \quad \sim\!\!\sim[SiR_2-O]_i-[SiR_2-O]_m\!\!\sim\!\!\sim$$

$$\sim\!\!\sim[O-SiR_2]_j \qquad [O-SiR_2]_n\!\!\sim\!\!\sim \qquad + \qquad \sim\!\!\sim[O-SiR_2]_j-[O-SiR_2]_n\!\!\sim\!\!\sim$$

Exchange reactions do not modify the number-average molar mass since the number of molecules remains constant. Two molecules with the same degree of polymerization are, however, converted into one molecule with a higher molar mass and one with a lower one, etc. Narrow molar mass distributions are thus broadened by exchange reactions until finally the equilibrium value of $\overline{X}_w = 2\overline{X}_n - 1$ is obtained. Very broad molar mass distributions are correspondingly narrowed.

Exchange reactions are always catalyzed, either by catalyst residues from polycondensations or by catalytic groups on the macromolecules themselves. Polyamides rarely undergo exchange reactions by transamidation (analog to Eqn.(4-18)); they rather equilibrate via an acid-catalyzed aminolysis of amide groups by amino end groups:

$$(4\text{-}19) \quad \sim\!\!\sim X-CO-NH-Y\sim\!\!\sim \quad \rightleftharpoons \quad \sim\!\!\sim X-CO \quad + \quad H_2N-Y\sim\!\!\sim$$

$$+ \qquad\qquad\qquad\qquad\qquad\qquad |$$

$$\sim\!\!\sim X-NH_2 \qquad\qquad\qquad\qquad\qquad \sim\!\!\sim X-NH$$

This exchange reaction is absent if the amino end groups are capped by regulators, e.g., by acetic acid, acetic anhydride, or ketene.

4.1.7. Cyclizations

Two types of cyclization reactions exist: intramolecular and intramerar. Intramolecular reactions occur between the *chain ends* of linear oligomer and polymer molecules; they lead to cyclic *molecules*. Intramerar reactions proceed within a *mer* or a *repeating unit* of a molecule; they generate cyclic *units* within otherwise linear macromolecules.

Intramolecular Cyclizations

These reactions are usually the most important side reactions in AB and AA/BB polycondensations to linear macromolecules. The proportion of ring molecules is controlled by the average distance between A and B groups.

At an infinitely high concentration of very long, infinitely thin chains, only B-groups of other molecules reside near A-groups: The reaction is exclusively inter-molecular to linear chains. Real chains are however not infinitely thin: the concentrations of A- and B-groups are not "infinitely high", not even in the melt. The monomeric units act rather as diluents. Melt condensations therefore always deliver small proportions of ring molecules (see Section 3.2.5.) that increase with increasing temperature. An example is the polycondensation of AH salt to PA 66 where the proportion of cyclic compounds is 1.9 % at 275°C, 4.3 % at 297°C, and 5.9 % at 310°C.

At infinitely high dilution, on the other hand, B-groups next to A-groups are always those from the same molecule. Reactions are exclusively *intra*molecular if ring closures are sterically possible. This Ruggli–Ziegler dilution effect is utilized for cyclopolycondensations and for the syntheses of cyclic oligomers and polymers.

Cyclic dimers and oligomers are of interest as monomers for ring-opening polymerizations since such polymerizations lead to the same polymer structures as the corresponding polycondensations but they (1) do not generate leaving molecules, (2) provide higher polymerization rates, and (3) lead to polymers with narrower molar mass distributions. Industrially even more advantageous is the direct synthesis of cyclic polymers with high molar masses for example, cyclic bisphenol A polycarbonates. Such polymers do not have end groups that may lead to **chemical aging**, i.e., the undesired change of useful properties by chemical reactions.

***Intra*molecular ring formation** is achieved by high *local* dilution, i.e., by the dropwise addition of AB monomers to a large excess of solvent. This way, the bis-chloroformic acid ester of bisphenol A delivers, with $(C_2H_5)_3N$ as catalyst in CH_2Cl_2 at 0°C, cyclic oligocarbonates:

Degrees of polymerization of these cyclic oligomers vary between 2 and 26. Yields are controlled by the bases: $(C_2H_5)_3N$ delivers 85 %, $(C_4H_9)_3N$ 50 %, $(C_2H_5)_2NCH_3$ 27 %, and pyridine 0 %.

Intramerar Cyclizations

Cyclopolycondensations are polycondensations of monomers with functionalities of three or more that result in "linear" (often slightly branched) polymers by intra-merar cyclizations, i.e., cyclizations within a mer. A suitable process control avoids *inter*molecular reactions to cross-linked polymers that cannot be processed:

The first steps of cyclopolycondensations are always polyadditions to linear oligo-mers (in Eqn.(4-21) shown as step I of the dimer formation). In highly polar solvents such as N,N-dimethyl formamide, dimethyl sulfoxide, etc., pyromellitic acid an-hydride reacts with 4,4'-diaminodiphenyl ether to form a polyamic acid, mainly in the *para* position. In order to avoid premature cross-linking, solids contents of solutions are restricted to ca. 15 % and polymer yields to 50 % in step I.

(4-21)

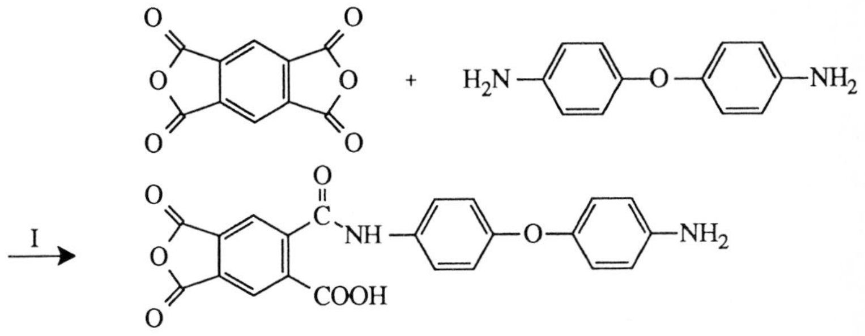

The resulting polymer solutions are raw materials for laminating resins. Step II con-sists of intramerar cyclizations to imide units (a condensation reaction, shown in Eqn.(4-21a) for the dimer), further polyadditions and polycondensations, and simul-taneous shaping of the material. Since some of the amic acid groups react intermole-cularly, the resulting polyimides are not linear (as shown) but cross-linked; they are thermosets:

(4-21a)

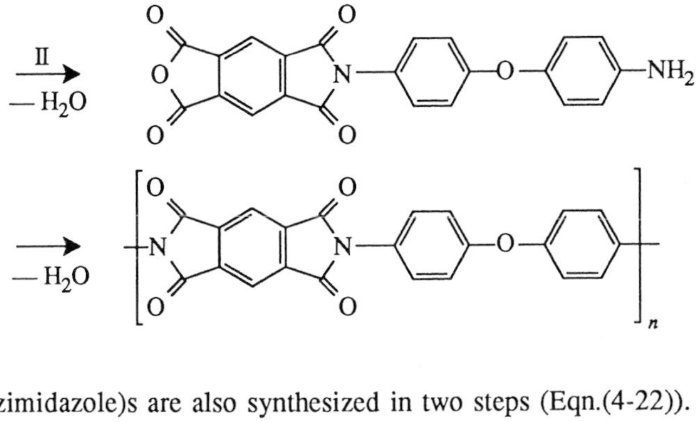

Poly(benzimidazole)s are also synthesized in two steps (Eqn.(4-22)). Diphenyliso-phthalate and 3,3'-diaminobenzidine tetrahydrochloride are the monomers preferred by industry. The tetrahydrochloride is employed instead of the free amine because the former is more stable against oxidation. The diphenyl ester is used because (a) free acids decarboxylate at high reaction temperatures, (b) acid chlorides react too fast, and (c) methyl esters will partially methylate amino groups. The reaction mixture foams due to the formation of volatile phenol. It delivers a prepolymer that is isolated and pulverized. The prepolymer is then polycondensed in nitrogen at $(265-425)°C$ to poly(benzimidazole), using $(5-50)$ % phenol as plasticizer.

(4-22)

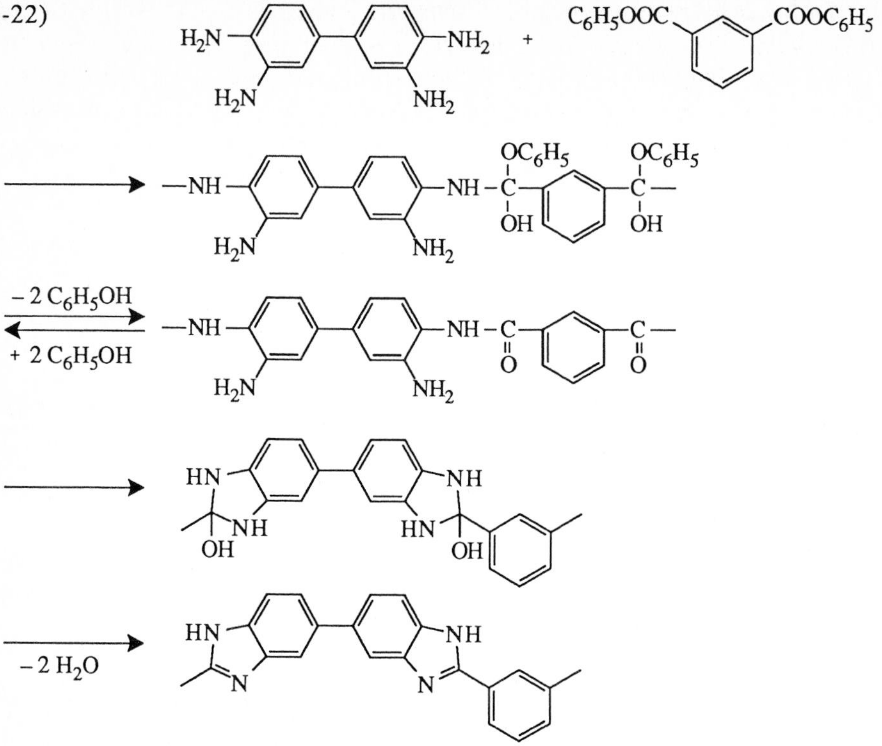

Poly(benzimidazole)s are stable up to ca. 500°C because they (a) contain only small proportions of oxidizable hydrogen ligands, (b) possess only small proportions of hydrolyzable groups, and (c) have the advantage that a scission of one bond in a heterocycle does not lead to cleavage of the whole chain.

The same principles have been utilized for other, even more thermally stable polymers. Examples are "Pyrron" = Poly(imidazopyrrolone) and BBB = Poly(benzimidazobenzophenanthroline); both are thermally stable up to ca. 600°C:

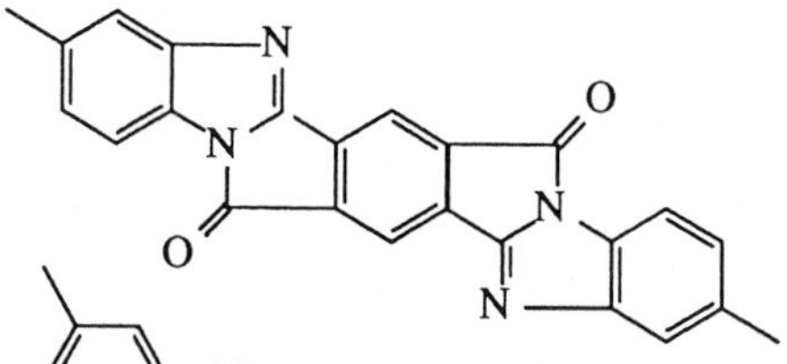

Pyrron

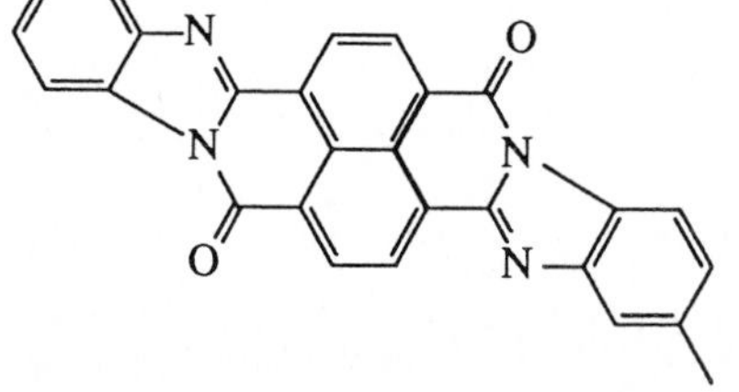

BBB

4.1.8. Polyadditions

Polyadditions consist of repeated additions of functional groups to double bonds or rings ("potential double bonds"). An **AB polyaddition to double bonds** is the self-addition of acrylic acid:

(4-23) $n\, CH_2=CH-COOH \longrightarrow +CH_2-CH_2-CO-O+_n$

The resulting poly(β-propionic acid) reverts to acrylic acid upon heating; this process is used for the purification of acrylic acid.

Another linear AB polyaddition to double bonds is the reversible **Diels–Alder polymerization**, for example:

(4-24)

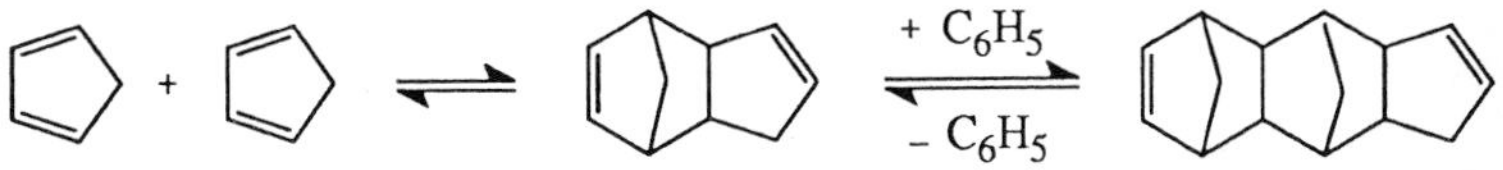

Bismaleimides add A-X-A monomers in linear **AA/BB polyadditions** to form "thermoplastic polyimides", i.e., polyimides that are not cross-linked as are the polyimides of Eqn.(4-21). Depending on functional groups A, polyimines (from NH_2), polythioethers (from SH), polyethers (from phenolic OH), or polyamides (from aldoxime groups CH=NOH) are obtained, for example, from diamines:

(4-25)

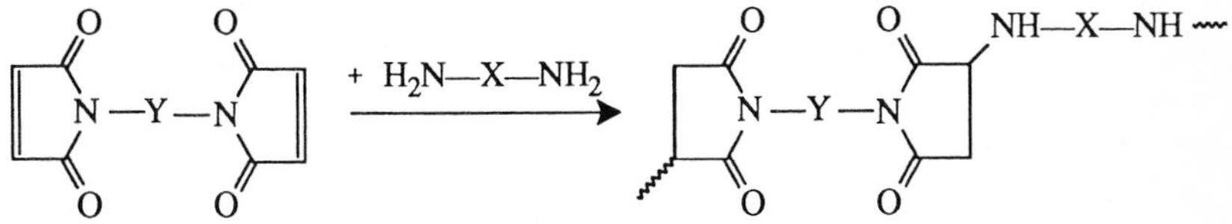

Industrially much more important than linear AA/BB polyadditions are polyadditions of multifunctional monomers to branched or cross-linked polymers, especially to polyurethanes and epoxy polymers.

Polyurethanes PUR
Polyurethanes are obtained by repeated addition of hydroxyl-group-containing monomers to isocyanates (~OH + O=C=N~ → ~O–OC–NH~). This reaction was originally conceived in 1937 by Otto Bayer as *diisocyanate addition* for the synthesis of linear polyurethanes $+O(CH_2)_4O-OC-NH(CH_2)_6NH-CO+_n$ from hexamethylenediisocyanate $OCN(CH_2)_6NCO$ and 1,4-butanediol $HO(CH_2)_4OH$. These linear polyurethanes were supposed to compete with PA 66, $+NH(CH_2)_6NH-OC(CH_2)_4CO+_n$, which had been discovered shortly before by W.H.Carothers; they are no longer produced. Linear polyurethanes are presently manufactured from MDI, hydroxyl-terminated polyesters, and glycol chain extenders with a 1:1 ratio of NCO and OH groups. These polyurethanes are thermoplastic elastomers that can be processed like elastomers and have properties like elastomers (see Chapter 12).

The majority of industrial polyurethanes are however branched or cross-linked polymers from diisocyanates and polyols. Toluylendiisocyanate TDI (an 8/20 mixture of the 2,4- and 2,6-diisocyanates) and diphenylmethanediisocyanate MDI comprise 95 % of the industrially used isocyanate monomers. Other isocyanates are used less frequently, especially 1,5-naphthalenediisocyanate NDI, *m*-xylylenediisocyanate XDI and isophoronediisocyanate IPDI, the "polymeric" MDI (= PMDI), and other multifunctional isocyanates that contain urea or biuret groups.

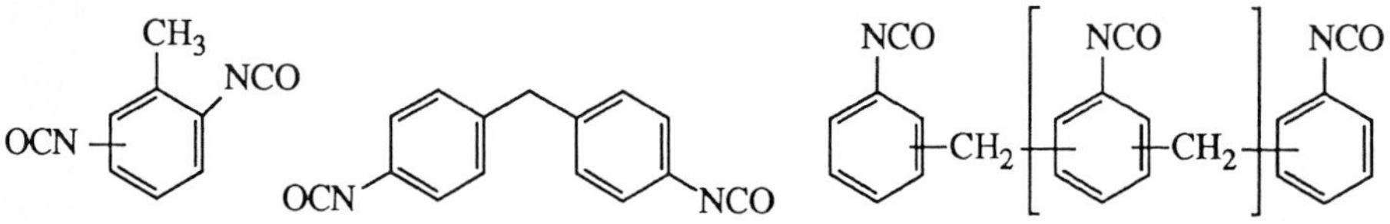

TDI   MDI   PMDI

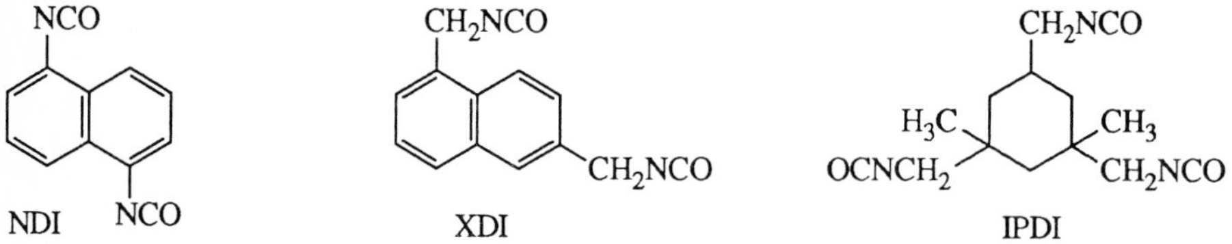

NDI   XDI   IPDI

More than 90 % of the industrially used polyols are so-called polyether-polyols. Simple bifunctional polyether-polyols are bifunctional polyethers with hydroxy end groups, for example, poly(ethylene glycol) $H+OCH_2CH_2\}_n OH$, poly(propylene glycol) $H+OCH(CH_3)CH_2\}_n OH$ and poly(tetrahydrofuran) $H+O(CH_2)_4\}_n OH$, all usually with molecular weights up to a few thousand. Higher functional polyols comprise glycerol ($f = 3$), pentaerythritol ($f = 4$), and saccharose ($f = 8$). The so-called modified polyether-polyols consist of hydroxyl-group-containing compounds that are grafted onto polymer particles. Polyesters with hydroxy end groups are also used.

Expanded plastics (foamed plastics) consume most of the isocyanates. Rigid PUR foams possess only short, rigid segments between cross-links. They are manufactured from short-chain diols, e.g., PMDI, and a physical blowing agent (often $CFCl_3$, nowadays CH_2Cl_2 or even CO_2). The heat of polyaddition causes the blowing agent to vaporize. The polymer and its cell structure are formed simultaneously.

Flexible PUR foams are prepared by chemical foaming, i.e., by addition of carefully controlled amounts of water to the mixture of (mainly) TDI and polyether-polyols. Water reacts with isocyanate groups according to ~NCO + H_2O → ~NH_2 + CO_2. Carbon dioxide foams the polyurethane; amine groups react with isocyanate groups and give urea groups according to ~NCO + H_2N~ → ~NH-CO-NH~.

An excess of isocyanate groups converts urea groups into biuret groups, urethane groups into allophanate groups, and amide groups into acylurea groups:

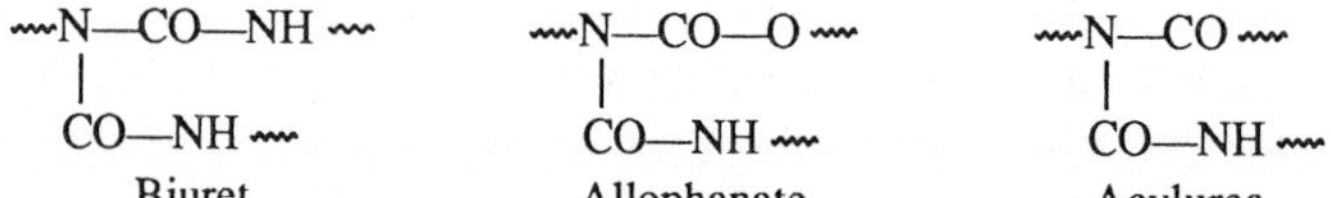

Biuret   Allophanate   Acylurea

Polyurethane elastomers are segmented polymers containing rigid and flexible segments. Rigid ("hard") segments have structures similar to those of foamed PUR plastics. Flexible ("soft") segments are polyether-polyols with low glass temperatures.

PUR elastomers are synthesized in two steps. In the first step, polyols are reacted with excess diisocyanates to give "extended diisocyanates" (i.e., polyols with isocyanate end groups).

The second step is the cross-linking of these extended diisocyanates:

a. Short chain diols added in less than stoichiometric amounts convert isocyanate groups into urethane groups which then react with the remaining isocyanate groups to form allophanate groups. At temperatures above 150°C, allophanate formation is reversed: a polymer that is cross-linked at room temperature can thus be processed at elevated temperatures like a thermoplastic. On cooling, a thermoset is formed again.

b. Addition of aromatic diamines in less than stoichiometric amounts converts isocyanate groups into urea groups which react with the remaining isocyanate groups to give biuret groups.

c. A careful control of stoichiometry (small excess of isocyanate groups because of side reactions) results in "linear" and weakly cross-linked polymers that serve as elastic fibers (Spandex fibers), for example:

Polyurethanes are chemically stable, flexible, and abrasion resistant. Besides foams and elastomers, they also furnish excellent surface coatings and good adhesives.

Epoxy Resins EP

Epoxy resins are oligomers with two or more epoxide groups per molecule, they often have hydroxyl groups. More than 90 % of the world production consists of epoxies from bisphenol A and epichlorohydrin with the idealized structure:

Epoxy resins with *average* repeating units of $0.1 < n < 0.6$ are liquids; they consist of a mixture of trimers (1 bisphenol unit and 2 epoxy units; $n = 0$) and higher oligomers ($n \geq 1$). Epoxy resins with an average of $n > 2$ are solids. Other epoxy resins contain cycloaliphatic or heterocyclic units instead of bisphenol A units. Industrial epoxy resins are always formulated, i.e., they contain diluents, plasticizers, fillers, pigments, etc. (see Chapter 11).

Epoxy resins are cured (i.e., cross-linked) at room temperature by polyfunctional amines and at elevated temperatures by carboxylic acid anhydrides. The *cold hardening* by diethylene diamine, 4,4'-diaminodiphenylmethane, etc., opens the epoxy ring:

(4-26) $R—NH_2$ + H_2C——$CH—CH_2$ ⌇⌇⌇   ⟶ $R—NH—CH_2—CH—CH_2$ ⌇⌇⌇
                              \ / |
                               O   OH

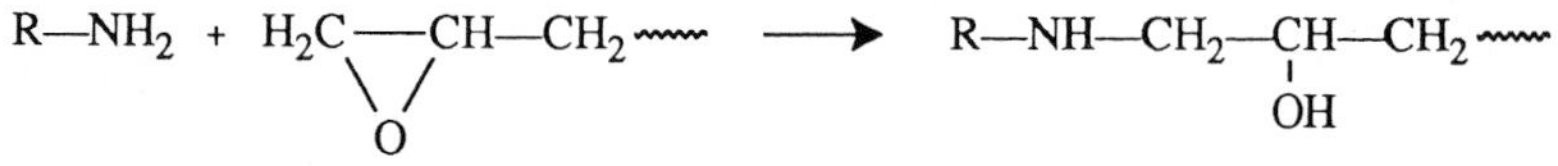

which is followed by the intermolecular cross-linking reaction between hydroxyl and epoxide groups.

The *warm hardening* by, e.g., phthalic acid anhydride at (80-100)°C, delivers ester structures:

(4-27)

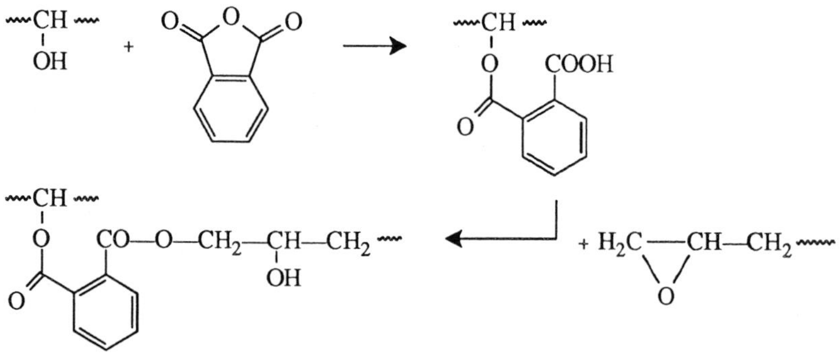

and, by addition of ~OH to protonated oxirane rings and subsequent ring-opening, also ether bonds:

(4-28)

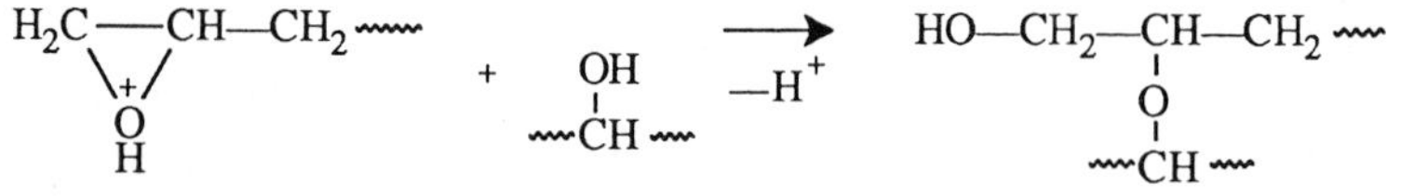

4.1.9. Branching Polycondensations

Multifunctional polycondensations and polyadditions are polymerization reactions in which at least one of the monomeric species has a functionality of three or higher. Examples are the self-condensation or self-addition of A groups of monomers A_3, the self-polycondensation or self-polyaddition of A and B groups of AB_2, the polycondensation or polyaddition of A_2 with B_3, A_3 with B_3, etc. Depending on the functionality of the monomer, the stoichiometry of the reaction, and the monomer conversion either branched or cross-linked polymers are obtained. No cross-linking at any monomer conversion is obtained if AB_f monomers alone or AB_f + AB is reacted where $f \geq 2$.

Hyperbranched Polymers

Monomers of the type AB_f (with $f \geq 2$) deliver alone or in a mixture with AB so-called **hyperbranched molecules** with irregular structure of the arms. Polycondensation of, e.g., AYB_2 with a trifunctional branch points Y and reactive groups A and B results in (schematically):

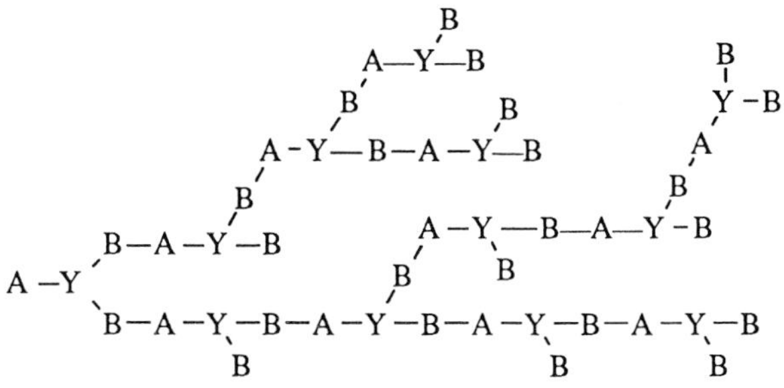

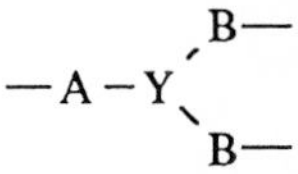

The reaction delivers branched molecules at *all* monomer conversions; the polymers never become cross-linked. The molecules possess many end groups B and are non-uniform with respect to composition and constitution and thus also to molar mass. They can be imagined as irregular comb polymers with subsequent branching of the side chains; the molecule shown above can also be depicted as:

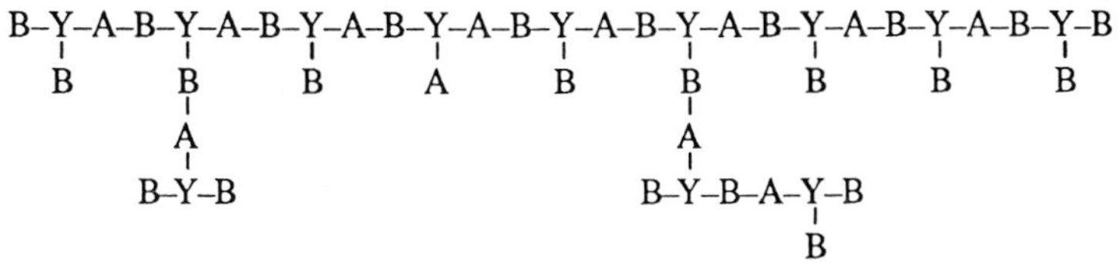

Dendritic Polymers

Dendritic polymers, on the other hand, are not synthesized by one-pot reactions as the hyperbranched polymers but by stepwise condensation or addition of one type of species after the other (see Section 2.1.5): they are not multifunctional polycondensations or polyadditions but a succession of bifunctional reactions of bifunctional or multifunctional monomers. For divergent syntheses of each new generation, monomers are usually reacted in great excess and the unreacted monomers subsequently removed. This strategy generates equal branch lengths and equal segments between branch points and few constitutional mistakes: dendritic polymers are very uniform with respect to composition and constitution.

Gel Point

All other monomer species $A_f B_g$ ($f \geq 2$; $g \geq 2$), $A_f B_g$ + AB, etc., deliver cross-linked polymers. The viscosity increases strongly with time if such species undergo polycondensation (Fig. 4-3). The system finally becomes so viscous that the stirrer stops and bubbles of water vapor no longer rise to the surface. The transition from a viscous liquid to a "solid" gel occurs in such a small time interval that one is entitled to speak of a **gel point**. At this gel point, not all functional groups have reacted ($p <$ 1). The gel not only contains a cross-linked polymer but also soluble branched or linear oligomers until the extent of reaction becomes unity.

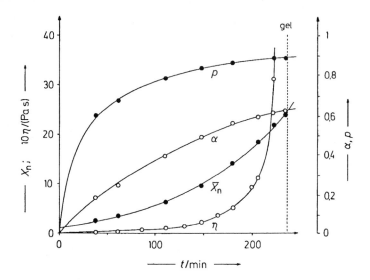

Fig. 4-3   Time dependence of viscosity η, number-average degree of polymerization of the soluble part, extent of reaction p, and branching coefficient α during the polycondensation of diethylene glycol and a mixture of succinic acid and tricarballylic acid at 109°C, catalyzed by p-toluene sulfonic acid. $x_{A,3} = 3$ [Tricarballylic acid]/[COOH] = 0.404; $r_0 = $ [COOH]/[OH] = 1.002. gel = Experimental gel point at $p = 0.894$. With permission of the American Chemical Society [3].

The reaction first generates branched molecules. With increasing extent of reaction, the average molar mass of these molecules increases and so does the number of functional groups per molecule. The more functional groups present per molecule, the greater is the likelihood of reaction of that molecule. Only a very small incremental reaction of the functional groups is necessary at the gel point to generate an "infinitely large", insoluble polymer molecule that extends from one reactor wall to the other. Not all initially present molecules are united to such a giant molecule at the gel point, however. Some soluble, low molar mass reactants remain (in Fig. 4-3, $\overline{X}_n$ of the soluble part is only 24.2 at the gel point).

Gel points are critical points that can be evaluated by a number of different methods. The classical statistical theory calculates the probability of certain bonds or units. The reaction of B-Z-B + A-X-A + YA_3 (corresponds to Fig. 4-3) leads to segments between branching points that possess the general structure (X is here bifunctional!)

$$\begin{matrix} \text{\tiny www} A \\ \diagdown \\ \diagup \\ \text{\tiny www} A \end{matrix} Y-(A-B)-[Z-(B-A)-X-(A-B)]_i-Z-(B-A)-Y \begin{matrix} A \text{\tiny www} \\ \diagup \\ \diagdown \\ A \text{\tiny www} \end{matrix} \quad ; \quad 0 < i < \infty$$

$$ \text{I} \qquad\qquad \text{II} \qquad\qquad \text{III} \qquad\qquad \text{IV}$$

The probability of the presence of bonds I equals the extent of reaction of A groups, p_A. The probability of the presence of bonds II is determined by the extent of reaction of B groups, p_B, and the mole fraction $x_{A,2} \equiv 1 - x_{A,3}$ of A groups that reside in molecules A-X-A ($x_{A,3}$ = mole fraction of A groups in molecules YA$_3$). The total probability for i bonds of type II is thus $[p_B(1-x_{A,3})]^i$. Correspondingly, probabilities are $(p_A)^i$ for type III bonds and $p_B x_{A,3}$ for type IV bonds. The probability for the total segment (the "repeating unit" rep) is given by the product of the probabilities for the various bond types:

(4-29) $\quad p_{A,rep} = p_A \, [p_B(1 - x_{A,3})]^i \, (p_A)^i \, p_B x_{A,3}$

The *branching coefficient* α is defined as the probability that a functional group at a branching unit is connected with another branching unit by other groups. This branching coefficient refers to the total system; it is the sum of the various segment probabilities, $p_{A,rep}$:

(4-30)

$$\alpha = \sum_{i=0}^{i=\infty} (p_{A,rep})_i = \sum_{i=0}^{i=\infty} p_A [p_B(1-x_{A,3})]^i (p_A)^i p_B x_{A,3} = Q \sum_i Z^i = Q(1 + Z + Z^2 + ...)$$

setting $Q \equiv p_A p_B x_{A,3}$ and $Z \equiv p_A p_B(1 - x_{A,3})$. The series $1 + Z + Z^2 + ...$ can be converted into a closed expression since Z is always smaller than unity. Introduction of the group ratio $r_o \equiv p_B/p_A \leq 1$ delivers:

(4-31) $\quad \alpha = \dfrac{p_A p_B x_{A,3}}{1 - p_A p_B(1 - x_{A,3})} = \dfrac{r_o p_A^2 x_{A,3}}{1 - r_o p_A^2(1 - x_{A,3})} = \dfrac{p_B^2 x_{A,3}}{r_o - p_B^2(1 - x_{A,3})}$

A branching unit has f_{br} arms. The number of arms increases by f–1 if another functional molecule is added. The probability is $\alpha(f_{br} - 1)$ that one of these arms carries another branching unit. If this probability becomes unity, all molecules will interconnect. The critical value for cross-linking is thus $\alpha_{crit} = 1/(f_{br} - 1)$:

(4-32) $\quad \alpha_{crit} = \dfrac{1}{f_{br} - 1} = \dfrac{r_o p_{A,crit}^2 x_{A,3}}{1 - r_o p_{A,crit}^2(1 - x_{A,3})}$

The critical extent of reaction, α_{crit}, depends on four parameters: functionality f_{br} of branching monomers, initial mole ratio r_o of B and A groups, and mole fractions $x_{A,3}$ of A groups in branching molecules relative to all A groups. α_{crit} and thus the gel point is independent of reaction time and reaction temperature.

Eqn.(4-32) delivers a critical extent of reaction of $p_{A,crit} = 0.843$ for the system shown in Fig. 4-3. Experimentally, a higher value of 0.894 is found because Eqn.(4-32) neglects intramolecular ring formations. The probability of such cyclizations increases with increasing dilution. If one extrapolates gel points obtained at various concentrations c to infinite concentration ($1/c \rightarrow 0$), then indeed correct gel points are observed. The higher than theoretical, experimental gel points thus provide a safety margin for industrial cross-linking polycondensations and polyadditions.

Gel points from theory and experiment (at infinite dilution) agree, however, only if the cross-linking reaction is controlled by thermodynamics. For kinetically controlled cross-linkings, gel points and structures of the cross-linked polymers depend on the reaction path, i.e., on prior history.

An important example is the **sol–gel process** for the synthesis of silicate glasses. In this process, tetramethylorthosilicate $Si(OCH_3)_4$ is first hydrolyzed to silicic acid $Si(OH)_4$ which immediately polycondenses to an amorphous gel. After drying and heating, optical glasses of $(SiO_2)_n$ result (see Section 5.7.3.).

Cross-linking polyadditions are used for polyurethanes and epoxides (see Section 4.1.8.). Important industrial cross-linking polycondensations comprise those leading to alkyd resins and those of formaldehyde with either phenol, urea, or melamine.

Alkyd Resins

Alkyd resins result from the polycondensation of multifunctional *al*cohols (al...) with bifunctional *acids* (...kyds) or their anhydrides. Examples are the glyptal resins from glycerol and phthalic acid anhydride. The reaction is conducted to just short of the gel point; cross-linking occurs after application, e.g., as paints.

Phenol–Formaldehyde Resins (Phenolic Resins) PF

Phenolic resins are condensation products of phenols (mainly phenol itself) with formaldehyde or (sometimes) other aldehydes. The first phenolic resins served as substitutes for shellac, the purified secretions (lac) of the larvae of the insect *Kerria lacca*. The synthetic phenolic resins were thus called **novolac** (L: *novo* = new; Sanskrit: *laksha* = hundred thousand) because a hundred thousand insects are needed to produce one ounce (28.4 g) of shellac.

Novolacs result from the **acid catalysis** of the reaction of formaldehyde with an excess of phenol. Formaldehyde CH_2O is protonated to the methylol cation $^{\oplus}CH_2OH$. This cation then reacts with phenol to form protonated *para-* and/or *ortho-*methylolphenols which cannot be isolated. In the presence of protons, they react with each other to form the methylene compound by releasing water. Two methylol groups can furthermore react with one molecule of formaldehyde to an open-chain formal:

o-Methylolphenol   Methylene compound   Open-chain formal
(protonated)

These reactions proceed to form novolacs, i.e., oligomers with average molecular weights of ca. 1000. Novolacs are soluble. They are cross-linked by hexamethylene-tetramine (often called hexa or urotropin) which generates methylene bridges and, at higher hexa concentrations, also nitrogen-containing bridges. *Para*-positions react faster than *ortho*-positions:

(4-33)

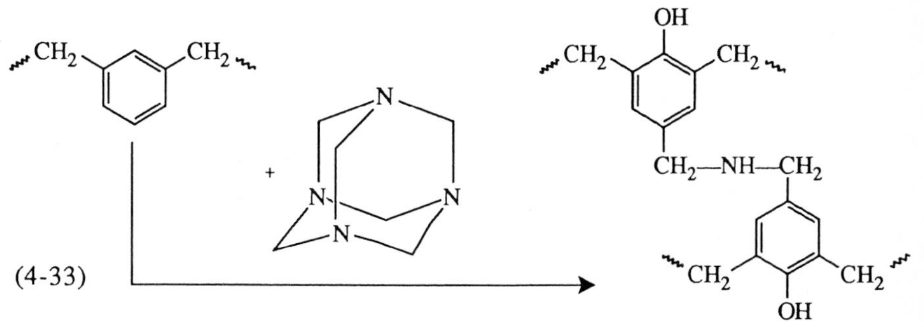

In **base catalyses**, phenolate ions are added nucleophilically to formaldehyde. Multiple additions are possible, for example in the *ortho* position to methylol anions:

(4-34)

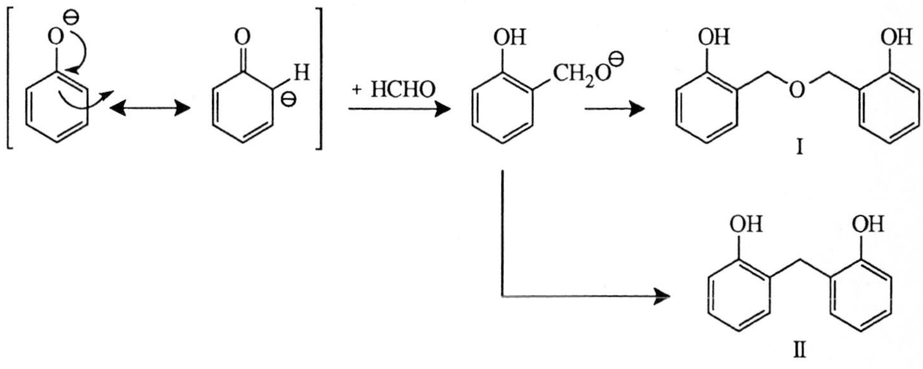

Two methylol groups each etherify by releasing water (I). At higher temperatures, methylol groups react with aromatic nuclei and form methylene bridges (II). The resulting **resols** (A-stage) condensate to **resitols** (B-stage) and finally cross-link to **resits** (C-stage). The hardened products are called **phenoplasts**. Resitols can be isolated by stopping the reaction short of the gel point. These resitols can then be hardened by acid catalysis.

Phenolic resins are mainly laminated with paper or fabrics to prepare laminated plastics or are filled with, e.g., wood flour to form thermosets. They also serve as raw materials in the paint industry, as tanning agents, binders for sand shell moldings, and heat shields for interplanetary rockets. Amine-hardened resins are yellowish-brown;

this color results from a dehydrogenation of nitrogen–bridges (see Eqn.(4-33)) to azomethine structures. Novolacs are also spun into flame-resistant fibers that are cross-linked by formaldehyde. The white fibers yellow with time because *p*-methylenephenol end groups ~(p-C$_6$H$_4$)–CH$_2$–(p-C$_6$H$_4$)–OH are oxidized to quinone methids by oxygen in air. This oxidation can be prevented by esterification of the phenolic end groups.

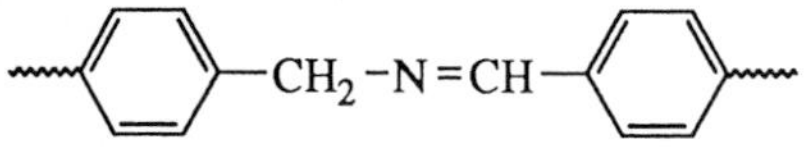

Azomethine 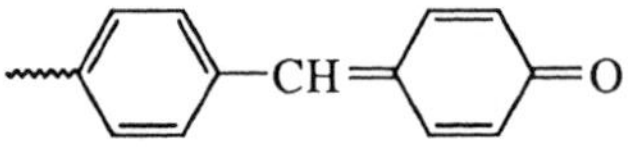   Quinone methide

Melamine Resins MF and Urea Resins UF

These are the most important representatives of the **amino resins**. They result from a kind of Mannich reaction between NH-containing compounds, nucleophilic molecules and carbonyl-containing compounds:

(4-35)   ~Y—H   +   Carbonyl   +   NH- $\xrightarrow{-H_2O}$

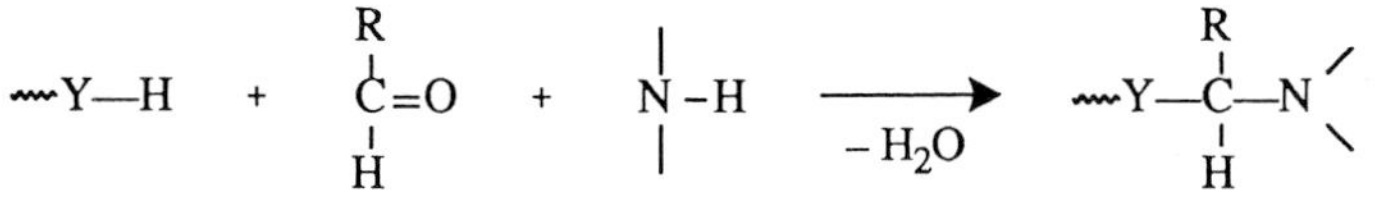

Nucleophilic   Carbonyl   NH-
component   component   component

NH-Components are mainly urea H$_2$N-CO-NH$_2$ or melamine (2,4,6-triamino-1,3,5-triazine). The carbonyl component is predominantly formaldehyde (rarely ketones or other aldehydes). Nucleophilic components may be H-acidic (halogen acids), OH compounds (alcohols, carboxylic acids), or NH compounds (urea, melamine, amines, etc.). For example, urea reacts with formaldehyde to *N*-methylol urea which is stabilized by an intramolecular hydrogen bond:

(4-36)   H$_2$N—CO—NH$_2$ $\xrightarrow{+ \text{HCHO}}$ H$_2$N—CO—NH—CH$_2$OH $\rightleftharpoons$

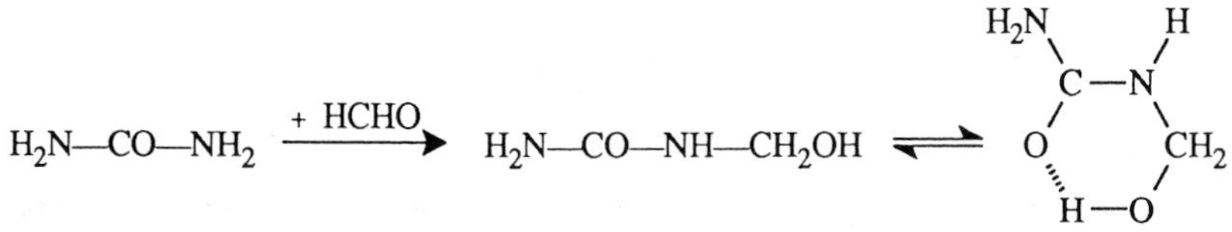

The base-catalyzed reaction stops at the methylol urea stage. Acids convert the methylol urea into a resonance-stabilized carbonium-immonium ion:

(4-37)   H$_2$N—CO—NH—CH$_2$OH $\xrightarrow[- H_2O]{+ H^{\oplus}}$

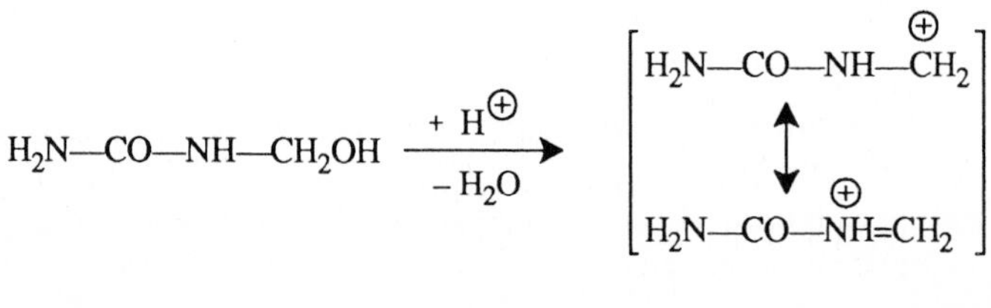

α-Ureidoalkyl(carbonium-immonium) ions subsequently react with suitable nucleophilic compounds. With urea as a partner, polymer chains are extended to give $H_2N-CO-NH-CH_2-NH-CO-NH_2$. Newer investigations indicate the existence of linear polymers, i.e., one methylol group per amino group. A dispersion is generated whose particles finally aggregate and precipitate. The particles are "hardened" by hydrogen bonds.

The resulting **aminoplastics** are colorless and less sensitive to light than phenolic resins but are also less stable against humidity and temperature. More than 85 % of urea resins are used as binders for wooden materials, e.g., particle board. Other applications include wrinkle-proofing of cotton, paints, expanded plastics, and wet-proof paper. Melamine resins are used for shatter-proof tableware.

4.2. Biological Polymerizations

4.2.1. Introduction

Biopolymers are macromolecular compounds that are generated by biochemical reactions. The monomeric units of **homologous biopolymers** consist of a single class of species, for example, different amino acid residues. The most important classes of homologous biopolymers are nucleic acids, proteins, polysaccharides, polyprenes, lignins, and aliphatic polyesters. **Heterologous biopolymers** comprise various classes of monomeric species. Examples are nucleoproteins composed of different nucleic acid residues and different α-amino acid residues. In general, heterologous biopolymers are block copolymers or comb copolymers.

Biopolymers are formed *in vivo* in cells with enzymes acting as catalysts. Cells (L: *cella* = chamber) consist of a nucleus (L: *nucleus* = kernel) and the surrounding cytoplasm (G: *kytos* = cavern; *plassein* = to form). In plants, cells are surrounded by a cell wall; in animals, by a cell membrane composed of lipids. Cells are subdivided into procaryotic and eucaryotic cells (L: *pro* = before; G: *eus* = good; G: *karyon* = kernel, nut).

The terms "procaryotic" and "eucaryotic" are linguistically wrong. "Procaryotic" has both Greek and Latin roots in one word. "...caryotic" is wrongly written with a "c" instead of a "k" (there is no "c" in the Greek alphabet). It should furthermore read "karyontic" and not "caryotic".

Blue algae and bacteria each consist of only one **procaryotic cell** of ca. (300-2000) nm diameter. The cytoplasma of these primitive cells is not subdivided; the nuclei are not surrounded by a hull.

All other living beings contain many **eucaryotic cells**. Most of these cells are ca. 10 times larger than the procaryotic ones but some can be very large (several meters in rubber trees!). The nuclei of eucaryotic cells are always surrounded by a membrane. The interior of the cell is subdivided by a series of membranes (endoplasmic reticulum). The cytoplasm contains not only the nucleus but also a number of organelles ("little organs"), i.e., special parts of cells that resemble a cell itself and function like one. These organelles are responsible for various biochemical reactions: ribosomes for protein syntheses; mitochondria for the oxidation of sugars, proteins and fats; lysosomes for the digestion of foodstuffs and dead cells; the Golgi apparatus for the encapsulation of carbohydrates and digesting enzymes.

Enzymes are either dissolved in certain cell compartments or in cell organelles (e.g., cytosol enzymes) or are bound to membranes or particles. One cell contains ca. 10^5 enzyme molecules which catalyze ca. 1000-2000 different reactions. Only 50-100 enzyme molecules are thus involved in each type of reaction!

All biological polymerizations are highly specific in the selection of molecules (but not always substrate specific), regioselective and stereoselective in chain propagation, and fast at the usual temperatures of (0-40)°C. Many polymerizations (if not all) proceed at matrices. Probably all biological polymerizations can be considered poly-eliminations ("condensative chain polymerizations"): one monomer after the other is connected to the growing chain with the formation of leaving molecules. These leaving molecules are usually complicated, energy-rich molecules, not the simple molecules such as water, methanol, etc., in synthetic polycondensations.

4.2.2. Nucleic Acids

Two types of polymeric esters of phosphoric acid exist in nature: teichoic acids and nucleic acids. **Teichoic acids** are polyesters of phosphoric acid and glycerol or ribitol; they reside in the cell walls of bacteria (G: *teichos* = wall). Glycerol and ribitol are partially substituted with D-alanine or sugars:

R = H, D-alanyl, sugar residues   R' = H, D-alanyl; R" = H, *N*-acetylglucosamine

Teichoic acids are primitive predecessors of **nucleic acids**, linear copolyesters from phosphoric acid and substituted ribose (**ribonucleic acids, RNA**) or substituted 2'-deoxyribose (**deoxyribonucleic acids, DNA**). Ribose units are present as furanoses; they are substituted by purine or pyrimidine bases.

2'-Deoxyribose   Ribose   Ribose nucleoside   Ribose nucleotide

Compounds of sugars and bases are called **nucleosides**; their individual names carry the prefix d if the sugar is deoxyribose. Phosphoric acid esters of nucleosides are called **nucleoside phosphates** or **nucleotides**. Phosphoric acid residues are symbolized by a p which is written in front of the nucleoside symbol if the phosphoric acid is in 5' position and after the symbol, if the acid is in 3' position. Nucleotides are always interconnected in the 5',3' position.

Ribonucleic acids contain the purine bases adenine and guanine and the pyrimidine bases cytosine and uracil. Deoxyribonucleic acids also possess adenine, guanine, and cytosine; uracil is however replaced by thymine (3-methyluracil). In RNAs, small proportions of 1-methylguanine and dihydrouracil are found; DNAs contain some 4-methylcytosine units.

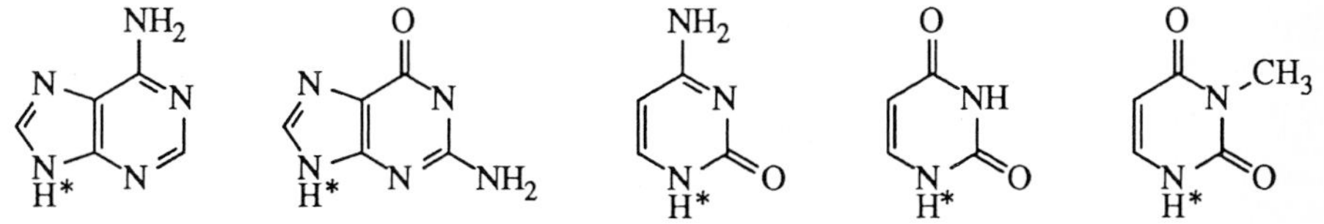

Names of bases
Adenine   Guanine   Cytosine   Uracil   Thymine
Names of corresponding nucleosides (H* of bases replaced by ribose or 2'-deoxyribose)
Adenosine   Guanosine   Cytidine   Uridine   Thymidine
Names of corresponding nucleotides (sugars of nucleosides substituted by phosphoric acid)
Adenylic acid   Guanylic acid   Cytidylic acid   Uridylic acid   Thymidylic acid

Ribonucleic acids are linear single-strand polymers; certain RNA segments form intramolecular helix structures with other segments. The DNA of the bacteriophage ɸX 174 is also single-stranded but it forms a ring and not a linear chain. Cyclic DNAs are likewise found in bacteria where they form intermolecular helices from double strands. DNAs of higher life forms are double strands of linear chains.

Double strand DNAs consist of two DNA chains with molecular weights between millions and (American) billions that are complementary to each other. They form intermolecular hydrogen bonds between the bases of thymidine (T) and adenosine (A) on one hand and between cytidine (C) and guanosine (G) on the other:

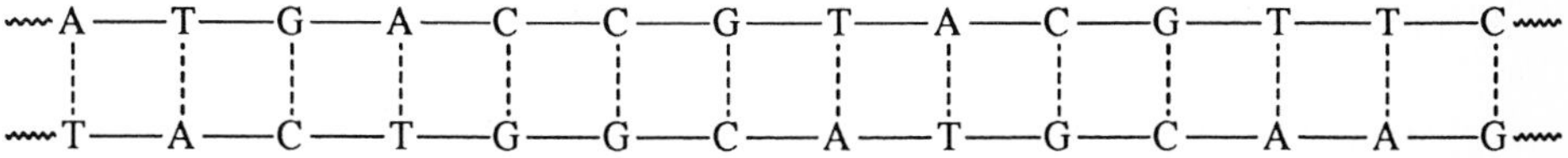

Three successive nucleotides of DNA (a **codon**) form the code for one amino acid (**triplet code**). The sequence of these triplets determines the genetic code. DNA is always generated by **replication**, never *de novo* from nucleosides. During replication, double strands are opened; the intermediate single strands serve as matrices for the synthesis of new DNA chains. After replication, each double helix contains one old and one new strand each (semi-conservative mechanism).

Genetically relevant segments of DNA can be multiplied *in vitro* by the **polymerase chain reaction** (PCR). DNA dissociates into two single strands on heating to 94°C. After cooling to 50°C, preformed DNA oligomers are placed on both strands. These oligomers act as primers (i.e., initiators) for the subsequent polyelimination reactions which proceed on the single strands as matrices by successive addition of nucleoside triphosphates to the primers under the action of the enzyme polymerase as catalyst. This very important reaction allows one to multiply the DNA initially present by a factor of 10^5 after 25 cycles of 7 minutes each.

Ribonucleic acids control protein syntheses. They exist in at least three types. **Transfer RNAs** (t-RNA) of all life forms have similar structures; they have relatively low degrees of polymerization of only 73 to 93. **Messenger RNAs** (m-RNA) have considerably higher molar masses of ca. 500 000 g/mol. They are copied by DNA in ways similar to the replication of RNA. The base sequences of m-RNA thus correspond to the base sequences of DNA in nuclei; they are therefore also called **template RNAs**. **Ribosomal RNAs** (r-RNA) possess even higher molar masses of $(6 \cdot 10^5 - 2 \cdot 10^6)$ g/mol; they exist in ribosomes as complexes between nucleic acids and proteins, i.e., as nucleoproteins.

4.2.3. Proteins

Structure

Proteins (G: *protos* = first) are natural copolymers composed of peptide residues, i.e., α-amino acid units -NH-CHR-CO- and imino acid units -NR'-CH(R")-CO- with different R, R' and R"; R' and R" may also be part of a ring. Imino acids are also called secondary or heterocyclic amino acids. Synthetic copolymers of α-amino acids are called **polypeptides. Poly(α-amino acid)**s are homopolymers consisting of only one species of α-amino acids.

About 240 different α-amino acid units are known to exist in nature. The genetic code, however, provides programs for only 19 α-amino acids and the imino acid proline (Scheme I). All other natural amino acid units are generated by postreactions of primary formed proteins. Proteins of higher living beings contain only L-α-amino acid units; they are practically always in [S] configuration. Proteins of lower life forms such as cell walls of bacteria or lower plants may include up to 15 % D-α-amino acid residues.

Several amino acid residues of native proteins are generated by after-reactions. SH groups of cysteine units are often oxidized to the -S-S- bridges of cystine units. In collagen, hydroxyproline residues are generated from proline units in procollagen.

$$
\begin{array}{cccc}
-\text{NH} & & & \text{NH}- \\
| & & & | \\
\text{HC}-\text{CH}_2-\text{S}-\text{S}-\text{CH}_2-\text{CH} & & & \text{cystine unit} \\
| & & & | \\
-\text{OOC} & & & \text{COO}-
\end{array}
$$

Scheme I   Substituents R of α-amino acids H₂N-CHR-COOH

H	gly	Glycine	CH_2CONH_2	asn	Asparagine
CH_3	ala	Alanine	$(CH_2)_2CONH_2$	gln	Glutamine
$CH_2C_6H_5$	phe	Phenylalanine	$(CH_2)_4NH_2$	lys	Lysine
$CH(CH_3)_2$	val	Valine	$(CH_2)_2N=C(NH_2)_2$	arg	Arginine
$CH_2CH(CH_3)_2$	leu	Leucine	CH_2OH	ser	Serine
$CH(CH_3)CH_2CH_3$	ile	Isoleucine	$CH(OH)CH_3$	thr	Threonine
CH_2COOH	asp	Aspartic acid	CH_2SH	cys	Cysteine
$(CH_2)_2COOH$	glu	Glutamic acid	$(CH_2)_2SCH_3$	met	Methionine
$CH_2(p\text{-}C_6H_4)COOH$	tyr	Tyrosine			

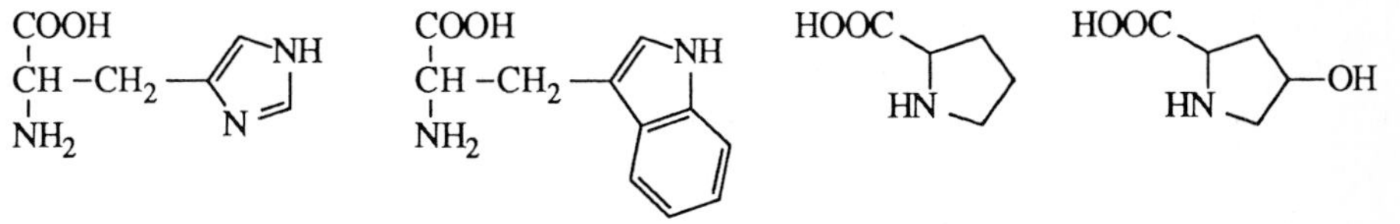

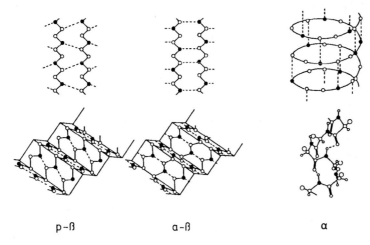

Histidine his   Tryptophan trp   Proline pro   Hydroxyproline hyp
          α-Amino acids   Imino acids

The sequence of peptide units in proteins is controlled by the genetic code; it is called the **primary structure**. Sequences are usually not repeated in enzyme molecules; other protein chains may however have periodic structures (silk, collagen, etc.).

Hydrogen bonds often exist between different peptide units; these bonds create **secondary structures** (Fig. 4-4). Intramolecular hydrogen bonds lead to helix structures, mainly right-handed helices with 3.6 peptide units per turn. **Pleated sheet** struc-

p-ß   a-ß   α

Fig. 4-4 Pleated sheet and helix structures of polypeptides. p-ß = Parallel pleated sheet (ß-structure), a-ß = antiparallel pleated sheet (ß-structure), α = α-helix. O Methyl groups in helices, o carbon atoms in helices and pleated sheets, ∘ hydrogen atoms in helices. ● Nitrogen atoms in helices or NH groups in pleated sheets, - - - hydrogen bonds between >CO and HN<.

tures may be intermolecular between two chains as well as intramolecular between
segments of the same chain. These sequence-dependent interactions often generate
spheroidal shapes of molecules (Fig. 4-5, below).

Macroconformations of protein molecules, the so-called **tertiary structures**, are
often stabilized by interactions between side groups, for example, by covalent S-S
bridges of cystine and ionic and/or hydrophobic interactions. Like or unlike protein
molecules often associate to defined **quaternary structures**.

Secondary, teriary, and quaternary structures often break up at elevated temperatures or by specific interactions with certain solvents. The protein chains then adopt
the shape of a random coil (Chapter 5). This **denaturation** is reversible. It is however
often followed by an irreversible aggregation of the coils which is also called a denaturation in industry.

Proteins are divided according to their native shapes into spheroproteins and scleroproteins (G: *skleros* = hard). **Spheroproteins** have spherical or ellipsoidal shapes.
Due to their shape, they diffuse rapidly (see Section 7.2.2.) and form low viscosity
solutions (see Section 5.8.). Nature uses them for transport (e.g., oxygen by
hemoglobin) and as rapid deployment catalysts (extracellular enzymes).

Scleroproteins have fibrillar structures; they are thus also called **fiber proteins** or
linear proteins ("linear" in contrast to "spherical" and not to "not branched" or "not
cross-linked"). Scleroproteins form supports; examples are keratins (hair, feathers,
silks), collagen (skin), and elastins (connective tissue).

Protein Syntheses

Poly(α-amino acid)s can be obtained by polymerization of *N*-carboxy anhydrides
of α-amino acids (Leuchs anhydrides) (see Eqn.(3-5)). *Peptide syntheses* require
three steps: (1) protection of NH_2 or COOH end groups of amino acids and peptides,
respectively, (2) peptide bond formation via the unprotected end groups:

(4-38)   ©–NH–CHR'–COQ + H_2N–CHR"–COOR^1 ⟶

©–NH–CHR'–CO–NH–CHR"–COOR^1 + QH

and (3) selective removal of one of the protecting groups. Typical protecting groups
© are *t*-butyloxycarbonyl (R^1 = $(CH_3)_3$C–O–CO) (BOC) for amino groups and ester
groups (e.g., R^1 = CH_3) for COOH groups. In step 2, COOH itself is not used as the
reactive group; rather, it is converted into activated groups –COQ, where Q may be,
for example, the azide group –N_3, a carbodiimide residue –O–C(=NR^2)(–NR^3), a
mixed anhydride –O–CO–OR^4, or a nitrophenyl ester –O–CO–(p-C_6H_4)NO_2.

These "free" syntheses require the removal of unreacted molecules before the next
step because otherwise non-regular polypeptides would result. A solid matrix is thus
used as the protecting group © by the **Merrifield method**, for example, functionalized porous glass. Peptide groups remain bound to this matrix throughout the total sequence of reactions; unreacted materials can be easily washed out. Computer-control-
led synthesis machines thus allowed the synthesis of the enzyme ribonuclease from
124 amino acid units in 369 chemical reactions and 11 911 individual steps.

Protein syntheses occur *in vivo* in two steps. In the **transcription**, the 4-letter alphabet of DNA (A, G, C, T) is transcribed into the complementary (but different) 4-letter alphabet of RNA (U, C, G, A). Transcriptions are copolyeliminations: DNA double strands are locally unraveled into segments of single strands. The single strands act as matrices for the successive addition of monomers to the primer. This reaction takes place in the nucleus; it is catalyzed by an enzyme complex. Each addition results in the release of diphosphoric acid. The monomers are not the nucleosides themselves but the nucleoside triphosphates, for example, adenosine triphosphate ATP:

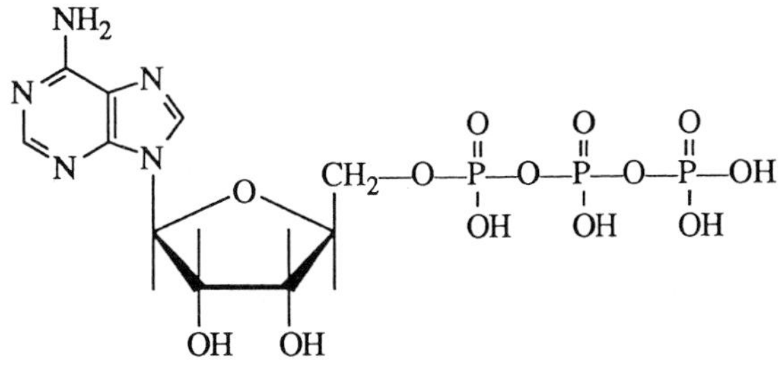

Nucleoside triphosphates are more energy-rich than the monophosphates. Biochemistry calls those chemical compounds "energy-rich" that are easy to hydrolyze (not those having high thermochemical dissociation energies!). ATP is such a compound according to quantum mechanics: its five successive positively charged chain atoms of the phosphate residue can be easily attacked by the negative ions of phosphatases (enzymes that hydrolyze phosphoric acid residues).

The newly synthesized m-RNA relocates from the nucleus into the cytoplasm which is the locus for the synthesis of all proteins. In the cytoplasm, the 4-letter code of m-RNA is converted by **translation** into the 20-letter code of the proteins. The protein code is a triplet code. A doublet code would generate only $4^2 = 16$ commands, too few for 20 amino acids. A quadruplet code would create $4^4 = 256$ commands, far too many; four bonds instead of three would also be too stable. A triplet code has $4^3 - 20 = 44$ more commands than needed. Several codons therefore code for the same amino acid which helps to stabilize the species. The control is usually exercised by the first two bases of a codon; GCU, GCC, GCA and GCG all code for alanine, for example.

The monomers are not the amino acids themselves but their esters with the 3' end of the t-RNAs. Each amino acid species requires a special t-RNA; each reaction is catalyzed by a special aminoacetyl t-RNA synthetase.

The resulting aminoacetyl t-RNAs are the true monomers. They possess special base sequences with which they bind specifically via hydrogen bridges to the bases of the complementary codons of the m-RNA. The synthesis of protein chains starts from the free, N-terminal end of the amino acid. The polyelimination step generates t-RNA as leaving molecule.

Some of the synthesized proteins remain in the cytoplasm. Their chains fold easily to tertiary structures. Local secondary structures originate in less than 10^{-2} seconds. These structures then reorganize themselves in less than 1 second via local tertiary folds to "molten globules" (water-containing fluid particles) that are held together by hydrophobic interactions. Native three-dimensional tertiary structures are then created within $(1-10^3)$ seconds.

Extracellular enzymes do not fold in the cytoplasm because folded (spherical, etc.) molecules cannot pass through the cell membrane. They are stabilized by other proteins as molten globules which are chaperoned through the membrane and move to organelles, e.g., mitochondria.

Enzymes (G: *enzumos* = leaven (sour dough))

Enzymes are a class of biochemical catalysts. With the exception of the so-called RNA enzymes (catalytically acting RNA molecules), they are all spheroidal proteins with different chemical structures. The names of enzymes reflect the name of the attacked substrate and the name of the catalyzed reaction; they always have the ending "ase". A hydrolase hydrolyses, a glucose polymerase polymerizes glucose, etc.

Simple enzymes consist of one protein chain with 60 to 2500 peptide units (molecular weights between ca. 6000 and 250 000). Many enzymes are composed of more than one protein chain, however. These protein chains are called **subchains**.

Subchains may be interconnected by covalent bonds (e.g., S-S bridges in insulin), by hydrogen bonds, or by other bonds. Hemoglobin has four subchains with molecular weights of ca. 16 000 each, glutamine dehydrogenase has eight subchains with $M_r \approx 250\,000$ each, etc. Subchains may have identical structures or different ones.

Protein chains of enzymes contain helical and non-helical sequences that cluster together to form spherical or ellipsoidal shapes (Fig. 4-5). All spheroproteins exhibit clefts in which the catalytically active center and the receptor, respectively, reside.

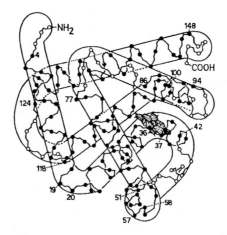

Fig. 4-5 The protein myoglobin of the sperm whale (schematic). The protein consists of 153 amino acid units in 8 helical sequences comprising the peptide units (•) 4-19, 20-35, 36-42, 51-57, 58-77, 86-94, 100-118, and 124-148. The active heme group (dotted) is in a cleft.

These clefts can accommodate several substrates; only some of those may be there to place their reactive sites in positions that are sterically and/or electronically favorable for an interaction with the catalytically active groups of the enzyme. These structures enable enzymes to be highly specific. The binding of the substrate to the catalytic center furthermore increases the effective substrate concentration. As a result, reaction rates may be enhanced by factors up to 10^{12} relative to similar non-enzymatic reactions. Since dissolution of an enzyme in certain organic solvents often does not change the macroconformation of the enzyme molecule, enzymatic reactions may not only be performed in aqueous solutions but also in organic solvents.

About 2100 different enzymes are known but only 150 are routinely prepared in milligram to kilogram amounts, mainly from microorganisms and for medical or biochemical purposes. Only 19 enzymes are used industrially, predominantly in the food industry. They are usually applied in very dilute solution which makes their recovery difficult. Enzymes are thus increasingly **immobilized**, for example, by incorporation into a matrix, micro-encapsulation by polymers, covalent bonding to or adsorption on a carrier, or intermolecular cross-linking of enzyme molecules.

Silk

– more exactly: natural silk – is produced by caterpillars and spiders. Economically most important is the silk of the cocoons of the mulberry silk spinner. These cocoons consist of 78 % silk fibroin and 22 % sericin (silk glue). They are treated with hot water or water vapor which kills the pupae. The sericin is softened by dipping the cocoon into hot water. Rotating brushes grasp the silk fibers which are wound up and dried. Only ca. 900 m fibers are recovered from the ca. 4000 m making up the cocoon. The rest of the fibers are broken and not clean enough; this raw silk waste is processed to fibers in so-called Schappe mills.

The silk fibers are then debasted, i.e., liberated from sericin by alkali-free soap. Silk loses weight by this treatment and is subsequently charged with $SnCl_4$ and Na_2HPO_4. The resulting tin phosphate is converted by water glass (sodium silicate) into the corresponding silicate. This mineral "weighing" improves the gloss and the hand as well as the balance sheet.

Silk fibers have a triangular cross-section (see Fig. 13-4) which causes the shine of the fibers (light refraction similar to that of a diamond). Fibers consist of microfibrils of ca. 10 nm width. In these fibrils, molecules are arranged as pleated sheets. The molecules themselves are segment copolymers: 10 crystallizable hexapeptide segments ser-gly-ala-gly-ala-gly are separated from each other by non-cystallizable segments comprising a total of 33 peptide units. The hexapeptide segments bundle parallel to crystalline domains which are responsible for the high strength of silk fibers. The non-crystalline domains provide the high extensibility of silk fibers.

Wool

Wool is the cut hair of sheep, goats, etc. Raw wool contains (20-50) % fiber and grease, suint, proteins, and plant residues. The ingredients are removed by mechanical means, washing, treatment with dilute sulfuric acid, or heating to (100-120)°C.

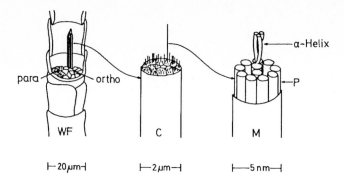

Fig. 4-6   Schematic representation of wool fibers WF (left) with their cortex cells C (center) composed of microfibrils M (right). Each microfibril contains 11 protofibrils P, each protofibril consists of 2-3 α-helices. para = Paracortex, ortho = orthocortex.

Wool is composed of ca. 200 different chemical compounds. Ca. 80 % are keratins, 17 % other proteins, and 3 % polysaccharides, nucleic acids, lipids, and inorganic molecules. Most peptide units of keratins possess bulky substituents which prevent the formation of pleated sheets. Keratins exist as α-helices which are intermolecularly cross-linked by either disulfide bridges $-CH_2-S-S-CH_2-$ or by N_ε-(γ-glutamyl)-lysine residues $-CH_2CH_2CONH(CH_2)_4-$. Wool is therefore insoluble in all solvents.

Two to three α-helices are united in a protofibril. Eleven of these protofibrils form a microfibril and many microfibrils a cortex cell (Fig. 4-6). The center of each cortex cell is composed of a matrix of proteins that are very rich in sulfur. Cortex cells are bundled to wool fibers which are bicomponent fibers: one half of the fibers is made up by the paracortex and the other half by the orthocortex. These two parts possess different chemical compositions. They thus absorb different amounts of water which causes the fibers to crimp. The scaly surface of the fibers produces the typical wool-like hand and the excellent voluminosity of wool.

Collagen and elastin

Collagen and elastin are the main constituents of skins, hides, tendons, cartilage, blood vessels, etc. Elastin provides the elasticity of connective tissue at small deformations, collagen prevents the fracture of skins and hides.

Hides consist of (top to bottom) the epidermis, cell layer (L: *corium minor*), dermis (L = *corium, cutis*), and hypodermis (L: *subcutis*). Epidermis, corium minor and subcutis are removed mechanically so that the dermis remains. The dermis is chemically or physically cross-linked to give **leather**.

The basic structural unit of collagen is **tropocollagen**, a protofibril composed of a triple helix of two α_1 chains and one α_2 chain. These chains have basically the same structure in all species. α_1 chains from calf hides consist of 1052 peptide units. 1011 of these units reside in triplets gl-X-Y where X is pro, leu, phe or glu and Y is mainly hydroxyproline or arg. Triplets are united in polar or apolar segments. The ends of the collagen chains carry oligopeptides without a triplet structure. These **telopeptides** are rich in lysine; they cause inter- and intramolecular cross-linking of chains.

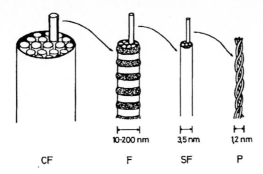

10-200 nm   3,5 nm   1,2 nm

CF   F   SF   P

Fig. 4-7   Structure of collagen fibers CF composed of aggregated fibrils F, subfibrils SF, and protofibrils P of tropocollagen. Protofibrils are 300 nm long triple helices from two α_1 chains and one α_2 chain. On staining by uranyl salts, fibrils show stripes which indicate alternating amorphous (dark) and crystalline (light) domains.

α_1 and α_2 chains form left-handed helices that are united in the triple superhelix of tropocollagen (Fig. 4-7). Protofibrils are joined to subfibrils in such a way that each unordered domain with polar segments and predominantly positively charged amino acid side groups faces a polar domain with predominantly negative charges. In subfibrils, protofibrils are also intermolecularly cross-linked by carbohydrate residues. Subfibrils are assembled to collagen fibrils and these to collagen fibers.

Biosynthesis first generates proline units containing pro-α_1 and pro-α_2 chains in the ratio 2:1. Each chain consists of ca. 1300 peptide units. Some of the proline and lysine units are subsequently hydroxylated. Hydroxylysine units are then bonded to glucose or galactose. Two pro-α_1 molecules and one pro-α_2 molecule associate side-by-side and their ends are then cross-linked via disulfide bridges. The resulting procollagen molecule leaves the cell and is transported to its final destination where most of the coil-like segments at the ends are removed. The residual coil-like peptide segments form the telopeptides. The procollagen molecules form tropocollagen molecules which agglomerate to subfibrils and then to collagen fibrils and collagen fibers. Tropocollagen does not metabolize; it remains in the body for life.

Heating of water-swollen collagen (G: *colla* = glue) to temperatures in excess of (40-60)°C causes tropocollagen to dissociate into α_1 and α_2 chains. These chains are subsequently partially hydrolyzed to **gelatin(e)** (L: *gelatum* = frozen). Gelatin is prepared from hog skins in the USA and from demineralized bones in Europe. It is used for photographic films, as a thickener and for sausage casings in the food industry, and for encapsulations in the pharmaceutical industry.

4.2.4. Lignins

Lignins are a group of high molar mass, amorphous compounds with high methoxyl contents that are present in wood and some other plants (L: *lignum* = wood). They derive from the three monolignols cumaryl alcohol, coniferyl alcohol and sinapyl alcohol:

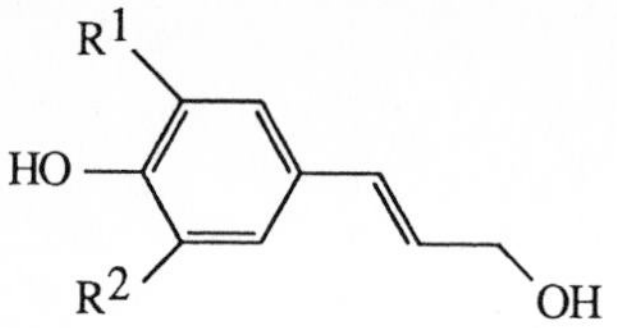

R^1	R^2	
H	H	*p*-cumaryl alcohol
H	OCH_3	coniferyl alcohol
OCH_3	OCH_3	sinapyl alcohol

Peroxidases polymerize monolignols free radically by dehydrogenation reactions. These random reactions generate lignins whose very complex cross-linked structures vary from plant to plant (Fig. 4-8). In wood, lignins are combined with celluloses. About 50 million tons of lignin sulfonates or sulfate lignins are produced each year by chemical cleavage of wood. Practically all of these compounds are burned in order to improve the energy balance of factories.

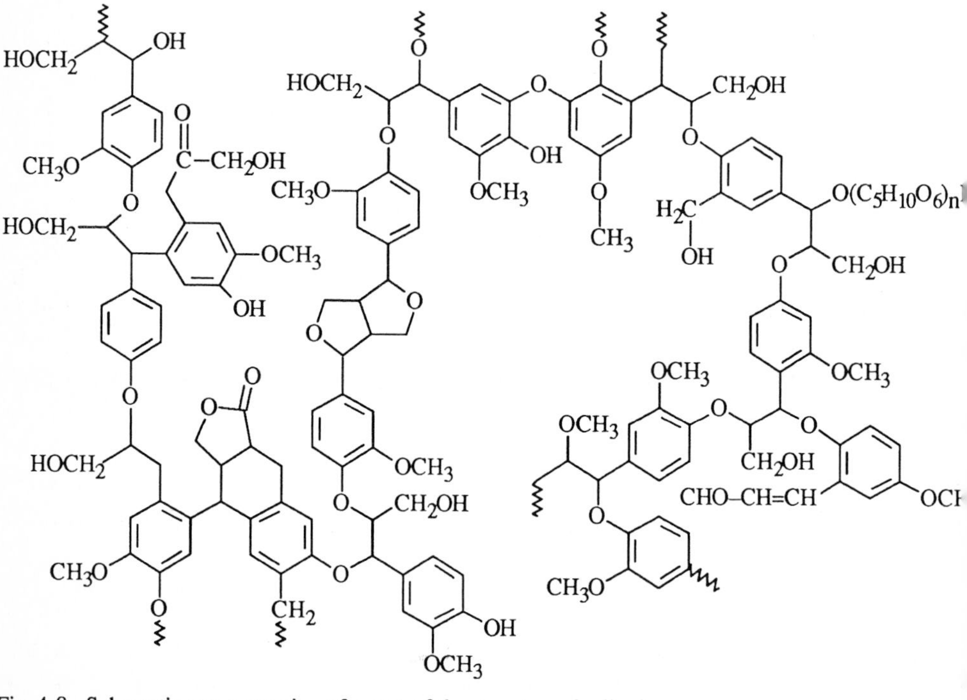

Fig. 4-8 Schematic representation of a part of the structure of a lignin.

4.2.5. Polysaccharides

Of all synthetic and natural polymers, polysaccharides are generated and consumed in the largest amounts, either directly (wood, cotton) or after physical preparation (paper, starch), chemical isolation (rayon), and chemical transformation (cellulose esters, etc.). Polysaccharides are usually subdivided into fiber-forming linear **structural polysaccharides** (e.g., cellulose, chitin), moderately to strongly branched

reserve polysaccharides (e.g., amylose, pectin), and physically cross-linked, **gel-forming polysaccharides** (e.g., gums, mucopolysaccharides) (G: *sakcharon* = sugar).

Nomenclature

Polysaccharides are homopolymers or copolymers of various sugars, almost exclusively hexoses and pentoses. IUPAC has proposed replacing "polysaccharide" by "polyglycane" because of the glycosidic bonds between sugar residues. In the food industry, polysaccharides are known as "complex carbohydrates" and as "fiber" if they cannot be digested (Section 13.1.). Homopolymers from a single species of sugar monomer are called homoglycanes, those from two or more species, heteroglycanes. "Copolyglycane" and "copolysaccharide" are not used.

The constitution of polysaccharides is characterized by poly(monomer) names if the types of bonds between monomeric units are unknown. Poly(glucose) is thus a polymerized glucose. If glucose molecules were joined with release of water, a poly-(anhydroglucose) will result. Systematic names also indicate the type of anomerism. Amylose is thus a poly[α-(1→4)-D-anhydroglucopyranose] with glucose as the repeating unit and cellulose a poly[β-(1→4)-D-anhydroglucopyranose] with β-cellobiose as the repeating unit:

Amylose   Cellulose

Polyglycanes with $\bar{X}_w/\bar{X}_n > 1$ are called "polymolecular", similar to synthetic polymers that are non-uniform with respect to molar masses. Polyglycanes comprised of molecules of similar (but not identical) structure are called "polydisperse" if they differ in small constitutive details, for example, in the number and length of side chains. Mixtures of polysaccharides with different constitutions of monomeric units or repeating units are called "polydiverse".

Biosyntheses

Biosyntheses of polysaccharides are not reversals of hydrolyses; the hydrated monomeric units (sugars) are not the monomers. Sugars S and adenosine triphosphate ATP first react to form sugar-6-phosphate S-6-P with adenosine diphosphate ADP as the leaving molecule. S-6-P is transformed by enzymes into sugar-1-phosphate S-1-P. This compound reacts in the presence of another enzyme with a nucleoside triphosphate NTP to a nucleoside diphosphate sugar NDP-S; phosphoric acid is released. NDP-S is then transformed into the true monomer, a phosphorylated lipid NDP-S-lipid. Depending on the life form and the polysaccharide, various sugars, nucleosides and enzymes are used by nature (Table 4-2).

Table 4-2 Sugars and primer nucleosides in biochemical polysaccharide syntheses.

Polysaccharide	Sugar	Nucleoside	Occurrence
Amylose	D-Glucose	Adenosine	Plants
Glycogen	D-Glucose	Adenosine	Bacteria
Glycogen	D-Glucose	Uridine	Liver
Cellulose	D-Glucose	Uridine	*Acetobacter xylinum*
Cellulose	D-Glucose	Guanosine	Plants
Callose	D-Glucose	Uridine	Higher plants
Chitin	*N*-Acetylglucosamine (as dextrin)	Uridine	Crustacea
Hyaluronic acid	*N*-Acetylglucosamine + glucuronic acid	Uridine	Synovial liquids
Slime polysacch.	Glucuronic acid + D-glucose	Uridine	Type III *pneumococci*
Pectic acid	D-Galacturonic acid	Uridine	Higher plants
Xylane	D-Xylose	Uridine	Grasses

Polysaccharide syntheses always seem to start from a protein. The start uses a nucleoside diphosphate that is different from the nucleoside diphosphate residue that is part of the true monomer. The start reaction of the cellulose synthesis requires uridine diphosphate glucose as primer (starter, initiator); one end of the cellulose chain probably carries protein-D-xylose units. The propagation reaction involves guanosine diphosphate glucose (lipid) as monomer, however. The ends of polysaccharide chains thus carry proteins and other sugar units that are often not completely removed in the work-up.

The propagation reaction is a polyelimination, not a polycondensation. Monomers NDP-S (or NDP-S-lipid?) are enzymatically attached to the non-reducing end of the growing chain S_n; nucleoside diphosphate NDP is released:

$$(4\text{-}39) \quad S_n\text{--O--Q} + S\text{--NDP} \longrightarrow S_{n+1}\text{--O--Q} + NDP$$

The new sugar unit S is inserted into a bond ~O–Q; it is not clear whether Q is a sugar unit or another group. NDP is then reconverted into NTP by ADP according to NDP + ATP → NTP + ADP. Oxidative phosphorylation converts ADP to ATP.

In vivo polysaccharide syntheses seem to be living polymerizations to polymers that are either molecularly uniform polymers or have narrow molar mass distributions. The true uniformity or non-uniformity of native polysaccharides is difficult to establish since polysaccharides often degrade during work-up. For example, cellulose from conventionally harvested cotton has a degree of polymerization of ca. 7000 but one of ca. 18 000 if it is obtained by opening seed capsules in the dark under oxygen exclusion.

Celluloses

"Cellulose" is a compound of defined chemical composition for the chemist but a group of substances for botanists, fiber technologists, and crystallographers. Only very few celluloses possess the constitution of a pure poly[β(1→4)-anhydro-D-glucopyranose], e.g., the cellulose of the algae *Valonia*. Cotton contains (1.5-2) % other sugar units. The cellulose of the red algae *Rhodumenia palmata* is even composed of 50 % xylose units and 50 % glucose units. Also, some CH_2OH units are always replaced by COOH units in native celluloses.

Celluloses exist in various crystal modifications. Chains are parallel in cellulose I from cotton, ramie and *Valonia* but antiparallel in cellulose II from rayon. Celluloses III and IV are also known.

In plants, celluloses are accompanied by pectins, polyoses, waxes, lignins, etc. They form fibers that can often be separated by mechanical means from other plant material; examples are seed hairs (cotton), stalks (flax) and leaves (sisal hemp).

In wood, cellulose fibers are intimately intermixed with lignin. The lignin is removed by cooking wood with SO_2 containing aqueous solutions of calcium hydrogensulfite $CaHSO_3$ (**sulfite process**), with $NaOH + Na_2S + Na_2CO_3 + Na_2SO_4$ (**sulfate process**), with NaOH, or with HNO_3. The sulfite process generates soluble lignin sulfonates. The remaining "sulfite cellulose" cell–OH is converted by NaOH solutions into the sodium alcoholate cell-ONa (alkali cellulose) and then by CS_2 into cellulose xanthogenate cell–O–C(=S)SNa with 0.5-0.6 xanthogenate groups per glucose unit (**viscose process**).

The resulting "viscose" is a highly viscous solution that is colored reddish-yellow by the byproduct sodium trithiocarbonate Na_2CS_3. Viscose is spun into a precipitation bath of sulfuric acid and sodium sulfate which generates $NaHSO_4$ and regenerates CS_2 and the viscose fiber cell-OH (**rayon**). Regenerated celluloses are also converted into cellulose esters and ethers by the respective chemical reactions.

Amylose and Amylopectin

Starches are intimate mixtures of amylose and amylopectin. They are obtained from corn in the United States, from corn and potatoes in Europe, and as tapioca from the roots of manioc (cassava) in tropical countries. Tapioca contains 17 % amylose (83 % amylopectin), potato starch 22 %, normal corn starch ca. 26 %, hybrid amylo maize 85 %, and hybrid waxy maize only 1 %.

Starch is consumed in great amounts by the food, pharmaceutical, paper, and textile industries as thickeners, puddings, encapsulations (pharmaceuticals), binders, wet strength improvers (paper) or sizes (textiles). Starch derivatives, e.g., acetates and phosphates, find similar applications.

Enzymes hydrolyze starches to the dimer maltose. Oligomeric maltoses may be obtained as intermediates, for example, maltopentaose, the pentamer. Acid hydrolysis of amyloses and amylopectins generates cyclic oligomers (**dextrins**) with 6, 7 or 8 glucose units that are preformed by the helical structures of the starch molecules.

Amyloses are practically linear poly[α-(1→4)-anhydro-D-glucopyranose]s with molecular weights of up to ca. one million. The stable macroconformation of amylose is that of a helix. These helices form inclusion complexes with iodine; the blue color of these complexes is due to the linear arrangement of iodine atoms in the channel provided by the helix. Amyloses are also obtained synthetically from glucose-1-phosphate as monomer, e.g., maltopentaose as starter, and the enzyme potato phosphorylase as catalyst.

Amylopectins are branched poly[α-(1→4)-anhydro-D-glucopyranose]s with Christmas tree-like structures. They possess 1 branching unit per 18-27 glucose units; the branching is via 1,6-positions. The distances between the branching points are large enough to allow a crystallization of the network chains. Starches containing exclusively amylopectin crystallize equally well or even better than starches from 100 % amylose. **Glycogens** are even more strongly branched than amylopectins; they reside in the liver and in the brain.

In **dextrans**, chain bonds are mainly in (1→6') positions and branches via α-(1→4') bonds. Dextrans are obtained by enzymatic polymerization of saccharose. They serve as blood plasma expanders and as columns in chromatography.

Other Polysaccharides

Some other naturally occuring polysaccharides are also poly(glucopyranose)s. In **chitin**, OH groups of C^2 atoms of β-cellobiose repeating units are replaced by N-acetylamino groups NHCOCH$_3$. Chitin is thus a poly[β-(1→4)-N-acetyl-2-amino-deoxyglucopyranose]. Partially deacetylated chitin is called **chitosan**.

Chitin is the structural polysaccharide of arthropods (insects, crustacea, etc.; G: *chitin* = armor). It is always associated with calcium carbonate and/or proteins. In order to obtain chitin, lobster shells are treated with cold hydrochloric acid to remove CaCO$_3$, then with proteolytic enzymes or 4 % caustic soda solution to destroy proteins, and finally with 40 % caustic soda to eliminate acetyl groups. Chitosan is soluble in dilute acids but chitin is not. Chitosan serves for the coverage of wounds, as an ion exchanger in water purification plants, to improve the wet-strength of papers, and as a biodegradable film for food packaging.

Xanthan is produced from D-glucose by the bacterium *Xanthomonas campestris* NRRL B-1459. This high molar mass polymer ($M_r \approx 5 \cdot 10^6$) has a cellulosic main chain at which every other glucose group carries a three-membered side chain of β-D-mannopyranosyl-α-(1→4)-D-glucopyranosyl-β-(1→2)-dimannopyranoside-6-O-acetate. The 4- and 6-positions of the ends of these side chains are esterified with pyruvic acid H$_3$C–CO–COOH. The periodic spacing of the side chains forces the main chain to adopt the macroconformation of a stiff helix. Aqueous xanthan solutions are lyotropic at concentrations above ca. 2.5 g/L. They serve as thickeners, gelation agents for explosives, pushers in secondary and tertiary oil recovery, etc.

Mucopolysaccharides are found in skin, connective tissue, cartilage, sweat, and mucous secretions (L: *mucus* = slime). They are glycosaminoglycanes of the general structure $\leftarrow$(1→3)Y-(1→4)Z$\rightarrow_n$. Y is either a glucose or a galactose unit and Z either a glucose or idose unit. Y carries in the 2-position either $NHCOCH_3$ or $NHSO_3H$, in the 4-position either OH or OSO_3H, and in the 6-position CH_2OH or CH_2OSO_3H. Z contains in the 2-position OH or $NHCOCH_3$ and in the 6-position COOH. Mucopolysaccharides comprise hyaluronic acid (synovial liquid of joints), chondroitin sulfate (matrix of cartilage), heparin (blood anticoagulant), dermatane sulfate (matrix of the skin), and many others.

Another large group of polysaccharides are gums. Agar-agar, carrageen, tragacanth, pectins, and gum arabic contain mainly D-galactose und D-galacturonic acid units. The most important carbohydrate units of pullulan, guaran, and alginates are D-mannose and D-mannuronic acid. Gums are used industrially as thickeners, emulsifiers, or protecting colloids (see also Table 1-2).

4.2.6. Polyesters

Poly(β-D-hydroxybutyrate) $+O–CH(CH_3)–CH_2–CO+_{\overline{n}}$ is produced by bacteria which store this polymer as reserve food in their cytoplasms. The same microbiological process is used in industry, emplying glucose as raw material. The polymer possesses similar mechanical properties as it-poly(propylene) but is biodegradable. An addition of propionic acid to the glucose solution results in copolymers with up to 47 mol% 3-hydroxyvaleric acid units.

Nature synthesizes a number of other aliphatic polyesters. The main component of cork is suberine, a polymer with ester and lactone units, especially phellogenic acid $HOOC(CH_2)_{20}COOH$ and phloionol acid $HOOC(CH_2)_{17}OH$. Earth bees coat their nests with copolyesters from lactones of the acids $HOOC(CH_2)_nOH$ ($n = 17, 19$).

4.2.7. Polyprenes

Polyprenes are oligomers and polymers of isoprene that occur in plants. They are synthesized in nature by an enzymatically catalyzed polymerization of isopentenyl diphosphate as monomer with dimethylallyl pyrophosphate as starter. The first step:

(4-40) $(CH_3)_2C=CH–CH_2–O–P_2O_6^{3-}$ + $CH_2=C(CH_3)–CH_2–CH_2–O–P_2O_6^{3-}$
$\longrightarrow$ $(CH_3)_2C=CH–CH_2–CH_2–C(CH_3)=CH–CH_2–O–P_2O_6^{3-}$ + H^+ + $P_2O_7^{4-}$

is followed by polyelimination reactions of other isopentenyl diphosphate molecules. It is unknown why some plants deliver cis-poly(isoprene)s and others trans polymers.

Natural rubber NR is a poly(isoprene) composed of ca. 95 % cis-1,4-units and ca. 3 % 3,4-units; it also contains aldehyde and epoxy groups. The tapping of rubber trees provides an aqueous latex with (20-60) % poly(isoprene) and smaller proportions of proteins, fats and antioxidants. Only a small portion of the latex is used directly for dip coating (e.g., rubber gloves). The major portion is coagulated to bales and processed to elastomers, mainly to rubber tires (Chapter 12).

Balata (Cariban: ?) and **gutta-percha** (Malay: *getah* = sap; *percha* = strip of cloth) are trans-1,4-poly(isoprene)s that are used for transmission belts, cable jackets, and golf balls. Some chewing gums are still manufactured from **chicle**, a mixture of 1,4-trans-poly(isoprene)s with terpenes; most chewing gums use poly(vinyl acetate) as their base, however.

4.3. Reactions of Macromolecules

Reactions of macromolecules may proceed with conservation, increase or decrease of the degree of polymerization. They may be desired (synthesis of new polymers or recycling of plastics waste) or undesired (chemical aging).

4.3.1. Polymer Analog Reactions

Polymer analog reactions transform ligands of a polymer completely and without side reactions into other ligands; the initial degree of polymerization remains constant. H.Staudinger used this strategy to prove the macromolecular character of polymers: if a polymer A can be converted into a differently structured polymer B and if both polymers have the same degree of polymerization by physical methods, then these polymers cannot be micelle colloids because of the difference in polymer–solvent interactions.

By reaction with acetic acid, hydroxyl groups of cellulose are completely esterified to **cellulose triacetate** with three acetate groups per glucose unit. At incomplete conversion of OH groups, totally acetylated chains are found besides completely unmodified cellulose chains: a direct acetylation of cellulose does not deliver *partially* acetylated cellulose molecules.

Cellulose triacetate could not be processed in the past due to lack of inexpensive solvents. It was thus partially saponified to **cellulose-2 1/2-acetate** (commonly called **cellulose acetate**). In industry, cellulose triacetate is therefore also called **primary acetate** and cellulose-2 1/2-acetate, **secondary acetate**. Cellulose-2 1/2-acetate was widely used as textile fiber but today is mainly processed into cigarette filters. Reaction of cellulose acetate with butyric acid results in **cellulose(acetate-co-butyrate)**, a special purpose thermoplastic.

Cellulose can be partially nitrated by a mixture of concentrated nitric acid and concentrated sulfuric acid (nitrating acid) to **cellulose nitrates** ("nitrocelluloses") with various degrees of substitution, DS. The desired DS can be obtained by varying the mixing ratio of the acids. DS = 2.8 is required for gun cotton and DS = 2.3 for **celluloid**, a cellulose nitrate plasticized by camphor.

Cellulose ethers result from the reaction of cellulose with oxiranes (ethylene oxide and/or propylene oxide). Oxiranes are multifunctional: more than one molecule of oxirane can be added to one hydroxyl group. One has thus to distinguish between the **degree of substitution**, DS, as the average number of reacted groups and the **degree of reaction**, DR, as the average number of reacted molecules, both with respect to a monomeric unit. The cellobiose unit shown below has a DR = 3; the glucose unit on the left has DS = 3 but the glucose unit on the right has DS = 2.

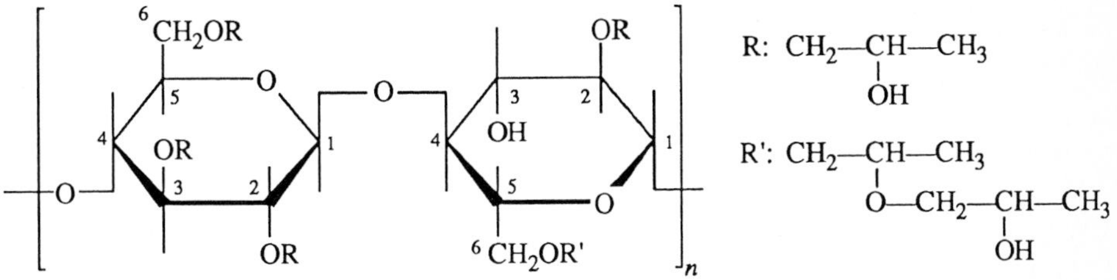

Poly(vinyl acetate) $+CH_2–CH(OOCCH_3)\frac{1}{n}$ is transesterified (but not saponified) industrially by butanol or methanol to **poly(vinyl alcohol)** $+CH_2–CH(OH)\frac{1}{n}$; the byproduct, butyl acetate, is an industrial solvent. This reaction is not 100 % polymer analog since the free radical polymerization of vinyl acetate is accompanied by chain transfer to ester groups. Branches $–O–CO–CH_2–[CH(OCOCH_3)–CH_2]_nR$ result; they are removed on transesterification. Poly(vinyl alcohol) dissolves in water. It is used as a size, emulsifier, adhesive, working material, copy layer for offset printing, etc., also, after cross-linking by formaldehyde as industrial fiber in East Asia.

Poly(vinyl alcohol) PVAL reacts with butyraldehyde C_3H_7CHO to **poly(vinyl butyral)**. Plasticized poly(vinyl butyral) constitutes the adhesive between the double panes of safety glass; on fracture, broken pieces of glass stick to the adhesive layer.

About 20 % of the OH groups of PVAL remain unreacted because some of the OH groups remain isolated upon reaction of adjacent groups in kinetically controlled ring closures. Theory predicts an unreacted proportion of $1/e^2 = 0.135$ if bifunctional reagents (such as aldehydes) are reacted with homopolymers that have head-to-tail structures and one functional group per monomeric unit. Different unreacted proportions are predicted for head-to-tail units with two functional groups per monomeric unit $(1/(2e))$, alternating head-to-tail and head-to-head units $(1/2)$, etc.

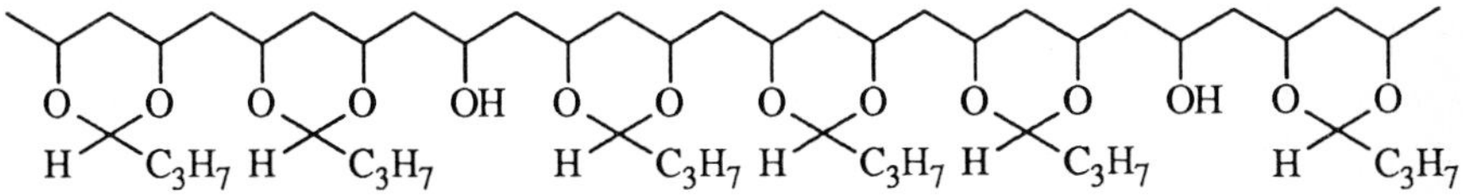

4.3.2. Chain Extensions

Block Formation

Chain extension reactions increase the degree of polymerization of the parent macromolecule. They are either block formations or grafting reactions.

Block copolymers are obtained by block coupling, block polymerization or block copolymerization. In **block coupling**, separately synthesized blocks are united according to $RA_n* + *B_mR' \rightarrow RA_n\text{-}B_mR'$. The active ends * may be functional groups (polycondensations, polyadditions); care must here be taken to avoid trans reactions. Coupling may also proceed between a macroanion $RA_n{}^\ominus$ and a macrocation $^\oplus B_mR'$. Poly(tetrahydrofuran) macrocations couple, for example, 100 % with poly(styrene) macroanions but only 20 % with poly(α-methyl styrene) macroanions. Anionic ends $RA_n{}^\ominus + {}^\ominus B_mR'$ can be coupled with the help of an agent $^\oplus X^\oplus$. Multiblock copolymers $*A_n\text{-}B_m...A_n\text{-}B_m*$ are obtained from bifunctional primary blocks $*A_n* + *B_m*$.

Block polymerizations consist of the successive addition of monomers to growing chains. These reactions usually proceed in two or three steps. Three-step processes start with monofunctional initiators, e.g., anions $R^\ominus$. First, an A block is formed by $R^\ominus + m\, A \rightarrow RA_m{}^\ominus$. Sufficiently basic macroanions $RA_m{}^\ominus$ then initiate the polymerization of B according to $RA_m{}^\ominus + n\, B \rightarrow R\text{-}A_m\text{-}B_n{}^\ominus$. Addition of p A (or p C) generates the desired triblock copolymers $R\text{-}A_m\text{-}B_n\text{-}A_p{}^\ominus$ (or $R\text{-}A_m\text{-}B_n\text{-}C_p{}^\ominus$). The polymerization is terminated by addition of $D^\oplus$. This strategy increases the probability of undesired termination reactions (and therefore the proportion of homopolymers and diblock copolymers) since three separate propagation reactions are involved.

Two-step processes are much more easy to control since they work with bifunctional initiators and require only two propagation reactions. An example is the sequence $^\ominus X^\ominus + 2\, m\, B \rightarrow {}^\ominus B_nXB_n{}^\ominus$ and $^\ominus B_nXB_n{}^\ominus + 2\, m\, A \rightarrow {}^\ominus A_mB_nXB_nA_m{}^\ominus$. This strategy is used in the industrial synthesis of poly(styrene)-*block*-poly(butadiene)-*block*-poly(styrene), a thermoplastic elastomer. The anionic polymerization is here initiated by hydrocarbon solutions of aromatic dilithium compounds in the presence of aromatic ethers. Sodium naphthalene cannot serve as initiator because it requires tetrahydrofuran THF as solvent. The anionic polymerization of butadiene proceeds however in THF with the formation of trans-1,4-butadiene units. These units are undesirable because *trans*-1,4-poly(butadiene) has a high glass temperature and insufficient elastic properties.

Block copolymerizations start with a mixture of both monomers. In anionic polymerizations, mixtures of monomers do not form random copolymers, however, because of the great differences in reactivity. They rather polymerize one after the other and form gradient copolymers (see also Fig. 3-18). This strategy is used industrially to synthesize poly(styrene)-*block*-poly(isoprene)-*block*-poly(styrene).

Grafting Reactions

Grafting creates side chains on a parent polymer chain of different constitution. It is performed to modify polymer surfaces (e.g., to increase hydrophilicity), to synthesize thermoplastic elastomers (e.g., isobutylene onto poly(ethylene)), or to prepare compatibilizers for polymer blends.

Grafting is either *to* or *from* the parent polymer. Grafting from the parent polymer is especially easy if the latter contains reactive groups, e.g., hydroxyl groups. Unreactive polymers are irradiated with γ-rays which generates radical sites that can initiate free radical graft polymerizations. Irradiation of a parent polymer in the presence of monomers also causes undesired homopolymerizations. In this case, grafting to parent polymers is preferred, especially by chain transfer from initiator radicals. The effectiveness of such radicals is not only a function of the initiator but also depends on the parent polymers. AIBN generates, e.g., polyradicals of poly(vinyl acetate) but not those of poly(styrene). Grafting is used industrially to produce impact-resistant polymer composites by grafting a mixture of styrene and acrylonitrile onto copolymers of butadiene and acrylonitrile (ABS polymers, see Chapter 14).

4.3.3. Cross-Linking Reactions

Chemical cross-links interconnect macromolecules via covalent bonds to "infinitely large" molecules. Cross-linking reactions occur simultaneously with chain formation if multifunctional monomers are used. Examples are polycondensations leading to phenolic resins and amino resins (Section 4.1.9.), polyadditions to polyurethanes (Section 4.1.8.), free radical polymerizations of diallyl compounds (Section 3.7.6.), and radical copolymerizations of styrene and divinyl benzenes or glycol methacrylate and glycol dimethacrylate, respectively (Table 3-6).

Oligomers and polymers may also be cross-linked by post-polymerization reactions. Epoxy resins are cured by amines or acids (Section 4.1.8.). Unsaturated polyesters (Eqn.(4-5)) are hardened by free radical copolymerization with styrene or methyl methacrylate. The cross-linking of rubbers is known as **vulcanization** since it was first applied to natural rubber by sulfur and heat, the attributes of the Roman God of Fire, Vulcanus (see Chapter 12).

Primary molecules are polymer molecules before cross-linking. A complete cross-linking unites *all* primary molecules to a network; this requires two trifunctional or one tetrafunctional cross-link per primary molecule:

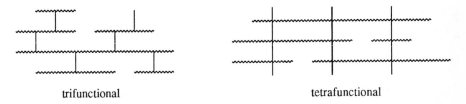

trifunctional   tetrafunctional

The probability P of finding, for example, a tetrafunctional cross-link in a primary molecule is given by the probability p_u that a monomeric unit carries a cross-linking site. The higher it is, the larger is the degree of polymerization, X, of the primary molecule. Since $X = 1$ leads to a linear molecule, values of $X \geq 2$ are required. The probability is thus $P = p_u(X - 1)$. In non-uniform polymers, primary molecules with higher degrees of polymerization have a greater chance of being cross-linked than those with lower ones. The average probability $\overline{P}$ must thus depend on the mass fractions w_i of the various primary molecules i, i.e., $\overline{P} = \Sigma_i P_i = \Sigma_i p_u w_i(X_i - 1) = p_u(\Sigma_i w_i X_i - \Sigma_i w_i) = p_u(\overline{X}_w - 1)$. Cross-linking occurs if each primary molecule possesses one cross-link, i.e., at $\overline{P} = 1$. It results in $p_u = 1/(\overline{X}_w - 1) \approx 1/\overline{X}_w$.

4.3.4. Degradation

In the polymer literature, "degradation" refers to a decrease of the degree of polymerization, an uncontrolled change of constitution of monomeric units, or a combination of both. Reactions that lower degrees of polymerization are reverse polymerization reactions. In analogy to polymerizations (see Section 3.1.1.), one can distinguish four different types:

Chain scissions at random sites along the polymer chains
   - without participation of low molar mass molecules ("retro-polyadditions"),
   - with participation ("retro-polycondensations", e.g., hydrolyses of polyesters).
Depolymerization from the chain ends leading to monomer
   - without participation of low molar mass molecules ("retro-chain polymerization"),
   - with participation ("retro-polyelimination").

Chain Scission

One scission per linear chain reduces the number-average degree of polymerization to one-half of its original value $\overline{X}_{n,o}$. If N_b bonds are broken per chain, the number average-degree of polymerization $\overline{X}_{n,o}$ is lowered to $\overline{X}_n = \overline{X}_{n,o}/(1 + N_b)$. Degrees of polymerization, here, relate to *all* molecules, including monomers ($X_i = 1$). The **degree of degradation** is defined as the fraction f_b of broken bonds:

$$(4\text{-}41) \quad f_b \equiv \frac{N_b}{\overline{X}_{n,o}-1} = \left(\frac{\overline{X}_{n,o}}{\overline{X}_n}-1\right)\left(\frac{1}{\overline{X}_{n,o}-1}\right) = \frac{\overline{X}_{n,o}-\overline{X}_n}{\overline{X}_n(\overline{X}_{n,o}-1)} \approx \frac{1}{\overline{X}_n} - \frac{1}{\overline{X}_{n,o}}$$

where the last expression is an approximation for high degrees of polymerization.

The rate of chain scission is controlled by the concentrations of chain bonds B and catalyst C, i.e., $-d[B]/dt = k_b[B][C]$ or $[B] = [B]_o \exp(-k_b[C]t)$. The initial scission is small at high degrees of polymerization: $[B]/[B]_o \approx 1 \approx \exp(-k_b[C]t) \approx 1 - k_b[C]t$. Since also $f_b \equiv ([B]_o - [B])/[B]_o$, one obtains with Eqn.(4-41):

$$(4\text{-}42) \quad 1/\overline{X}_n = (1/\overline{X}_{n,o}) + k_b[C]t$$

Inverse number-average degrees of polymerization thus decrease linearly with time. The time dependence of the mass-average degrees of polymerization, however, is controlled by the molar mass distribution. A molecularly uniform polymer first becomes non-uniform upon chain scission but will finally be uniform again at $X = 1$.

Rate constants k_b are independent of X at high degrees of polymerization, X. At low degrees of polymerization, they depend on the number N_i of bonds, the rate constant k_m of the main chain bonds, and the rate constant k_e of bonds at the end of the chains according to $k_b = k_m + (k_e/N_i)$. If only one bond at each end is different from the main chain bonds of a linear chain, then $k_e = 2(k_t - k_m)$. The difference $k_t - k_m$ may be positive (acid hydrolysis of poly(1,4-D-anhydroglucose)s at 18°C) or negative (alkali hydrolysis of poly(glycine)s at 20°C).

Not all main chain bonds may be cleaved with the same rate because polymer chains may contain a few "wrong" structures which provide **weak bonds**. These structures may be other sugar units in the poly(glucose) chains of cellulose (Section 4.2.5.), peroxy units -O-O- from traces of oxygen in free radical vinyl polymerizations (Section 3.7.), or regioisomerism. For example, head-to-tail and head-to-head structures are oxidized with different rates in poly(vinyl alcohol).

Depolymerization

In depolymerizations, one monomeric unit after the other is successively split off. They can proceed only at active chain ends P_i^*, for example, at living anionic polymers according to $P_i^\ominus \rightarrow P_{i-1}^\ominus + M$. Dead polymers must first undergo homolytic chain scission $P_{i+j} \rightarrow P_i^\bullet + {}^\bullet P_j$. The radical chain ends make the chain active for depolymerizations $P_i^\bullet \rightarrow P_{i-1}^\bullet + M$, etc. The number-average degree of polymerization of the remaining polymer stays constant during such depolymerizations if the initial molar mass is low. It decreases linearly with the mole fraction x_M of the leaving monomer, however, if the initial molar mass is high (Fig. 4-9).

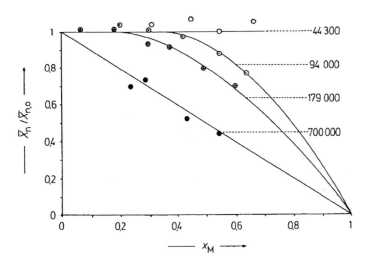

Fig. 4-9 Variation of number-average degrees of polymerization with the mole fraction x_M of the released monomer in the system during the thermal degradation of poly(methyl methacrylate). With permission of the Royal Society of Chemistry [4].

This behavior is caused by the presence of chain transfer and termination reactions in addition to depropagation. In analogy to the kinetic chain length v of chain polymerization, a **zip length** Ξ can be defined. This zip length gives the number of monomer molecules that can be split off from a macroradical before that radical is annihilated by either termination or chain transfer. Small initial degrees of polymerization and large zip lengths allow total depolymerization of activated macromolecules while the number-average degree of polymerization of the non-activated macromolecules remains constant (see $\overline{M}_n$ = 44 300 g/mol in Fig. 4-9).

Another relationship between $\overline{M}_n$ and x_M is obtained if the zip length is far smaller than the initial degree of polymerization. A polymer contains initially $N_{u,o}$ monomeric units in $N_{P,o}$ molecules, i.e., the number-average degree of polymerization is $\overline{X}_{n,o} = N_{u,o}/N_{P,o}$. A depolymerization in a closed system (no removal of monomer molecules) preserves the total number $N_{u,o} = N_{u,P} + N_M$ of monomeric units in polymers ($N_{u,P}$) and monomers (N_M). The number of molecules is increased to $N = N_P + N_M = N_o(1 + N_b) = N_{P,o}(1 + N_b)$ by depolymerization where N_b is the number of broken bonds per initial polymer molecule. Each broken bond generates a monomer molecule, $N_b = N_M/N_{P,o}$, if initial (random) chain scissions are neglected. With the mole fraction $x_M = N_M/N_{u,o}$ of monomer molecules with respect to all monomeric units in monomers and polymers, it follows:

$$(4\text{-}43) \quad \overline{X}_n = \frac{N_{u,P}}{N_P} = \frac{N_{u,o} - N_M}{N - N_M} = \frac{N_{u,o} - N_M}{N_{P,o}\{1 + N_b\} - N_M} = \frac{N_{u,o} - N_M}{N_{P,o}\{1 + (N_M/N_{P,o})\} - N_M}$$

$$\overline{X}_n = \overline{X}_{n,o}(1 - x_M)$$

The number-average degree of polymerization of the remaining polymer thus decreases linearly with the mole fraction of generated monomer molecules (see $\overline{M}_n = 700\ 000$ g/mol in Fig. 4-9). More complicated expressions are obtained for non-uniform polymers and for those with intermediate molar masses.

Zip lengths vary between several hundred for poly(methyl methacrylate) and poly(α-methylstyrene) to 3.1 for poly(styrene) and 0.01 for poly(ethylene). The thermal decomposition of polymers in vacuum at 300°C thus delivers 100 % monomer from poly(α-methylstyrene) but only 65 mol-% (= 42 wt-%) from poly(styrene) and 21 mol-% (= 3 wt-%) from poly(ethylene). Recycling polymers to raw materials by depolymerization is thus efficient for only a few monomers.

The easy depolymerization of poly(1-butene-*alt*-sulfur dioxide) is utilized in the manufacture of printing plates, printed circuit boards, and microelectronic chips. In the latter case, a thin layer of SiO_2 is deposited on silicon wafers, followed by a layer of the polymer. Irradiation by X-ray or electron beams through a mask causes the polymer to depolymerize at the exposed areas; the depolymerized material is removed by vacuum (dry process). SiO_2 is etched away at the exposed areas and the liberated Si is subsequently doped with Ga, As, etc. During etching and doping, the polymer protects the unexposed areas; it thus acts as a **positive resist**.

Poly(methyl methacrylate) is also a positive resist; the depolymerized material, however, is removed by solvents (wet process). Other polymers are used as **negative resists**, for example poly(styrene) which cross-links on irradiation through a mask. The unexposed areas can then be removed by solvents.

Literature

G.Allen, J.C.Bevington, eds., Comprehensive Polymer Science, Pergamon Press, Oxford 1989; Vols. **3-6**, G.C.Eastmond, A.Ledwith, S.Russo, P.Sigwalt, eds., Vol. 3: Chain Polymerization I ; Vol. 4: Chain Polymerization II; Vol. 5: Step Polymerization; Vol. 6: Polymer Reactions.

4.1. POLYCONDENSATIONS AND POLYADDITIONS
R.W.Lenz, Organic Chemistry of Synthetic High Polymers, Interscience, New York 1967
F.Millich, C.E.Carraher, Eds., Interfacial Synthesis, Dekker, New York 1977 (2 vols.)
P.Rempp, E.W.Merrill, Polymer Synthesis, Hüthig and Wepf, Basel, 2nd ed. 1991
H.R.Kricheldorf, Ed., Handbook of Polymer Synthesis, Dekker, New York 1992 (2 vols.)
G.Odian, Principles of Polymerization, Wiley, New York, 3rd ed. 1992
W.J.Mijs, Ed., New Methods for Polymer Synthesis, Plenum, New York 1992
G.B.Butler, Cyclopolymerization and Cyclocopolymerization, Dekker, New York 1992
M.Kucera, Mechanism and Kinetics of Addition Polymerizations, Elsevier, Amsterdam 1992
R.K.Sadhir, R.M.Luck, Eds, Expanding Monomers: Synthesis, Characterization and Applications, CRC Press, Boca Raton, FL 1992
D.J.Brunelle, Ed., Ring-Opening Polymerization, Hanser, Munich 1993
Y.Yamashita, Ed., Chemistry and Industry of Macromonomers, Hüthig and Wepf, Basel 1993

4.2. BIOLOGICAL POLYMERIZATIONS
R.E.Dickerson, I.Geis, The Structure and Action of Proteins, Benjamin, New York 1981
J.V.F.Vincent, Structural Biomaterials, Wiley, New York 1982
T.E.Creighton, Proteins: Structures and Molecular Properties, Freeman, New York 1984
R.W.Stoddart, The Biochemistry of Polysaccharides, Croom Helm, Beckenham, UK, 1984
R.A.Young, R.M.Rowell, Cellulose. Structure, Modification and Hydrolysis, Wiley, New York 1986
W.Saenger, Principles of Nucleic Acid Structure, Springer, Berlin 1988
O.Kramer, Ed., Biological and Synthetic Polymer Networks, Elsevier, Applied Science, New York 1988
G.M.Blackburn, M.J.Gait, Nucleic Acids in Chemistry and Biology, Oxford University Press, Oxford 1990
R.H.Pain, eds., Mechanisms of Protein Folding, Oxford University Press, New York 1994
J.F.Kennedy, G.O.Phillips, P.A.Williams, Eds., Cellulosics: Chemical, Biochemical and Materials Aspects, Ellis Horwood, New York 1993
R.L.Whistler, J.N.BeMiller, Eds., Industrial Gums, Academic Press, San Diego, 3rd ed. 1993
R.Gilbert, Ed., Cellulosic Polymers, Hanser, Munich 1994

4.3. REACTIONS OF MACROMOLECULES
P.E.Cassidy, Thermally Stable Polymers, Dekker, New York 1980
N.Grassie, G.Scott, Polymer Degradation and Stabilisation, Cambridge University Press, Cambridge, UK, 1988
O.Guven, Crosslinking and Scission in Polymers, Kluwer, Dordrecht 1990
G.Griffin, Ed., Chemistry and Technology of Biodegradable Polymers, Chapman and Hall, New York 1994
E.A.Bekturov, S.E.Kudaibergenov, Polymeric Catalysis, Hüthig and Wepf, Zug, Switzerland, 1994

References

[1] P.J.Flory, J.Am.Chem.Soc. **61** (1939) 3334, Table I
[2] H.Dostal, R.Raff, Mh. Chemie **68** (1936) 188, Table 7
[3] P.J.Flory, J.Am.Chem.Soc. **63** (1941) 3083, Fig. 2
[4] N.Grassie, H.W.Melville, Proc.R.Soc. [London] **A 199** (1949) 14, Fig. 2

5.   Size and Shape of Molecules

5.1. Introduction

Atoms of macromolecules can be interconnected in various ways which in turn leads to an extraordinary variety of chemical structures (Chapter 2) and thus to different sizes and shapes of single polymer molecules. Dimensions and shapes of *single* polymer molecules can only be studied in solution because these molecules degrade at the temperatures required for their vaporization (Section 4.3.4.). Solvents may or may not interact with polymer molecules which generates additional varieties of sizes and shapes. In melts, macromolecules are "solvents" for their own kind.

Depending on constitution, configuration, solvent and temperature, single linear chains may form unperturbed coils (Section 5.3.), wormlike coils (Section 5.4.), perturbed coils (Section 5.5.) and even various more or less compact structures that resemble Euclidian bodies (Section 5.6.). Sizes and shapes of these basic physical structures are modified by various types of branching as well as by intramolecular and intermolecular association. The external shape, the size and the internal physical structure of single molecules determine physical properties of polymers much more than constitution and configuration.

Sizes and shapes of single macromolecules can be evaluated by a number of experimental methods. Compact spheroidal molecules (spheres, ellipsoids) as well as rods and coiled molecules with large chain diameters can be seen by electron microscopy or by atomic force microscopy. Pictures of molecules do not necessarily reveal the true dimensions of the "free", three-dimensional molecules, however, because they may be affected by experimental techniques (collapse of internal structures by loss of solvent, spreading of three-dimensional structures on a two-dimensional plane, etc.).

Dimensions of molecules are obtained directly from their scattering of electromagnetic radiation (Section 5.2.). They can also be obtained indirectly from chromatography (Section 2.2.4.), from the viscosity of dilute solutions (Section 5.8.), and from other hydrodynamic properties (Chapter 7). These data also allow conclusions about the shapes of molecules since dimensions of homologous macromolecules often change systematically with the degree of polymerization,

5.2. Scattering Methods

Scattering methods are the most important methods for the direct determination of polymer shapes and dimensions. In particles, incoming electromagnetic waves keep shifting electrons and atomic nuclei relative to each other. This shifting creates induced dipoles which follow the oscillating electric field with the same frequency and produce scattered radiation (see physics textbooks). The more dipoles are generated per particle, i.e., per its mass or molar mass, the higher is the scattering intensity. The angular dependence of the scattered radiation depends on the distribution of dipoles, i.e., dimensions and shapes of particles.

The theory of scattering applies to any wavelength. Most important for polymer science are visible light (wavelengths between 300 nm and 700 nm), neutron beams (0.1 nm – 1 nm), X-rays (0.02 nm – 2 nm) and electron beams (ca. 0.01 nm). *Static light scattering* depends on the difference in polarizabilities of polymers and solvents. *Small angle X-ray scattering*, SAXS, registers differences in electron densities of molecules and *small angle neutron scattering*, SANS, differences in coherent neutron scattering lengths of monomeric units and solvent molecules. Static light scattering is the most affordable method, small angle neutron scattering the most expensive.

5.2.1. Static Light Scattering

Infinite Dilution

In gases at low pressure, each molecule scatters independently of the other molecules. At the scattering angle ϑ relative to the incident beam, the scattering intensity of a molecule is i_ϑ.. The relative scattering intensity i_ϑ/I_0 is defined as the ratio of scattering intensity i_ϑ to the intensity I_0 of the incident light. Since intensities are given by the energy that falls on a unit area at a unit distance, an intensity $I_\vartheta = i_\vartheta/L^2$ is observed at a distance L from the light source. The ratio $R_\vartheta \equiv i_\vartheta/I_0 = I_\vartheta L^2/I_0$ is called the **Rayleigh ratio**.

From the electromagnetic theory for unpolarized (natural) light, it follows that the Rayleigh ratio of a system with the number concentration $N/V = cN_A/M$ of particles can be obtained from the polarizability α of the particles:

$$(5\text{-}1) \quad R_\vartheta = (I_\vartheta L^2/I_0) = (8\ \pi^4 \alpha^2/\lambda_0^4)[\sin^2 \vartheta_h + \sin^2 \vartheta_v](N/V)$$

where $c = m/V$ = mass concentration, M = molar mass, N_A = Avogadro number, λ_0 = wavelength of incident light. The angular term in square brackets relates to the different scattering of the horizontal (h) and vertical (v) components of unpolarized incident light; at a scattering angle ϑ, $[\sin^2 \vartheta_h + \sin^2 \vartheta_v] = (1 + \cos^2\vartheta)$.

According to **Clausius–Mosotti**, the polarizability of a gas can be calculated from the relative permittivity ε_r of the gas via $\alpha = \{3/(4\ \pi\ N/V)\}[(\varepsilon_r - 1)/(\varepsilon_r + 2)]$. Permittivities ε_r of dilute solutions are not referred to the vacuum (relative permittivity 1) but to the solvent (relative permittivity $\varepsilon_{r,1}$). For solutions, the square-bracket term of the Clausius–Mosotti equation thus has to be replaced by $[(\varepsilon_r - \varepsilon_{r,1})/(\varepsilon_r + 2\varepsilon_{r,1})]$. The relative permittivity of a dilute solution, however, is only slightly higher than that of the solvent itself, thus $\varepsilon_r + 2\varepsilon_{r,1} \approx 3\ \varepsilon_r$.

According to **Maxwell**, relative permittivities can be replaced by the squares of refractive indices n, i.e., $\varepsilon_r - \varepsilon_{r,1} = n^2 - n_1^2$. Refractive indices n of dilute solutions are concentration dependent which can be expressed by $n = n_1 + (dn/dc)c$, where dn/dc is the **refractive index increment**. The square of this equation can be approximated by $n^2 \approx n_1^2 + 2\, n_1 (dn/dc)c$ since the quadratic term $(dn/dc)^2 c^2$ is very small compared to the other terms. Insertion of all these expressions into the Eqn.(5-1) delivers:

(5-2)

$$R_\vartheta = \frac{2\,\pi^2 n_1^2 (dn/dc)^2 (1+\cos^2 \vartheta)cM}{N_A \lambda_o^4} = \left(\frac{4\,\pi^2 n_1^2 (dn/dc)^2}{N_A \lambda_o^4}\right)\left(\frac{(1+\cos^2 \vartheta)}{2}\right)cM = K_\vartheta cM$$

Eqn.(5-2) allows the *absolute* calculation of molar masses M of molecules. No model needs to be assumed since M is the only unknown and all other parameters R_ϑ, n_1, dn/dc, ϑ, λ_o and c can be determined by experiment. Note that Eqn.(5-2) applies to infinite dilution (it is based on the behavior of a gas at $p \to 0$): the molar mass M is an apparent molar mass that has to be extrapolated to infinite dilution (see below).

Molar masses by Eqn.(5-2) depend also on the scattering angle if the dimensions of particles are larger than ca. $0.05\, \lambda_o$. Thus, they have to be extrapolated to the scattering angle $\vartheta \to 0$ (see below). Modern scattering photometers employ laser beams as the light source; the coherent laser light allows one to determine R_ϑ at very small angles without an extrapolation to $\vartheta \to 0$ (LALLS = low-angle laser light scattering), thus giving $R_\vartheta \approx R_0$. The scattering intensity R_0 depends only on the concentration. It is not sensitive to dust particles which, due to their large sizes, strongly affect the scattering behavior at angles $\vartheta > 0$ (see Fig. 5-1), even at small concentrations.

The refractive index increment dn/dc is independent of the molar mass at large degrees of polymerization because it is no longer affected by end groups. The **optical constant** K_ϑ is thus a system-dependent constant. Since the scattering by molecules is additive, one obtains $R_\vartheta = \Sigma_i (R_\vartheta)_i = \Sigma_i K_\vartheta c_i M_i = K_\vartheta \Sigma_i c_i M_i = K_\vartheta c \overline{M}_w$ for polymers that are nonuniform with respect to molar mass: the M in Eqn.(5-2) is the mass-average molar mass (see Eqn.(2-6)).

For copolymers that are nonuniform with respect to the composition of molecules, the refractive index increment is no longer the same for all copolymer molecules. Since dn/dc is now a variable, it can no longer be part of the optical constant K_ϑ but has to be included in the sum $\Sigma_i c_i M_i$. The derivation shows that the molar mass of Eqn.(5-2) is then an apparent mass-average molar mass $\overline{M}_{w,app}$ that depends on the average refractive index increment of the copolymer, $(dn/dc)_{copolymer}$, and the refractive index increments $(dn/dc)_a$ and $(dn/dc)_b$ of the two homopolymers:

(5-3) $$\overline{M}_{w,app}/\overline{M}_w = 1 + 2\, v_z^{(1)}\left(\frac{(dn/dc)_a - (dn/dc)_b}{(dn/dc)_{copolymer}}\right) + v_z^{(2)}\left(\frac{(dn/dc)_a - (dn/dc)_b}{(dn/dc)_{copolymer}}\right)^2$$

$v_z^{(1)}$ and $v_z^{(2)}$ are the first and second moment of the z-distribution of the differences $w_{a,i} - w_{a,cp}$ of the mass fractions of "a" units in molecules i to the average mass fraction of "a" units in the copolymer. True mass-average molar masses are obtained only if the second and third terms of Eqn.(5-3) are zero.

Concentration Dependence

The outline above refers to small, isotropic molecules that are distributed at random and move independently of each other. In a first approximation, fluctuations of polymer concentration and solvent density caused by Brownian motion are independent of each other in space and time. The scattering intensity I_ϑ of very dilute solutions is simply the difference of the scattering intensities of solution and solvent, i.e., $I_\vartheta = I_{soln} - I_{solv}$.

This is no longer true for higher polymer concentrations. Here, the instantaneous polarizability α of a volume element deviates by $\Delta\alpha$ from the average polarizability $\bar{\alpha}$ of the solution. In Eqn. (5-1), α^2 becomes $\alpha^2 = (\bar{\alpha} + \Delta\alpha)^2 = \bar{\alpha}^2 + 2\,\bar{\alpha}\Delta\alpha + (\Delta\alpha)^2$. Neither $\bar{\alpha}$ nor $\Delta\alpha$ add to the fluctuation of polarizabilities, however, since the average polarizability $\bar{\alpha}$ is the same for all volume elements and the average fluctuation $\Delta\alpha$ equals zero. Only the last term, $(\Delta\alpha)^2$, is important. This term is an average, $\langle(\Delta\alpha)^2\rangle$. Its calculation is cumbersome.

The essential result of the calculation is that Eqn.(5-2) has to be extended by the average of the square of concentration fluctuations, $\langle(\Delta c)^2\rangle$. These fluctuations are proportional to the concentration: Eqn.(5-2) has to be normalized by c^2 in order to arrive at the correct physical units. Since $(1 + \cos^2\vartheta)/2 = 1$ for $\vartheta = 0$, one gets:

$$(5\text{-}4) \qquad R_0 = K_0[\langle(\Delta c)^2\rangle/c^2]c\,\overline{M}_w$$

Thermodynamics traces fluctuations of concentrations to the concentration dependence of the Gibbs energy G of the system. Introducing the chemical potential μ_1 of the solvent, one obtains $\langle(\Delta c)^2\rangle = k_B T/(\partial^2 G/\partial c^2)_{p,T} = k_B T^* V_{1,m} c(N/V)/(-\partial\mu_1/\partial c)$, where $^*V_{1,m}$ is the partial molar volume of the solvent. The concentration dependence of the chemical potential is given by $-\partial\mu_1/\partial c = RT^* V_{1,m}(M^{-1} + 2\,A_2 c + 3\,A_3 c^2 + ..)$ (see Eqn.(6-23)). With $k_B = R/N_A$ and $M = \overline{M}_w$ (see above), Eqn.(5-4) thus becomes:

$$(5\text{-}5) \qquad K_0 c/R_0 = (1/\overline{M}_w) + 2\,A_2 c + 3\,A_3 c^2 + ...$$

At $c \to 0$, a plot of $K_0 c/R_0 = f(c)$ furnishes the inverse mass-average of molar mass. The initial slope of this function delivers $2\,A_2$ where A_2 is the so-called second virial coefficient (see Section 6.3.1.). Second virial coefficients from light scattering experiments on molecularly nonuniform polymers are complex average quantities that differ from those obtained by osmotic measurements.

Angular Dependence

Small molecules possess only one scattering center. However, a large molecule contains many scattering centers if the dimension of the molecule is greater than ca. $0.05\ \lambda_0$. Light scattered by two centers at the same angle ϑ differs in its path length by $\Delta = \overrightarrow{DB} = \overrightarrow{AB} - \overrightarrow{AD} = \overrightarrow{AB}(1 - \cos\vartheta)$ (Fig. 5-1). Waves emanating from different scattering centers all originate from the same light source; thus they can interfere. The greater the scattering angle, the larger is the interference. Thus, the ratio of Rayleigh ratios at two scattering angles is a measure of the interference.

This ratio is called **dissymmetry** z; it usually refers to angles of 45° and 135° ($z \equiv R_{45}/R_{135}$). Dissymmetries increase with increasing size of the molecule. It is for this reason that light scattering experiments are sensitive to dust at large scattering angles because dust particles are much larger than molecules (hence the use of LALLS for molar mass determinations (see above).

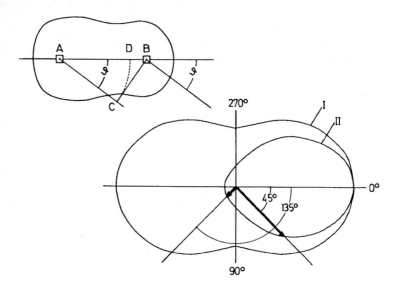

Fig. 5-1   Above left: Schematic representation of the phase shift by light scattering from two scattering centers A and B in a large particle. Below right: Scattering diagram for unpolarized incident light. I: Small particles; II: large particles (here calculated for a dilute solution of monodisperse spheres with a diameter $\lambda_o/2$).

Theory allows one to calculate the **radius of gyration**, s, of molecules from dissymmetries if the particle shape is known (sphere, rod, coil, etc.). This radius is more easily obtained without any assumptions about the particle shape if one measures the complete angular dependence of Rayleigh ratios. The mathematical functions describing this angular dependence are complicated, however. One thus uses approximations, usually the **Zimm scattering function** $P(\vartheta)$ for the limiting case $\vartheta \to 0$:

(5-6) $P(\vartheta) = R_\vartheta/R_0 \ = 1 - (1/3) \ [(4 \ \pi n_1)/\lambda_o]^2 \ [\sin^2(\vartheta \ /2)] \ \langle s^2 \rangle_z = 1 - (Q_\vartheta^2/3) \ \langle s^2 \rangle_z$

where $Q_\vartheta \equiv 4 \ \pi n_1 [\sin (\vartheta \ /2)]/\lambda_o$ and $\langle s^2 \rangle_z$ = z-average of the mean-square radius of gyration (see below).

The Zimm scattering function applies only to infinite dilution. At concentrations $c > 0$, the scattering function $P(\vartheta)$ has to be modified. In the case of low concentrations, small scattering angles, and large molecules, Eqn.(5-5) becomes:

(5-7) $\dfrac{K_\vartheta c}{R_\vartheta} = \dfrac{1}{\overline{M}_w P(\vartheta)} + \dfrac{2 \ A_2 c}{P''(\vartheta)} + \ ...$

$K_\vartheta c/R_\vartheta$ is plotted against $[\sin^2(\vartheta/2) + kc]$ in a **Zimm diagram** (Fig. 5-2) and extrapolated to both zero scattering angle (Eqn.(5-6)) and zero concentration (Eqn.(5-5)). k is an arbitrarily chosen constant whose sole purpose is to give a good spread to the grid-shaped plot (Fig. 5-2).

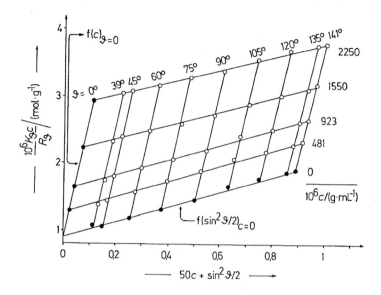

Fig. 5-2 Zimm diagram of a poly(vinyl acetate) in butanone at 25°C; k = 50 mL/g. The slope $f(c)$ for $\vartheta \to 0$ indicates A_2, the slope $f(\sin^2 \vartheta /2)$ for $c \to 0$ relates to the radius of gyration.

For polymolecular polymers, s^2 appears as a z-average $\langle s^2 \rangle_z$. Since $\langle s^2 \rangle \sim 1 - P(\vartheta)$ and, at $c \to 0$, also $P(\vartheta) = R_\vartheta /(K_\vartheta c \, \overline{M}_w)$ according to Eqn.(5-7), summation of the contributions by all components i delivers with $c_i = m_i/V$ and Eqn. (2-7):

$$(5\text{-}8) \qquad P(\vartheta) = \frac{R_\vartheta}{K_\vartheta c \overline{M}_w} = \frac{\sum_i K_\vartheta c_i (\overline{M}_w)_i P_i(\vartheta)}{\sum_i K_\vartheta c_i (\overline{M}_w)_i} = \frac{\sum_i z_i P_i(\vartheta)}{\sum_i z_i} \equiv \overline{P}_z(\vartheta)$$

5.2.2. Small Angle X-Ray and Neutron Scattering

Static scattering experiments can determine radii of gyration of $s \geq 0.05 \, \lambda_o$, i.e., radii greater than ca. 22 nm for blue light (λ_o = 436 nm). Visible light is thus a fairly insensitive probe. X-rays have much smaller wavelengths and can therefore measure much smaller radii of gyration of ca. 1 nm. This, however, requires experiments at very small scattering angles:

Small angle X-ray scattering SAXS is sensitive to the differences in electron densities of solvents and polymers. The optical constant (at ϑ = 0°) here becomes $K_{SAXS} = N_A i_e (d\rho_e/dc)^2$ instead of $K_{LS} = [4\,\pi^2 n_1^{\,2}(dn/dc)^2]/[N_A \lambda_o^{\,4}]$ (light scattering); i_e = scattering power of an electron and $d\rho_e/dc$ = specific electron density increment.

In order to get the same effect, the quantity $(1/3)\,[(4\,\pi n_1)/\lambda_0]^2\,[\sin^2(\vartheta/2)]\,\langle s^2\rangle$ (Eqn. (5-6)) must be the same for small angle X-ray scattering and static light scattering. What is observed by light scattering at $\lambda_0 = 436$ nm and $90°$ in a solvent of $n_1 = 1.45$ (i.e., $\lambda = 436$ nm/$1.45 \approx 300$ nm) has to be measured by SAXS with $\lambda = 0.1$ nm at an angle of $\vartheta = 0.027°$! At these small angles, the scattering function can be approximated by the **Guinier function** $P(\vartheta) = \exp(-\,Q^2\langle s^2\rangle/3)$; the radius of gyration, however, refers to the distribution of electrons and not to that of masses.

At $Qs \approx 1$, the numerical value of the radius of gyration is underestimated by the Guinier function $P(\vartheta) = \exp(-\,Q_\vartheta^2 s^2/3)$ and overestimated by the Zimm function $P(\vartheta) = 1 - (Q_\vartheta^2 s^2/3)$. SAXS delivered $s = 11.4$ nm (Zimm) and $s = 8.7$ nm (Guinier) for a 32-arm star poly(butadiene) polymer ($M_r = 280\,000$) whereas static light scattering provided $s = 10.0$ nm (Zimm).

Small angle neutron scattering measures the scattering of neutrons by atoms. The optical constant is given by $K_{\mathrm{SANS}} = N_A[a_u - a_1(*V_{u,m}/*V_{1,m})]^2/M_u^2$ where $*V_m = $ partial molar volume (monomeric unit u, solvent 1) and $M_u = $ molar mass of a mer. The coherent scattering lengths a_u (monomeric units) and a_1 (solvent molecules) are, respectively, the sums of scattering lengths b of atoms in mers and solvent molecules. In general, one measures undeuterated ("protonated") polymers in deuterated solvents (or vice versa) because of the great difference of coherent scattering lengths of hydrogen atoms ($-\,0.374\cdot10^{-12}$ cm) and deuterium atoms ($0.667\cdot10^{-12}$ cm).

5.3. Ideal Coil Molecules

5.3.1. Macroconformations

The simplest physical structure of a macromolecule consists of a linear chain composed of N_c chain atoms with $N = N_c - 1$ bonds of identical length b. The *mathematically* possible maximum length of such a chain is $L_{\mathrm{chain}} = Nb$ (Fig. 5-3). This length was correctly called "contour length" in the older literature because it paces off the contour of the chain.

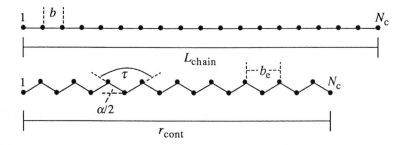

Fig. 5-3   Chain length L_{chain} and contour length r_{cont} of a chain in all-trans conformation. $N_c = 19$ chain atoms, $N = 18$ chain bonds, $N_e = 9$ effective bonds, $b = $ bond length, $b_e = $ effective bond length (= crystallographic length for vinyl polymers), $\tau = $ bond angle (valence angle), $\alpha = 180° - \tau$ = complementary angle to the bond angle τ.

IUPAC designates the *physically* possible maximum length as the **contour length** r_{cont} of a chain. The contour length of a chain in all-trans conformation with bond angles τ is given by geometry as:

(5-9) $r_{cont} = Nb \sin(\tau/2) = N_e b_e$

The contour length thus equals the end-to-end distance of such a fully extended chain (Fig. 5-3). The projection of the bonds onto the end-to-end vector is the effective bond length b_e which, for chains $-\!\!\!+\!CH_2\!-\!CHR\!+\!\!\!-_{\overline{n}}$, usually refers to the monomer unit $-CH_2\!-\!CHR\!-$, i.e., to two bonds ($N_e = N/2$). In this case, the effective bond length is identical to the crystallographic length of a monomeric unit (see Chapter 8). Vinyl polymers $-\!\!\!+\!CH_2\!-\!CHR\!+\!\!\!-_{\overline{n}}$ have effective bond lengths of $b_e = 0.254_6$ nm since $b = 0.154$ nm and $\tau = 111.5°$.

Fully extended chains represent a special type of polymer chain with periodic microconformations (see Section 2.4.), for example, all-trans conformations in Fig. 5-3. These and other periodic sequences of microconformations are observed in crystalline chain assemblies (Chapter 8). Chain molecules in solution usually adopt aperiodic sequences of microconformations which lead to the overall shape of a random coil (Fig. 2-21).

Gauche conformations have higher energies than trans conformations in chains with monomeric units $-CH_2\!-\!CHR\!-$. The energy difference is $\Delta E_{TG} = 1.46$ kJ/mol in a partially deuterated it-poly(methyl methacrylate). The gauche conformations G of each chain are either G^+ or G^- (see Section 8.3.4.). At 30°C, ca. twice as many trans conformations than gauche conformations are present ($x_T/x_G \approx 2$) whereas $x_T = x_G$ at infinitely high temperature (Fig. 5-4).

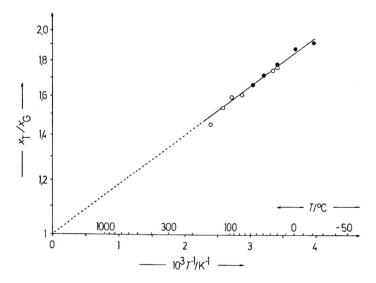

Fig. 5-4   Temperature dependence of the ratio x_T/x_G of mole fractions of trans and gauche conformations ($x_T + x_G \equiv 1$ where G is either G^+ or G^-) according to measurements of the geminal NMR coupling constants of an it-poly(methyl acrylate) with $-CHD\!-\!CH(COOCH_3)-$ units ($x_i = 0.93$) in deuterated chloroform ($\bullet$) and o-dichlorobenzene ($\circ$) [1].

A polymer chain can adopt many different macroconformations. If three energetically equal types of microconformations are present, then a poly(methylene) chain $H[CH_2]_XH$ with a degree of polymerization $X = 20\ 002$ would exist in $3^{X-2} = 3^{20\ 000} \approx 10^{9452}$ macroconformations! Macroconformations are rapidly converted into other macroconformations. For example, 5 % of a given type of microconformation with a rotational barrier of $\Delta E_{TG}{}^{\ddagger} = 10$ kJ/mol are converted at $T = 298.2$ K within $t_{5\%} = 0.05/k \approx 1.4 \cdot 10^{-16}$ s into another microconformation according to the **Eyring equation** $k = (k_B T/h) \cdot \exp(-\Delta E^{\ddagger}/RT)$. Each microconformation exists for only a very short time: the observed proportions of microcoformations, and thus the resulting macroconformations, are *temporal averages* over *all* molecules.

Polymer chains with the same constitution, configuration and degree of polymerization thus exhibit different macroconformations at any given time (Fig. 5-5). Practically none of these macroconformations corresponds to a simple geometric shape: chain molecules in solution are neither rods nor rotational ellipsoids or spheres. The instantaneous shape, averaged over many conformations, resembles a kidney.

Molecules with these macroconformations are called **random coils** or simply **coils** in polymer science. In the biological sciences, the word "coil" has a different meaning: it denotes a helix. "Super coils" are helical chains in the macroconformation of a random coil. "Coiled coils" in the parlance of molecular biologists are helices that are wound into another helix structure.

A linear chain can be characterized by its end-to-end distance r and its radius of gyration s. The **end-to-end distance** is the spatial distance between the end groups of a linear chain (graph in Eqn.(5-11)). It has no meaning for a branched chain. However, particles and molecules can always be characterized by their **radii of gyration**.

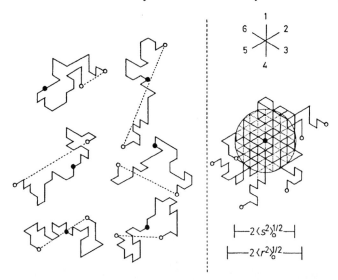

Fig. 5-5   Left: Snapshots of two-dimensional macroconformations of 6 chains with $N_b = 30$ chain bonds each obtained by rolling a die to determine the direction of the next bond. The central chain atom of each chain (chain atom no. 16) is indicated by ●. Chains have different end-to-end distances - - - - between the two end groups O. Right: Superposition of the six chains at their center atoms shows that a considerable number of chain atoms resides outside the area determined by the average radius of gyration, $s \equiv \langle s^2 \rangle_o^{1/2}$. With permission by Springer-Verlag [2].

Particles (spheres, rods, coils, etc.) consist of entities i (atoms, groups, units, etc.) with masses m_i at distances R_i from the center of gravity (graph in Eqn.(5-11)). The mean-square radius of gyration, $\langle s^2 \rangle$, of a particle is defined as the mass-average of R_i^2 of all entities; non-rigid particles require an additional averaging over all conformations (see Eqn.(5-11), right). Thus:

$$(5-10) \quad \langle s^2 \rangle \equiv \langle \Sigma_i\, m_i R_i^2 \rangle / \Sigma_i\, m_i \equiv s^2$$

By definition, averages of spatial quantities are denoted by $\langle\ \rangle$ and not by a line atop the symbol. Radii of gyration are always measured as mean-square quantities $\langle s^2 \rangle$. For reasons of simplicity, square roots $\langle s^2 \rangle^{1/2}$ are often written as s.

5.3.2. Random Coils

End-to-end distances of single linear chains can be calculated by various models. In simple cases, only **short-range interactions** exist within a chain segment, i.e., over a distance of 2, 3, 4, or 5 adjacent chain atoms (Table 5-1). **Long-range interactions** are defined as those between spatially proximate groups that are separated *along* the chain by many segments (Fig. 5-6). "Long-range" and "short-range" do not refer to the range of forces but rather to the number of segments along the chain that separate interacting groups.

For single chains, both long-range and short-range interactions are intramolecular but the former is intersegmental and the latter intrasegmental. The positions of segments of coils without long-range interactions are distributed at random in space and time. Such coils are thus called **random coils** or **statistical coils**; they are present in so-called theta solutions (see Chapter 6). The spatial distributions of segments are only approximately random in coils with long-range interactions.

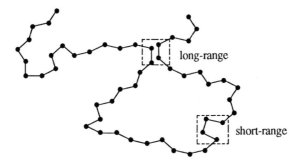

Fig. 5-6 Long-range and short-range (local) interactions between chain segments.

The various models for random coils with short-range interactions differ in the number of interacting chain atoms and in the restrictions applied to valence angles and torsion angles (Table 5-1). All models are nonspecific with respect to the type of interactions (van der Waals, dipole–dipole, etc.).

Table 5-1 Models for random coils differ in the number of adjacent chain atoms that are sufficient to describe the model. The higher the number of these chain atoms, the more realistic is the model. — Covalent bonds between participating chain atoms ●.

	Freely jointed chain	Freely rotating chain	Chain with restricted rotation	Rotational isomeric state model
Chain atoms implied	2	3	4	5
Controlling factor	bond length	valence angle	conformer	conformer pair
Valence angle τ	any	defined	defined	defined
Torsion angle θ	any	any	one defined	two defined
End-to-end distance	r_{oo}	r_{of}	r_{α}	r_o

The **freely jointed chain** describes a hypothetical chain with a number N_{seg} of infinitely thin linear segments of uniform length L_{seg} that can adopt any spatial position with equal probability. The "segment" remains unspecified. In the simplest case, it is composed of two covalently bound chain atoms. Freely jointed chains with segments of different lengths have been called **random-walk chains** or **random-flight chains**.

Freely jointed chains may have any valence angle and torsion angle. The controlling factor is the chain bond composed of two chain atoms (see Table 5-1). The end-to-end distance r_{oo} of such a chain is calculated by vector analysis. The essential feature can be obtained by a simplified approach:

Two bonds of length b form a valence angle τ. The distance r_{oo} between the two ends of this "chain" is given by the cosine rule:

(5-11) $\quad r_{oo}^2 = 2\,b^2 - 2\,b^2 \cos \tau$

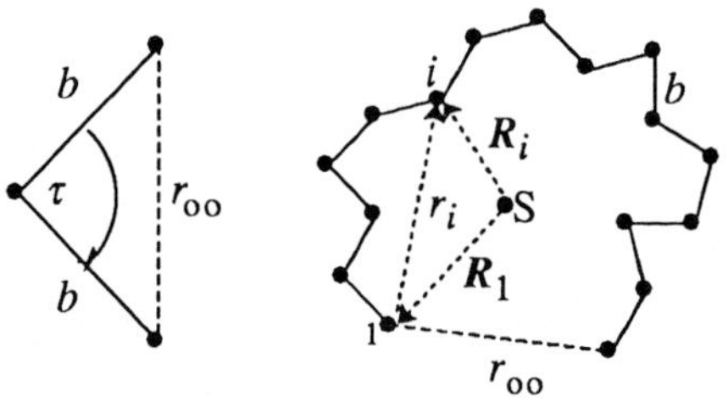

Eqn.(5-11) also applies to the corresponding averages $\langle r^2 \rangle_{oo}$ and $\langle \cos \tau \rangle$ so that $\langle r^2 \rangle_{oo} = 2\,b^2 - 2\,b^2 \langle \cos \tau \rangle$. Since all directions are equally probable, $\langle \cos \tau \rangle = 0$. Replacement of 2 bonds by N bonds results in:

(5-12) $\quad \langle r^2 \rangle_{oo} = Nb^2$

The **freely rotating chain** contains N bonds of length b that form identical valence angles τ. Chain segments rotate freely around bonds. Local interactions are now controlled by two adjacent chain bonds (three chain atoms; Table 5-1). The end-to-end distance is calculated for the limiting case of infinitely high degree of polymerization ($X \to \infty$) as (see Appendix A 5-1):

$$(5\text{-}13) \quad \langle r^2 \rangle_{of} = Nb^2(1 - \cos\tau)/(1 + \cos\tau) = N(b')^2 \quad ; \quad N \to \infty$$

The transition from arbitrary to defined valence angles results in an expansion of the coil if the valence angle is greater than $\tau = 90°$. For freely rotating carbon chains with $\tau = 109.5°$, Eqn.(5-13) reduces to $\langle r^2 \rangle_{of} \approx 2\,Nb^2$: The mean-square end-to-end distance of freely rotating chains is just twice as large as that of freely jointed chains. The constant angular term $(1 - \cos\tau)/(1 + \cos\tau)$ can be merged with the square of the bond lengths b which then become effective bond lengths b'.

In reality, however, the rotation around chain bonds is not free but restricted since chain atoms reside preferentially in discrete microconformations such as T, G$^+$ and G$^-$. On average, they occupy a torsion angle θ: Such a chain with local interactions between three bonds (4 chain atoms, Table 5-1) is called a **chain with restricted rotation**. According to vector analysis, its end-to-end distance is given by:

$$(5\text{-}14) \quad \langle r^2 \rangle_{or} = Nb^2 \left(\frac{1 - \cos\tau}{1 + \cos\tau} \right)\left(\frac{1 + \cos\theta}{1 - \cos\theta} \right) = N(b'')^2 \quad ; \quad N \to \infty$$

Eqns.(5-13) and (5-14) are derived by stochastic arguments. They diverge for certain limiting conditions ($\tau = 180°$ and $\theta = 0°$) and thus do not apply to rods ($\theta = 0°$ for all-T conformations).

The **rotational isomeric state model (RIS model)** assumes that the probability of a microconformation is controlled by previous microconformations. In the simplest RIS model, the controlling factor is a *pair* of microconformations (see the pentane effect, Fig. 2-20) which requires the introduction of a new parameter, Q_{pair}. More sophisticated RIS models assume control by three microcoformations, etc. The dimensions are calculated by matrix methods. They deliver the **unperturbed mean–square end-to-end distance**; for $N \to \infty$:

$$(5\text{-}14a) \quad \langle r^2 \rangle_0 = Nb^2 \left(\frac{1 - \cos\tau}{1 + \cos\tau} \right)\left(\frac{1 + \cos\theta}{1 - \cos\theta} \right) Q_{pair} = Nb^2 \left(\frac{1 - \cos\tau}{1 + \cos\tau} \right)\sigma^2 = \langle r^2 \rangle_{of}\sigma^2$$

5.3.3. Steric Factor and Hindrance Parameter

Q_{pair} and the torsional angle θ both relate to microconformations. They are thus often united in a new parameter $\sigma^2 \equiv Q_{pair}(1 + \cos\theta)/(1 - \cos\theta)$ which is introduced as a square in order to make it comparable to b^2. The parameter σ is called the **steric factor, hindrance parameter,** or **conformational factor**. It is the ratio of the end-to-end distance $\langle r^2 \rangle_0^{1/2}$ of coil molecules with **unperturbed dimensions** to the end-to-end distance $\langle r^2 \rangle_{of}^{1/2}$ of a freely rotating chain. These dimensions are called *unperturbed* because they are not perturbed by long-range interactions as caused by thermodynamically good solvents. Unperturbed dimensions can be obtained from measurements at the so-called **theta temperature** (p. 238).

The end-to-end distance $\langle r^2 \rangle_{of}^{1/2}$ of a freely rotating chain is calculated from the number N of chain bonds, bond lengths b and valence angles τ (the latter two usually from crystallographic data). The valence angle of compounds in crystals, however, is not necessarily identical with that of dissolved compounds. The energy required for the deformation of a C-C-C valence angle by 5.6° is ca. 2 kJ/mol and thus in the same range as conformational energies (see Section 2.4.). Since the hindrance parameter and the valence angle term are not quantities that can be assumed to be the same as in the solid state, they are combined in a **characteristic ratio** C_N:

$$(5\text{-}15) \quad C_N = \sigma^2 (1 - \cos \tau)/(1 + \cos \tau) = \langle r^2 \rangle_o/(Nb^2) = \langle r^2 \rangle_o/\langle r^2 \rangle_{oo}$$

The characteristic ratio is given by the ratio of the unperturbed dimension $\langle r^2 \rangle_o$ to the dimension $\langle r^2 \rangle_{oo} = Nb^2$ of a freely jointed chain. The characteristic ratio and the steric factor are material constants for apolar polymers. They often vary with the solvent for polar polymers (see amylose in Table 5-2) because the interaction of solvent and polymer may change the population of microconformations. Both σ and C_∞ (C_N for $N \rightarrow \infty$) increase with increasing bulkiness of substituents (Table 5-2).

Table 5-2   Steric factors σ and characteristic ratios C_∞.

Polymer	Monomeric unit	Solvent	$T/°C$	σ	C_∞
Poly(butadiene), 1,4-trans	$-CH_2CH=CHCH_2-$	various	50	1.23	5.8
Poly(butadiene), 1,4-cis	$-CH_2CH=CHCH_2-$	decalin	55	1.63	4.9
Poly(ethylene)	$-CH_2CH_2-$	1-chloronaphthalene	140	1.77	6.87
Poly(propylene), at	$-CH_2-CH(CH_3)-$	cyclohexane	92	1.76	6.8
Poly(isobutylene)	$-CH_2C(CH_3)_2-$	benzene	24	1.80	6.5
Poly(styrene), at	$-CH_2CHC_6H_5-$	cyclohexane	35	2.2	10.2
Poly(methyl methacrylate), at	$-CH_2C(CH_3)COOCH_3-$	butanone	25	1.89	7.9
		butyl chloride	25	1.87	7.7
		benzene/cyclohexane	25	2.14	10.1
Poly(decyl methacrylate), at	$-CH_2C(CH_3)COOC_{10}H_{21}-$	butanone	25	2.4	12.7
Poly(docosyl methacrylate), at	$-CH_2C(CH_3)COOC_{22}H_{45}-$	butanone	25	3.3	23.9
Cellulose	$-C_6H_{10}O_5-$	Cadoxene®	25	2.0	
Amylose	$-C_6H_{10}O_5-$	dimethylsulfoxide	25	1.8	
Amylose	$-C_6H_{10}O_5-$	nitromethane	23	2.8	

End-to-end distances increase with longer chain bonds b, larger valence angles τ and/or greater hindrance parameters (Eqns.(5-12)-(5-14a)). These factors act as if larger segments of length L_K are present in smaller numbers N_K (**Kuhn's equivalent coil**, also called **ersatz coil** (from G: *ersatz* = substitute, replacement)). Eqn.(4-14a) can therefore be written $\langle r^2 \rangle_o = N_K L_K^2$. The **Kuhn length** L_K can be calculated from $L_K = \langle r^2 \rangle_o/r_{cont}$ since the product $N_K L_K$ must equal the contour length $r_{cont} = N_e b_e$ (see Fig. 5-3). The smaller the Kuhn length, the more flexible is the chain. Chains with large Kuhn lengths no longer form random coils (except in the limit $N_K \rightarrow \infty$) but adopt a worm-like structure (Section 5.4.).

5.3.4. Radius of Gyration

End-to-end distances are easy-to-grasp parameters that can be calculated by simple theoretical means. However, they cannot be measured directly and they do not make sense if a molecule has no ends (rings) or more than two ends (branched polymers). But all bodies do possess a radius of gyration that is accessible by experiment (Section 5.2.). It turns out for $M \rightarrow \infty$ that a general relationship exists between the radius of gyration, s, and the end-to-end distance, r, for all linear chains with only short-range interactions, regardless of the model (Appendix A 5-2): $\langle r^2 \rangle_0 = 6 \langle s^2 \rangle_0$ for unperturbed coils (RIS model), $\langle r^2 \rangle_{oo} = 6 \langle s^2 \rangle_{oo}$ for freely jointed chains, etc.

It follows from Eqn.(5-14a) with $N \approx N_c = M/M_c$ that the radius of gyration increases with the square root of the molar mass for *unperturbed coils of linear chains*:
$$\langle s^2 \rangle_0^{1/2} = \{[b^2(1 - \cos \tau)\sigma^2]/[6 \, M_c(1 + \cos \tau)]\}^{1/2} M^{1/2} \equiv K_s M^{1/2}.$$

This square root dependence $s_0 \sim M^{1/2}$ is often found for an astonishingly broad range of molar masses, for example, for the range $5 \cdot 10^3 \le M/(\text{g mol}^{-1}) \le 5 \cdot 10^7$ of linear poly(styrene)s PS in cyclohexane at 34.5°C where this solvent becomes a theta solvent for PS (Fig. 5-7). The exponent v of the molar mass increases for PS from $v = 0.50$ in theta solvents to $v = 0.59$ in the thermodynamically good solvent toluene at 15°C. The exponent becomes independent of the solvent quality for molar masses of $M \le 5000$ g/mol but no longer follows the simple relationship $s \sim M^v$.

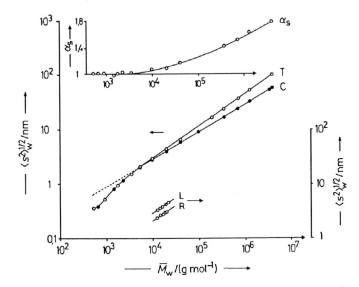

Fig. 5-7   Mass-average radius of gyration, $\langle s^2 \rangle_w^{1/2}$, as function of the mass-average of molar masses for practically molecularly uniform linear "atactic" poly(styrene)s (mole fraction of syndiotactic diads $x_s = 0.59$) in toluene (T) at 15°C and cyclohexane (C) at 34.5°C [3] and for linear (L) and cyclic (R) poly(styrene)s in toluene-d$_8$ at 22°C [4]. Insert: Expansion factor α_s in toluene at 15°C. Molar masses were determined by static light scattering for high molar masses and by small angle X-ray scattering [3] or small angle neutron scattering [4] for low molar masses.

Radii of gyration are independent of solvent quality for low molar mass poly-(styrene)s because short chains cannot exert long-range interactions but only short-range ones. The strong decrease of s below $\overline{M}_w \leq 5000$ g/mol ($X \leq 50$) of poly-(styrene) is caused by short helical segments which make the whole chain worm-like (Section 5.4.). Only for $X \geq 50$, chains behave like random coils (cf. Fig. 2-21).

Cyclic molecules are only half as long in the completely extended state as linear ones with the same molar mass. On average, this can also be assumed for all other macroconformations, i.e., $s_{0,lin} = K_{s,o}M^{1/2}$ and $s_{0,ring} = K_{s,o}(M/2)^{1/2}$. The radius of gyration of cyclic molecules should thus be smaller than that of linear chains by a factor of $(1/2)^{1/2} = 0.707$ (Fig. 5-7). These two types of molecules assume their unperturbed states at different temperatures, however: the theta temperatures are 34.5°C (linear) and 28°C (cyclic) for high molar mass poly(styrene)s in cyclohexane.

Short-range interactions depend on the local structure and therefore on *differences in tacticities*. Unperturbed radii of gyration may differ by up to 20 % for isotactic and syndiotactic polymers with the same constitution and molar mass.

5.3.5. Coil Density

Single polymer coils have very low densities. A compact sphere from a poly-(methylene) molecule $H(CH_2)_{20001}H$ of density $\rho = 1$ g/mL would have a volume of $V = m/\rho = M/(\rho N_A)$ and thus a radius of $R_{sph} = (3\,V/(4\,\pi))^{1/3} = 4.82$ nm and a radius of gyration of $s_{sph} = (3/5)^{1/2}\,R_{sph} = 3.74$ nm (see Section 5.6.1.). The same molecule would have average radii of gyration of $s_{oo} = 8.9$ nm as freely jointed chain and $s_o = 23.3$ nm as unperturbed coil.

Segments are not evenly spaced in such coils. Their number concentration $C = N_{seg}/V$ depends on the distance R from the center of gravity. The distribution of C can be approximated by a Gaussian function $C = A \cdot \exp(-B^2R^2)$ where A and B are model constants. Coils are often assumed to possess a spherical symmetry. A shell of the sphere contains $C(4\,\pi R^2 dR)$ segments in a volume $4\,\pi R^2 dR$. The number N_{seg} of all segments in a coil is obtained by integration from $R = 0$ to $R = \infty$:

$$(5\text{-}16) \quad N_{seg} = \int_0^\infty 4\pi R^2 C dR = 4\pi A \int_0^\infty R^2 \exp(-B^2R^2) dR = \pi^{3/2}A/B^3$$

$dN = 4\,\pi R^2 C dR$ segments reside in a spherical shell of radius R and thickness dR. Since the average of all R^2 for all N_{seg} segments equals the mean-square radius of gyration, $\langle s^2 \rangle_0$, one obtains:

$$(5\text{-}17) \quad \langle s^2 \rangle_0 = \int_0^N NR^2 / \int_0^N N = \int_0^\infty 4\pi R^4 C dR / \int_0^\infty 4\pi R^2 C dR = 3\pi^{3/2}A/(2N_{seg}B^5)$$

The combined Eqns.(5-16) and (5-17) deliver $A = N_{seg}[3/(2\,\pi\,\langle s^2 \rangle_0)]^{3/2}$ and $B^2 = 3/(2\,\langle s^2 \rangle_0)$. Furthermore, N_{seg} equals the degree of polymerization, X, if the segments are the monomeric units. The number concentration of segments is thus:

(5-18) $C = X \, [3/(2 \, \pi \, \langle s^2 \rangle_0)]^{3/2} \exp[- \, 3 \, R^2/(2 \, \langle s^2 \rangle_0)]$

The number concentration C of segments is largest at the center of gravity ($R = 0$) (Fig. 5-8). It decreases with increasing distance R from the center of gravity and with increasing molecular weight M_r. According to Eqn.(5-18), $9.77 \cdot 10^{19}$ mer per milliliter reside at the center of gravity of an unperturbed coil of a poly(α-methyl styrene) molecule with $M = 1.19 \cdot 10^6$ g/mol. The molar volume of a mer is $V_m = 74$ mL/mol and the volume of a mer is $V_m/N_A = 1.23 \cdot 10^{-22}$ mL. The volume fraction of a mer at the center of gravity of the poly(α-methyl styrene) molecule is thus only $\phi_u = (1.23 \cdot 10^{-22}$ mL$) \cdot (9.77 \cdot 10^{19}$ mL$^{-1}) = 0.012$. Only 1.2 % of the space at the center of gravity is occupied by monomer units and 98.8 % by solvent molecules.

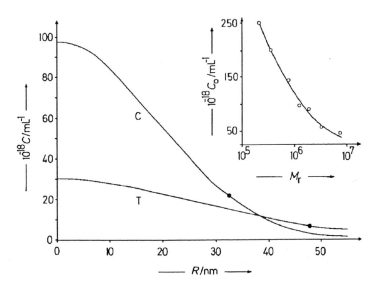

Fig. 5-8   Number concentration C of monomeric units as function of the distance R of these units to the center of gravity of single poly(α-methyl styrene) molecules with $M = 1.19 \cdot 10^6$ g/mol. C = Unperturbed coils in the theta solvent cyclohexane at 34.5°C; T = expanded coils in the good solvent toluene at 25°C. ● Root mean-square average radii of gyration (32.4 nm in cyclohexane, 48.0 nm in toluene). Insert: Number concentrations C_o at the center of gravity as a function of the molecular weight M_r for the theta solvent cyclohexane at 34.5°C.

5.4. Worm-Like Chains

Segments of a freely rotating chain can adopt any spatial position since the torsional angles can adopt any value although the valence angle is constant (e.g., $\tau = 111.5°$). So-called *rigid chains* have angles τ of nearly 180° between two adjacent segments of length b. The segments following the first segment thus cannot occupy any possible position in space. The chain direction has a *persistence* which causes a resistance against coiling. The **persistence length** a is defined as $a \equiv b/(1 + \cos \tau)$. Such a chain is continuously curved like a worm; it is often said to be semiflexible.

The mean-square end-to-end distance of such a **worm-like chain (Kratky–Porod chain)** is calculated as a limiting case for a freely rotating chain with the following restrictions: the segment lengths are vanishingly small ($b \to 0$), the bond angles approach 180° ($\tau \to \pi$), and the number of bonds N approaches infinity ($N \to \infty$). Chain lengths $L_{chain} = Nb = N_a a$ (see Fig. 5-3) and persistence lengths a should be constant. The worm-like chain thus consists of $N_a = Nb/a$ segments of length a. The mean-square end-to-end distance of the chain is obtained as (see Appendix A 5-3):

$$(5\text{-}19) \quad \langle r^2 \rangle_{worm} = 2\,N_a a^2 [1 - (1/N_a) + (1/N_a) \cdot \exp(-N_a)]$$
$$= 2\,aL_{chain} - 2\,a^2 \{1 - \exp(-L_{chain}/a)\}$$

Random coils with flexible chains: The chain length $L_{chain} = Nb$ is much larger than the persistence length a. The condition $Nb/a = N_a \gg 1$ leads to $\exp(-N_a) \to 0$. Eqn.(5-19) is simplified to $\langle r^2 \rangle_{of} = 2\,N_a a^2$. Since $N_a = N_K L_K/a$ and $\langle r^2 \rangle_o = N_K L_K^2$, one obtains $L_K = 2\,a$: the Kuhn length is twice as large as the persistence length.

Rigid chains: The persistence length a is much greater than the chain length Nb. The exponential $\exp(-N_a)$ can be expanded into a Taylor series for $a/Nb \ll 1$. The series $\exp(-N_a) = 1 - N_a + (N_a^2/2) - \ldots$ is terminated after the third term. Insertion into Eqn.(5-19) delivers $\langle r^2 \rangle_o = N_a^2 a^2$ and $r_o \equiv \langle r^2 \rangle_o^{1/2} = N_a a$, respectively. The end-to-end distance is identical with the chain length $N_a a$. An infinitely rigid chain is a rod (see also Section 5.6.2.).

The Kratky–Porod chain thus describes the transition from a rigid rod ($N_a = 0$) to a flexible coil ($N_a = \infty$). For example, the term in square brackets in Eqn.(5-19) adopts values of 0.3679 at $N_a = 1$, 0.8013 ($N_a = 5$), 0.9000 ($N_a = 10$), 0.9900 ($N_a = 100$), and 1.0000 ($N_a = \infty$).

5.5. Real Coil Molecules

5.5.1. Excluded Volume

The freely jointed chain, the freely rotating chain, the chain with restricted rotation, the RIS model, and the Kuhn coil all lead to the same result: the mean-square end-to-end distance $\langle r^2 \rangle_y$ is always given by the product of the number N_y of units y and the square of their lengths L_y, i.e., $\langle r^2 \rangle_y = N_y L_y^2$. Such a relationship is characteristic for all processes with random walk statistics, e.g., diffusion processes (Section 7.2.1.).

The length L_y characterizes short-range (local) effects. It depends on the chosen model: if it increases at constant $\langle r^2 \rangle_y$, numbers N_y of units become smaller and vice versa. The relationship $\langle r^2 \rangle_y = N_y L_y^2$ between r, N_y and L_y is however independent of the local effects along the chain, i.e., length of bonds (or segments), valence angles, and steric hindrance, as long as only short-range interactions are present.

In coils with long-range interactions (Fig. 5-6), the space occupied by one segment is excluded for all other segments. This intramolecular **excluded volume** expands the coil as one may see for two-dimensional chains (Fig. 5-9). Intramolecular excluded volumes in particular are present in coils in very dilute solutions where the effects of intermolecular excluded volumes are absent, e.g., those of solid or hollow spheres.

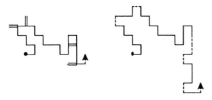

Fig. 5-9 Schematic representation of two-dimensional chains on a cubic lattice. Left: unperturbed chain with overlapping segments (double lines). Right: the same chain with intramolecular excluded volume. Segment overlaps are avoided by shifting the rest of the chain (these shifts are indicated by broken lines).

Space requirements of segments are not only determined by their physical volumes (as for solid spheres) but also by their interactions. Excluded volumes increase due to repulsion between segments and decrease due to attraction. Interactions between two segments furthermore compete with those between segments and solvent molecules. The magnitude of the excluded volume effect thus depends on the thermodynamic quality of the solvent. In so-called theta solvents, all effects of interactions and physical volume requirements cancel each other and the excluded volume becomes zero. The coils adopt their unperturbed dimensions.

5.5.2. Dependence on Molar Mass

Coil expansions caused by excluded volumes can be described by an **expansion factor** $\alpha_s \geq 1$:

$$(5\text{-}20) \quad \langle s^2 \rangle \equiv \alpha_s^2 \langle s^2 \rangle_o$$

The expansion causes the spatial distribution of segments to deviate from the segmental distribution of unperturbed coils: excluded volumes lead to **perturbed coils**. The better the solvent, the more is the coil expanded and the larger is α_s which is a temporal average over the expansion factors of all macroconformations.

The expansion factor is a function of the segment-segment excluded volume u_{seg}. This function is defined for a parameter z, the so-called **binary cluster integral**:

$$(5\text{-}21) \quad z \equiv (4\,\pi)^{-3/2}[M/\langle s^2 \rangle_o]^{3/2}\{u_{seg}/M_{seg}^2\}M^{1/2}$$

(often written in terms of the end-to-end distance, not the radius of gyration). u_{seg}, M_{seg} and $M/\langle s^2 \rangle_o = (K_{s,o})^2$ are material constants. u_{seg} and M_{seg} are not unambiguously defined; z can also not be measured directly. Neither Eqn.(5-21) alone nor the calculated theoretical function $\alpha_s = f(z)$ can therefore be checked by experiment.

However, since $\alpha_s = f(z)$ and $z = f'(M)$, one also has $\alpha_s = f''(M)$. From $\langle s^2 \rangle_o = K_s^2 M$ and Eqn.(5-20) one obtains $\langle s^2 \rangle = \alpha_s^2 K_s^2 M$. Assuming $\alpha_s^2 = M^{2v-1}$, the molar mass dependence of the radii of gyration in thermodynamically good solvents is given by

$$(5\text{-}22) \quad \langle s^2 \rangle = K_s^2 M^{2v} \qquad \text{or} \qquad s \equiv \langle s^2 \rangle^{1/2} = K_s M^v$$

The exponent v can be calculated by the **mean field theory** which assumes that the same average force field acts on each segment (actually, no mean field is explicitly applied in the calculation which averages over all energetic contributions). This force field is composed of two opposite effects. Repulsive forces between chain segments expand coils and lead to excluded volumes. The larger the expansion, the fewer microconformations are available. The repulsion is thus opposed by an elastic retractive force which tries to increase the number of microconformations (elastic effect). The Gibbs energy is therefore composed of two parts, $\Delta G = \Delta G_{rep} + \Delta G_{elast}$:

The *repulsion term* is the product of four parameters. (1) It increases with the average number concentration $C_K = N_K/V_{coil}$ of Kuhn segments. The coil is here replaced by an **equivalent sphere** which confines all segments to the volume $V_{coil} = (4 \pi s^3)/3$ that is prescribed by the radius of gyration. The repulsion term must also increase with (2) the number N_K of segments and (3) the thermal energy $k_B T$. One must furthermore consider (4) the polymer-solvent interaction relative to the polymer–polymer and solvent–solvent interactions. This interaction term must have the physical unit of an effective volume $V_{eff} = K_{eff} L_K^3$ so that the physical unit of the repulsion term equals the physical unit of the retractive term.

The repulsion term is thus given by the product of all these factors, resulting in $\Delta G_{osm} = C_K N_K k_B T V_{eff} = (3 K_{eff}/4 \pi) k_B T N_K^2 L_K^3 s^{-3}$.

The *retractive energy* is assumed to be the Gibbs energy of rubber elasticity (see Eqn.(10-14)), i.e., $\Delta G_{elast} = (k_B T/2)(3 \alpha_s^2 - 3)$ for a coil from one chain (one network chain; $N_c = 1$) of functionality $f = 2$ (two end groups) that expands equally into all three spatial directions $(\lambda_x = \lambda_y = \lambda_z = \alpha_s)$ with *high* values of α_s (good solvent). Insertion of $\alpha_s^2 = s^2/s_o^2$ and $s_o^2 = r_o^2/6 = N_K L_K^2/6$ into this equation results in $\Delta G_{elast} = (k_B T/2)(18 N_K^{-1} L_K^{-2} s^2 - 3) \approx 9 k_B T N_K^{-1} L_K^{-2} s^2$.

The total Gibbs energy is thus:

$$(5\text{-}23) \quad \Delta G = \Delta G_{osm} + \Delta G_{elast} = (3 K_{eff}/4 \pi) k_B T N_K^2 L_K^3 s^{-3} + 9 k_B T N_K^{-1} L_K^{-2} s^2$$

The minimum of the energy is obtained by taking the first derivative of the Gibbs energy with respect to the radius of gyration and setting it to zero:

$$(5\text{-}24) \quad \partial \Delta G/\partial s = -3 (3 K_{eff}/4 \pi) k_B T N_K^2 L_K^3 s^{-4} + 18 k_B T N_K^{-1} L_K^{-2} s = 0$$

Solving for s, introduction of $N_K = M/M_K$, and collection of all constants into a factor K_s results in:

$$(5\text{-}25) \quad s \equiv \langle s^2 \rangle^{1/2} = [6^{-1/5}(3 K_{eff}/4 \pi)^{1/5} L_K M_K^{-3/5}] M^{3/5} = K_s M^{3/5} = K_s M^v$$

The radius of gyration thus increases with the 3/5th power of the molar mass for perturbed coils in *good solvents*. The exponent $v = 3/5$ is often called the **Flory exponent**. It characterizes a fractal dimension (see Section 5.7.3.).

The same value of $v = 3/5$ is also obtained from random walk statistics for self-avoiding walks. Renormalization theory delivers a slightly smaller value of $v = 0.588$ instead of $v = 3/5 = 0.600$. This method is a mathematical procedure for the calculation of properties by successive doubling of parameters (here: segment lengths) near critical points.

Experiments in good solvents always deliver v values of ca. 0.59-0.60 in the limit $M \to \infty$ regardless of the particular system polymer–solvent–temperature. In theta solvents, one always observes $v = 0.50$ (Table 5-3), even for random coils composed of DNA double helices (see Fig. 5-11). The mean field theory thus describes very well the dependence of radii of gyration on molar masses.

Table 5-3   Exponents v. CD = Coefficient of determination = square of correlation coefficient. x_s = Mole fraction of syndiotactic diads. * Ditto for Poly(D,L-β-methyl-β-propiolactone).

Polymers	$10^{-3}M/(\text{g mol}^{-1})$	x_s	Solvent		$T/°C$	v	CD	Ref.
Poly(styrene)	5.4 - 3900	0.59	cyclohexane	Θ	34.5	0.501	1.000	[3]
Poly(α-methyl styrene)	342 - 7500	0.40	cyclohexane	Θ	34.5	0.499	0.999	[5]
Poly(α-methyl styrene)	768 - 7500	0.40	trans-decalene	Θ	9.5	0.492	0.998	[5]
Poly(methyl methacrylate)	5.5 - 2830	0.79	acetonitrile	Θ	44	0.501	1.000	[6]
Poly(styrene)	5.4 - 3900	0.59	toluene		15	0.590	1.000	[3]
Poly(α-methyl styrene)	204 - 7500	0.40	toluene		25	0.577	0.997	[5]
Poly(D-β-hydroxybutyrate)*	86.5 - 9100	it+at	trifluoroethanol		25	0.603	0.999	[7]

Long-range interactions dominate in good solvents; radii of gyration are thus scarcely affected by local structures. Polymers with the same constitution and molar mass but different tacticity have the same dimensions in good solvents. This is the reason why the isotactic poly(D-β-hydroxybutyrate) and the constitutionally identical, either atactic or stereoblock-containing polymer from racemic D,L-β-methyl-β-propiolactone have an identical function $s = K_s M^{0.603}$ (Table 5-3) and the same K_s.

5.5.3. Branched Polymers

Branched molecules occupy smaller volumes than linear molecules with the same molar mass (cf. dendrimers in Section 2.1.5.). They thus possess smaller radii of gyration. The contraction is described by a **branching index** $g_s = \langle s^2 \rangle_{br}/\langle s^2 \rangle_{lin}$. Large star molecules with f equally long arms and Gaussian distributions of intersegmental distances are predicted to show $g_s \approx (3f - 2)/f^2$ for theta solvents; if the lengths of arms follow a Gaussian statistics, $g_s = 6 f/[(f + 1)(f + 2)]$ is calculated. Stars with equally long arms should obey $g_s \approx 3/f$ in good solvents. Star molecules with a small number of arms should thus contract more strongly in theta solvents than in good solvents. The branching index g_s is calculated to be independent of the solvent quality for stars with a large number of arms, however.

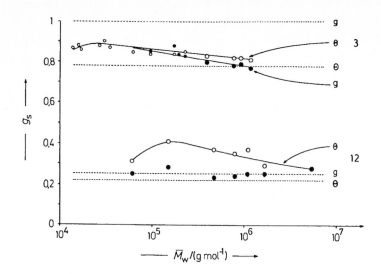

Fig. 5-10 Branching index g_s as function of the mass-average molar mass of narrow distribution star poly(styrene)s with 3 or 12 arms (see text) in the good solvent toluene (●) or the theta solvent cyclohexane (○), both at 35°C. - - - Theory for good (g) and theta solvents (Θ). Radii of gyration were determined by static light scattering or neutron scattering (●,○) or by hydrodynamic measurements (●,○). Data of [8].

Poly(styrene)s $Si\{CH_2CH_2Si[(CH_2CH=CHCH_2)_{2\text{-}3}(CH_2CHC_6H_5)_N]_3\}_4$ with $f = 12$ arms of length N show indeed a $g_s = 0.25 \pm 0.02$ in the good solvent toluene as predicted by theory (Fig. 5-10). Far higher than theoretical values are observed for the same polymers in the theta solvent cyclohexane. Since the values of g_s in cyclohexane pass through a maximum with increasing molar mass and approach the theoretical value only at very high molar masses, two effects must work against each other:

Very short arms possess the same random distribution of segments as linear chains with the same degree of polymerization. The decrease of dimensions is only due to the functionality of the star molecule: low molar masses lead to the theoretical g_s value. The segment distribution of somewhat longer arms is however impeded near the core due to the adjacent arms. The coiled arms stretch a little and g_s increases. For very long arms, the constraints near the core can be neglected; these arms can adopt almost the same segment distribution as linear chains.

Values of g_s of stars with constant number of arms decrease with increasing $\overline{M}_w$ and approach the theoretical value. This effect is less pronounced in good solvents because perturbed coils have lower segment densities near the center of gravity than unperturbed ones (see Fig. 5-8).

The maximum of the function $g_s = f(M)$ is located at $M = 150\ 000$ g/mol for the 12-arm polymer (corresponds to $N \approx 120$ for each arm). It is at a lower molar mass ($M \approx 25\ 000$ g/mol) and a shorter arm ($N \approx 80$) for 3-arm polymers with the constitution $HSi\{[CH_2CH=CHCH_2)_{2,5}(CH_2CHC_6H_5]_n\}_3$ because fewer arms compete for space near the core.

Combs and dendrimers have more complicated branching indices than stars.

5.5.4. Tethered Chains

Chains bound to surfaces are also experiencing restrictions with respect to their local conformations near the surface. These **tethered chains** possess macroconformations that are different from those of "free" chains, regardless of whether the surface is macroscopic or molecular and the bond chemical or physical. Examples are arms of star or comb polymers, side chains grafted to polymer molecules, or chain molecules adsorbed on solid surfaces. Tethered chains have always more stretched macroconformations than isolated chains. Because space restrictions force the chains to become more or less parallel, they resemble **brushes**. Stretched macroconformations are also found if linear chains are forced to flow through narrow pores, for example, in size-exclusion chromatography (Section 2.2.4.).

5.6. Euclidian Bodies

Single macromolecules may also exist in physical structures other than ideal and real coils. The external shape of these other structures can resemble Euclidian bodies but the internal segment distribution may range from coil-like arrangements to dense packing with or without internal order.

Solid bodies with Euclidian shapes (spheres, ellipsoids, rods, etc.) have interiors that are homogeneously filled with matter. Such bodies are thus isotropic: the same density prevails at each point of their interior. **Non-solid Euclidian bodies** have shapes like Euclidian bodies but the matter in their interior has a density distribution.

5.6.1. Spheres and Spheroids

Spherical molecules are fairly rare in polymer science. Seemingly *solid spheres* are formed by hyperbranched lysine polymers (but see end of Section 5.9.5.). Practically solid spheres of single molecules are generated if random coils are rapidly cooled below the theta temperature; these entities are unstable, however. The protein apoferritin is a *hollow sphere*; its interior is filled with water (Section 2.1.5.).

Other macromolecules have external shapes of spheres or ellipsoids but their interior is composed of polymer segments and solvent. These **solvated spheroids** are not solid bodies. They are usually generated by the particular constitution of the polymer chain. The chain of the protein myoglobin contains several segments that are rich in helicogenic α-amino acid residues (Fig. 4-5). Other segments are non-helical which allows the chain to fold to a spheroid. It is unclear whether the folding starts with the formation of α-helices by hydrogen bonds or with hydrophobic bonding. Hydrophobic "bonds" between apolar groups are created in an aqueous environment because very strong attractions exist between water molecules that must be severed on contact with a hydrophobic group. A folding caused by hydrophobic bonding seems to be the most likely effect because hydrogen bonds between amide groups and water molecules are energetically favored over hydrogen bonds between two amide groups.

The total mass is accumulated at a distance R from the center of gravity in hollow spheres with infinitely thin shells. Radius R and radius of gyration s are identical here. This is not true for hollow spheres with shells of finite thickness, i.e., with external radii R_e and internal radii R_i. The square of the radius of gyration of hollow spheres is given by $s^2 = (3/5)(R_e^5 - R_i^5)/(R_e^3 - R_i^3)$.

For solid spheres ($R_i = 0$, $R_e \equiv R$), one obtains $s^2 = (3/5) R^2$. Since $V = (4 \pi R^3/3)$ for solid spheres and $V = m/\rho = M/(\rho N_A)$ for all rigid bodies, the molar mass dependence of the radii of gyration is given by $s = (3/5)[3/(4 \pi \rho N_A)]^{1/3} M^{1/3}$. The radius of gyration of solid spheres increases with the cube root of the molar mass (Fig. 5-11).

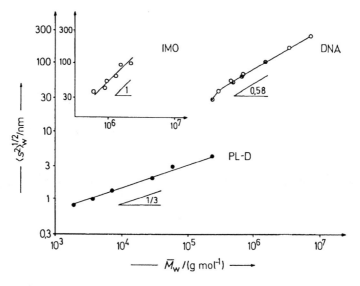

Fig. 5-11 Dependence of mass-average radii of gyration on the mass-average of molar masses.
DNA: Coils of double helices of high molar mass deoxyribonucleic acids in dilute aqueous salt
   solutions (O: [9]; ⊙: [10]); for the structure of DNA see explanation to Fig. 5-12.
PL-D: Nearly compact spheres of hyperbranched poly(lysine)s in DMF [11]; for the constitution
   of PL-D see legend to Fig. 5-21.
IMO (insert): Rods of the mineral imogolite in dilute salt solution [12] (see also Fig. 5-19).

5.6.2. Rods

Helical conformations are either stabilized by attractions due to internal bonds (Sections 4.2.2. and 4.2.3.) or by steric repulsions caused by large substituents (Section 2.4.). Helices generated by steric repulsion between ligands survive dissolution processes only if ligands are very large or if chains are very rigid. An example of the former is poly(triphenylmethyl methacrylate) $+CH_2CH(CH_3)(COOC(C_6H_5)_3)\frac{}{n}$, whereas poly(butyl cyanide) $+N=C(C_4H_9)\}_{\overline{n}}$ is an example of the latter. Most other sterically generated helices exist only in molecular assemblies (see Chapter 8.3.4.).

Examples of helices generated by internal attractive forces include (Fig. 5-12)
- single helices of poly(γ-benzyl-L-glutamate) which are stabilized by hydrogen bonds between amide groups;
- double helices of deoxyribonucleic acids with stabilization by hydrogen bonds and π–π interactions between bases; and
- protein-studded helices of ribonucleic acid in tobacco mosaic virus.

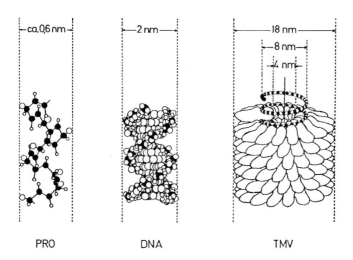

Fig 5-12   Rod-like segments of molecules with right-handed helices. From left to right: α-helix of proteins and poly(L-α-amino acid)s, double helix of deoxyribonucleic acids, and tobacco mosaic virus with a single strand of ribonucleinc acid and hull proteins. The tobacco mosaic virus is a hollow cylinder of 300 nm length and 18 nm width; its central hole is ca. 4 nm wide. The structures are not to scale.

Compact cylindrical rods with radius R and length L have volumes of $V = \pi R^2 L$. The center of gravity of a rod is at $r = 0$ and $L/2$. The number of mass elements residing in the longitudinal distance from x to $(x + dx)$ is dx; x can adopt values between 0 and $L/2$ (or 0 and $-L/2$). If the rods are not infinitely thin, radial distances y between 0 and R also have to be considered. After integration between the limits $L/2$ and R, the square of the radius of gyration is given by:

$$(5\text{-}26) \quad s^2 = \int_0^{L/2} \int_0^R \frac{2\pi y(x^2 + y^2)\mathrm{d}y\mathrm{d}x}{\pi R^2 L/2} = \frac{L^2}{12} + \frac{R^2}{2}$$

Introduction of $L = V/(\pi R^2)$ and $V = M/(N_A\rho)$ leads to $s = (12^{1/2}\pi R^2 N_A\rho)^{-1}M$ if $R \to 0$: The radius of gyration of infinitely thin, rigid rods is directly proportional to the molar mass and not to the square root of M as for unperturbed coil molecules. For $L/R > 25$, one can use $s^2 \approx L^2/12$ with less than 1 % error.

Helical molecules in solution are not necessarily rigid rods. Similar to garden hoses, they resemble rods if they are short but they bend and form random coils if they are very long. The extent of bending of molecular rods is determined by the flexibility of chain segments and this in turn by the oscillation of torsion and valence angles around the potential minima. Large helical molecules with the same diameter may thus behave quite differently in solution. The volcanic material imogolite, $SiO_2 \cdot Al_2O_3 \cdot 2\ H_2O$, and the double helices of DNA have almost the same diameters (2.3 nm vs. 2.0 nm). Yet, the molar mass dependence of the radius of gyration at $M \approx 10^6$ g/mol is given by $s \sim M^{1.0}$ for imogolite (typical for a truly rigid rod) and by $s \sim M^{0.58}$ for DNA (characteristic for random coils in good solvents) (Fig. 5-11).

5.6.3. Excluded Volume

With one exception, identical Euclidian bodies cannot completely fill a given volume. The exception is a cube (a hexahedron having six congruent square faces). If all faces are arranged parallel to each other, cubes can pack cubic densest and the volume fraction of cubes is unity ($\phi_{max} = 1$). If the side faces of cubes are oriented at random to each other, then the maximum volume fraction is only $\phi_{max} \approx 0.524$. The maximum volume fraction ϕ_{max} is called the **maximum packing density**. Its value depends on the shape and arrangement of Euclidian bodies (Table 5-4). Parallel cylindrical rods can be packed to $\phi_{max} = 0.907$ if their cross-sections are arranged hexahedrally but only to $\phi_{max} = 0.820$ if the cross-sections are placed at random.

Euclidian bodies begin to affect each other before they attain their maximum packing densities. The centers of gravity of two spheres with radii R and volumes $V = 4\ \pi R^3/3$ can approach each other only to a distance of $d = 2\ R$ (Fig. 5-13). A certain volume of one sphere is excluded to the other one. This **excluded volume** of *spheres* (solid or hollow) is easily calculated as $u = 4\ \pi d^3/3 = 32\ \pi R^3/3 = 8\ V$.

Table 5-4   Maximum packing densities of identical Euclidian bodies. z = Number of nearest neighbors. * Bimodal distribution of two sizes of spheres with $d_1/d_2 = 47$ and $d_1/d_2 = 100$, respectively. For comparison: Perturbed coils of poly(styrene) of molar mass $M = 1 \cdot 10^6$ g/mol begin to touch each other in good solvents at $\phi \approx c_s^* = 0.0036$ (see Sections 5.7.2. and 6.1.2.).

Type of packing	z	Maximum packing density of Cubes	Spheres	Rods (cylindrical)
Hexahedral densest	12	-	0.754	-
Cubic face-centered	12	-	0.741	-
Cubic body-centered	8	-	0.605	-
Hexahedral parallel	6	-	-	0.907
Cubic simple	6	1	0.524	0.785 (parallel)
Random parallel	-	-	-	0.820
Random densest, $L/d = 1$	-	0.524	0.637	0.704
Random densest, $L/d = 47$	-	-	0.814*	0.108 (3-dimensional)
Random densest, $L/d = 100$	-	-	0.868*	0.041 (3-dimensional)

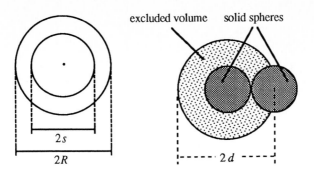

Fig. 5-13   Left: Radius R and radius of gyration, s, of a solid sphere.
Right: External excluded volume of spheres (solid or hollow)

External excluded volumes of rods are more difficult to calculate than those of spheres because the longitudinal axes may be oriented to each other at different angles γ ($\gamma = 0$ for parallel arrangement). The volume excluded by one rigid rod for all others with the same length L and the same circumference U has been calculated as $u = 8\,V[1 + (L/U)\cdot\sin\gamma]$. The circumference equals $U = 2\,\pi R$ for rods with circular cross-sections and radius R and $U = 4\,a$ for square cross-sections with side length a.

5.7. Scaling

5.7.1. Self-Similarity

Euclidian bodies are self-similar: their characteristic features remain the same on enlargement or diminution of the body. An isotropic sphere of density ρ and radius R has a mass of $m = \rho V = (4\,\pi\rho/3)\,R^3$. Doubling the radius increases the mass eightfold because $m = (4\,\pi\rho/3)\cdot(2\,R)^3 = 8\,(4\,\pi\rho/3)\cdot R^3$ but the mass is still proportional to the third power of the radius. The exponent 3 in $m \sim R^3$ indicates the scaling dimensionality $d = 3$ of the sphere with respect to the mass. The smooth surface of a sphere scales with the square of the radius ($A \sim R^2$; i.e., $d = 2$).

The dimensionalities of Euclidian bodies are thus positive integers (1, 2, 3 ...). Other bodies may possess dimensionalities which are fractional numbers. For example, the surface of infinitely thin circular discs with rough surfaces increases more strongly than with the square of the radius of the disc: $A \sim R^d$ and d is now between 2 and 3 although the discs are still two-dimensional bodies.

Random coils are also self-similar which allows one to propose models such as the freely rotating chain or the Kuhn coil without any reference to specific chemical structures. Self-similarity disappears, however, if the scale is too fine (see Fig. 5-14, II → III). The applicability of models to real structures thus depends on the length scale.

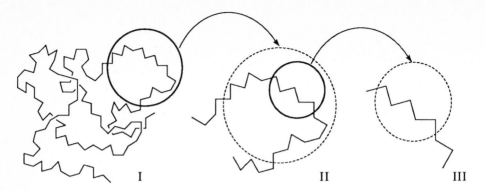

Fig. 5-14   Self-similarity of random coils. The encircled segment of I appears on magnification (broken circle of II) *on average* like the total chain in I. On further magnification to a segment of the order of a Kuhn length L_K, self-similarities disappear.

Dimensionalities are not always positive integers for self-similar entities. Eqn.(5-25) was derived, for example, by assuming that the volume of a random coil is that of an equivalent sphere and that the radius of gyration s is identical with the radius R of the equivalent sphere. The dimensionality of the random coil was thus set to be d = 3. Since the self-similarity of random coils need not be identical with the self-similarity of solid spheres, a more general expression would be $V_{coil} = (4 \pi/3) s^d$ instead of $V_{coil} = (4 \pi/3) s^3$. A recalculation of Eqn.(5-25) then provides:

(5-27) $\langle s^2 \rangle^{1/2} = K_s M^{3/(d+2)} = K_s M^\nu$

The exponent ν is predicted to be ν = 0.600 by the mean-field theory (see Eqn.(5-22), thus d = 3. Renormalization theory delivers ν = 0.588 and thus d = 3.102 in the limit of high molar masses. Experimentally, one cannot distinguish between ν = 0.600 and ν = 0.588 within limits of error (see Table 5-3).

5.7.2. Overlap of Coils

Random coils of polymer molecules are very loose entities; they occupy a large volume per mass (Section 5.3.4.). At very low polymer concentrations, they are separated from each other by many solvent molecules. With increasing polymer concentration, less and less space is available between random coils until they finally begin to touch each other and then to overlap. If random coils are assumed to be equivalent spheres with radii equaling those of radii of gyration of coils, then the overlap concentration is $c_s^* = m_{mol}/V_{mol} = (3 M)/(4 \pi s^3 N_A)$ (see also Section 6.1.2.).

For the following discussion, it is convenient to express the overlap concentration as volume fraction ϕ^* instead of mass concentration c_s^*. Since $c_2^* = m_2^*/V = V_2^* \rho_2/V = \phi_2^* \rho_2$ and $M = M_u X$, one can write $\phi^* = [(3 M_u)(4 \pi N_A \rho_2)^{-1}][X s^{-3}] = K_c' X s^{-3}$. At the overlap concentration, the same relationship must also apply to blobs (see below), thus $\phi^* = K_c X_{bl} s_{bl}^{-3}$.

At concentrations $\phi > \phi^*$, overlapping of molecules becomes so prominent that segments of different molecules begin to entangle (Fig. 5-15, left). Chain sections are large between two adjacent **entanglements**; thus, they behave as if they are individual coils (Fig. 5-15, right). These **blobs** are large enough to be self-similar to the total coil; they are "coils within a coil".

A polymer chain with molar mass $M = N_{bl}M_{bl} = X_u M_u$ composed of X_u monomeric units of molar mass M_u can thus be depicted as a string of pearls comprising N_{bl} blobs of molar mass M_{bl} and degree of polymerization $X_{bl} = M_{bl}/M_u$, respectively. The diameter d_{bl} of a blob is taken as the distance between two adjacent entanglements; it is the mesh size of the network generated by entanglements.

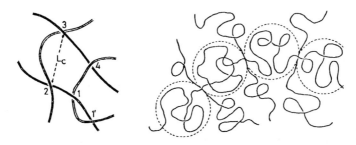

Fig. 5-15   Schematic representation of entanglements in large chain molecules. Contact points 1 and 1' define the entanglements between segments of different chains. Contact points 2, 3, 4 ... can likewise form entanglements with contact points 2', 3', 4' ... (not plotted). Molecule sections between 2 entanglements behave as quasi-independent blobs (right, broken circles).

Near the overlap concentration ϕ^*, blob diameters d_{bl} must be proportional to the radii of gyration of the blobs, i.e., $d_{bl} = Q_{bl}s_{bl}$ where Q_{bl} is a proportionality factor. At this concentration, blobs are relatively large because of the low entanglement density. *Within* these large blobs, volumes must be excluded for monomeric units in good solvents similar to the excluded volumes of units within whole coil molecules. The radii of gyration of blobs thus follow the general relationship $s_{bl} = K_{bl}'M_{bl}^{3/5}$ for perturbed coils (Eqn.(5-25)). Again, it is convenient to use the degree of polymerization of the blob instead of the molar mass, so that $s_{bl} = K_{bl}X_{bl}^{3/5}$.

However, the proportionality factor Q_{bl} in $d_{bl} = Q_{bl}s_{bl}$ cannot be a concentration-independent constant. At volume fractions ϕ of the polymer greater then the overlap fraction ϕ^*, blobs in good solvents are more and more compressed. The compression is more difficult at higher polymer concentrations; the proportionality factor must therefore be $(\phi/\phi^*)^y$ where y is a negative number. It results in $d_{bl} = (\phi/\phi^*)^y s_{bl}$.

The exponent y in $d_{bl} = (\phi/\phi^*)^y s_{bl}$ can be evaluated as follows. Introduction of $\phi^* = K_c X_{bl}s_{bl}^{-3}$ and $s_{bl} = K_{bl}X_{bl}^{3/5}$ leads to $d_{bl} = [K_c^{-y}K_{bl}^{3y+1}]\phi^y X_{bl}^{(3/5)+(4y/5)}$. The blob diameter is by definition independent of the degree of polymerization of the blob; thus $d_{bl} \sim X_{bl}^0$ and $(3/5) + (4y/5) = 0$. The exponent becomes $y = -3/4$. Setting $L_u = K_c^{-y}K^{3y+1}$, one arrives at $d_{bl} = L_u\phi^{-3/4}$. The blob diameter thus decreases rapidly with increasing volume fraction of the polymer. In the melt ($\phi = 1$), the blob diameter becomes the length of a monomeric unit.

Since excluded volumes are present within blobs, blob diameters must follow $d_{bl} = L_u X_{bl}^{3/5}$ where L_u is the length of a monomeric unit ($d_{bl} = L_u$ for $X = 1$). Introduction of $d_{bl} = L_u \phi^{-3/4}$ leads to $X_{bl} = \phi^{-5/4}$.

No excluded volume effects exist *between* blobs because the total random coil consists of only a few blobs. The pearl string composed of blobs thus represents a Kuhn coil where the blob diameter d_{bl} portrays the Kuhn length L_K and the number $N_{bl} = X_{chain}/X_{bl}$ depicts the number of Kuhn elements. The blob diameter d_{bl} is consequently the distance L which screens the excluded volume within the blobs. The natural tendency to exclude other segments is correlated for $L < d_{bl}$ but not for $L > d_{bl}$. The blob diameter is therefore also called the **screening length** or the **correlation length.**

The absence of excluded volume effects *between* blobs lets the total string of blobs behave like a Kuhn coil composed of N_{bl} blobs of "segment length" d_{blob}. The radius of gyration of such a coil is $\langle s^2 \rangle = \langle r^2 \rangle / 6 = (1/6)(N_{bl} d_{bl}^2)$. The number of blobs is given by the ratio of the degree of polymerization of the molecule to the degree of polymerization of the blob, i.e., $N_{bl} = X/X_{bl}$. Introduction of $X_{bl} = \phi^{-5/4}$ and $d_{bl} = L_u \phi^{-3/4}$ results in:

$$(5\text{-}28) \quad \langle s^2 \rangle = (1/6) X L_u^2 \phi^{-1/4} \quad \text{or} \quad s \sim \phi^{-1/8}$$

The radius of gyration of polymer coils in semidilute solutions of *good solvents* should thus decrease with the $-1/8$th power of the volume fraction ϕ. A value of -0.078 instead of -0.125 is found experimentally (Fig. 5-16) probably because the theory applies to infinitely high molar masses whereas the experiment was performed with a polymer of rather low molar mass. For polymers in theta solvents, theory derives $\langle s^2 \rangle_o \sim M$, i.e., concentration-independent radii of gyration.

5.7.3. Fractals

For a Euclidian body, the mass m is proportional to the dth power of a characteristic length L where d is a positive integer ($m \sim L^d$ with d = 1, 2, 3, etc.). Another example is the dependence of the radius of gyration of coils on the molar mass, $s \sim M^\nu = M^{3/(d+2)}$ (see Eqn.(5-27)). Such a correlation is called **scaling**.

The scaling of the radius of gyration of perturbed chains leads to $M \sim s^{1/\nu}$ where $1/\nu$ is a fraction. Such interrelationships are characteristic of irregular objects which are therefore called **fractals** (L: *fractus*, from *frangere* = to break). Macroscopic examples of fractals are mountain ranges, snow flakes, and meandering rivers. Crosslinked polymers, dendrimers, and perturbed coils are examples of geometric fractals in polymer science.

Fractals are characterized by fractal dimensions. Geometric fractal dimensions $\bar{d}$ play a similar role as spatial dimensions d. The distribution of masses within an object is described by a fractal dimension $\bar{d}_m$ (**Hausdorff dimension**) which is defined as a limiting value. For perturbed coils, it is given as $\bar{d}_m = \lim_{s \to \infty} (d \log_e M / d \log_e s)$.

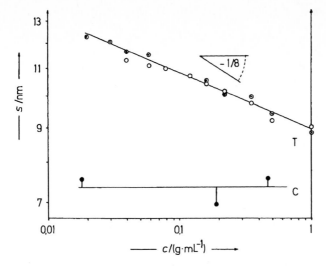

Fig. 5-16 Double-logarithmic plot of radii of gyration s (by small angle neutron scattering) as a function of polymer concentration c for semidilute concentrations of mixtures of poly(styrene)s with deutero-poly(styrene)s of the same molar mass. Both polymers T and C deliver the same reduced radius of gyration of $s/M^{1/2} = 0.027$ for the melt (density ρ = concentration c = 1.054 g/mL), independent of molar mass, temperature and solvent.

    T: $\overline{M}_w$ = 114 400 g/mol in the good solvent deutero-toluene at room temperature [13].
    C: $\overline{M}_w$ = 75 700 g/mol in the theta solvent deutero-cyclohexane at 34°C [14].

The dependence of molar masses on radii of gyration is therefore given by:

$$(5\text{-}29) \quad M \sim s^{\overline{d}_m}$$

Comparison of Eqns.(5-25) and (5-27) shows that polymer coils in good solvents have a fractal dimension of $\overline{d}_m = 1/\nu = 1.667$.

Objects that follow Eqn.(5-29) are called **mass fractals**. A similar expression can be written for **surface fractals**:

$$(5\text{-}30) \quad A \sim L^{\overline{d}_a}$$

Surface fractals are homogeneous with respect to density (i.e., d = 3) but have a rough surface ($2 \leq \overline{d}_a \leq 3$). Whether a fractal object is a mass fractal or a surface fractal can be decided by the dependence of scattering intensities $I \sim (Q_\vartheta)^P$ on the scattering angle; $Q_\vartheta = q \cdot \sin(\vartheta/2) = (4\,\pi/\lambda) \cdot \sin(\vartheta/2)$ and λ = wavelength of radiation in the system (see Eqn.(5-6)). The exponent P in $I \sim (Q_\vartheta)^P$ is the **Porod slope**.

In sol–gel processes (Section 4.1.9.), fractals are formed as intermediates. The cross-linking reaction is controlled by the structure of already formed particles once the gel point is passed. The resulting gels are either mass fractals (usually called "polymers") or surface fractals (called "colloids"). The structure of these fractals determines the structure of the ceramic masses which result after drying and firing (heating).

Porod slopes can distinguish between mass fractals and surface fractals. The mass fractal $\overline{d}_m$ of "colloids" equals the dimensionality d = 3 of Euclidian bodies, whereas the surface fractal $\overline{d}_a$ ranges between 2 (smooth surface) and 3 (very rough surface). The Porod slope is thus $-4 \leq P \leq -3$.

The situation is different for "polymers". The mass fractal $\bar{d}_m = 1/v$ is determined by the molar mass exponent v of the function $s = f(M)$, i.e., by the shape of the polymer particles. At the gel point (onset of cross-linking), polymers start to precipitate. A precipitating linear polymer would be at least in the theta state ($v = 1/2$). The true value of v must be smaller because the polymer is branched which introduces a shrinking factor $g_s < 1$ (see Section 5.5.3.). The polymer can also collapse to a more dense structure while passing through the gel point. These two effects will make the polymer behave more like a sphere ($v = 1/3$). The mass fractal is thus $2 \le \bar{d}_m \le 3$. The surface fractal of polymers equals the dimensionality of Euclidian bodies ($\bar{d}_a = d = 3$). The Porod slope must therefore be in the range $-3 \le P \le -1$.

According to investigations of sol–gel processes, "polymers" are obtained by acid-catalyzed 2-step reactions of $Si(OR)_4$ in the presence of a little water. "Colloids" are produced by the base-catalyzed reaction of the same compound if the molar ratio $[H_2O]/[Si(OR)_4] \approx 4$.

5.8. Viscosimetric Dimensions

5.8.1. Introduction

Dimensions of polymer molecules are not only obtainable by scattering experiments but also from the response of molecules to hydrodynamic forces, for example, during diffusion and viscous flow (Chapter 7). Viscometry of dilute solutions allows one to determine specific volumes of macromolecules that are directly related to the shape and dimensions of the dissolved molecules and indirectly to their molar masses.

In the simplest case, the flow behavior of a material is described by **Newton's law**, $\eta = \sigma_{21}/\dot{\gamma}$. The viscosity η is the ratio of shear stress σ_{21} to shear rate $\dot{\gamma} = \partial v/\partial y$, i.e., the change of the rate of flow, v, of the liquid with the distance y perpendicular to the flow direction (Chapter 7). Liquids that follow Newton's law are called **Newtonian liquids**; their viscosities $\eta = \sigma_{21}/\dot{\gamma}$ do not depend on the shear rate and they do not vary with time.

Viscosities of dilute solutions of low and medium molar mass polymers are very often Newtonian. They are usually determined by capillary viscometry. The product of the flow time of a specified volume and the density of the liquid is proportional to its viscosity. The proportionality constant need not be determined since one is usually only interested in the relative viscosity η/η_1, the ratio of the viscosity η of the solution to the viscosity η_1 of the solvent which allows one to calculate a reduced viscosity $\eta_{red} = \{(\eta/\eta_1) - 1\}/c$. The extrapolation of reduced viscosities to zero concentration furnishes the intrinsic viscosity $[\eta]$. This fast and simple method is the most important technique for the determination of molar masses from $[\eta]$ (see below).

Capillary viscometers usually generate shear rates of ca. 1000 s^{-1} which are high enough to cause non-Newtonian behavior of dilute solutions of high molar mass polymers. These **non-Newtonian viscosities** can be studied by considerably more expensive rotational viscometers (see Chapter 7). Non-Newtonian viscosities must be extrapolated to zero shear rates in order to obtain relative viscosities. For the purpose of Sections 5.8. and 5.9., all viscosities are assumed to be Newtonian.

5.8.2. Concentration Dependence

The viscosity η of dispersions of small spheres (glass, gutta percha) in solvents of viscosity η_1 can be described by a power series with respect to the volume fraction ϕ_2 = V_2/V of the spheres:

(5-31) $\eta = \eta_1[1 + B_1\phi_2 + B_2\phi_2^2 + ...]$

The coefficient B_1 was calculated by **Albert Einstein** as $B_1 = 5/2$ for unsolvated, rigid spheres which is confirmed by experiment. The extension of the theory by **Eugen Guth** furnished $B_2 = 14.1$ in accordance with experimental data.

The **Einstein–Guth equation** (5-31) can be applied to any particle shape and size albeit with different values of B_1 and B_2. It is convenient to define the **viscosity ratio** $\eta/\eta_1 \equiv \eta_{rel}$ as **relative viscosity** (IUPAC recommends η_r). The **specific viscosity** $\eta_{sp} \equiv$ $\eta_r - 1 \equiv (\eta - \eta_1)/\eta_1$ has been renamed **relative viscosity increment** η_i by IUPAC.

The historic name "specific viscosity" is incorrect since it is not a viscosity related to a mass (see Chapter 15). This book retains the symbol η_{sp} in order to avoid confusion between η_i = relative viscosity increment and η_i = viscosity of the component i.

The volume fraction $\phi_2 = V_2/V$ can be replaced by the mass concentration c. The volume $V_2 = N_2V_h$ of all particles is given the product of their number N_2 and the volume V_h of one particle. The number concentration N_2/V of all particles is related to the mass concentration c and the molar mass M via $N_2/V = cN_A/M$ where N_A is the Avogadro number. Introduction of the **reduced viscosity** (or **viscosity number**) η_{red} $\equiv \eta_{sp}/c = (\eta - \eta_1)/(\eta_1 c)$ into Eqn.(5-31) leads to:

(5-32) $\eta_{red} = (5/2)(V_hN_A/M) + B_2(V_hN_A/M)^2 c + ... = [\eta] + B_2(2[\eta]/5)^2 c + ...$

Extrapolation of viscosity numbers η_{red} measured at various concentrations c to infinite dilution ($c \rightarrow 0$) delivers the **limiting viscosity number** $[\eta]$, most commonly called **intrinsic viscosity**. $[\eta]$ is now reported in mL/g, also in 100 mL/g in the USA, and in 1000 mL/g in the old literature. $[\eta]$ has also been known as the **Staudinger index** because H.Staudinger was the first to recognize the importance of this quantity.

Note the misleading names: η_{red} is not a viscosity *number* and $[\eta]$ is not a limiting viscosity *number* since both physical quantities have the physical unit of a specific volume (volume per mass). $[\eta]$ is also not an intrinsic *viscosity* since it does not have the unit Pa s.

Eqn.(5-32) is identical with the empirical **Huggins equation** for nonelectrolytes:

(5-33) $\eta_{red} \equiv \eta_{sp}/c = [\eta] + k_H[\eta]^2 c + ...$

if $[\eta] = (5/2)(V_hN_A/M)$ and $k_H = 4 B_2/25$. The Huggins constant k_H usually has values between 0.33 and 0.8.

Viscosity numbers η_{red} of **non-electrolytes** often show a *positive* curvature with the polymer concentration c (Fig. 5-17). Other empirical terms and functions have therefore been proposed to make plots more linear. Such a term is the inherent viscosity $\eta_{inh} \equiv (\log_e \eta_{rel})/c$. The best known functions are:

(5-34) η_{red} $= [\eta] + k_{SB}[\eta]\eta_{sp} + ...$ **Schulz–Blaschke equation**
(5-35) $(\log_e \eta_{rel})/c$ $= [\eta] + k_K[\eta]^2 c + ...$ **Kraemer equation**
(5-36) $\log_{10} \eta_{red}$ $= \log_{10} [\eta] + k_M[\eta]c + ...$ **Martin equation** or
                                                         Bungenberg–de Jong equation

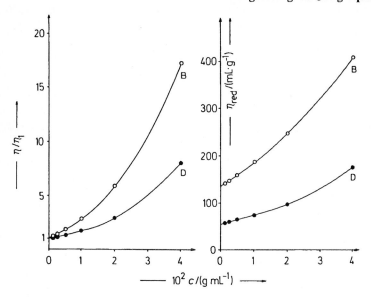

Fig. 5-17 Concentration dependence of relative viscosities (left) and viscosity numbers (right) of a poly(styrene) in benzene B or decalin D at 25°C. Data of [15].

The concentration dependence of viscosities η of aqueous solutions of **polyelectrolytes** differs from that for non-electrolyte polymers. Viscosities of polyelectrolyte solutions have a *negative* curvature with the polymer concentration c if salts are absent or only present in small concentrations (Fig. 5-18, left).

Polyelectrolytes are *water-soluble* polymers with *many* electrically charged groups per molecule. They form **polyions** on dissociation. These polyions may be **polyanions** with negative charges as in dissociated poly(acrylic acid)$-\{CH_2\text{-}CH(COO^\ominus)\}_{\overline{n}}$ or **polycations** as in protonated poly(vinylamine)$-\{CH_2\text{-}CH(NH_3^\oplus)\}_{\overline{n}}$; they may also be **polysalts** such as the sodium salt of poly(acrylic acid)$-\{CH_2\text{-}CH(COONa)\}_{\overline{n}}$. Polyions, polyanions, etc. are to be distinguished from **macroions, macroanions**, etc., which carry only *one* ionic group, usually as an end group (macromolecules with two ionic groups are called macrodiions, etc.). *Water-insoluble* polymers with relatively few ionic groups *in the chain* are known as **ionomers** (see also Section 8.6.).

The viscosity number η_{red} of aqueous polyelectrolyte solutions increases with decreasing polymer concentration until it passes through a maximum at c_{max} and then declines (Fig. 5-18). This phenomenon is explained by an increasing dissociation of polyelectrolytes to charged molecules. Charges repel each other and the polymer molecules become more rod-like. Rigid chains, however, have higher reduced viscosities than flexible ones (see below). The origin of the maximum is controversial.

Reduced viscosities of dissociating polyelectrolytes follow the **Fuoss equation:**

(5-37) $1/\eta_{red} = (1/[\eta]) + Ac^{1/2} + ...$

in the concentration range $c > c_{max}$. Addition of salt suppresses the dissociation. Polyelectrolytes are not dissociated at high salt concentrations; they then follow the Huggins equation (5-33).

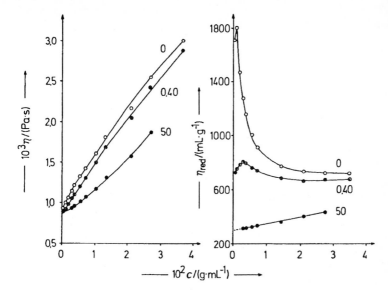

Fig. 5-18 Dependence of viscosities η (left) and viscosity numbers η_{red} on the concentration c of a sodium pectinate at 27°C in water and in aqueous NaCl solutions of 0.40 mmol NaCl/L and 50 mmol NaCl/L. Data of [16].

5.8.3. Molar Mass Dependence

The dependence of intrinsic viscosities $[\eta]$ on molar masses can often be described for surprisingly wide molar mass ranges by the empirical **Kuhn–Mark–Houwink–Sakurada equation** (KMHS equation) (Fig. 5-19):

$$(5-38) \quad [\eta] = K_\eta M^\alpha$$

K_η and α are system-specific constants that depend on the constitution, configuration, and molar mass distribution of the polymer as well as on the solvent and temperature. The KMHS equation is also called the Mark–Houwink–Sakurada equation, Mark–Houwink equation, or Staudinger equation (Staudinger used only $\alpha = 1$!).

Exponents α depend on molecule shapes and segment distributions. Theory predicts $\alpha = 0$ for spheres, $\alpha = 1/2$ for unperturbed coils (in theta solvents), $\alpha = 0.764$ for perturbed coils (in good solvents) and $\alpha = 2$ for infinitely thin, rigid rods (see Section 5.9.). In principle, worm-like chains may exhibit $1/2 \leq \alpha \leq 2$ with higher values for *more rigid* chains at *lower* molar mass and vice versa. The following exponents are observed for the indicated molar mass ranges (α) and asymptotes to high molar masses (α_∞) of the polymers of Fig. 5-19 (see Section 5.8.4.):

PL-D	$512 < M_r < 234\ 000$	$\alpha = 0$	$\alpha_\infty = 0$	spheres
PS-C$_6$H$_{12}$	$10\ 000 < M_r < 3\ 800\ 000$	$\alpha = 0.50$	$\alpha_\infty = 0.50$	unperturbed coils
PS-toluene	$20\ 000 < M_r < 3\ 800\ 000$	$\alpha = 0.72$	$\alpha_\infty = 0.74$	perturbed coils
Amylose	$10\ 800 < M_r < 1\ 720\ 000$	$\alpha = 0.68$	$\alpha_\infty = 0.74$	perturbed coils
Imogolite	$590\ 000 < M_r < 2\ 600\ 000$	$\alpha = 1.77$	$\alpha_\infty = 2.00$	rods

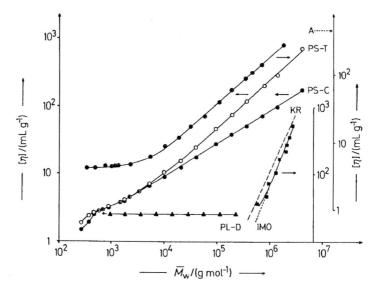

Fig. 5-19   Intrinsic viscosity–molar mass relationships of polymers with narrow molar mass distributions (double-logarithmic plot).

IMO: Imogolites in dilute acetic acid (pH = 3; + 0.02 wt-% NaN$_3$) at 30°C [12].
A: Maltodextrins ($342 < M_r < 2448$) and synthetic amyloses in DMSO at 25°C [17].
PS-T: Poly(styrene)s ($x_s = 0.59$) in toluene T ($\theta = 15$°C) [18].
PS-C: Poly(styrene)s ($x_s = 0.59$) in cyclohexane C ($\theta = 34.5$°C) [18].
PL-D: Hyperbranched poly(lysine)s in DMSO (+ 1 % LiCl) at 25°C [19]; see also Fig. 5-11.
The broken line to the left of IMO indicates the theoretical values for rods according to the Kirk-wood–Riseman theory KR (see below). Chemical structure of PL-D: see legend to Fig. 5-21.

Constants K_η and α must be determined for each system polymer-solvent-temperature by calibration with molecularly uniform polymers of the same structure. Instead of using polymers with $\overline{M}_w/\overline{M}_n \approx 1$, one can also calibrate with polymers having broad molar mass distributions if the so-called viscosity-average molar masses are known. Viscosity-average molar masses are the averages corresponding to intrinsic viscosities as the following derivation shows.

The intrinsic viscosity $[\eta]$ of a polymer is a **mass-average** of the intrinsic viscosities of all macromolecules. According to Eqn.(5-33), one gets $\eta_{sp} \approx [\eta]c$ at small concentrations. As confirmed by experiment, it follows that specific viscosities $(\eta_{sp})_i$ of homologs i of non-electrolyte polymers are additive: $\eta_{sp} = \Sigma_i\ (\eta_{sp})_i = \Sigma_i\ [\eta]_ic_i$. Intrinsic viscosities are defined as $[\eta] \equiv \lim_{c\to 0}\ (\eta_{sp}/c)$. Division of this equation by $c = \Sigma_i\ c_i$ and introduction of the mass fraction $w_i = c_i/c$ delivers $[\eta] = [\Sigma_i\ [\eta]_ic_i]/[\Sigma_i\ c_i] = \Sigma_i\ w_i[\eta]_i \equiv \overline{[\eta]}_w$: intrinsic viscosities are mass-averages.

The molar mass of Eqn.(5-38) is however an exponential average. Solving Eqn.(5-38) for the molar mass, introducing first $[\eta] = \Sigma_i w_i[\eta]_i$ and then $[\eta]_i = K_\eta (M_i)^\alpha$ for each component i, results in the so-called **viscosity-average molar mass**:

$$(5\text{-}39) \quad M = \{[\eta]/K_\eta\}^{1/\alpha} = \{(\Sigma_i w_i[\eta]_i)/K_\eta\}^{1/\alpha} = \{\Sigma_i w_i M_i^\alpha\}^{1/\alpha} \equiv \overline{M}_\eta$$

The viscosity-average molar mass can be calculated if the molar mass distribution is known. It is identical with the mass-average molar mass for $\alpha = 1$; for $\alpha < 1$, it is smaller than the mass-average. Molar masses calculated with Eqn.(5-38) are always viscosity-average molar masses if this equation was calibrated with molecularly uniform polymers or polymers with known viscosity-average molar masses. Calibrations of Eqn.(5-38) with polymers of unknown molar mass distributions or molar mass-averages other than viscosity-averages deliver undefined molar mass averages.

5.8.4. Hydrodynamic Volumes

The KMHS equation is usually called the **viscosity–molar mass relationship**. However, intrinsic viscosities are not viscosities but specific volumes = volume/mass (see Eqn.(5-32)). If $[\eta]$ is measured in mL/g, then it indicates that the hydrodynamic volume in mL is that occupied by 1 g of macromolecules in the limit of infinite dilution. Intrinsic viscosities directly measure hydrodynamic dimensions of macromolecules and only indirectly molar masses because of well-defined relationships between hydrodynamic volumes and molar masses for polymers of identical shapes and segment distributions. These relationships can be elucidated as follows:

The hydrodynamic volume V_{sph} of a *sphere* equals the volume of the sphere: $V_{sph} = (4 \pi/3) R_{sph}^3$. The radius $R_{sph} = Q_{sph}s$ is however greater than the radius of gyration s by a factor $Q_{sph} = (5/3)^{1/2}$ (Section 5.6.1.). Eqn.(5-32) becomes:

$$(5\text{-}40) \quad [\eta] = \frac{5 N_A V_h}{2 M} = \frac{10 \pi N_A R_{sph}^3}{3 M} = \Phi_{sph,R} \frac{R_{sph}^3}{M} = \frac{10 \pi N_A (5/3)^{3/2} s^3}{3 M} = \Phi_{sph,s} \frac{s^3}{M}$$

where $\Phi_{sph,s} \approx 13.57 \cdot 10^{24}$ mol^{-1} and $\Phi_{sph,R} \approx 6.306 \cdot 10^{24}$ mol^{-1}.

The hydrodynamic volume $V_h = m/\rho = M/(\rho N_A)$ can be expressed by the molar mass M and the density ρ of the sphere. Eqn.(5-40) becomes $[\eta] = 5/(2 \rho)$, i.e., intrinsic viscosities of spheres are independent of the molar mass. They depend only on densities ρ. Such spheres are formed by the hyperbranched poly(lysine)s (Fig. 5-19).

The situation is more complicated for *random coils*. Application of Eqn.(5-40) to random coils with hydrodynamic volumes V_η delivers the viscosimetric hydrodynamic radius R_η (**Einstein radius**) of a viscosimetric equivalent sphere:

$$(5\text{-}41) \quad [\eta] = \frac{5 N_A V_\eta}{2 M} = \frac{10 \pi N_A R_\eta^3}{3 M} = \frac{10 \pi N_A Q_\eta^3 s^3}{3 M} = \Phi \frac{s^3}{M} \quad ; \ \Phi \equiv 10 \pi N_A Q_\eta^3/3$$

The factor $Q_\eta = R_\eta/s$ for the conversion of the Einstein radius R_η of equivalent spheres to the radius of gyration of random coils differs from $Q_{\mathrm{sph}} = (5/3)^{1/2}$. The radius of gyration can thus only be calculated if Q_η is known (see Section 5.9.).

According to Eqn.(5-22), $\langle s^2 \rangle = K_s M^{2\nu} = s^2$. Consolidation of the respective constants as $K_\eta = \Phi K_s^{3/2}$ and $\alpha = 3\nu - 1$ converts Eqn.(5-41) into:

$$(5\text{-}42) \quad [\eta] = \Phi \langle s^2 \rangle^{3/2}/M = \Phi K_s^{3/2} M^{3\nu - 1} = K_\eta M^\alpha$$

The exponent α of the KMHS equation is thus related to the exponent ν of the dependence of the radii of gyration on molar masses.

The cube of the radius of gyration, $s^3 = \langle s^2 \rangle^{3/2}$, in Eqn. (5-41) is a number-average! The $\eta_{\mathrm{sp},i}$ values of all i components are additive, thus $\eta_{\mathrm{sp}} = \Sigma_i \, \eta_{\mathrm{sp},i}$. Insertion into Eqn.(5-41) results in $\eta_{\mathrm{sp}} = \Sigma_i \, \Phi(s_i^3/M_i)c_i$. Successive introduction of $c_i = w_i c$, $w_i = m_i/m$, $m_i = n_i M_i$, $m = n \overline{M}_n$, and $n = \Sigma_i \, n_i$ as well as $[\eta] \approx \eta_{\mathrm{sp}}/c$ delivers $[\eta] = (\Phi/\overline{M}_n)[\Sigma_i \, n_i s_i^3]/n = (\Phi/\overline{M}_n)[\Sigma_i \, n_i \langle s^2 \rangle_i^{3/2}]/[\Sigma_i \, n_i]$, i.e., the number-average of s^3.

5.9.   Interpretation of $[\eta] = f(M)$

5.9.1. Unperturbed Coils

Theory (Section 5.3.3.) predicts that the exponent ν becomes 1/2 for *unperturbed coils*; thus, α should also become 1/2 according to Eqn.(5-42). $\alpha = 1/2$ has been found for wide molar mass ranges of many different polymers, for example, for 2000 $< \overline{M}_w/(\mathrm{g\ mol}^{-1}) < 3\,000\,000$ of narrowly distributed poly(styrene)s at 34.5°C in the theta solvent cyclohexane (Fig. 5-19).

The same system deviates from the square root relationship $[\eta] = K_\eta M^{1/2}$ for molar masses below $\overline{M}_w \approx 1000$ g/mol. There may be three reasons for this effect. (1) The influence of end groups with constitutions that differ from those of monomeric units is relatively great at low molar masses. (2) Polymer chains are no longer self-similar since chains are too short to adhere to ideal coil statistics. (3) Self-similarity can no longer be achieved because short chains are dominated by certain local conformations.

The latter influence can be seen for poly(methyl methacrylate) PMMA in acetonitrile at the theta temperature of 44°C. Deviations from $[\eta] = K_\eta M^{1/2}$ are observed here for $M < 50\,000$ g/mol. At this degree of polymerization, $X \approx 500$, ideal coil statistics should be achieved according to point (2). Short sequences of PMMA units, however, form helical segments which dominate the $[\eta] = f(M)$ relationships at low molar masses (α similar to that of rod-like molecules). At high molar masses, these helical segments are separated by coil-like segments. The molecule adopts the global shape of a random coil while locally segments are still helical (see also Fig. 2-21).

Table 5-5 Exponents α of the intrinsic viscosity–molar mass relationships for polymers with narrow molar mass distributions; x_s = mole fraction of syndiotactic diads. Theoretical values α_{theor} are calculated from $\alpha = 3v - 1$ with $v = 1/2$ for unperturbed coils and $v = 0.588$ for perturbed coils. Experimental values α_{exp} refer to sufficiently high molar masses (no effects of end groups and local conformations) and, for good solvents, also to the asymptote at high molar masses. CD = Coefficient of determination. Cf. Table 5-3 for the dependence of radii of gyration on molar masses.

Polymer	x_s	Solvent	$T/°C$	α_{theor}	α_{exp}	CD	Ref.
Unperturbed coils (theta solvents)							
Poly(styrene)	0.59	cyclohexane	34.5	0.500	0.499	1.000	[18]
Poly(α-methyl styrene)	0.40	cyclohexane	34.5	0.500	0.505	0.999	[20]
Poly(α-methyl styrene)	0.40	trans-decalin	9.5	0.500	0.482	0.999	[20]
Poly(methyl methacrylate)	0.79	acetonitrile	44	0.500	0.502	0.999	[21]
Poly(methyl methacrylate)	0.79	butyl chloride	40.8	0.500	0.499	0.999	[21]
Perturbed coils (good solvents)							
Poly(styrene)	0.59	toluene	15	0.764	0.752	0.998	[18]
Poly(α-methyl styrene)	0.40	toluene	25	0.764	0.766	0.997	[20]
Poly(methyl methacrylate)	0.79	benzene	30	0.764	0.660	1.000	[21]
Amylose	–	dimethylsulfoxide	25	0.764	0.740	0.998	[17]
Poly(D-β-hydroxybutyrate)	it+at	trifluoroethanol	25	0.764	0.751	1.000	[22]

Intrinsic viscosities of low molar mass polymers ought to be identical in good and theta solvents because short chains have no excluded volumes. This is indeed found (Fig. 5-19) albeit only for a small range because at still lower molar masses, effects of end groups and local conformations prevail. At higher concentrations in good solvents, excluded volume effects cause increasing deviations from $[\eta] \sim M^{1/2}$. The function $\log_{10}[\eta] = f(\log_{10}M)$ becomes slightly curved until it finally approaches an asymptote at high molar masses. This asymptote is predicted by the mean-field theory to be $\alpha = 3v-1$ (Section 5.5.2.).

Experiments confirm $\alpha = 1/2$ for unperturbed coils ($v = 1/2$). For perturbed coils, the mean-field theory predicts $\alpha = 0.800$ ($v = 0.600$) and renormalization theory $\alpha = 0.764$ ($v = 0.588$) (Table 5-5). Values of $\alpha < 0.764$ are often found for perturbed coils because some solvents are only moderately good and/or the asymptote at high molar masses has not been reached. An example of the former is PMMA in benzene, one of the latter is amylose in DMSO. The ten highest molar masses of amylose (Fig. 5-19) furnish $\alpha = 0.68$ (CD = 0.999) whereas the three highest provide $\alpha = 0.74$ (CD = 0.998). Values of $\alpha > 0.764$ are caused either by insufficient polymolecularity corrections for flexible chains or by worm-like characteristiscs of chains. However, the latter do approach $\alpha = 1/2$ for very high molar masses (see Fig. 5-21).

5.9.2. Flory Constant of Flexible Molecules

Eqn.(5-41) predicts that one can calculate radii of gyration, s, from intrinsic viscosities $[\eta]$ if the conversion factor $Q_\eta = R_\eta/s$ is known. Since this conversion factor relates hydrodynamic radii R_η to radii of gyration, it must depend on the segment distribution in the coil which is known to be different for flexible molecules in good and bad solvents (Fig. 5-8). The **mean-field theory** holds that this difference is caused by perturbations due to long-range interactions (Fig. 5-6) that are equally effective for the coil molecule under static conditions (e.g., static light scattering) and in a flowing solution. This theory predicts for flexible coils that the exponent α can vary only between 0.50 and 0.80 (renormalization theory: 0.764). **Draining theories**, on the other hand, assume that long-range interactions are absent and that the different behavior in good and bad solvents is caused by differences in the draining behavior of molecules. They predict α to vary between 0.50 and 1.00.

Draining theories are intuitively appealing. In good solvents, polymer coils are more expanded than in theta solvents; coils in good solvents may thus be better draining than those in theta solvents. Since polymer concentrations are very low in polymer coils, solvent molecules may flow relatively easily through a coil.

In order to calculate the effects, draining theories represent polymer chains by beads that are interconnected by massless springs (see Section 7.2.3.). The **Rouse theory** assumes that there are no hydrodynamic interactions between the beads so that the solvent can flow freely through the coil whereby it causes friction. The viscosity $\eta = \xi_{seg}F_\eta$ is predicted to be the product of the friction coefficient ξ_{seg} of a segment and a global factor $F_\eta = (\rho N_A/6)(\langle s^2\rangle_o/M)N_{seg}$ that describes the effect of the macroconformation. In dilute solutions, densities $\rho = m_{coil}/V_{coil}$ of coils become concentrations $c = m_2/V$ where m_2 is the polymer mass and V the solution volume. Viscosities can be written as $\eta = \eta_1(\eta/\eta_1) \approx \eta_1[(\eta - \eta_1)/\eta_1] = \eta_1\eta_{sp}$. Since $\eta/\rho \approx \eta_1\eta_{sp}/c \approx \eta_1[\eta]$ in dilute solutions, one also obtains for free-draining coils without excluded volume:

$$(5\text{-}43) \qquad [\eta] = \frac{N_A\xi_{seg}N_{seg}}{6\,\eta_1} \cdot \frac{\langle s^2\rangle_o}{M} = \frac{N_A\xi_{seg}}{6\,\eta_1 M_{seg}} \cdot \frac{\langle s^2\rangle_o}{M} \cdot M = K_\eta M \quad ; \quad \langle s^2\rangle_o/M = const.$$

The Rouse theory predicts $\alpha = 1$ for free-draining coils without excluded volume.

The **Kirkwood–Riseman theory** (KR theory) also assumes the absence of excluded volumes. The beads should however interact (Section 7.2.3.) so that the friction factor varies with the extent of hydrodynamic interaction. A free-draining coil obviously has no hydrodynamic interactions. Very strong interactions, on the other hand, lead to non-draining coils. The KR theory results in:

$$(5\text{-}44) \qquad [\eta] = \pi^{3/2}N_A[Q\cdot f(Q)]\frac{\langle s^2\rangle_o^{3/2}}{M} = \Phi\frac{\langle s^2\rangle_o^{3/2}}{M} = \Phi\left(\frac{\langle s^2\rangle_o}{M}\right)^{3/2}M^{1/2}$$

The Kirkwood–Riseman function $[Q\cdot f(Q)]$ depends on the segmental friction coefficients. It ranges from very small values for freely draining coils to $[Q\cdot f(Q)] = 1.259$ for non-draining coils (Auer–Gardner revision of the KR function). The factor $\Phi = \pi^{3/2}N_A[Q\cdot f(Q)]$ thus becomes $\Phi_\Theta \approx 4.22\cdot10^{24}$ mol^{-1} for non-draining coils.

Eqn.(5-44) from the Kirkwood–Riseman theory is identical with an equation that can be derived from the **mean-field theory**. This theory writes Eqn.(5-41) for theta conditions as $[\eta]_\Theta = \Phi_\Theta(\langle s^2 \rangle_0^{3/2}/M)$ where $\Phi_\Theta = 10\ \pi N_A(Q_{\eta,\Theta})^3/3$. The factor $Q_{\eta,\Theta} = R_{\eta,\Theta}/s_0$ for the conversion of the unperturbed radius of gyration, s_0, into the unperturbed radius of a viscosimetrically equivalent sphere is independent of the polymer structure since the distribution of segments in unperturbed coils does not depend on the polymer structure (see Eqn.(5-18)). According to Section 5.3.4., the ratio $\langle s^2 \rangle_0/M$ is furthermore independent of molar mass and solvent for high-molar-mass, unperturbed coils. All constants can be united in a system-dependent constant $K_{\eta,\Theta}$ and Eqn.(5-44) becomes:

$$(5\text{-}45) \quad [\eta]_\Theta = \Phi_\Theta \frac{\langle s^2 \rangle_0^{3/2}}{M} = \Phi_\Theta \left(\frac{\langle s^2 \rangle_0}{M} \right)^{3/2} M^{1/2} = K_{\eta,\Theta} M^{1/2}$$

Φ_Θ is thus predicted to be a *universal constant* for unperturbed coils of high molar masses. This **Flory constant** adopts a value of $\Phi_\Theta = 4.22 \cdot 10^{24}$ mol^{-1} according to the KR theory. This value leads to $Q_{\eta,\Theta} = R_{\eta,\Theta}/s_0 = 0.874$ for unperturbed coils whereas it is $R_{sph}/s_{sph} = (5/3)^{1/2} \approx 1.291$ for solid spheres (Section 5.6.1.).

Various numerical values for Φ_Θ can be found in the literature because either the characteristic length is related to the end-to-end distance instead of the radius of gyration and/or the various quantities are used with different physical units. Φ_Θ is also often given without any physical unit. If the Flory constant is based on the end-to-end distance instead of the radius of gyration, then its numerical value is $\Phi_{\Theta,r} = 2.87 \cdot 10^{23}$ mol^{-1} because $\Phi_{\Theta,r} = [\eta]M/\langle r^2 \rangle_0^{3/2} = [\eta]M/[6^{3/2}\langle s^2 \rangle_0^{3/2}]$.

As predicted, Flory constants are independent of molar masses for *unperturbed coils* in the high molar mass limit (Fig. 5-20). They are however lower than the KR value of $10^{-24}\Phi_\Theta/\text{mol}^{-1} = 4.22 \cdot 10^{24}$.

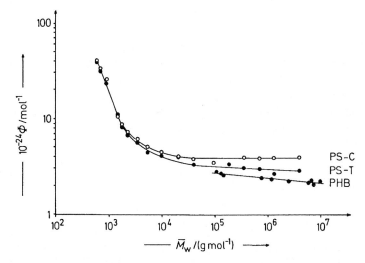

Fig. 5-20 Dependence of the Flory constant $\Phi = [\eta]M/s^3$ on the mass-average of molar masses. PS-C: Poly(styrene) in the theta solvent cyclohexane (34.5°C) [18]; PS-T: Poly(styrene) in the good solvent toluene (15°C) [18]; PHB: Poly(D-β-hydroxybutyrate) in trifluorethanol (25°C) [22].

Flory constants also depend on the polymer–solvent system:

poly(styrene) in cyclohexane at $\Theta = 34.5°C$	3.94 ± 0.06
synthetic amylose in dimethyl sulfoxide at $\Theta = 25°C$	3.59 ± 0.50
poly(isobutylene) in isoamyl valerate at $\Theta = 25°C$	3.58 ± 0.12
poly(methyl methacrylate) in acetonitrile at $\Theta = 44°C$	3.33 ± 0.11
poly(α-methyl styrene) in cyclohexane at $\Theta = 34.5°C$	2.99 ± 0.13

These differences may be caused by different unperturbed dimensions in various theta solvents (different specific polymer–solvent interactions) and may also be due to different degrees of draining.

5.9.3. Perturbed Coils

Number concentrations of segments are very small in coils (Fig. 5-8). They decrease rapidly with increasing molar mass of the polymers at the center of gravity of coils. One would thus expect that the degree of draining increases with increasing molar mass of the polymer. If draining is indeed important, this should cause the exponent α in $[\eta] = K_\eta M^\alpha$ to increase continuously with increasing molar mass from $\alpha = 1/2$ (no draining) to $\alpha = 1$ (freely draining). One does indeed find $\alpha = 1/2$ for small molar masses (if end effects are negligible) and then an increase of α with M albeit only to $\alpha \approx 0.76$ (see Table 5-5). Since flexible chains behave in theta solvents like unperturbed coils, it is generally assumed that draining is negligible even in good solvents and that values of $\alpha > 1/2$ for *flexible chains* are practically due only to effects of excluded volumes.

The mean-field theory predicts $\langle s^2 \rangle^{1/2} = K_s M^\nu$ for the radius of gyration of such perturbed chains (Eqn.(5-25)). One can thus write in analogy to Eqn.(5-45):

$$(5\text{-}46) \quad [\eta] = \Phi \langle s^2 \rangle^{3/2}/M = \Phi K_s^3 M^{3\nu-1} = K_\eta M^\alpha$$

The upper limit of $\nu = 0.588$ leads to an upper limit of $\alpha = 0.764$ for flexible coils with excluded volume. The mean-field theory also predicts that Φ is not a universal constant for perturbed chains. Φ is not only different for each system (like Φ_Θ), it is also continuously decreasing with increasing molar masses and never becomes constant (Fig. 5-20).

Since radii of gyration are expanded by $\alpha_s = [\langle s^2 \rangle/\langle s^2 \rangle_o]^{1/2}$ in good solvents (Section 5.5.2), expansions of hydrodynamic radii can be assumed by analogy to follow $\alpha_\eta = ([\eta]/[\eta]_\Theta)^{1/3}$. Eqn.(5-46) converts into:

$$(5\text{-}47) \quad [\eta] = \alpha_\eta^3 [\eta]_\Theta = \alpha_\eta^3 \Phi_\Theta \langle s^2 \rangle_o^{3/2}/M = \Phi_\Theta \langle s^2 \rangle^{3/2} M^{-1} (\alpha_\eta^3/\alpha_s^3)$$

Scattering methods and viscosities "see" different radii ($\alpha_s \neq \alpha_\eta$). Theoretical calculations furnished $\alpha_\eta^3 = \alpha_s^q$ with q = 2.43 for equivalent spheres and q = 2.18 for equivalent ellipsoids.

These two quantities lead to $10^{24}\Phi/\text{mol}^{-1} = 3.76 \pm 0.17$ (equivalent spheres) and 4.21 ± 0.26 (equivalent ellipsoids) for poly(styrene)s in the good solvent toluene at 15°C. Both values are identical within limits of error with the value $10^{24}\Phi_\Theta/\text{mol}^{-1} = 3.94 \pm 0.06$ for the theta solvent cyclohexane. One thus cannot distinguish between equivalent spheres and ellipsoids using this method.

5.9.4. Rods

Radii of gyration of infinitely thin rigid rods are directly proportional to their molar masses: $s = (12^{1/2} \pi R^2 N_A \rho)^{-1} M$ (Section 5.6.2.). Insertion into Eqn.(5-41) delivers $[\eta] \sim M^2$: the intrinsic viscosity of rigid rods is proportional to the square of the molar mass which is indeed observed for imogolite at *high molar masses* (Fig. 5-19). These polymers are so rigid that the transition to random coils is outside the measured range of molar masses.

Hydrodynamic theories model rod-like molecules as prolate ellipsoids. They obtain for infinitely long rods of length L and diameter d:

$$(5\text{-}48) \quad [\eta] = \frac{2\pi N_A L^3}{45 M[\log_e(L/d) + K]}$$

where $K = 0$ (Kirkwood–Riseman theory) or $K = (2 \log_e 2) - (7/3)$ (Doi–Edwards theory). Both theories predict remarkably well the function $[\eta] = f(M)$ for the observed lengths and diameters of imogolites (Fig. 5-19).

5.9.5. Branched Polymers

The function $[\eta] = f(M)$ for branched polymers is determined by the type of branching. Linear high-density poly(ethylene)s by Ziegler–Natta polymerization follow the KMHS equation $[\eta] = K_\eta M^{0.74}$ in good solvents as one would expect for perturbed coils (Fig. 5-21). The same function is also obtained for linear low-density poly(ethylene)s from ethylene and 7.7 wt-% 1-butene or 9 wt-% 1-decene. Low-density poly(ethylene)s by free radical polymerization do not adhere to the KMHS equation, however, because the proportion of randomly distributed short-chain and long-chain branches varies with the molar mass. These polymers are not self-similar: at lower molar masses, they behave like perturbed coils ($\alpha \approx 0.74$) but at high molar masses, they act like highly solvated spheres ($\alpha \approx 0$; $[\eta]_\infty = 210$ mL/g).

Comb polymers with poly(methyl methacrylate) main chains and poly(styrene) side chains behave at 25°C in the good solvent toluene (Fig. 5-21) similar to amylose in a good solvent (Fig. 5-19). At low molar masses ($X < 58$), they appear as slightly swollen spheres ($\alpha = 0$), at high molar masses ($X > 58$) as perturbed coils ($\alpha \approx 0.75$). The transition from spheres to coils is fairly sharp; it occurs if the main chain is about twice as long as the side chains ($X = 58$; $N = 28$).

$$PMMA{-}c{-}PS \qquad CH_3{-}\overset{\overset{\displaystyle CH_2}{|}}{\underset{\displaystyle }{C}}{-}COO(CH_2)_2{-}(CH{-}CH_2)_N{-}CH_2CH(CH_3)_2$$

with the $\overset{|}{C}_6H_5$ on the side chain and the $\Big]_x$ backbone bracket.

Intrinsic viscosities of **hyperbranched** poly(α,ε-lysine)s PL-D are independent of molar masses (Fig. 5-21). This sphere-like behavior is probably caused by a dense packing of lysine units in the spheres due to unequal lengths of the two branching units -(CH$_2$)$_4$NH- and -NH- in the lysine unit -NH-CO-CH(CH$_2$CH$_2$CH$_2$CH$_2$NH-)-.

The experimental value of $[\eta] = 2.5$ mL/g $\neq f(M)$ is however greater than the intrinsic viscosity of $[\eta] = 2.12$ mL/g that can be calculated from $[\eta] = 5/(2\,\rho)$ with the density $\rho = 1.18$ g/mL of poly(α,ε-lysine). These hyperbranched poly(lysine)s are therefore slightly solvated and not solid spheres.

PL—D   (C$_6$H$_5$)$_2$CH—NH—   —CO—CH—(CH$_2$)$_4$—NH—   —CO—O—C(CH$_3$)$_3$
                                                                |
                                                              NH —

    core   monomeric unit   end group

Poly(benzylidenedioxide) **dendrimers** PPO-D show a completely different behavior. These dendrimers possess a trifunctional core and 3,5-dioxybenzylidene units as chain segments between branching points. Their intrinsic viscosities first increase with increasing molar mass, pass through a maximum, and finally decrease (Fig. 5-21).

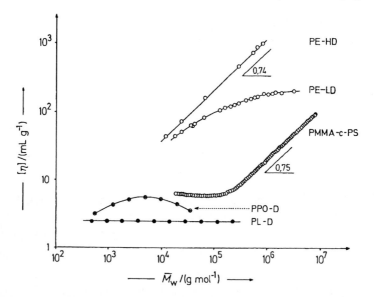

Fig. 5-21   Viscosity–molar mass relationships for polymers with various types of branches.
 PE-HD: Unbranched high-density poly(ethylene) in tetralin at 120°C [23];
 PE-LD: Randomly branched low-density poly(ethylene) in tetralin at 120°C [23];
 PMMA-c-PS: Comb polymers of poly(styrene) on PMMA in toluene at 25°C [24];
 PPO-D: Dendrimers with 3,5-dioxybenzylidene units in tetrahydrofuran at 30°C [25];
 PL-D: Hyperbranched α,ε-L-lysine polymers in *N,N*-dimethylformamide at 25°C [26].

This behavior follows from the different dependence of molecular volumes and molar masses with increasing number of generations (increasing molar mass). Intrinsic viscosities are proportional to the ratios of hydrodynamic volumes V_h to molar masses M (Eqn.(5-40)). Molar masses of PPO-D increase exponentially with the number of generations g according to 2^g-1 whereas their volumes V_h increase with g^3. The ratios $V_h/M \sim g^3/(2^g-1)$ are 2.67 (at $g = 2$), 3.86 (at $g = 3$), 4.27 (at $g = 4$), 4.03 (at $g = 5$), 3.43 (at $g = 6$), 0.98 (at $g = 10$) and 0.103 (at $g = 15$). Intrinsic viscosities $[\eta] \sim V_h/M$ thus pass through a maximum with increasing M.

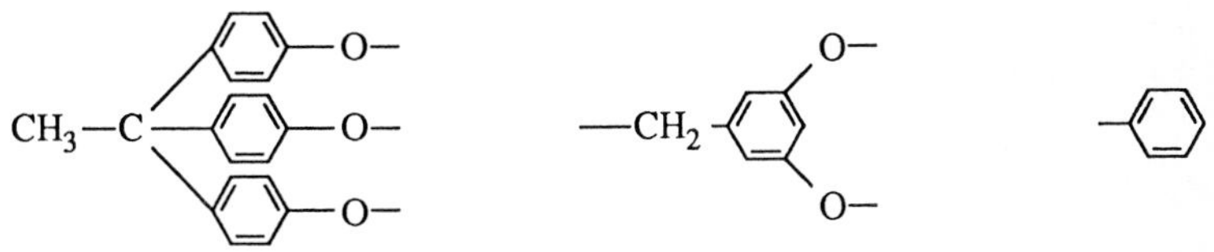

| core | dendrimer unit | end group |

Appendix to Chapter 5

A 5-1: End-to-End Distances of Freely Rotating Chains

The length and direction of each bond i is determined by a vector $|b_i| = const.$ The vectorial distance between chain ends is

$$r_{oo} = b_1 + b_2 + ... b_N = \sum_{i=1}^{i=N} b_i$$

(see figure in Eqn.(5-11)). The mean-square average of end-to-end distances is given by the scalar product:

(A 5-1) $\langle r^2 \rangle_{oo} = \langle r_{oo} r_{oo} \rangle = \langle \Sigma_i\, b_i \cdot \Sigma_j\, b_j \rangle = b_1 b_1 + b_2 b_2 + ... b_N b_N + 2\, \Sigma_j \Sigma_{j<i} \langle b_i b_j \rangle$

where the index j has the same meaning as the index i and only indicates that each term of the first sum has to be multiplied by each term of the second sum. Vectors b_i and b_{i+1} define an angle $\alpha = 180° - \tau$ that is complimentary to the valence angle τ. Their scalar product is $b_i b_{i+1} = |b_i||b_{i+1}|\cdot\cos(180°-\tau)$. Eqn.(A 5-1) thus becomes for chains with N chain units:

(A 5-2) $\langle r^2 \rangle_{of} = Nb^2 + 2(N-1)\langle b_1 b_2 \rangle + 2(N-2)\langle b_1 b_3 \rangle + ... + 2\langle b_1 b_N \rangle$

All chain bonds are of equal length for vinyl polymers $+CH_2\text{-}CHR\}_{\overline{n}}$ and all valence angles of equal size. After introduction of the averages $\langle b_i b_{i+j} \rangle = b^2 \cos^j(180° - \tau) = b^2 \cos^j \alpha$ of the scalar products, Eqn.(A 5-2) converts into:

(A 5-3) $\langle r^2 \rangle_{\text{of}} = b^2[N + 2(N-1)\cos\alpha + 2(N-2)\cos^2\alpha + ... + 2(N-j)\cos^j\alpha + ... + 2\cos^{N-1}\alpha]$

The two series in Eqn.(A 5-3) can be solved. The result is:

(A 5-4) $\langle r^2 \rangle_{\text{of}} = Nb^2 \left[\dfrac{1-\cos\tau}{1+\cos\tau} + \dfrac{2\cos\tau}{N} \left(\dfrac{1-(-\cos\tau)^N}{(1+\cos\tau)^2} \right) \right]$

For poly(methylene), $\alpha = 180° - \tau = 180° - 111.5° = 68.5°$ (Section 2.4.1.) and $\cos\tau = -0.3665...$. Since the second term in Eqn.(A 5-4) is less than 2 % of the first term for $N = 100$, higher order terms can be neglected in Eqn.(A 5-3).
Eqn.(A 5-4) thus reduces to:

(A 5-5) $\langle r^2 \rangle_{\text{of}} = Nb^2(1 + \cos\alpha)(1 - \cos\alpha)^{-1} = Nb^2(1 - \cos\tau)(1 + \cos\tau)^{-1}$

for long vinyl chains. More complicated expressions are obtained for chains other than vinyl polymers because of the differences in bond lengths and valence angles.
    A similar derivation can be performed for the more complicated case of a chain with restricted rotation.

A 5-2: End-to-End Distance and Radius of Gyration

End-to-end distances and radii of gyration are interrelated in the same manner for all chains without excluded volume. The masses of chain atoms are assumed to be concentrated in N_c mass points that are interconnected by bonds of length b. $\boldsymbol{R}_1$ is the vector from the center of gravity S to the first mass point, $\boldsymbol{R}_i$ the corresponding vector to the ith mass point, and $\boldsymbol{r}_i$ the vector between these two mass points (see figure in Eqn.(5-11)). $\boldsymbol{R}_i = \boldsymbol{R}_1 + \boldsymbol{r}_i$ applies to each mass point. One obtains $\Sigma_i \boldsymbol{R}_i = N_c\boldsymbol{R}_1 + \Sigma_i \boldsymbol{r}_i = 0$ (for $1 \leq i \leq N_c$) for all mass points and therefore also:

(A 5-6) $\boldsymbol{R}_1 = -(1/N_c) \Sigma_i \boldsymbol{r}_i$

For a freely jointed chain, the mean-square radius of gyration is the second moment of the mass distribution of all radii, i.e., $\langle s^2 \rangle_{oo} = \langle \Sigma_i m_i R_i^2 \rangle / \Sigma_i m_i$. All masses are equal, the number of all chain units is thus $N_c = (\Sigma_i m_i)/m$. One can average first over all sums and then the products or first over the products and then the sums. It follows:

(A 5-7) $\langle s^2 \rangle_{oo} = \dfrac{\langle \Sigma_i m_i R_i^2 \rangle}{\Sigma_i m_i} = \dfrac{\Sigma_i m_i \langle R_i^2 \rangle}{m} = \dfrac{\Sigma_i \langle R_i^2 \rangle}{N_c}$

(A 5-8) $\langle s^2 \rangle_{oo} = (1/N_c) \Sigma_i (\boldsymbol{R}_1 + \boldsymbol{r}_i)(\boldsymbol{R}_1 + \boldsymbol{r}_i) = R_1^2 + (1/N_c) \Sigma_i r_i^2 + (2/N_c) \boldsymbol{R}_1 \Sigma_i \boldsymbol{r}_i$

According to Eqn.(A 5-6) one has:

(A 5-9) $R_1^2 = (\Sigma_i \Sigma_j \, r_i r_j)/N_c^2$

and, according to Eqns.(A 5-6) and (A 5-1):

(A 5-10) $(2/N_c) \, R_1 \, (\Sigma_i \, r_i) = -(2/N_c^2) \, \Sigma_i \Sigma_j \, r_i r_j$

Introduction of Eqn.(A 5-9) and (A 5-10) into Eqn.(A 5-8) delivers:

(A 5-11) $\langle s^2 \rangle_{oo} = (1/N_c) \, \Sigma_i \, r_i^2 - (1/N_c^2) \, \Sigma_i \Sigma_j \, r_i r_j$

The scalar product is solved by the cosine rule $r_i r_j = r_i r_j \cdot \cos \omega = [r_i^2 + r_j^2 - r_{ij}^2]/2$. The sum of the squares of distances is identical since the indexes i and j have the same meaning. Introduction of this expression into Eqn.(A 5-11) delivers:

(A 5-12) $\langle s^2 \rangle_{oo} = [1/(2 \, N_c^2)] \, \Sigma_i \Sigma_j \, \langle r_{ij}^2 \rangle$

where $\langle r_{ij}^2 \rangle$ is the end-to-end distance of a chain of $|j-i|$ elements of length b, i.e., $\langle r_{ij}^2 \rangle = |j-i| \cdot b^2$. Eqn.(A 5-12) lets one express b^2 by $b^2 = \langle r^2 \rangle_{oo}/N$. With $N \approx N_c$, Eqn.(A 5-12) converts into:

(A 5-13) $\langle s^2 \rangle_{oo} = [1/(2 \, N^2)] \, \Sigma_i \Sigma_j \, |j-i| \langle r^2 \rangle_{oo}/N$

The double sum of the absolute differences $|j-i|$ can be solved consecutively for each sum. The summation over all j values results in:

(A 5-14)

$$\sum_{j=1}^{j=N} |j-i| = \sum_{j=1}^{i}(i-j) + \sum_{j=i+1}^{N}(j-i) = i^2 - (1/2)i(i+1) + (1/2)(N-i)(N+i+1) - i(N-i)$$

$$= i^2 - iN + (1/2)N^2 + (1/2)N - i$$

The sum of all squares of i is $\Sigma_i \, i^2 = 1^2 + 2^2 + ...N^2 = N(N+1)(2\,N+1)/6$. Thus:

(A 5-15) $\displaystyle\sum_{i=1}^{i=N} \sum_{j=1}^{j=N} |j-i| = (N^3 - N)/3 \approx N^3/3$

Introduction of Eqn.(A 5-15) into Eqn.(A 5-13) delivers:

(A 5-16) $\langle s^2 \rangle_{oo} = \dfrac{1}{2\,N^2} \cdot \dfrac{N^3}{3} \cdot \dfrac{\langle r^2 \rangle_{oo}}{N} = \dfrac{\langle r^2 \rangle_{oo}}{6}$

Eqn.(A 5-16) was derived for a freely jointed chain. Calculations for the more complicated cases of freely rotating chains, chains with restricted rotation, or other unperturbed chains (e.g., by the RIS model) always deliver the same result: The mean-square radius of gyration is always smaller by a factor of 6 than the mean-square end-to-end distance of unperturbed chains, i.e., $\langle s^2 \rangle_o = \langle r^2 \rangle_o/6$, as long as $N \to \infty$.

A 5-3: Chains with Persistence

Eqn.(A 5-4) is applied to a freely rotating chain where N segments of length b are joined through angles τ. The first segment b_1 of this chain should point toward the z direction. The average of the z-component of the vector r_{1N} between the chain ends is thus $\langle r_{1N} e_z \rangle = b^{-1} \Sigma_i \langle r_1 r_i \rangle$. The scalar product is obtained from Eqn.(A 5-2) as $\langle r_1 r_i \rangle = b^2(- \cos \tau)^{i-1}$. Thus:

(A 5-17) $\langle r_{1N} e_z \rangle = b[1 - (- \cos \tau)^N][1 + \cos \tau]^{-1}$

This equation becomes $\langle r_{1N} e_z \rangle_\infty = N/(1 + \cos \tau) \equiv 1/a$ in the limiting case of long chains ($N \to \infty$), where a is the persistence length. For the limiting case of an infinitely long ($N \to \infty$) and very rigid chain ($\tau \to 180°$, i.e., $(1 - \cos \tau) \to 2$), one obtains from Eqns.(A 5-4) and (A 5-17) with $b/(1 + \cos \tau) = 1/a$ and $L_{chain} = Nb = N_a a$

(A 5-18) $\langle r_{1j} e_z \rangle = a^{-1}[1 - \exp(- L_{chain}/a)]$

(A 5-19) $\langle r^2 \rangle_{worm} = 2\, a L_{chain} - 2\, a^2[1 - \exp(- L_{chain}/a)]$

Literature

H.-G. Elias, Makromoleküle (in German), Hüthig and Wepf, Basel, 5th ed. (1990), Vol I: Fundamentals (1990); Chapter 16, Shape and Dimensions of Single Macromolecules

5.2. SCATTERING METHODS
H.Brumberger, ed., Small Angle X-Ray Scattering, Gordon and Breach, New York 1967
M.B.Huglin, ed., Light Scattering from Polymer Solutions, Academic Press, London 1972
O.Glatter, O.Kratky, Small Angle X-Ray Scattering, Academic Press, New York 1982
P.Kratochvil, Classical Light Scattering from Polymer Solutions, Elsevier, Amsterdam 1987
B.Chu, Laser Light Scattering, Academic Press, San Diego 1990
J.S.Higgins, H.C.Benoit, Polymers and Neutron Scattering, Clarendon Press, Oxford 1994

5.3.-5.5. COIL MOLECULES
P.J.Flory, Statistical Mechanics of Chain Molecules, Interscience, New York 1969
W.C.Forsman, ed., Polymers in Solution, Theoretical Considerations and Newer Methods of Characterization, Plenum, New York 1983
J.A.Semlyen, ed., Cyclic Polymers, Elsevier Sci. Publ., New York 1986
J.des Cloizeaux, G.Jannink, Les Polymères en Solution: Leur Modélisation et Leur Structure, Les éditions de physique, Les Ulis Cedex (France) 1987
H.Fujita, Polymer Solutions, Elsevier, Amsterdam 1990
A.Halperin, M.Tirrell, T.P.Lodge, Tethered Chains in Polymer Microstructures, Adv. Polym.Sci. **100** (1992) 31
E.A.Colbourn, ed., Computer Simulation of Polymers, Longman Higher Education, Harlow (Essex) 1994
B.R.Gelin, Molecular Modeling of Polymer Structures and Properties, Hanser, München 1994
W.L.Mattice, U.W.Suter, Conformational Theory of Large Molecules, Wiley, New York 1994
H.Dautzenberg, W.Jaeger, J.Kötz, B.Philipp, C.Seidel, D.Stscherbina, Polyelectrolytes, Hanser, Munich 1994
A.Yu.Grosberg, A.R.Khokhlov, Statistical Physics of Macromolecules, American Institute of Physics, New York 1994

K.Binder, ed., Monte Carlo and Molecular Dynamics Simulations in Polymer Science, Oxford
   University Press, New York 1995

5.6. EUCLIDIAN BODIES
R.H.Pain, ed., Mechanisms of Protein Folding, Oxford University Press, New York 1994

5.7. SCALING
P.G. de Gennes, Scaling Concepts in Polymer Physics, Cornell University Press, Ithaca (NY)
   1979
D.Stauffer, Introduction to Percolation Theory, Taylor and Francis, London 1985
K.R.Freed, Renormalization Group Theory of Macromolecules, Wiley, New York 1987
D.W.Schaefer, Polymers, Fractals, and Ceramic Materials, Science **243** (1989) 1023-1027
C.J.Brinker, G.W.Scherer, Sol-Gel Science. The Physics and Chemistry of Sol-Gel Processes,
   Academic Press, San Diego (CA) 1990
F.F.Cassidy, T.Vicek, The Dynamics of Fractal Surfaces, World Scientific, Singapur 1992

5.8. VISCOSIMETRIC DIMENSIONS
M.Bohdanecky, J.Kovar, Viscosity of Polymer Solutions, Elsevier, Amsterdam 1982
K.S.Schmitz, Macroions in Solution and Colloidal Suspension, VCH, Weinheim 1992
M.Hara, ed., Polyelectrolytes. Science and Technology, Dekker, New York 1993
H.Dautzenberg, W.Jaeger, J.Kötz, B.Philipp, C.Seidel, D.Stscherbina, Polyelectrolytes, Hanser,
   Munich 1994

References

[1]   K.Matsuzaki, F.Kawazu, T.Kanai, Makromol.Chem. **183** (1982) 185, Fig. 5
[2]   H.-G.Elias, Grosse Moleküle, Springer-Verlag, Berlin 1985; Mega Molecules, Springer-
          Verlag, Berlin 1987; Fig. 14
[3]   F.Abe, Y.Einaga, T.Yoshizaki, H.Yamakawa, Macromolecules **26** (1993) 1884
[4]   M.Ragnetti, D.Geiser, H.Höcker, R.C.Oberthür, Makromol.Chem. **186** (1985) 1709
[5]   T.Kato, K.Miyaso, I.Noda, T.Fujimoto, M.Nagasawa, Macromolecules **3** (1970) 777
[6]   Y.Tomai, T.Konishi, Y.Einaga, M.Fujii, H.Yamakawa, Macromolecules **23** (1990) 4067
[7]   Y.Miyaki, Y.Einaga, T.Hirosye, H.Fujita, Macromolecules **10** (1977) 1356
[8]   N.Khasat, R.W.Pennisi, H.Hadjichristidis, L.J.Fetters, Macromolecules **21** (1988) 1100
[9]   P.Doty, B.McGill, S.A.Rice, Proc.Natl.Acad.Sci. **44** (1958) 432
[10]   R.Pecora, Science **251** (1992) 893
[11]   S.M.Aharoni, N.S.Murthy, Polym.Commun. **24** (1983) 132
[12]   N.Donkai, H.Inagaki, K.Kanjiwara, H.Urakawa, M.Schmidt, Makromol.Chem. **186**
          (1985) 2623
[13]   J.S.King, W.Boyer, G.D.Wignall, R.Ullman, Macromolecules **18** (1985) 709, Fig. 9
[14]   R.W.Richards, A.Maconnachie, G.Allen, Polymer **19** (1978) 266, Table 3
[15]   D.J.Streeter, R.F.Boyer, Ind.Engng.Chem. **43** (1951) 1790
[16]   D.T.F.Pals, J.J.Hermans, Rec.Trav. **71** (1952) 433, Fig. 1
[17]   T.Nakanishi, T.Norisuye, A.Teramoto, S.Kitamura, Macromolecules **26** (1993) 4220
[18]   F.Abe, Y.Einaga, H.Yamakawa, Macromolecules **26** (1993) 1891
[19]   S.M.Aharoni, C.R.Crosby III, E.K.Walsh, Macromolecules **15** (1982) 1093
[20]   I.Noda, K.Mizutani, T.Kato, T.Fujimoto, M.Nagasawa, Macromolecules **3** (1970) 787
[21]   Y.Fujii, Y.Tamai, T.Konishi, H.Yamakawa, Macromolecules **24** (1991) 1608
[22]   Y.Miyaki, Y.Einaga, T.Hirosye, H.Fujita, Macromolecules **10** (1977) 1356
[23]   R.Kuhn, H.Krömer, G.Rossmanith, Angew.Makromol.Chem. **40/41** (1974) 361, Fig. 3
[24]   M.Wintermantel, M.Schmidt, Y.Tsukahara, K.Kajiwara, S.Kohjiya, Macromol.Rapid
          Commun. **15** (1994) 279, Fig. 3
[25]   T.H.Mourey, S.R.Turner, M.Rubinstein, J.M.J.Fréchet, C.J.Hawker, K.L.Wooley, Macro-
          molecules **25** (1992) 2401
[26]   S.M.Aharoni, N.S.Murthy, Polym.Commun. **24** (1983) 132

6. Solution Thermodynamics

6.1. Phenomena

6.1.1. Introduction

Solutions of polymers are important for the molecular characterization of polymers (Chapters 2, 5-7). Solvents also serve as media for polymerizations or polymer reactions (Chapters 3 and 4). Polymer solutions are furthermore essential in industry as vehicles for the processing of polymers to fibers (Chapter 13), polymer blends (Chapter 14), coatings, and adhesives. Here, the choice of the solvent is dictated primarily by the constitution of the polymer and secondarily by technical, economic or ecological considerations.

Some polymer solutions are used, however, because of their specific solution properties, especially their high viscosities. Examples are paints, motor oils, and certain additives to foodstuffs. Here, the solvent dictates the choice of polymer, not the other way around. Water is the most common solvent for this type of application, followed by liquid hydrocarbons (oils). Aqueous solutions of starch and starch derivatives, some cellulose derivatives, gums, poly(vinyl alcohol) and poly(acrylamide) are used for the manufacture of paper and textiles, in mineral oil production, in water treatment, for laundry purposes, for paints, and in the cosmetic and pharmaceutical industries. The world consumption of water-soluble polymers for industrial purposes is ca. $9 \cdot 10^6$ tons per year. It is dwarfed by the consumption of water-soluble thickeners in the food industry, especially of starch and its derivatives (ca. $100 \cdot 10^6$ tons per year), some plant gums, and several cellulose derivatives (see Table 1-2).

Present-day motor oils always contain (0.5-3) % of polymers which increase the viscosity index (a measure of the temperature dependence of viscosity) and improve the pour point (exception: fully synthetic motor oils). These polymers are methacrylate copolymers, ethylene–propylene bipolymers, ethylene–propylene-diene terpolymers, hydrogenated styrene–isoprene block copolymers, and hydrogenated star-poly(isoprene)s; poly(isobutylene)s are practically no longer used. The annual world consumption of these polymers is estimated as ca. 500 000 tons per year.

Polymers are added to water or oils as "thickeners", i.e., to increase the viscosity. Economy demands that the required high viscosities should be generated by as low as possible polymer concentrations. This goal can be achieved by various means. The higher the molar mass of the polymer, the greater are the intrinsic viscosities (Section 5.8.3.) and the viscosities themselves. Polymers must be selected in such a way that the required solvents, water or oil, are thermodynamically good solvents for the polymers. Solution viscosities are also often enhanced by intermolecular association of chemical macromolecules to "physical macromolecules" with greater apparent molar masses (Section 6.4.). Since polymer association is caused by polymer–polymer attraction and coil expansion requires segment–segment repulsion and strong polymer–solvent attraction, it is often assumed that polymer association cannot occur in thermodynamically good solvents. Polymer association *is* however possible in good solvents since specific interactions between certain groups (end groups, comonomer units, etc.) may promote association without much change of the dimensions of individual coils. Many polymer chains furthermore contain short helical sequences that may cluster at higher polymer concentrations and lead to intermolecular association and physical networks (gels).

Physical molecules are composed of chemical molecules; depending on their structure and the scientific field, they are also called multimers, associates, aggregates, micelles, etc. The chemical molecules in such physical molecules are held together by intermolecular forces such as hydrogen bonds, dipole–dipole interactions and dispersion forces or by entropy effects of the surrounding solvent molecules (iceberg effect of water molecules causing the "hydrophobic bonding" between apolar chemical molecules). Individually, these intermolecular (physical) bonds between groups or atoms are much weaker than chemical bonds (covalent, coordinative, electron-deficient). However, since there may be many such bonds per chemical molecule, the combined bond strengths *per molecule* may exceed that of a single chemical bond. Physical molecules may thus *appear* to be stable chemical molecules. An example is the protein "molecule" hemoglobin which is composed of 4 "subunits" that are the true chemical molecules. In equilibrium at infinite dilution, all physical molecules are dissociated into chemical molecules.

It is therefore of interest to study why polymer solutions are formed (Section 6.1.3.), which models can be used to describe their properties (Section 6.2.), and how they behave in dilute (Section 6.3.) and semidilute solutions (Section 6.4.). Solution properties depend not only on the interaction between polymer and solvent but also on the spatial distribution of segments and therefore on the polymer concentration. It is thus useful to define various concentration ranges (Section 6.1.2.).

6.1.2. Concentration Ranges

Macromolecules are separated from each other in very dilute solutions. Their physical structures are determined by their constitution, configuration and molar mass (Chapter 2) and their microconformations and macroconformations (Chapter 5) which in turn are affected by interaction with the solvent. At somewhat higher polymer concentrations, specific interactions between polymer molecules may operate, for example, hydrogen bonds. They lead to association of molecules, i.e., "physical molecules" and, at even higher concentrations, to gelation.

Nonspecific effects may also be present. Polymer coils have low coil densities (Section 5.3.5.); they require plenty of space. With increasing polymer concentration, coils approach each other and start to overlap at a certain concentration c^*. At and above this concentration, the same volume has to accommodate more segments:

Coils *in good solvents* are compressed and the radius of gyration decreases (Fig. 5-16). They finally attain their unperturbed dimensions at a concentration c^{**}.

Coils in *theta solvents* are not compressed further since they have already their unperturbed dimensions. Their radii of gyration remain constant throughout the whole concentration range (see Fig. 5-16).

One can thus distinguish between three concentration ranges for polymer coils in good solvents: dilute solutions in the range $c < c^*$, semidilute (moderately concentrated) solutions in the range $c^* < c < c^{**}$, and concentrated solutions in the range $c > c^{**}$. The transition dilute $\rightleftarrows$ semidilute cannot be sharp since coils do not possess solid surfaces. Another question is which coil dimension best describes this transition: radius of gyration, Einstein radius (viscometry), Stokes radius (diffusion), etc.

If coils are approximated as equivalent spheres and if the radius of gyration is taken as the radius controlling the overlap, then the overlap concentration is given by $c_s^* = m_{mol}/V_{mol} = (3\ M)/(4\ \pi\ s^3 N_A)$, where $m_{mol} = M/N_A$ is the mass of the molecule and V_{mol} the volume of a molecule. Insertion of $s = K_s M^\nu$ (Eqn.(5-25)) leads to:

$$(6\text{-}1)\qquad c_s^* = \frac{3}{4\pi K_s^3 N_A} \cdot M^{1-3\nu}$$

In good solvents, $\nu \approx 3/5$: the critical overlap concentration c_s^* decreases with the 0.8 power of the molar mass. A poly(styrene) of molar mass $M = 1\cdot 10^6$ g/mol, $K_s = 1.2\cdot 10^{-9}$ cm·(mol/g)$^{0.6}$ and $\nu = 0.6$ (PS in CS$_2$ at 25°C) thus has an overlap concentration of $c_s^* \approx 3.6\cdot 10^{-3}$ g/mL.

Alternatively, overlap concentrations may also be calculated from intrinsic viscosities. To get comparable results, however, the hydrodynamic radii (Einstein radii) $R_\eta = [(3\ M[\eta])/(10\ \pi\ N_A)]^{1/3}$ from intrinsic viscosities (see Eqn.(5-41)) have to be converted into radii of gyration s in order to obtain static overlap concentrations. The radius of gyration of unperturbed coils is greater by $s/R_d = 1.28$ than the Stokes radius R_d (from diffusion; Section 7.2.2.) and the Stokes radius is greater by $R_d/R_\eta = 1.055$ than the Einstein radius (from viscosity; Eqn.(5-41)) according to experimental data, thus $s = 1.28\cdot 1.05_5\ R_\eta$. Assuming that these relationships also apply to perturbed coils and inserting the Einstein radius $R_{sph} = R_\eta = [(3\ M[\eta])/(10\ \pi\ N_A)]^{1/3}$ of an equivalent sphere (see Eqn.(5-40)), one finds that the critical overlap concentration c_η^* is approximately given by the inverse of the intrinsic viscosity:

$$(6\text{-}2)\qquad c_\eta^* = \frac{3M}{4\pi N_A s^3} = \frac{3M}{4\pi N_A (1.28\cdot 1.055 R_\eta)^3} = \frac{1.015}{[\eta]} \approx \frac{1}{[\eta]}$$

6.1.3. Solubility Parameters

The solubility of an amorphous polymer 2 in a low molar mass solvent 1 can often be estimated by the concept of solubility parameters:

Solvent molecules (degree of polymerization $X_1 \equiv 1$) are held together in liquids by intermolecular forces. This **cohesion** is caused by dispersion forces, dipole–dipole interactions, and/or hydrogen bonds. The cohesion energy is ε_{11} per binary contact and $z\varepsilon_{11}/2$ per solvent molecule since each molecule is surrounded by z others. The cohesion energy per unit volume $V_{1,mol}$ is called the **cohesion energy density** Υ_1:

(6-3) $\Upsilon_1 = z(\varepsilon_{11}/2)/V_{1,mol} = N_A z\varepsilon_{11}/(2\ V_{1,m}) = \delta_1^2$

where $V_{1,m} = N_A V_{1,mol}$ is the molar volume of the solvent. The square root of the cohesion energy density is the **solubility parameter** δ_1.

The solubility parameter of a solvent is calculated from the molar vaporization energy $E_{1,m} = V_{1,m}\delta_1^2 = N_A z\varepsilon_{11}/2$ which is the separation energy *per mole molecule* (i.e., per half mole of binary contacts). Molar vaporization energies are obtained as the difference of experimental vaporization enthalpies and the work required to overcome the external pressure.

Solubility parameters are often seen as "three-dimensional" parameters with contributions by dispersion forces (δ_d), dipole forces (δ_p), and hydrogen bonds (δ_h): $\delta_1^2 = \delta_d^2 + \delta_p^2 + \delta_h^2$. The contributions by dispersion forces are by and large independent of the chemical structure (Table 6-1). The usefulness of separating polar forces into δ_p and δ_h, however, is questionable since both parameters depend on interactions between Lewis acids and Lewis bases.

Table 6-1   Solubility parameters δ_1, δ_d, δ_p and δ_h of solvents with the traditional unit of 1 Hildebrand = 1 $(cal/cm^3)^{1/2}$ = 2.046 $(J/cm^3)^{1/2}$ (often reported without units!).

Solvent	Solubility parameters in $(cal/cm^3)^{1/2}$			
	δ_1	δ_d	δ_p	δ_h
Heptane	7.4	7.4	0	0
Carbon tetrachloride	8.65	8.65	0	0
Benzene	9.05	8.99	0.5	1.0
Chloroform	9.33	8.75	1.65	2.8
Tetrahydrofuran	9.51	8.22	3.25	3.5
Acetone	9.69	7.58	5.7	2.0
Dimethylacetamide	11.12	8.2	5.6	5.0
Acetonitrile	11.95	7.50	8.8	3.0
N,N-Dimethylformamide	12.14	8.5	6.7	5.5
Dimethylsulfoxide	13.04	9.0	8.0	5.0
Methanol	14.6	7.42	6.1	11.0
Water	23.43	7.0	8.0	20.9

The mixing of solvent molecules 1 and monomeric units 2 of a polymer is treated as a quasichemical reaction between one pair each of 1-1 and 2-2 to two pairs of 1-2. The **interchange energy** per pair is thus:

(6-4) $\Delta\varepsilon = \varepsilon_{12} - (1/2)(\varepsilon_{11} + \varepsilon_{22})$

Quantum chemical calculations for *dispersion forces* between spherical molecules show that the cohesion energy ε_{12} is given by the geometric mean of the homocohesion energies, i.e., by $\varepsilon_{12} = (\varepsilon_{11}\varepsilon_{22})^{1/2}$. Furthermore, $\varepsilon_{11} = [2\ V_{1,m}/(N_A z)]\delta_1^2 = K\delta_1^2$ (Eqn.(6-3)) and $\varepsilon_{22} = K\delta_2^2$ if solvent molecules and monomeric units have the same molar volume and the same number of neighbors. Introduction into Eqn.(6-4) gives:

(6-5) $\Delta\varepsilon/K = -(\delta_1 - \delta_2)^2/2$

The difference of the solubility parameters, $\delta_1 - \delta_2$, becomes zero at identical interactions 1-1 and 2-2 and so does the enthalpy of mixing, $\Delta H_{mix} \sim -\Delta\varepsilon$. The Gibbs energy of mixing remains negative, however, since the entropy of the system increases on mixing. The less similar the interactions 1-1 and 2-2, the more positive ΔH_{mix} becomes until it finally can no longer be compensated by the positive term $-T\Delta S_{mix}$ and $\Delta G_{mix} = \Delta H_{mix} - T\Delta S_{mix}$ becomes positive. There should be a difference $|\delta_1 - \delta_2|$ beyond which a polymer 2 cannot be mixed with a solvent 1. The range of $|\delta_1 - \delta_2|$ for mixing is fairly narrow for apolar polymers in apolar or polar solvents but broad for polar polymers in polar solvents (Table 6-2).

Table 6-2   Experimental solubility ranges. 1 $(cal/cm^3)^{1/2} = 2.046$ $(J/cm^3)^{1/2}$.

Polymer	Polymer $\delta_2/(cal\ cm^{-3})^{1/2}$	Solvent $\delta_1/(cal\ cm^{-3})^{1/2}$ apolar	polar
Poly(dimethylsiloxane)	7.5	8.3 ± 0.8	8.9 ± 0.7
Poly(isobutylene)	8.0	7.9 ± 0.6	9.0 ± 0.9
Poly(styrene)	9.1	9.3 ± 1.3	9.0 ± 0.9
Poly(methyl methacrylate)	9.1	10.8 ± 1.2	10.9 ± 2.4
Poly(vinyl acetate)	9.4	10.8 ± 1.9	11.6 ± 3.1
Cellulose trinitrate	10.8	11.9 ± 0.8	11.2 ± 3.4
Poly(acrylonitrile)	12.5	—	13.1 ± 1.4

Deviations are expected for the following reasons. (1) The condition $\Delta G_{mix} \leq 0$ is necessary but not sufficient. (2) $\Delta H_{mix} < 0$ is never true. (3) The theory applies only to dispersion forces, not to polar ones. (4) Only *mixing* of two fluid compounds is considered, not the *dissolution* of a solid polymer in a solvent: on dissolution, a heat of melting has to be considered for a crystalline polymer and a freezing-in energy for an amorphous polymer (see Chapter 9).

A cohesion energy ε_{22} exists accordingly between every two monomeric units 2 in liquid polymers. Solubility parameters of polymers cannot be calculated from vaporization energies, however, because macromolecules do not evaporate but rather degrade at the temperatures required for evaporation.

Such solubility parameters are either estimated from those of low molar mass model compounds or determined via intrinsic viscosities $[\eta]$ of soluble polymers or degrees of swelling Q of cross-linked polymers. The better the interaction between polymer and solvent, the more are polymer coils expanded and the higher is the intrinsic viscosity. The same is true for network chains and the degree of swelling; the volume fraction ϕ_2 of the polymer in swollen gels declines accordingly. The solubility parameter δ_2 of a polymer is thus given by the maximum in the functions $[\eta] = f(\delta_1)$ and the minimum in the function $\phi_2 = f(\delta_1)$, respectively (Fig. 6-1).

Solvents that interact strongly with polymers are called thermodynamically **good solvents**. They cause large radii of gyrations and intrinsic viscosities of soluble polymers and strong swelling of cross-linked polymers. In the paint industry, such solvents are called *poor* because they lead to unwanted high viscosities of paints.

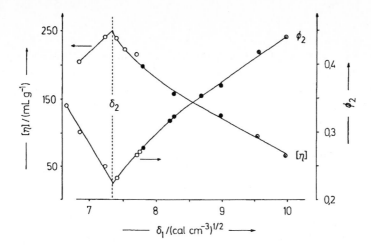

Fig. 6-1 Intrinsic viscosities [η] of a natural rubber and volume fractions ϕ_2 of swollen gels of the same but cross-linked polymer as a function of the solubility parameters δ_1 of aliphatic hydrocarbons (O), esters (●) and ketones (⊙). Data of [1].

6.2. Statistical Thermodynamics

6.2.1. Types of Solutions

Enthalpy, entropy and volume may all change if two chemical compounds 1 and 2 are mixed, be it a polymer 2 in a solvent 1 or two polymers 1 and 2 in a blend. For an isothermal process, these changes result in a change of the Gibbs energy of mixing:

$$(6\text{-}6) \qquad \Delta G_{mix} = \Delta H_{mix} - T\Delta S_{mix} = \Delta U_{mix} + \Delta(pV) - T\Delta S_{mix} = \Delta A_{mix} + \Delta(pV)$$

where ΔH_{mix} = enthalpy of mixing, T = thermodynamic temperature, ΔS_{mix} = entropy of mixing, ΔU_{mix} = internal energy of mixing, p = pressure, V = volume, ΔA_{mix} = Helmholtz energy of mixing. Volumes are often additive for isobaric processes in condensed systems. In this case, $\Delta G_{mix} \approx \Delta A_{mix}$.

Solutions are subdivided into ideal, athermal, regular, irregular (or real), and theta solutions:

Ideal solutions. The heat of mixing is zero and the entropy of mixing is simply given by the ideal combinatorial entropy that results from the mixing of two types of solid spheres of the same size. An ideal solution is ideal at all temperatures.

Athermal solutions. The heat of mixing is zero but the entropy of mixing is larger than the ideal combinatorial entropy by an excess entropy of mixing.

Regular solutions. The heat of mixing does not equal zero but the entropy of mixing equals the ideal combinatorial entropy.

Irregular solutions (or *real solutions*). The heat of mixing is not zero and the entropy of mixing is composed of both ideal and excess terms.

Theta solutions. In dilute solutions, the heat of mixing just compensates the excess entropy of mixing at a certain temperature. At this *theta temperature*, solutions behave like ideal solutions and hence are also called *pseudo-ideal solutions*. A theta temperature thus corresponds to the Boyle temperature of real gases.

6.2.2. Lattice Theory

Statistical thermodynamics tries to calculate by statistical means the contributions of enthalpy and entropy of mixing to the Gibbs energy. In **lattice theories**, solutions are depicted as three-dimensional lattices with a total of $N_g = N_1 X_1 + N_2 X_2$ lattice sites (Fig. 6-2). Each lattice site is occupied by either a segment (e.g., a monomeric unit) or a solvent molecule. The solution comprises N_1 solvent molecules of degree of polymerization $X_1 = 1$ and N_2 polymer molecules with $X_2 > 1$ each. The theory can also be applied to a polymer blend of two polymers 1 and 2 with N_1 polymer molecules of $X_1 > 1$ and N_2 polymer molecules of $X_2 > 1$.

The number N_{12} of contact pairs is calculated by the **Flory–Huggins theory** from the number N_g of all lattice sites, the number z of nearest neighbors of a unit, and the probability that adjacent lattice sites are occupied by either solvent molecules 1 or monomeric units 2. These probabilities are identical with the volume fractions ϕ_1 of solvent molecules and ϕ_2 of monomeric units.

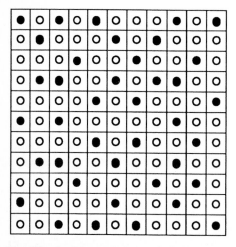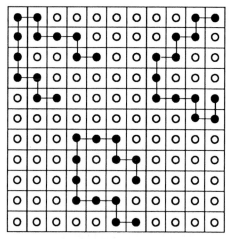

Fig. 6-2 Arrangement of dissolved low molar mass molecules (●, left) or monomeric units of macromolecules (●, right) and low molar mass solvent molecules (○) in a two-dimensional lattice. The volume fraction of the solute is always $\phi_2 = 0.322$; the degree of polymerization of the polymer is $X_2 = 13$ (right). In dilute polymers solutions, polymer coils with relatively high segment concentrations swim in a sea of solvent molecules. The assumption of the mean field theory of an even distribution of all monomeric units can thus only be fulfilled at high polymer concentrations.

Enthalpy of Mixing

The enthalpy of mixing is calculated from the exchange energy $\Delta\varepsilon$ per pair 1-2 (Eqn.(6-4)) and the number N_{12} of pairs:

$$(6\text{-}7) \qquad \Delta H_{mix} = N_{12}\Delta\varepsilon = N_g z \phi_1 \phi_2 \Delta\varepsilon$$

The product of the exchange energy and the number z of neighbors is divided by the thermal energy $k_B T$. The resulting quantity χ is called the **Flory–Huggins interaction parameter**:

$$(6\text{-}8) \qquad \chi \equiv z\Delta\varepsilon/(k_B T) = \Delta H_{mix}/[N_g \phi_1 \phi_2 k_B T] \; ;$$

(the original Flory definition of this parameter has the right side of Eqn.(6-8) multiplied by the number X_1 of the segments per solvent molecule).

The Flory–Huggins interaction parameter describes the thermodynamic goodness of a solvent for a polymer. It is only slightly dependent on concentration for apolar or weakly polar polymers in apolar or weakly polar solvents (PIP in Fig. 6-3). The concentration dependence of χ for other systems can be described by $\chi = \chi_0 + K\phi_2 + K'\phi_2^2$ where K and K' are constants for a given polymer–solvent system at $T = $ const. This function may generate a maximum in the $\chi = f(\phi_2)$ curve (see PVMA in Fig. 6-3). K and K' are negative for cellulose nitrate CN; the decrease of χ with increasing polymer concentration here is caused by the formation of liquid crystals (Section 8.4.).

According to Eqn.(6-8), χ is supposed to be inversely proportional to the temperature T. The interaction parameter is thus assumed to be an enthalpic quantity according to the second law of thermodynamics, $\Delta G/T = (\Delta H/T) - \Delta S$. Experimentally, a temperature dependence $\chi = \chi_\infty + (K_T/T)$ was found (χ_∞, $K_T = $ constants). The interaction parameter must thus contain an entropic part χ_s. This part is of a non-combinatorial nature (see below); it is caused by orientations of chain segments, variations of vibrational frequencies of components on formation of 1-2 contacts, etc.

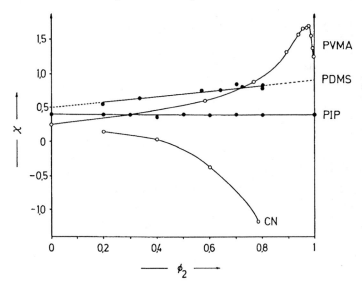

Fig. 6-3 Flory–Huggins parameter as a function of the volume fraction ϕ_2 of polymer. PVMA = Poly(vinylmethoxy acetal) in water at 25°C; PDMS = poly(dimethyl siloxane) in benzene at 20°C; PIP = cis-1,4-poly(isoprene) in benzene at 20°C; CN = Cellulose nitrate (DS = 2.6) in acetone at 20°C.

The **mixing enthalpy per mole lattice site** (index m) is obtained from Eqn.(6-8) with $R = k_B N_A$ and the mole fraction $n_g = N_g/N_A$ of lattice sites:

(6-9) $\Delta H_{mix,m} = \Delta H_{mix}/n_g = RT\phi_1\phi_2\chi$

Entropy of Mixing

The **molar entropy of mixing** $\Delta S_{mix,m} = \Delta S_{mix}/n_g$ is given by the entropy of mixing $\Delta S_{mix} = S_{comb}(N_1,N_2) - S_{comb,o}$ which in turn is the difference of combinatorial entropy $S_{comb}(N_1,N_2)$ and disorientation entropy $S_{comb,o}$.

Solvent molecules and monomeric units can be arranged in many different ways. These various combinations give rise to a **combinatorial entropy** $S_{comb}(N_1,N_2)$ on mixing. This entropy term is often called a *configuration entropy* because it relates to the various physical "configurations" (= macroconformations, see Section 2.4.3.) of a polymer chain.

Combinatorial entropies of polymer solutions ($X_1 = 1$; $X_2 >> 1$) are calculated via $S_{comb}(N_1,N_2) = k_B \ln \Omega$ according to statistical thermodynamics. The thermodynamic probability Ω is obtained by the Flory–Huggins theory from the v_i different arrangements for the *i*th chain:

The first monomeric unit can occupy any lattice site but the second one only one of those z sites that are adjacent to the first site. The third monomeric unit has to confine itself within one of the remaining $z-1$ sites. The first polymer chain with X_2 monomeric units thus has $v_1 = N_g z(z-1)^{X_2-2}$ possibilities. This approach neglects that there must be less than $z-1$ sites for the third unit (and so on) since a lattice site may already be occupied by one of the previously placed units.

After the lattice has been filled with $i-1$ polymer chains, $N_f = N_g - (i-1)X_2$ lattice sites remain. The probability of finding an empty site is approximately N_f/N_g. The ith polymer chain can thus be arranged in $v_i = N_f(N_f/N_g)^{X_2-1}z(z-1)^{X_2-2}$ different ways.

The thermodynamic probability Ω is proportional to $\Pi_i v_i$, i.e., the product of all values of v_i for the N_2 chains. These chains are all identical if the polymer is molecularly uniform. However, one is allowed to count only distinguishable arrangements. Thus, the $\Pi_i v_i$ has to be divided by the factorial of N_2 (i.e., $N_2!$).

Each of the chains can furthermore be placed head first or tail first. Depending on the symmetry number σ one counts σ^{N_2} too many combinations ($\sigma = 2$ for indistinguishable chains, $\sigma = 1$ for distinguishable ones) and one also has to divide by σ^{N_2}. It results in $\Omega = (N_2! \, \sigma^{N_2})^{-1} \Pi_i v_i$ for $1 \leq i \leq N_2$.

Introduction of $v_i = N_f(N_f/N_g)^{X_2-1}z(z-1)^{X_2-2}$ and $N_f = N_g - (i-1)X_2$ and applying Stirling's equation $x! \cong (x/e)^x$ (for x >> 1) delivers

(6-10) $\Omega = \dfrac{N_g!}{N_1!(N_2X_2)!}\left(\dfrac{N_2X_2}{N_g}\right)^{N_2(X_2-1)} \left[X_2 z(z-1)^{X_2-2}\{\sigma \exp(X_2-1)\}^{-1}\right]^{N_2}$

(6-11) $\Omega_{rel} \equiv X_2 z(z-1)^{X_2-2}[\sigma \exp(X_2-1)]^{-1}$

The combinatorial entropy of polymer solutions is obtained from Eqn.(6-10) by using Stirling's approximation $\log_e N_i = N_i(\log_e N_i) - N_i$ and the definitions of the volume fractions of lattice components ($\phi_1 \equiv N_1 X_1/N_g$; $\phi_2 \equiv N_2 X_2/N_g$):

(6-12) $\quad S_{comb}(N_1,N_2) = k_B \log_e \Omega = - k_B(N_1 \log_e \phi_1 + N_2 \log_e \phi_2) + k_B N_2 \log_e \Omega_{rel}$

Eqn.(6-12) still contains the disorientation entropy $S_{comb,o} = k_B N_2 \log_e \Omega_{rel} = S_{comb}(N_1,0) + S_{comb}(0,N_2)$ which describes the entropy of coiled polymer molecules relative to the entropy of polymer molecules in a perfect crystal. The value of $S_{comb,o}$ must be subtracted from the combinatorial entropy $S_{comb}(N_1,N_2)$ in order to obtain the entropy of mixing of amorphous polymers, $\Delta S_{mix} = - k_B(N_1 \cdot \log_e \phi_1 + N_2 \log_e \phi_2)$.

The disorientation entropy is a complex quantity for *polymer blends* because both the N_1 chains of polymer 1 as well as the N_2 chains of polymer 2 can adopt many different macroconformations. Its calculation is much more simple for *polymer solutions* since one only has to consider the N_2 polymer chains. Solvent molecules just fill the empty sites left by the polymer molecules, thus $N_1 = 0$ and $S_{comb,o} = S_{comb}(0,N_2) = k_B N_2 \log_e \Omega_{rel}$.

The entropy of mixing simply becomes $\Delta S_{mix} = - k_B(N_1 \log_e \phi_1 + N_2 \log_e \phi_2)$. Introduction of $N_1 = \phi_1 N_g/X_1$, $N_2 = \phi_2 N_g/X_2$; $N_g = n_g N_A$, and $N_A k_B = R$ into $\Delta S_{mix,m} = \Delta S_{mix}/n_g$ delivers the **entropy of mixing per mole of lattice sites**:

(6-13) $\quad \Delta S_{mix,m} = \Delta S_{mix}/n_g = - R(X_1^{-1}\phi_1 \log_e \phi_1 + X_2^{-1}\phi_2 \log_e \phi_2)$

Gibbs Energy of Mixing

The molar Gibbs energy of mixing results from Eqns.(6-6), (6-9) and (6-13) as:

(6-14)
$$\Delta G_{mix,m} = \Delta H_{mix,m} - T\Delta S_{mix,m} = RT[\phi_1\phi_2\chi + X_1^{-1}\phi_1 \log_e \phi_1 + X_2^{-1}\phi_2 \log_e \phi_2]$$

It is symmetrical around $\phi_2 = 1/2$ for $X_1 = X_2$ (Fig. 6-4) but unsymmetrical for $X_2 > X_1$. Solutions of polymers thus differ from those of low molar mass substances because macromolecules and small molecules differ greatly in molecular size.

The simple Flory–Huggins theory outlined above describes qualitatively the essential characteristics of polymer solutions but breaks down for more sophisticated analyses, especially for dilute solutions. In particular, the assumption is incorrect that the number N_f of free lattice sites is given by $N_f = N_g - (i - 1)X_2$. Interaction parameters are also often concentration dependent. The theory also neglects specific polymer-solvent interactions (solvation) and polymer–polymer associations.

6.2.3. Phase Separation

Solutions of Amorphous Polymers

The Flory–Huggins theory allows the calculation of equilibria between liquid phases. The change of the Gibbs energy of mixing, ΔG_{mix}, with the amount of substance, n_i, of component *i* is defined as the **chemical potential** μ_i of that component:

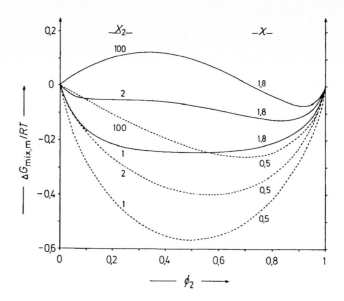

Fig. 6-4   Reduced molar Gibbs energy of mixing $\Delta G_{\text{mix,m}}/RT$ as a function of the volume fraction ϕ_2 of the solute for various interaction parameters χ and degrees of polymerization X_2 in solvents with $X_1 = 1$. Calculated from Eqn.(6-14).

$(\partial \Delta G_{\text{mix}}/\partial n_i)_{T,p} \equiv \mu_i$. Inserting the **Gibbs–Duhem relationship** $\Sigma_i\, n_i \mathrm{d}\mu_i = 0$ as well as $n_2 = N_2/N_A$, $N_2 = \phi_2 N_g/X_2$ and $\phi_1 = 1 - \phi_2$, one arrives at:

$$(6\text{-}15) \quad \Delta G_{\text{mix,m}} = n_1 \Delta\mu_1 + n_2 \Delta\mu_2 = (N_g/N_A)[\Delta\mu_1 - (\Delta\mu_1 - X_2^{-1}\Delta\mu_2)\phi_2]$$

The chemical potential of a component must be the same in each phase of a multi-phase system. A binary system with two components 1 and 2 and two phases ' and " must therefore obey $\mu_1' = \mu_1''$ and $\mu_2' = \mu_2''$ and thus $\Delta\mu_1' = \mu_1' - \mu_1^0 = \mu_1'' - \mu_1^0 = \Delta\mu_1''$ and $\Delta\mu_2' = \mu_2' - \mu_2^0 = \mu_2'' - \mu_2^0 = \Delta\mu_2''$. The chemical potentials are given by the intercepts of the function $\Delta G_{\text{mix,m}} = f(\phi_2)$ for $\phi_2 \to 0$ and $\phi_2 \to 1$, respectively (Eqn.(6-15)). They can only be identical in both phases if two points of the function $\Delta G_{\text{mix,m}} = f(\phi_2)$ possess a common tangent. This tangent determines the intercepts (Fig. 6-5).

At any temperature (e.g., 260 K in Fig. 6-5), three regions can be observed: (I) between $\phi_2 = 0$ and the point o at low ϕ_2, (II) between both points o, and (III) between point o at high ϕ_2 and $\phi_2 = 1$. Solutions in regions I and III are stable; they form one phase and do not demix.

Two phases coexist in region II, however. The border line B-P-B between the one-phase region and the two-phase region is called the **binodal**. Binodals are difficult to calculate for polymers with broad molar mass distributions because chemical potentials depend strongly on the degrees of polymerization (see below).

It is much more easy to calculate the **spinodal** S-P-S which divides the non-stable region below B-P-B into two metastable regions m and one unstable region u. The

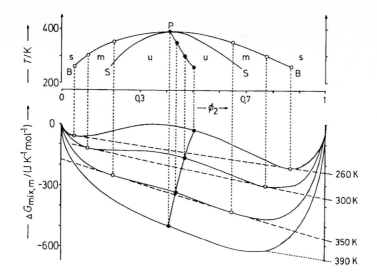

Fig. 6-5   Bottom: Molar Gibbs energy of mixing, $\Delta G_{mix,m}$, as function of the volume fraction ϕ_2 of the solute ($X_2 = 2$) in a solvent with $X_1 = 1$ at various temperatures. Calculations with Eqn.(6-14) and $\chi = 0.3 + \{(450 \text{ K})/T\}$. ○ Contact points of the tangent with the function $\Delta G_{mix,m} = f(\phi_2)$; ● ϕ_2 from Eqn.(6-18).
    Top: Binodal B and spinodal S with critical point P as calculated from these data and Eqn.(6-19). Ranges: s = stable, m = metastable, u = unstable.

spinodal is given by the two inflection points in the function $(\Delta G_{mix,m})_T = f(\phi_2)$, i.e., by $\partial^2(\Delta G_{mix,m})/\partial \phi_2^2 = 0$. Solutions in region u between the spinodal concentrations are unstable; they demix spontaneously into two continuous phases which form a physical interpenetrating network. The regions between the spinodals and the binodals are metastable. Their phase separations are controlled kinetically by nucleation which leads to dispersions of one phase in the other.

The difference $\Delta \mu_1$ of chemical potentials of the solvent in the solution and in the pure solvent is obtained by differentiation of the molar Gibbs energy of mixing with respect to the amount n_1 of the solvent. Using $X_1 = 1$, $\phi_2 = 1 - \phi_1$ and $\phi_1 = n_1 N_A X_1/N_g$, one obtains:

(6-16) $\partial \Delta G_{mix,m}/\partial n_1 = \Delta \mu_1 = RT[\chi \phi_2^2 + \log_e(1 - \phi_2) + (1 - X_2^{-1})\phi_2]$

The spinodal is characterized by the inflection points of the function $\Delta G_{mix,m} = f(\phi_2)$ and the extreme values of the function $\Delta \mu_1 = f(\phi_2)$:

(6-17) $\partial \Delta \mu_1/\partial \phi_2 = RT[2 \chi \phi_2 - (1 - \phi_2)^{-1} + (1 - X_2^{-1})] = 0$

At the critical point, the second derivative of the function $\Delta \mu_1 = f(\phi_2)$ becomes zero; here, the maximum, minimum and inflection point unite:

(6-18) $\partial^2 \Delta \mu_1/\partial \phi_2^2 = RT[2 \chi - (1 - \phi_2)^{-2}] = 0$

Solving both Eqns.(6-17) and (6-18) for χ and equating the result indicates that the critical volume fraction of the polymer decreases with increasing degree of polymerization:

$$(6\text{-}19) \quad \phi_{2,\text{crit}} = 1/(1 + X_2^{1/2})$$

The critical interaction parameter is obtained from Eqns.(6-18) and (6-19)

$$(6\text{-}20) \quad \chi_{\text{crit}} = [(1 + X_2^{1/2})^2]/[2\,X_2] \approx (1/2) + (1/X_2)^{1/2} \quad ; \text{if } \chi \neq f(\phi_2)$$

It depends only on the degree of polymerization, provided that the interaction parameter χ is independent of the volume fraction ϕ_2. At infinite degrees of polymerization, it approaches 1/2. χ_{crit} differs from 1/2 at $X_2 \to \infty$ if $\chi = f(\phi_2)$).

The dependence of critical volume fractions on degrees of polymerization and temperature is utilized in the **precipitation fractionation** of polymers. Practically all synthetic polymers possess molar mass distributions; with solvents, they form *quasi-binary* systems and not true binary systems as assumed by simple theory. On lowering the temperature of endothermic solutions, polymer fractions with the highest degree of polymerization phase-separate first; they form a physical gel composed of polymer and solvent at the bottom of the supernatant solution (if $\rho_2 > \rho_1$) containing the remaining polymer. Successive lowering of the temperature produces fractions with successively lower degrees of polymerization. The temperatures required for this type of precipitation are however often outside convenient laboratory procedures (see below). Precipitation fractionation is therefore usually performed at constant temperature by successive addition of a non-solvent to the polymer solution. Such fractionations are more effective, the more dilute the solutions.

Polymer Blends

The same principles can also be applied to mixtures of two amorphous polymers 1 and 2 with $X_2 \gg 1$ and $X_1 \gg 1$. The presence of polymer 1 reduces the possible arrangements of monomeric units of polymer 2: the molar entropy of mixing, $\Delta S_{\text{mol,m}}$, can never become as positive as in polymer–solvent systems. The resulting entropy term $-T\Delta S_{\text{mix,m}} = RT[X_1^{-1}\phi_1 \ln \phi_1 + X_2^{-1}\phi_2 \ln \phi_2]$ is only slightly negative and can no longer compensate the positive enthalpy term $\Delta H_{\text{mix,m}} = RT[\phi_1\phi_2\chi]$ if the interaction parameter is positive (see Eqn.(6-14)). The molar Gibbs energy of mixing becomes positive; the polymer–polymer system cannot exist as one phase and demixes.

Most interaction parameters are indeed positive. As a result, mixtures of two polymers are usually immiscible. However, there are exceptions, i.e., systems with negative interaction parameters. Such parameters are produced by strong attractions between unlike polymers.

Mixtures of two polymers are called **polymer blends, polyblends** or **blends** in industry. They are prepared to improve the property of the products as well as to reduce the costs. **Homogeneous blends** are true (molecular) mixtures of two different polymers (see Chapter 14). **Heterogeneous blends** are thermodynamically immiscible in the important concentration ranges. However, they very often do not demix spontaneously into two macroscopic phases because of their high viscosity and rather remain as dispersions of microphases of component 1 in the matrix of component 2. Their mechanical properties behave *as if* they were generated by thermodynamically miscible components; such components are often called **compatible**.

The dispersion is promoted by the presence of **compatibilizers** which act similar to emulsifiers in low molar mass systems (see Fig. 8-16). Compatibilizers are often diblock copolymers: one block resides in a microphase 1 and the other block in the matrix 2 which provides a kind of anchor. Graft copolymers act similarly.

Critical Solution Temperatures

Various types of phase diagrams are observed for liquid systems depending on the type of temperature dependence of the interaction parameters (Fig. 6-6). Systems with $\chi = \chi_\infty + (K_T/T)$ and positive K_T are endothermic. Interaction parameters decrease with increasing temperature and the system demixes below the binodal into two liquid phases (Fig. 6-6, I). The maximum of the binodal is the **upper critical solution temperature UCST**. Exothermic systems demix with increasing temperature (Fig. 6-6, II). The minimum of the binodal is the **lower critical solution temperature LCST**.

LCSTs correspond to entropically induced phase separations and UCSTs to enthalpically induced ones. The terms UCST and LCST do *not* refer to the absolute position of critical temperatures since an LCST may be at a higher temperature than a UCST (Fig. 6-6, IV).

The function $\chi = f(T)$ may also exhibit maxima or minima (see Fig. 6-3). Functions with maxima lead to closed miscibility loops (Fig. 6-6, III) whereas those with minima create hour-glass diagrams (Fig. 6-6, V). The type of phase diagram depends on the polymer–solvent interaction: poly(oxyethylene) shows a closed miscibility loop in water but an hour-glass diagram in *t*-butyl acetate.

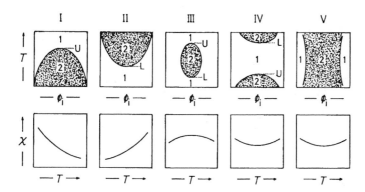

Fig. 6-6   Top: Types of phase diagrams $T = f(\phi_i)$ of fluid systems (polymer solutions or polymer melts) with one-phase (1) or two-phase (2) regions and upper (U) and lower (L) critical solution temperatures. III: Closed miscibility loop; V = hour-glass diagram. Bottom: Temperature dependence of interaction parameters.

Curves are drawn for idealized systems with $X_1 = X_2$ and $\chi \neq f(\phi_2)$. Real systems usually have asymmetric phase diagrams because $X_1 \neq X_2$ and $\chi = f(\phi_2)$. Examples:

Solutions	*Blends*
I: Poly(styrene)–cyclohexane	Poly(butadiene) (deuterated *vs.* nondeuterated)
II: Poly(ethylene)–hexane (5 bar)	Poly(styrene)–poly(vinylmethyl ether)
III: Poly(oxyethylene)–water	Poly(methyl methacrylate)–polycarbonate A
IV: Poly(styrene)–acetone (low molar mass)	
V: Poly(styrene)–acetone (high molar mass)	

The case UCST > LCST is often found for water-soluble polymers. These polymers desolvate at temperatures T > LCST whereupon the solution separates into two phases, except at very low and very high polymer concentrations (Fig. 6-6, III). The UCST is often above the boiling temperature of water.

Most amorphous polymer + solvent and amorphous polymer 1 + amorphous polymer 2 systems exhibit only upper critical solution temperatures UCST. Under pressure, such solutions often show an LCST above the normal boiling temperature of the solvent (Fig. 6-6, II and IV). The mixing of the dense polymer with the highly expanded solvent causes the system to contract; the entropy of mixing becomes negative. The solvent quality thus passes through a maximum between UCST and LCST: the temperature dependence of χ shows a minimum.

Solutions of Crystalline Polymers

Each of the liquid phases generated by phase separations of solutions of amorphous polymers contains both components of a binary mixture; pure phases consisting of only one component are almost never present. The situation is quite different for solutions of *crystalline polymers* where one phase may consist solely of the pure (crystallized) polymer depending on concentration and temperature.

A crystalline polymer in a thermodynamically *good solvent* forms a homogeneous solution phase S. A heterogeneous system develops below the concentration-dependent phase separation temperature. This heterogeneous system consists of the crystalline polymer C and a polymer solution S' (Fig. 6-7). The melting temperature of the polymer is lowered; one observes a freezing point depression. A polymer–solvent eutectic (if any) must exist at very low volume fractions ϕ_2.

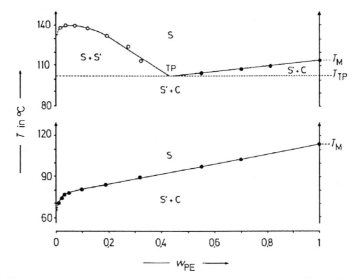

Fig. 6-7   Temperature T of phase separation as a function of the mass fraction w_{PE} of a poly-(ethylene) in the thermodynamically bad solvent amyl acetate AA (top) and the moderately good solvent xylene (bottom). O Binodal, ● crystallization curve, TP = triple point, T_M = melting temperature of the polymer, T_{TP} = temperature at the triple point. See text. After data of [2].

In *poor solvents*, a lowering of the temperature also leads to the formation of heterogeneous systems. The heterogeneous region consists of S' and C at mass fractions $w_2 \geq w_{2,\text{TP}}$ but of two liquid phases S and S' at $w_2 \leq w_{2,\text{TP}}$. Further cooling of the liquid heterogeneous region below a temperature T_{TP} causes the polymer to crystallize and the formation of a heterogeneous region S'+C. In thermal equilibrium, three phases (P = 3) S, S' and C thus exist at the triple point TP according to the **Gibbs phase rule** P + F = K + 2 for a system of K = 2 components with one degree of freedom F = 1 (pressure).

Semicrystalline polymers can thus be fractionated with respect to molar mass in thermodynamically bad solvents at concentrations below the triple point. The fractionation does not succeed in good solvents since the enthalpy of melting is independent of the molar mass for high molar mass polymers (Section 9.4.3.). However, polymers can be fractionated from good solvents according to chemical structure because enthalpies of melting depend on constitution (composition, branching) and configuration (tacticity).

This phenomenon is exploited for the characterization of short-chain branching SCB in poly(ethylene) by **temperature rising elution fractionation (TREF)**. A poly(ethylene) solution is introduced into a heated column with an inert carrier, e.g., glass beads. Controlled lowering of the temperature causes the polymer to deposit on the carrier particles: first the fractions forming the most perfect crystallites and later the least crystallizable ones. The degree of crystallinity is mainly controlled by the content of short chain branches; the least branched polymers crystallize best. Solvent is then passed through the column. A controlled increase of temperature causes first the elution of highly branched and then of lesser branched fractions. Fractions are collected as a function of temperature; the SCB content is monitored by infrared via the C-H stretch at 3.41 μm.

6.3. Osmotic Pressure

6.3.1. Fundamentals

The differential of the Gibbs energy is given by the change of the partial molar Gibbs energy $^{*}G_{i,\text{m}}$ with pressure p, temperature T and mole fraction x_i (see textbooks of chemical thermodynamics):

$$(6\text{-}21) \quad \begin{aligned} \text{d}^{*}G_{i,\text{m}} &= (\partial^{*}G_{i,\text{m}}/\partial p)\text{d}p &+ \quad (\partial^{*}G_{i,\text{m}}/\partial T)\text{d}T &+ \quad (\partial^{*}G_{i,\text{m}}/\partial x_i)\text{d}x_i \\ \text{d}^{*}G_{i,\text{m}} &= {}^{*}V_{i,\text{m}}\text{d}p &- \quad {}^{*}S_{i,\text{m}}\text{d}T &+ \quad RT\,\text{d}\log_{e} a_i \\ \text{d}^{*}G_{i,\text{m}} &= {}^{*}V_{i,\text{m}}\text{d}p &- \quad {}^{*}S_{i,\text{m}}\text{d}T &+ \quad \text{d}\Delta\mu_i \end{aligned}$$

In osmometry, the activity a_1 of the solvent in the solution is made to equal the activity of the pure solvent by applying a pressure differential $\text{d}p$. In an isothermal process ($\text{d}T = 0$) at equilibrium ($\text{d}^{*}G_{1,\text{m}} = 0$), this pressure difference equals the osmotic pressure differential $\text{d}\Pi$. Eqn.(6-21) thus converts to $^{*}V_{1,\text{m}}\Pi = -RT\log_{e} a_1 = -\Delta\mu_1$.

In ideal solutions, activities a_1 are identical with mole fractions $x_1 = 1 - x_2$ for the whole concentration range. For dilute ideal solutions ($n_1 > n_2$; $m_1 > m_2$; $V_1 > V_2$), one can furthermore write $\log_e(1-x_2) \approx -x_2$ and thus $*V_{1,m}\Pi = RTx_2$. The mole fraction x_2 can be replaced by the mass concentration $c_2 = m_2/(V_1+V_2) \approx m_2/V_1 = n_2M_2/V_1 = n_2M_2/(n_1*V_{1,m}) \approx x_2M_2/*V_{1,m}$. The result is the **van't Hoff law** which says that the **reduced osmotic pressure** Π/c_2 is inversely proportional to the molar mass M_2 at infinite dilution ($c_2 \rightarrow 0$):

$$(6\text{-}22) \quad \lim_{c_2 \rightarrow 0} \frac{\Pi}{c_2} = \frac{RT}{M_2}$$

Eqn.(6-22) applies to each component i of a polymolecular polymer, thus $\Pi_i = RT(c_i/M_i)$. The osmotic pressure Π of a dilute solution of a polymolecular polymer is the sum of the osmotic pressures Π_i of all components i, which results in $\Pi = \Sigma_i \Pi_i = RT \Sigma_i (c_i/M_i)$. Introducing this equation into Eqn.(6-22) delivers $M_i = c_i/[\Sigma_i (c_i/M_i)]$ and, with $c_i = \Sigma_i c_i$ and $c_i = n_iM_i/V$, also $M_i = \Sigma_i n_iM_i/(\Sigma_i n_i) \equiv \overline{M}_n$ (see Eqn.(2-5)). The molar mass M_i in Eqn.(6-22) is thus the *number-average* if the specimen has a molar mass distribution.

Eqn.(6-22) applies only to infinite dilution. At concentrations $c_2 > 0$, reduced osmotic pressures are usually concentration dependent: for ideal solutions because of the ideal entropy of mixing, for athermal solutions because of the nonideal entropy of mixing, for regular solutions because of the enthalpy of mixing and the ideal entropy of mixing, and for irregular solutions because of the enthalpy of mixing and the nonideal entropy of mixing (see Section 6.2.1.). Only at low concentrations in theta solutions is Π/c_2 independent of the concentration.

For all other solutions, activities can always be expressed as a power series with respect to mole fractions according to statistical mechanics: $-\log_e a_1 = x_2 + Bx_2^2 + Cx_2^3 + ...$ Introduction into $*V_{1,m}\Pi = -RT \log_e a_1$ and replacement of the mole fraction by $x_2 = *V_{1,m}c_2/M_2$ (see above) delivers:

$$(6\text{-}23) \quad \frac{\Pi}{c_2} = RT\left[\frac{1}{M_2} + \frac{B*V_{1,m}}{M_2^2}c_2 + \frac{C*V_{1,m}^2}{M_2^3}c_2^2 + ... \right]$$

$$= RT[A_1 + A_2c_2 + A_3c_2^2 + ...]$$

where $A_1, A_2, A_3 ...$ are the first, second, third ... **virial coefficients**. This equation is also often written as $\Pi/c_2 = (RT/M_2) + A_2'c_2 + A_3'c_2^2 + ...$ with $A_2' = RTA_2$ and $A_3' = RTA_3$. Virial coefficients are complicated averages, for example, $(A_2)_{OP} = \Sigma_i\Sigma_j w_iw_jA_{ij}$ for osmometry and $(A_2)_{LS} = \Sigma_i\Sigma_j w_iM_iw_jM_jA_{ij}/(\Sigma_i w_iM_i)^2$ for static light scattering where the coefficients A_{ij} are due to the interaction of molecules i and j in the presence of the solvent. Virial coefficients depend on the solvent (see Fig. 6-8), the molar mass (see Section 6.3.4.), and the temperature.

The term "virial coefficient" got its name from the virial theorem which was widely used in the nineteenth century. This theorem considered forces which were expressed by the so-called virial (L: *vis* = force). The virial was expanded into a power series whose coefficients were the virial coefficients. It was shown later that the second virial coefficient depends on the interaction between two bodies, the third virial coefficient on those between three bodies, etc.

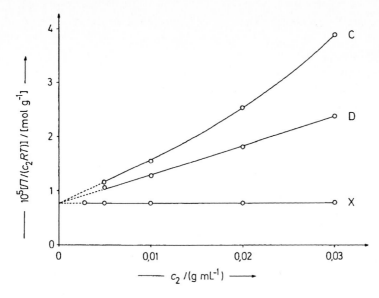

Fig. 6-8 Concentration dependence of reduced osmotic pressures $\Pi/(RTc_2) \equiv 1/\overline{M}_{n,\,\text{app}}$ of a poly-(methyl methacrylate) in various solvents at 20°C [3]. m-Xylene X is a theta solvent ($A_2 = 0$), 1,4-dioxane D leads to a positive second virial coefficient A_2 and $A_3 = 0$, and chloroform C furnishes both second and third virial coefficients. The common intercept at $c_2 \rightarrow 0$ equals $1/\overline{M}_n$.

The first virial coefficient is the inverse molar mass. It follows that $\Pi/(c_2RT)$ delivers a concentration-dependent inverse **apparent molar mass** for concentrations $c_2 > 0$ (Fig. 6-8). This apparent molar mass is not to be confused with the apparent molar mass obtained from light scattering measurements of copolymers (see Eqn.(5-3)). The initial slope of the function $\Pi/(c_2RT) = f(c_2)$ is the second virial coefficient A_2 (see Eqn.(6-23)) if association is absent (see Section 6.3.4.).

6.3.2. Membrane Osmometry

Eqn.(6-23) provides the foundation for the determination of number-average molar masses and osmotic second virial coefficients by membrane osmometry. In membrane osmometers, a membrane separates a solution chamber from a solvent chamber. The membrane is semipermeable: it can be permeated only by the solvent and (ideally) not by the solute. Membranes usually consist of regenerated cellulose for organic solvents, cellulose acetate for water, and porous glass for acids.

At the beginning of an osmotic experiment with solvent and solution levels equal, solution and solvent are not in thermodynamic equilibrium. Solvent thus flows from the solvent chamber into the solution chamber (or vice versa) until the equilibrium pressure Π is obtained. The time to reach equilibrium is very long in this **static osmometry**, sometimes days or weeks. Modern osmometers thus use a servo mechanism: a small increase in osmotic pressure is immediately and automatically balanced by a change of the hydrostatic pressure. Only very small solvent volumes need to pass through the membrane. In this **dynamic osmometry**, the equilibrium state is obtained after (10-30) min instead of days.

Since the osmotic pressure is inversely proportional to the molar mass, osmotic measurements become less accurate with increasing molar mass (see Eqn.(6-23)); the upper limit is at ca. $M_2 \approx 10^6$ g/mol. The lower limit is due to permeation of low molar mass macromolecules through the membrane; it is ca. 10^4 g/mol for membranes from regenerated cellulose.

It is often assumed that the permeation of low molar mass fractions through the membrane is negligible in dynamic osmometry because the time to reach equilibrium is short and the change in volume is small. This assumption is wrong; even if the loss of polymer from the solution chamber is vanishingly small, it still affects the measured osmotic pressure:

The total volume flow J_V from the solvent chamber into the solution chamber is caused by both the osmotic pressure Π and the hydrostatic pressure difference Δp: J_V = $L_p \Delta p + L_{pD} \Pi$ where L_p and L_{pD} are the **phenomenological coefficients (Onsager coefficients)**. The pressure difference obtained by dynamic osmometry at a volume flow of $J_V = 0$ thus corresponds to:

$$(6\text{-}24) \quad \Delta p = -(L_{pD}/L_p)\Pi \equiv s\Pi$$

The **Staverman coefficient** s becomes unity for truly semipermeable membranes since in this case $-L_p = +L_{pD}$. For leaky membranes, $s < 1$, however: even at zero volume flow ($J_V = 0$), i.e., infinitely fast measurements, one never obtains the true osmotic pressure Π by dynamic osmometry with leaky membranes, but always a smaller value $s\Pi$.

6.3.3. Vapor Phase Osmometry

Number-average molar masses can also be obtained by **vapor phase osmometry**. The sensitivity is not as high as in membrane osmometry, however, which restricts the application of this method to molar masses below ca. 50 000 g/mol.

Vapor phase osmometry is based on the following principle. A droplet of the solution of a nonvolatile solute in a volatile solvent resides on a temperature sensor (thermistor) and a droplet of the pure solvent is located on another sensor. The two thermistors are connected to a measuring unit so that their temperature difference can be measured. The large space surrounding the two sensors is saturated with solvent vapor; the two drops and the solvent vapor are initially in thermal equilibrium. Solvent vapor condenses on the solution droplet because of the lower vapor pressure of the latter. The condensation generates a heat of condensation which in turn causes the temperature of the solution droplet to rise until the temperature difference ΔT between solution droplet and solvent droplet balances the original difference in vapor pressures, and the chemical potentials equal each other. Similar to **ebullioscopy** (cf.

textbooks on thermodynamics), one obtains for infinitely dilute solutions of density ρ in solvents of molar mass M_1 and molar vaporization enthalpy $\Delta H_{v,m}$:

$$(6\text{-}25) \quad \lim_{c_2 \to 0} K \Delta T = RT c_2 / \overline{M}_n \quad ; \quad K = \rho \Delta H_{v,m}/(TM_1)$$

Vapor phase osmometry has a strict thermodynamic foundation if the solvent droplet and the solution droplet are thermally isolated from each other. In experiments, however, both droplets and vapor are not thermally isolated from each other. The temperature difference between solvent and solution slowly decreases due to convection, radiation and conduction. This, in turn, causes more solvent vapor to condense on the solution droplet until a steady state with a temperature difference $\Delta T_{st} = K_E \Delta T$ is approached. The proportionality constant K_E is difficult to calculate from theory; it is thus obtained by calibration with compounds of known molar mass.

6.3.4. Virial Coefficients

Second and third virial coefficients depend on excluded volumes u (A_2 disappears for theta solutions), i.e., on interaction parameters χ. These two quantities determine the dependence of virial coefficients on molar mass and temperature. Virial coefficients can be calculated by the simple lattice theory as follows.

The chemical potential of the solvent is given by $\Delta \mu_1 = -{}^*V_{1,m} \Pi$ (see above). It can also be obtained from Eqn.(6-14) as $\Delta \mu_1 = RT[\chi \phi_2^2 + \log_e (1-\phi_2) + (1-X_2^{-1})\phi_2]$ because $\Delta \mu_1 \equiv (\partial G/\partial n_1)_{T,p,n_2}$ and $\phi_1 \equiv 1-\phi_2$. The logarithmic term can be developed into a series $\log_e(1-\phi_2) = - \phi_2 - \phi_2^2 - \phi_2^3 - \ldots$ and the chemical potential becomes $\Delta \mu_1 = -{}^*V_{1,m}\Pi = - RT[X_2^{-1}\phi_2 + \{(1/2)-\chi\}\phi_2^2 + (1/3)\phi_2^3 + \ldots]$. Introduction of $X_2 = M_2/M_u$ and $\phi_2 = V_2/(V_1+ V_2) \approx V_2/V_1 = v_2 c_2$ delivers:

$$(6\text{-}26) \quad \Pi = RT\left(\frac{1}{M_2} \cdot \frac{M_u v_2}{{}^*V_{1,m}} \cdot c_2 + \frac{[(1/2)-\chi]v_2^2}{{}^*V_{1,m}} \cdot c_2^2 + \frac{v_2^3}{3 \, {}^*V_{1,m}} \cdot c_2^3 + \ldots \right)$$

The *first virial coefficient* is obtained as $A_1 = (M_u v_2)/(M_2 \, {}^*V_{1,m})$. The specific volume v_2 of the polymer is identical with the specific volume of the polymer segments ($v_2 = V_2/m_2 = V_u/m_u = v_u$). The molar mass of the segments is $M_u = m_u/n_u$. The product $M_u v_2$ is therefore $M_u v_2 = V_u/n_u = V_{u,m}$ where $V_{u,m}$ = molar volume of segments. The partial molar volume ${}^*V_{1,m}$ of the solvent furthermore equals the molar volume $V_{1,m}$ at small concentrations. Since lattice sites of polymer segments and solvent molecules have the same size, one obtains $V_{u,m} = V_{1,m}$ and thus also $M_u v_2/{}^*V_{1,m} = 1$. The first virial coefficient is thus correctly given as $A_1 = 1/M_2$ (van't Hoff equation).

The simple lattice theory delivers for the *second* and *third virial coefficients*:

$$(6\text{-}27) \quad A_2 = \frac{[(1/2)-\chi]v_2^2}{{}^*V_{1,m}} \quad \text{and} \quad A_3 = \frac{v_2^3}{3 V_{1,m}}$$

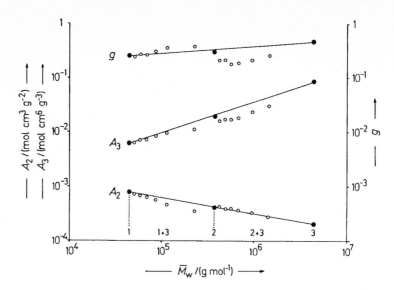

Fig. 6-9   Dependence of the second virial coefficient A_2, third virial coefficient A_3 and reduced third virial coefficient g on the molar masses of three narrowly distributed poly(styrene)s 1, 2 and 3 (●) and their mixtures 1+3 and 2+3 in various proportions (○) in benzene at 25°C. The functions are $A_2 \sim M^{-0.28}$, $A_3 \sim M^{0.58}$ and $g \sim M^{0.14}$. Data of [4].

According to theory, both A_2 and A_3 should be independent of the molar mass M_2 of the solute. Experiments, however, do show a molar mass dependence (Fig. 6-9). These molar mass dependences can be expressed by $A_2 \sim 1/M_2^h$ with $0,2 \le h \le 0,3$ and $A_3 \sim M_2^k$, respectively. The so-called **reduced third virial coefficient** $g = A_3/(A_2^2 M_2)$ also depends on the molar mass.

A solution with $A_2 = 0$ is called a **theta solution**. The polymer is said to be in the **theta state** which is obtained at a certain temperature (the **theta temperature** Θ) for a given solvent which becomes a **theta solvent** at this temperature.

The term 1/2 in the term $[(1/2) - \chi]$ of Eqn.(6-27) originates from the entropic terms $\log_e(1 - \phi_2) + (1 - X_2^{-1})\phi_2$ according to the derivation of Eqn.(6-26). Another entropic contribution is contained in the interaction parameter χ as shown in Section 6.2.2; the interaction parameter can thus be written as $\chi = \chi_H + \chi_S$. The two entropic terms 1/2 and χ_S are combined into a new "entropy term" $\psi = (1/2) - \chi_S$.

If the remaining term χ_H represents an enthalpy (thus an energy), then the "entropy term" ψ must have the physical unit (the "dimension") of an energy, too. Neither ψ nor χ_S is an entropy but ψ/T has the unit of an entropy. Gibbs energies become zero at a certain characteristic temperature Θ according to $\Delta G = \Delta H - \Theta \Delta S = 0$; thus also $\chi_H - \Theta(\psi/T) = 0$. One can therefore write $\chi_H = \Theta(\psi/T)$ and consequently also $[(1/2) - \chi] = \psi[1 - (\Theta/T)]$.

ψ and χ_H/ψ are usually positive in thermodynamically poor solvents; Θ must thus also be positive. The solution behaves at the theta temperature Θ as if it is ideal. Theta solutions are however only pseudo-ideal since they exhibit both excess entropies of mixing (here: $\psi/T \ne 0$) and enthalpies of mixing (here: $\chi_H \ne 0$).

The second virial coefficient is also a measure of the excluded volume u of a segment pair. Excluded volumes are not only controlled by the space requirements of the segments but also by the interactions between segments and between segments and solvent molecules (Section 5.5.1.). The thermodynamic probability Ω of the various arrangements of excluded segments therefore also includes the enthalpic contributions without considering them explicitly.

The first molecule has available a total volume of V. The second molecule can occupy only $(V-u)$, the third only $(V-2u)$, etc., and the ith molecule only $(V-iu)$. The product of all these possibilities is proportional to the probability Ω. It follows:

$$(6\text{-}28) \quad \Delta G = -T\Delta S = -k_B T \log_e \Omega = -k_B T \log_e \left[const \cdot \prod_{i=0}^{N_2-1}(V-iu) \right]$$

The logarithm of the products may be expressed as a sum of logarithms:

$$(6\text{-}29) \quad \Delta G = -k_B T \left[N_2 \log_e V + \sum_{i=0}^{N_2-1} \log_e (1 - iu/V) \right] + const'$$

The term iu/V is very much smaller than unity for dilute solutions ($iu/V \ll 1$). The logarithms can thus be expanded into a series $\log_e(1-y) = -y -...$ and $\log_e(1-(iu/V))$ becomes $-iu/V$. The summation only has to be performed over all i since u/V is a constant. The resulting sum becomes approximately $N_2^2/2$. It follows that $\Delta G = -k_B T[N_2\log_e V - (N_2^2/2)(u/V)] + const'$ and $\partial \Delta G/\partial V = -k_B T N_2/V - k_B T N_2^2 u/(2 V^2)$, respectively. Insertion of $N_2/V = c_2 N_A/M_2$ and $R = k_B N_A$ into the latter equation leads to $\partial \Delta G/\partial V = -RTc_2/M_2 - RTc_2^2 N_A u/(2 M_2^2)$. Because $^*V_{1,m} = \partial V/\partial n_1$, one also obtains $\partial \Delta G/\partial V = (1/^*V_{1,m})(\partial \Delta G/\partial n_1) = -\Pi$ (see the statements prior to Eqn.(6-22)). The result is $\Pi/c_2 = RT\{(1/M_2) + [(N_A u)/(2 M_2^2)]c_2$. Comparison of the coefficients of this equation with those of Eqn.(6-26) shows that the second virial coefficient is directly proportional to the excluded volume:

$$(6\text{-}30) \quad A_2 = (N_A u)/(2 M_2^2)$$

6.3.5. Overlap Concentrations

The second virial coefficient describes interactions between two molecules, the third virial coefficients between three, etc. Third virial coefficients are most likely to be observed for polymer coils in good solvents at higher concentrations (Fig. 6-8) since such coils are more expanded than coils in theta solvents (Fig. 5-8). At still higher concentrations, polymer coils begin to overlap (Section 5.7.2.). In a plot of the reduced osmotic pressure $(\Pi M)/(cRT)$ vs. the reduced overlap concentration c/c_v^* one does indeed see two ranges (Fig. 6-10). $(\Pi M)/(cRT)$ approaches unity at $c/c_v^* \ll 1$ but increases strongly with c/c_v^* at $c/c_v^* \gg 1$.

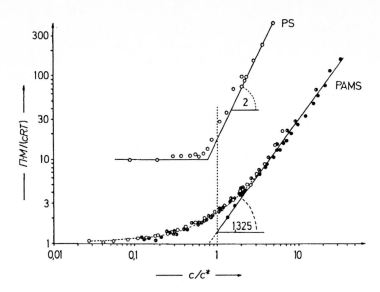

Fig. 6-10   Dependence of reduced osmotic pressures, $(\Pi M)/(cRT)$, on the normalized overlap concentration c/c^* from viscosities ($c^* = c_v^*$) for poly(styrene)s PS in the theta solvent cyclohexane at 34°C [5] (data shifted upwards by a factor of 10) and poly(α-methylstyrene)s PAMS in the good solvent toluene at 25°C [6]. The osmotic overlap concentration is at $c_\pi = 0.8\, c_v^*$.

If the transition from one range to the other is caused by overlapping of coils, then the coils must be increasingly compressed at higher concentrations. This compression can be measured by the change $dc/d\Pi$ of polymer concentration with the osmotic pressure. It is counteracted by the effect of external excluded volumes. The reduced osmotic pressure of polymer coils in good solvents must therefore increase more than proportional to the concentration, i.e., $\Pi/c = f(c^n)$ with n > 1 (Fig. 6-10).

At these higher concentrations, the reduced osmotic pressure can be approximated by $(M\Pi)/(cRT) = 1 + A_2 cM \approx A_2 cM$ (see Eqn.(6-26)). The second virial coefficient is proportional to the excluded volume (Eqn.(6-30)), thus $(M\Pi)/(cRT) = (N_A u/2)(c/M)$. Coils are practically isolated just below the critical concentration c^*. They can therefore be treated like equivalent spheres whose excluded volume u is just 8 times as large as the volume of the sphere (Section 5.6.1.). The radius of the equivalent sphere is assumed to be the radius of gyration of the coil, thus $u = 8\,V = (32\,\pi/3)\,s^3$. The radius of gyration is proportional to a power v of the molar mass ($s = K_s M^v$, see Eqn.(5-25)). It follows that $(M\Pi)/(cRT) = (16\,\pi/3)\,N_A K_s^3 M^{3v-1} c$. The exponent is $v = 3/5$ for perturbed coils. The critical overlap concentration is therefore

$$(6\text{-}31)\qquad c^* = \left(\frac{3\,M\Pi}{16\,\pi\,cRTN_A K_s^3}\right)\cdot\frac{1}{M^{3v-1}} = K_a M^{-4/5}$$

The reduced osmotic pressure $(M\Pi)/(cRT)$ can also be evaluated as a function of the extent of overlap, c/c^*. It must however possess about the same numerical value above and below c^*. The ratio c/c^* has thus to be scaled by the exponent $1/(3v-1)$:

(6-32) $(\Pi M)/(cRT) = f(c/c^*) = const\cdot(c/c^*)^{1/(3\nu-1)} = const\cdot(c/c^*)^m$

The exponent ν equals 1/2 for the unperturbed state (Section 5.3.3.); for $c > c^*$, the exponent m becomes m = 2 according to both theory and experiment (Fig. 6-10). For perturbed coils, the mean-field theory predicts $\nu = 0.600$ and thus m = 1.25 whereas renormalization theory leads to $\nu = 0.588$ and m = 1.309. Experimentally, values of $\nu = 0.585$ and m = 1.325 were found for poly(α-methyl styrene) in the good solvent toluene (Fig. 6-10). The critical overlap concentration c_{π}^* from osmometry is, however, lower than that from viscometry ($c_{\pi}^* \approx 0.8\ c_{\nu}^*$).

6.4. Association

Coil overlap is an unspecific event that is caused by the increased competition of segments of large coil molecules for space at higher polymer concentrations. An increase of concentration, however, may also induce specific interactions between certain groups or segments that lead to polymer association and, at even higher polymer concentrations, to polymer gelation. These specific interactions usually begin at polymer concentrations far below the critical overlap concentration c^*.

An **association** is the reversible formation of physical bonds between chemical groups (thermodynamic control), an **aggregation** is the corresponding irreversible process. Both terms are often used interchangeably in the literature, however.

Associations may be both intramolecular and intermolecular. *Intramolecular association* leads to a contraction of polymer coils. The coils become more compact and the viscosity decreases (Section 5.8.4.). Intramolecular associations are independent of the polymer concentration. The number concentration of solute molecules per volume does not change but the polymer–solvent interaction does. Since the polymer molecule becomes more compact by intramolecular association, the second virial coefficient decreases but remains positive in good solvents.

In *intermolecular association*, on the other hand, two or more chemical molecules are assembled into one physical molecule. Since the probability of intermolecular association increases with increasing polymer concentration, the average molar mass of the molecules (chemical and physical) in equilibrium increases too. The inverse apparent molar mass thus decreases with increasing concentration but the resulting negative slopes are not caused by negative second virial coefficients as a measure of excluded volume and polymer–solvent interactions, respectively, but by the increase of average apparent molar masses due to increasing association.

The type of the function $1/(\overline{M}_n)_{app} = f(c)$ depends on the type of association: open (Section 6.4.1.) or closed (Section 6.4.2.). In both types, one has furthermore to distinguish between molecule-based and segment-based associations:
– In *molecule-based* associations, the number of associogenic groups is constant per molecule; an example is an association via the two end groups of linear molecules.

The equilibrium constants are based on amount (molar) concentrations of physical and chemical molecules, i.e., $^nK_o = [M_N]/([M_{N-1}][M_I])$, where $[M_I]$ = molar concentration of unimers (chemical molecules), $[M_N]$ = concentration of physical molecules and N = number of chemical molecules per physical molecule. The superscript n to the left of K indicates that the equilibrium constant refers to number concentrations.
– In *segment-based* associations, the number of associogenic sites varies with the molar mass; an example is the intermolecular association of helical segments in random coil-forming molecules. The equilibrium constants have to be based on mass concentrations of molecules, i.e., $^wK_o = c_N/(c_{N-1}c_I)$.

6.4.1. Open Associations

In **open associations, unimers** (non-associated chemical molecules) of molar mass M_I are in equilibrium with their **multimers** (physical molecules: dimers, trimers, ..., N-mers of chemical molecules) with twofold, threefold, ..., N-fold molar masses M_{II}, M_{III}, ..., M_N. Two unimers associate to form a dimer according to $2\,M_I \rightleftarrows M_{II}$. Trimers are generated by $M_{II} + M_I \rightleftarrows M_{III}$ and tetramers by either $M_{III} + M_I \rightleftarrows M_{IV}$ or $2\,M_{II} \rightleftarrows M_{IV}$, etc. (cf. the similar situation for the formation of chemical molecules by polyaddition (Chapter 4)).

At higher polymer concentrations, not only are *more* multimers formed but these multimers are also of *higher molar mass*. The inverse apparent molar masses decrease with increasing concentration until the virial term starts to dominate and the inverse apparent molar masses increase (Fig. 6-11). In associating systems, negative initial slopes of $1/M_{app} = f(c)$ curves do *not* indicate negative second virial coefficients A_2 if one adheres to the physical meaning of A_2 as a measure of the excluded volume.

Open associations can be described by consecutive equilibria. Equilibrium constants $^nK_o = [M_N]/([M_{N-1}][M_I])$ can often be assumed to be independent of the degree of association, N, e.g., for association via the end groups (this is equivalent to the assumption of equal chemical reactivity in polymerizations). Since the condition $^nK_o[M_I] = [M_{II}]/[M_I] < 1$ must apply for molecule-based associations (see Section 3.2.2.), one obtains for the total molar concentration $[M]$ of all species (unimers, dimers, ..., N-mers): $[M] = [M_I] + [M_{II}] + [M_{III}] + ... = [M_I](1 - {}^nK_o[M_I])^{-1}$. A similar expression $c = [M_I]\,\overline{M}_n(1 - {}^nK_o[M_I])^{-2}$ is obtained for the total mass concentration c. From the definition of the molar concentration $[M] = c/(\overline{M}_n)_{app,\Theta}$, one obtains for molecule-based associations in the absence of virial coefficients:

$$(6\text{-}33) \quad (\overline{M}_n)_{app,\Theta} = (\overline{M}_I)_n + {}^nK_o(\overline{M}_I)_n\{c/(\overline{M}_n)_{app,\Theta}\} \qquad ; \text{molecule-based}$$

A plot of $(\overline{M}_n)_{app,\Theta}/(\overline{M}_I)_n = f(\{c/(\overline{M}_n)_{app,\Theta}\})$ should thus have an intercept of unity and an initial slope of K_o if an open association is present. This was indeed observed experimentally for poly(ethylene glycol)s $HO(CH_2CH_2O)_nH$ that associate in benzene at 25°C via the hydroxyl end groups. Equilibrium constants of $K_o = (4.1 \pm 0,14)$ L/mol were found to be independent of the molar mass of unimers.

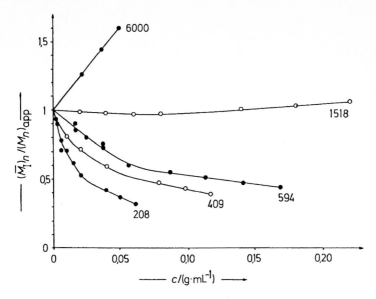

Fig. 6-11   Concentration dependence of normalized inverse apparent number-average molar masses of poly(ethylene glycol)s $HO(CH_2CH_2O)_nH$ in benzene at 25°C as measured by vapor-phase os-mometry. Numbers indicate the number-average molar masses $\overline{M}_{I,n}$ of unimers. Note that the initial slopes are *not* the scond virial coefficients A_2 (see Eqn.(6-33)), except for the polymer with $\overline{M}_{I,n}$ = 6000 g/mol. With permission of Gordon and Breach, Science Publishers [7].

Other expressions are obtained for the apparent mass-averages in molecule-based associations and for both apparent number-average and mass-averages in segment-based ones. They are given here for comparison with Eqn.(6-33) without further discussion:

$$(6\text{-}34) \quad (\overline{M}_w)_{app,\Theta} = (\overline{M}_I)_w + 2\,{}^nK_o(\overline{M}_I)_n\,\{c/(\overline{M}_n)_{app,\Theta}\} \quad ; \text{ molecule-based}$$

$$(6\text{-}35) \quad (\overline{M}_w)_{app,\Theta} = (\overline{M}_I)_w + {}^wK_o(\overline{M}_I)_w c \qquad\qquad ; \text{ segment-based}$$

$$(6\text{-}36) \quad (\overline{M}_n)_{app,\Theta} = M_I + {}^wK_o c[\log_e(1 + {}^wK_o c)]^{-1} \qquad ; \text{ segment-based}$$

Note that Eqn.(6-34) still requires the knowledge of apparent-number average molar masses for the evaluation of apparent mass-averages. Eqn.(6-36) applies only to molecularly uniform unimers; no closed expression has been reported for segment-based open associations of unimers with molar mass distributions.

6.4.2. Closed Associations

Only two species exist in **closed associations**: Unimers M_I and *N*-mers M_N are in equilibrium according to $N\,M_I \rightleftarrows M_N$. The higher the polymer concentration, the more multimers are formed. In contrast to open associations, the number N of unimers in *N*-mers stays constant in closed associations.

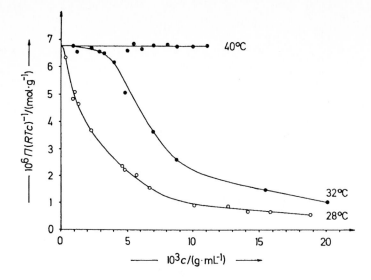

Fig. 6-12   Concentration dependence of reduced osmotic pressures, $\Pi/(cRT)$, of solutions of a di-block copolymer poly(styrene)-*block*-poly(2-vinylpyridine) with $\overline{M}_{\mathrm{I,n}} = 150\ 375$ g/mol and $w_{\mathrm{sty}} = 0.32$ in toluene. The polymer does not associate at 40°C. It undergoes a closed association at 32°C and 28°C because poly(2-vinylpyridine)-blocks are insoluble in toluene at these temperatures [8].

Closed associations follow the same formalism as micelle formation from low molar mass amphoteric molecules. They exhibit a more or less pronounced "critical micelle concentration" c_{crit} and a strong decrease of inverse apparent molar masses at $c > c_{\mathrm{crit}}$ (Fig. 6-12). However, a similar *formalism* indicates neither the same *mechanism* of formation nor the same *structure* of physical molecules. For example, the closed association of the block copolymers of Fig. 6-12 is enthalpically controlled whereas that of the low molar mass emulsifiers of Fig. 3-14 is governed by entropy.

The critical micelle concentration is often at such a low polymer concentration that it is no longer detectable by experimental methods (Fig. 6-12, 28°C). The concentration dependence of inverse apparent molar masses then resembles that of an open association (compare Fig. 6-11). Closed and open associations can however be distinguished by the dependence of $(\overline{M}_{\mathrm{w}})_{\mathrm{app,\Theta}}$ on $[(\overline{M}_{\mathrm{n}})_{\mathrm{app,\Theta}}]^{-1}$ which is linear for closed associations but not for open ones:

$$(6\text{-}37)\quad (\overline{M}_{\mathrm{w}})_{\mathrm{app,\Theta}} = [(\overline{M}_{\mathrm{I}})_{\mathrm{w}} + N(\overline{M}_{\mathrm{I}})_{\mathrm{n}}] + N(\overline{M}_{\mathrm{I}})_{\mathrm{n}}^{2}[(\overline{M}_{\mathrm{n}})_{\mathrm{app,\Theta}}]^{-1}$$

It follows from Eqn.(6-37) that the ratio $(\overline{M}_{\mathrm{w}})_{\mathrm{app,\Theta}}/(\overline{M}_{\mathrm{n}})_{\mathrm{app,\Theta}}$ is not a measure of the polymolecularity of the N-mers.

Associations of dissolved block copolymers are utilized industrially for so-called **polymer detergents**. Examples are multiblock copolymers that are composed of water-soluble ethylene oxide blocks $+\mathrm{OCH_2CH_2}\overline{\}_m$ and water-insoluble propylene oxide blocks $+\mathrm{OCH_2CH(CH_3)}\overline{\}_n$ which generate very high viscosities at low polymer concentrations.

6.4.3. Gels

A **gel** consists of a chemically or physically cross-linked polymer that is highly swollen by a solvent. The word "gel" is derived from "gelatin(e)" (see Section 4.2.3.) which needs only 0.6 % polymer to generate an aqueous gel (e.g., as Jello®). The solvent is tightly bound by the polymer in these gels; it does not exude.

Chemical cross-links result from the polymerization of most multifunctional monomers by either chain polymerizations or polycondensations and polyadditions. Physical cross-links are formed by macromolecules which contain segments that are incompatible with the rest of the molecule. Examples are long stereoregular sequences in otherwise atactic polymers, ionic structures in predominantly nonionic polymers, or hydrophobic regions in water-soluble polymers. These incompatible segments try to assemble themselves into domains but the formation of large perfect structures is prevented by the presence of the other segment types. The resulting junction zones or domains are thus crystallites arising from short stereoregular segments as in gelled poly(vinyl chloride), hydrophobic regions as in pectin gels, or bundles of short helical segments as in starch gels. The junction zones dissociate if the temperature is changed. Such **physical gels** are therefore also called **thermoreversible gels**.

Starch is composed of amylose and amylopectin (Section 4.2.5.). The physical network of the branched amylopectin molecules prevents the formation of large crystallites of helical amylose molecules. On dissolution of starch in hot water, amylose molecules enter the solution as random coils with unordered segments and short helical segments (cf. Fig. 2-21, III). The short helical segments can grow larger at the expense of the unordered segments in dilute solution; these larger segments form junction zones. The amylose crystallizes and becomes insoluble in water by this **retrogradation**. In concentrated starch solutions, longer helical segments are formed only slowly but these segments can still form intermolecular and intramolecular associations (though not large crystallites) because of the amylopectin network. This results in a physical network which contains a lot of water, a **hydrogel**.

Gels can be easily deformed by shearing but only slightly deformed by hydrostatic pressure. On the further addition of solvent, physically cross-linked gels can be converted into true solutions if the junction zones can be dissolved (e.g., melting of crystallites, dissociation of intermolecular associates, etc.). A chemically cross-linked gel cannot be converted into a true solution, however. Further addition of solvent causes the network chains to expand but this expansion is limited by the presence of chemical cross-links which try to retract the network chains.

In swelling equilibrium, the Gibbs energy of mixing for polymer and solvent equals the opposing Gibbs energy of elasticity ($\Delta G_{mix} = - \Delta G_{elast}$): a thermodynamically good solvent swells a weakly cross-linked polymer more strongly than a poor one. In strongly cross-linked polymers, the elasticity term dominates whereas negligible contributions come from the polymer–solvent interaction and the dilution by the solvent. The swelling of such networks is only slightly affected by the thermodynamic quality of the solvent; the networks swell very little.

Literature

H.Yamakawa, Modern Theory of Polymer Solutions, Harper and Row, New York 1971
H.Morawetz, Macromolecules in Solution, Interscience, New York, 2nd ed. 1975
H.Eisenberg, Biological Macromolecules and Polyelectrolytes in Solution, Clarendon Press, Oxford 1976
V.N.Tsvetkov, Rigid Chain Polymers. Hydrodynamic and Optical Properties in Solution, Consultants Bureau, New York 1989
G.Jannink, J. des Cloizeaux, Polymers in Solution, Oxford University Press, Oxford 1990
H.Fujita, Polymer Solutions, Elsevier, Amsterdam 1990

6.1.3. SOLUBILITY PARAMETERS
A.F.M.Barton, CRC Handbook of Solubility Parameters and Other Cohesion Parameters, CRC Press, Boca Raton (FL) 1983

6.2. STATISTICAL THERMODYNAMICS and 6.3. OSMOTIC PRESSURE
L.H.Tung, ed., Fractionation of Synthetic Polymers, Dekker, New York 1977
O.Olabisi, L.M.Robeson, M.T.Shaw, Polymer-Polymer Miscibility, Academic Press, New York 1979
M.Kurata, Thermodynamics of Polymer Solutions, Harwood Academic Publ., Chur 1982
K.Solc, ed., Polymer Compatibility and Incompatibility: Principles and Practices, Harwood Academic Publ., Chur 1982
L.Wild, Temperature Rising Elution Fractionation, Adv.Polym.Sci. **98** (1990) 1
M.M.Coleman, J.F.Graf, P.C.Painter, Specific Interactions and Miscibility of Polymer Blends, Technomic, Lancaster (PA) 1991
E.Frankuskiewicz, Polymer Fractionation, Springer, Berlin 1994
M.A.McHugh, V.J.Krukonis, J.A.Pratt, Supercritical Fractionation of Polymers and Copolymers, Trends in Polymer Science **2** (1994) 301

6.4. ASSOCIATION
H.-G.Elias, Association and Aggregation as Studied via Light Scattering, in M.B.Huglin, ed., Light Scattering from Polymer Solutions, Academic Press, London 1972
H.-G.Elias, Association of Synthetic Polymers, in K.Solc, ed., Order in Polymer Solutions, Gordon and Breach, New York 1975
Z.Tuzar, P.Kratochvil, Block and Graft Copolymer Micelles in Solution, Adv.Colloid Interface Sci. **6** (1976) 201
A.Ben-Naim, Hydrophobic Interactions, Plenum, New York 1980
O.Kramer, ed., Biological and Synthetic Polymer Networks, Elsevier Appl.Sci., New York 1988
W.Burchard, S.B.Ross-Murphy, eds., Physical Networks, Elsevier Appl.Sci., London 1990
J.M.Guenet, Thermoreversible Gelation of Polymers and Biopolymers, Academic Press, London 1992

References

[1]   G.M.Bristow, W.F.Watson, Trans.Faraday Soc. **54** (1958) 1567, Table 1; **54** (1958) 1731, Tables 1 and 2
[2]   R.B.Richards, Trans.Faraday Soc. **42** (1946) 10, Fig. 10
[3]   G.V.Schulz, H.Doll, Ber.Dtsch.Chem.Ges. **80** (1947) 232
[4]   T.Sato, T.Norisuye, H.Fujita, J.Polym.Sci.-Polym.Phys.Ed. **25** (1987) 1
[5]   P.Stepanek, R.Perzynski, M.Delsanti, M.Adam, Macromolecules **17** (1984) 2340, Fig. 2
[6]   I.Noda, N.Kato, T.Kitano, M.Nagasawa, Macromolecules **14** (1981) 669, Fig. 7
[7]   H.-G.Elias, Internat.J.Polym.Mat. **4** (1976) 209, Fig. 5
[8]   A.Sikora, Z.Tuzar, Makromol.Chem. **184** (1983) 2049; numeric data: private communication

7. Polymer Hydrodynamics

7.1. Introduction

Molecular dynamics is concerned with time-dependent fluctuations of structures and properties at equilibrium as well as with the approach of systems to equilibrium. A subfield is molecular hydrodynamics which studies the dynamics of fluid systems.

The hydrodynamic behavior of large rigid spheres in the continuum of a solvent is very well explored. It forms the basis for the determination of intrinsic viscosities of coil molecules (Section 5.8.2.) since a coil can be visualized as an equivalent sphere (Section 5.8.). This rather crude model neglects the local structure of a coil; such a modeling is therefore called **coarse-graining**. An example of coarse-graining is the Kuhn "ersatz coil" (G: *Ersatz* = substitute; Section 5.3.2.).

Time and length scales of probes used in coarse-graining are so large that they monitor only the dynamic motion of the *whole chain*. At somewhat shorter length and time scales, one would detect the dynamics inside coils. At even shorter time and length scales, only local changes would be observed, for example, the isomerization dynamics of conformers. An example is the lifetime of trans or gauche conformers which is ca. 10^{-9} s to 10^{-11} s in liquid hydrocarbons. Nuclear magnetic resonance works on a time scale of $(10^{-6}\text{-}10^{-7})$ s; individual types of conformers thus do not show up at these rather long times. Vibrational, infrared and Raman spectroscopy, however, work at much smaller time scales $(10^{-13}$ s); these techniques provide a kind of snapshot of the proportion of conformers.

There are therefore three different regimes for the dynamics of an isolated coil: (I) the long-time regime originating from the Brownian motion of the whole chain; (II) an intermediate regime where chain connectivity is important; and (III) the short-time regime covering the dynamics of events in monomeric units. The time scale in II is much larger than the time scale in III. The dynamic behavior of chains in regime II is thus independent of the chemical (local) structure of the chain; it describes the **global features** of chains.

7.2. Diffusion in Solution

7.2.1. Mutual Diffusion

Concentration differences between two solutions with concentrations $c' > c''$ of the solute contacting each other are diminished by **translational diffusion** of the solute from c' to c''. The transport of mass m of the solute through a unit area A per unit time is defined as the **flux** J_d. It is proportional to the concentration gradient dc/dr according to **Fick's first law**: $J_d \equiv \delta m/(A\delta t) = -D(\partial c/\partial r)$ (see textbooks of physical chemistry). The proportionality constant D is the **mutual diffusion coefficient** which is the same for solute and solvent.

Classical diffusion experiments measure the progress of the diffusion as the change of concentration c with time t per volume element, i.e., as $\partial c/\partial t = -\partial J_d/\partial r$. Introduction of this equation into Fick's first law delivers **Fick's second law** of diffusion: $\partial c/\partial t = \partial[D(\partial c/\partial r)]/\partial r$. This differential equation can be solved for various conditions. If a solution of concentration c' diffuses in an infinitely large space against its solvent ($c'' = 0$), then the concentration gradient will be $dc/dr = c[2\,(\pi Dt)^{1/2}]^{-1} \exp[-r^2/(4\,Dt)]$ at time t and distance r.

Diffusion coefficients D are generally concentration dependent which can be written as $D = D_0(1 + k_d c + ...)$. The concentration c is the averaged concentration $c = (c' + c'')/2$. The slope constant $k_d = 2\,A_2 M_2 - k_f - *v_2$ is controlled by both thermodynamics and hydrodynamics; the former via the second virial coefficient A_2 and the latter by a hydrodynamic interaction parameter k_f; $*v_2$ = partial specific volume of the polymer. Depending on the method of evaluation, diffusion coefficients D_0 are obtained as either mass-averages $\overline{D}_w = \Sigma_j\, w_j D_j$ (also called moment averages) or as so-called area averages $\overline{D}_A = (\Sigma_j\, w_j D_j^{-1/2})^{-2}$.

The equilibration of concentrations is caused by **Brownian motions** of molecules. These thermal motions generate local fluctuations of concentrations which last microseconds. The scattering behavior of these fluctuations can be measured by **dynamic light scattering (quasi-elastic scattering, photon correlation spectroscopy)** which is presently the most important method for the determination of mutual diffusion coefficients in solution.

Brownian motions cause scattering centers to move towards the detector or away from it. The former movement shifts the scattered light to higher frequencies, the latter to lower frequencies: a **Doppler effect** results. Molecules in solution, however, move in all three spatial directions and furthermore possess a velocity distribution. Instead of a sharp line, a distribution curve is obtained. The half-width $\Gamma = DQ_\vartheta^2$ of this curve is given by the product of the diffusion coefficient D and the square of the scattering factor $Q_\vartheta = (4\,\pi/\lambda)[\sin(\vartheta/2)]$ (see Eqn.(5-6)); it also contains contributions of rotation and vibration. The dependence of the reduced half-width Γ/Q_ϑ^2 on Q_ϑ^2 (i.e., on the scattering angle ϑ) delivers the z-average of the mean-square radius of gyration from the slope and (approximately!) the z-average of the diffusion coefficient from the intercept at $Q_\vartheta^2 \to 0$:

(7-1) $\Gamma/q^2 = \overline{D}_z\,(1 + K_q\langle s_z^2\rangle Q_\vartheta^2 + ...)$; K_q = optical constant

7.2.2. Frictional Coefficients

Theoretical calculations of diffusion coefficients require assumptions about the motions of particles in liquids. Assuming a liquid to be a continuum, the **Einstein–**

Sutherland equation predicts that the diffusion coefficient D_0 at infinite dilution ($c_2 \to 0$) is inversely proportional to the frictional coefficient ξ_d:

(7-2) $D_0 = k_B T/\xi_d = RT/(N_A \xi_d)$

Frictional coefficients depend on the types of bodies undergoing diffusion. *Solid spheres* with **Stokes radii** R_d in a continuous medium of viscosity η_1 move with a frictional resistance $\xi = \eta_1 \int(dv/dy)da$ if eddies are absent. The velocity gradient dv/dy of the medium perpendicular to the movement of the body has to be integrated over all area elements da (i.e., the total surface A of the sphere). It can be measured in fractions of v/R_d; the area elements da are then expressed as fractions of the surface $4\pi R_d^2$ of the sphere. The result is $\xi = const \cdot (v/R_d)(4\pi R_d^2)$ where the integration constant adopts a value of $const = 3/2$. Since the molecular frictional coefficient ξ_d is defined as the frictional coefficient per unit velocity, one obtains the **Stokes equation**:

(7-3) $\xi_d \equiv \xi/v = 6 \pi \eta_1 R_d$

Introduction of Eqn.(7-3) into Eqn.(7-2) leads to the **Stokes–Einstein equation**:

(7-4) $D_0 = RT/(N_A \xi_d) = RT/[N_A(6 \pi \eta_1 R_d)] = k_B T/(6 \pi \eta_1 R_d) = k_B T/\xi_d$

Diffusion coefficients deliver hydrodynamic radii R_d which are the true radii $R_{sph} = [(3 V_2/(4 \pi)]^{1/3}$ of *solid spheres*. Furthermore, since $M_2 = m_2 N_A = \rho_2 V_2 N_A$, one obtains $D_0 = [4 \pi N_A \rho_2/3]^{1/3}[k_B T/(6 \pi \eta_1)]M_2^{-1/3}$: diffusion coefficients D_0 of solid spheres decrease with the cube root of the molar masses M_2.

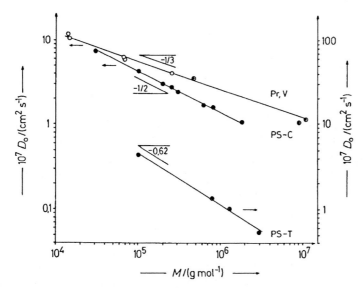

Fig. 7-1 Molar mass dependence of diffusion coefficients of spherical ($\bigcirc$) or spheroidal ($\odot$) proteins Pr and viruses V in dilute aqueous salt solutions (25°C, [1]) or of coils of poly(styrene)s PS in cyclohexane C ($\Theta = 35°C$; $\bullet$ [2]) and in the good solvent toluene T ($T = 20°C$; $\oplus$ [3]).

A similar function $D_0 = K_d M^{-1/3}$ is found for *solvated spheres*, e.g., for some proteins and viruses (Fig. 7-1). The proportionality constant K_d also depends on the degree of solvation in this case (the amount of solvent transported by the polymer molecule). For particles other than spheres (e.g., coils), Stokes radii represent radii of **equivalent spheres**, i.e., the hydrodynamic radii that particles would exhibit if they were solid spheres. The diffusion coefficients of coils follow $D_0 = K_d' M^{-(\alpha+1)/3} = K_d' M^\delta$. For $\alpha = 1/2$ (theta solvents), one obtains an exponent of $\delta = -1/2$ (see Fig. 7-1), for $\alpha = 0.8$ (good solvent), an exponent $\delta = -0.60$.

7.2.3. Spring-and-Bead Models

Various models have been developed for the calculation of the dependence of diffusion coefficients of coil molecules on their frictional coefficients. The simplest model is that of an elastic dumbbell where two beads are connected by an elastic spring (Fig. 7-2, I). In **spring-and-bead models**, segments of a chain are approximated by dumbbells. This model corresponds to unperturbed coils with Gaussian statistics of end-to-end distances.

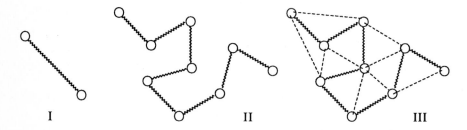

I   II   III

Fig. 7-2   Models for the calculation of polymer dynamics of coil molecules. I: Elastic dumbbell; II: spring-and-bead model without hydrodynamic interactions between segments (Rouse model); III: spring-and-bead model with (some) hydrodynamic interactions (- - -) between segments (as beads) that are interconnected by massless springs (Kirkwood–Riseman model).

Beads are assumed to contain all the mass in such models; no mass should reside in the springs. The polymer dynamics of such a model is described by a Hookean behavior (see Chapter 10) of springs. The resulting mathematical problem resembles that of the Brownian motion of a system of coupled harmonic oscillators.

The **Rouse theory** assumes that no hydrodynamic interaction exists between segments. Solvent can flow undisturbed through the coil which is therefore **free-draining**. The motion of a segment is however hampered by the presence of solvent molecules in the coil which causes friction. Each segment thus has a frictional coefficient ξ_{seg}. Since the N_{seg} segments of a chain are hydrodynamically independent of each other, molecular frictional coefficients become $\xi_d = N_{seg}\xi_{seg}$. The Rouse model delivers a diffusion coefficient:

$$(7\text{-}5) \quad D_0 = k_B T/(N_{seg}\xi_{seg}) = RT/(N_A N_{seg}\xi_{seg})$$

The **Zimm theory**, on the other hand, assumes very strong hydrodynamic interactions between segments. Solvent molecules can no longer move through the polymer coil which is completely **non-draining**.

The **Kirkwood–Riseman theory** allows variable hydrodynamic interactions. The higher the segment density of the coil, the more impeded is the flow of solvent molecules through the coil and the more the solvent moves with the center of the coil. The perturbation of solvent flow was calculated by Debye and Bueche for coils with homogeneous segment density and by Kirkwood and Riseman for coils with Gaussian segment distributions. The mathematically difficult derivation employs Oseen's method for the solution of the Navier–Stokes equations of hydrodynamics (see specialized literature). It delivers for N_{seg} segments with segmental frictional coefficients ξ_{seg} in solvents of viscosity η_1:

$$(7\text{-}6) \qquad D_0 = k_B T \left(\frac{[8/(3\pi)] + [6\pi\,\eta_1\langle s^2\rangle^{1/2} / (N_{seg}\xi_{seg})]}{6\pi\,\eta_1\langle s^2\rangle^{1/2}} \right)$$

The second term of the numerator dominates for short chains (small N_{seg}). One obtains $D_0 = k_B T/(N_{seg}\xi_{seg})$, i.e., the Rouse equation for free-draining coils (Eqn.(7-5)). Such chains should show $D_0 \sim N_{seg}^{-1} \sim M^{-1}$: the diffusion coefficient should be inversely proportional to the molar mass.

For long chains (large N_{seg}), the second term of the numerator is much smaller than $8/(3\pi)$ and the diffusion coefficient becomes $D_0 = (4\,k_B T)/(9\,\pi^2\eta_1\langle s^2\rangle^{1/2})$. Since radii of gyration increase with molar masses according to $\langle s^2\rangle^{1/2} = K_s M^\nu$ (see Eqn.(5-22) for coils and Eqn.(5-26) for rods), the Kirkwood–Riseman theory applied to long chains predicts:

$$(7\text{-}7) \qquad D_0 = \frac{4\,k_B T}{9\pi^2\,\eta_1 K_s} M^{-\nu}$$

For $\nu = 1/2$, one obtains $D_0 \sim M^{-1/2}$, i.e., the prediction of the Zimm theory for strong hydrodynamic interactions between segments, i.e., non-draining coils.

The Kirkwood–Riseman theory (KR) describes non-draining coils ($\nu = 1/2$), partially draining coils ($1/2 < \nu < 1$) and free-draining coils ($\nu = 1$). Strongly expanded coils should be better draining than non-expanded coils which is indeed found experimentally: higher values of ν are observed for a polymer in a thermodynamically good solvent (large coil dimensions, low segment densities) than in a theta solvent (small coil dimensions, high segment densities).

The coil density should also decrease strongly with increasing molar mass. The KR theory thus predicts an increase of ν with increasing molar mass. This prediction contradicts the experimental finding that ν values are practically constant for large ranges of molar masses (Fig. 5-7), regardless of solvent quality. Also, no values of $\nu > 0.6$ have ever been found for *flexible chains in good solvents* although values up to $\nu = 1$ are permitted by the KR theory.

It follows for flexible chains that exponents $v > 1/2$ cannot come from a partial draining of coils. Since $v = 1/2$ is caused by unperturbed coils without excluded volume and since the KR theory expects such coils to have strong hydrodynamic interactions, it is assumed that $v > 1/2$ does not originate from partial draining but exclusively from excluded volume effects (see also Section 5.9.3.). Indeed, experimental values of v never exceeded ca. 3/5 for flexible chains in good solvents (Table 5-3) which agrees with theories of the excluded volume.

The situation is different for rigid chains (Section 5.4.) and rods (Section 5.6.2.). The KR theory delivers for cylindrical rods of constant radius r and length L

$$(7\text{-}8) \qquad D_0 = \frac{k_B T}{3 \pi \eta_1} \cdot \frac{\log_e (L/2r)}{L} \sim L^{-v} \sim M^{-v}$$

The length L of rods is proportional to the aspect ratio $\Lambda = L/(2r)$ and the molar mass M. It follows from the logarithmic term of Eqn.(7-8) that the exponent v varies with the molar mass in the relationships $D_o \sim M^{-v}$ and $D_o \sim \Lambda^{-v}$, respectively. Aspect ratios of $10^2 < \Lambda < 10^3$ deliver $v \approx 0.81$ and aspect ratios of $10^8 < \Lambda < 10^9$ a value of $v \approx 0.96$. An exponent $v = 1$ is expected for rods of infinitely large molar mass ($\Lambda \to \infty$).

7.3. Sedimentation in Solution

7.3.1. Fundamentals

Sedimentation at high gravitational fields is a common method for the separation and characterization of biological macromolecules; it is less often used for synthetic polymers. Such sedimentation experiments are performed with special ultracentrifuges that achieve speeds of up to 150 000 rotations per minute and gravitational fields of up to 900 000 times the earth's gravity ($g \approx 981$ cm s^{-2}). For comparison: a jack rabbitt start of a car corresponds to ca. 1.5 g; astronauts have to suffer ca. 9 g.

Analytical ultracentrifuges possess optical devices which allow one to measure concentrations c and/or concentration gradients dc/dr via the adsorption, interference or refractive indices of polymer solutions.

A particle with a hydrodynamically effective mass m_h experiences a centrifugal force $F_z = m_h \omega^2 r$ if it is at a distance r from the center of rotation and if it is subjected to an angular velocity ω and thus an acceleration of $\omega^2 r$. The centrifugal force F_z is counteracted by the frictional force F_r and the buoyancy force F_b. The frictional force $F_r = \xi_s (dr/dt)$ is proportional to the sedimentation velocity dr/dt; the proportionality constant is the frictional coefficient ξ_s of sedimentation. The sedimentation velocity $dr/dt = s\omega^2 r$ is proportional to the acceleration $\omega^2 r$; the proportionality coefficient s is called the **sedimentation coefficient**. This system-specific quantity is usually reported in **Svedberg units** S = $1 \cdot 10^{-13}$ s.

The buoyancy force $F_b = V_h\rho_1\omega^2 r$ is generated by the solvent of density ρ_1 for particles with hydrodynamically effective volumes V_h. Particles sediment to the bottom of the ultracentrifuge cell if the density of particles is greater than the density of the solvent; they float to the top in the opposite case. In the steady state, one has $F_z = F_r + F_b$ and thus $m_h\omega^2 r = \xi_s(dr/dt) + V_h\rho_1\omega^2 r$.

The *hydrodynamically effective mass* is therefore $m_h = \xi_s[(dr/dt)/(\omega^2 r)] + V_h\rho_1 = \xi_s s + V_h\rho_1$. It is composed of the mass $m_2 = M_2/N_A$ of the "dry" macromolecule and the mass m_s of the solvent that is transported with the macromolecule. Defining a "degree of solvation" as $\Gamma_s \equiv m_s/m_2$, one obtains $m_h = M_2(1 + \Gamma_s)/N_A$.

The *hydrodynamically effective volume* is $V_h = V_2 + V_s$; it is calculated as $V_h = M_2[*v_2 + (\Gamma_s/\rho_1)]/N_A$ (see Appendix A 7-1). The last expression contains the partial specific volume $*v_2$ of the solute which is determined experimentally from $\rho = \rho_1 + (1 - *v_2\rho_1)c_2$ via the concentration dependence of the density of polymer solutions. Introduction of all these expressions into $m_h = \xi_s s + V_h\rho_1$ delivers

$$(7-9) \quad M_2 = \xi_s s N_A/(1 - *v_2\rho_1)$$

Eqn.(7-9) no longer includes explicitly the degree of solvation Γ_s, the hydrodynamic mass m_h, and the hydrodynamic volume V_h. These quantities are however implicitly contained in the frictional coefficient ξ_s which is controlled by particle shape and density. Frictional coefficients cannot be measured directly. In order to obtain an equation for the molar mass which contains only measurable quantities, they are therefore replaced by other quantities, for example, the frictional coefficient ξ_d of diffusion (see below).

Frictional coefficients and sedimentation coefficients both depend on concentration. The molar mass M_2 of Eqn.(7-9) is thus an apparent (concentration-dependent) molar mass M_{app} that has to be extrapolated to zero concentration, usually via $1/M_2 = f(c)$ in analogy to osmotic measurements (Eqn.(6-23)).

7.3.2. Sedimentation Rate

Sedimentation and flotation are opposed by the back diffusion of particles caused by Brownian motion. This back diffusion prevents the formation of a sharp boundary between the solution with sedimenting (or floating) macromolecules and the solvent volume that is already devoid of particles. A bell curve is observed instead of a sharp signal (Fig. 7-3). The sedimentation coefficient $s = (dr/dt)/(\omega^2 r)$ is obtained from the movement dr/dt of the peak. The broadening of the bell curve with time allows one to calculate the diffusion coefficient.

The distribution of sedimentation coefficients of polymolecular polymers can be calculated from the broadening of the bell curves after suitable corrections for the effects of back diffusion. This distribution can in turn be converted into the distribution of molar masses. In order to do so, the effects of concentrations on s have to be eliminated first. Eqn.(7-9) stipulates that sedimentation coefficients s are inversely pro-

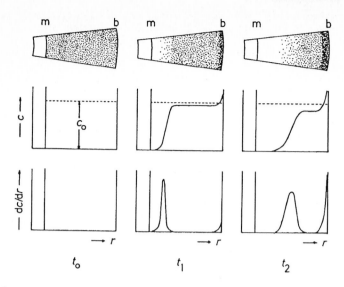

Fig. 7-3 Polymer concentrations c and concentration gradients dc/dr during the ultracentrifugation of particles in sector-shaped cells before the experiment (time t_o) and at times t_1 and t_2. m =
Meniscus, b = bottom of the cell.

portional to frictional coefficients ξ_s at $M = const$. Since frictional coefficients are
proportional to viscosities (see Eqn.(7-3)) and viscosities η increase with the concentration c_2 of the solute, one obtains $1/s \sim \xi_s \sim \eta \sim c_2$ and, to a first approximation, $1/s$
$= (1/s_o)(1 + k_s c_2 + ...)$.

Sedimentation coefficients s_o at infinite dilution deliver molar masses M via $s_o =$
$K_s M^\gamma$ if K_s and γ are known from calibration. This exponential expression results because frictional coefficients in Eqn.(7-9) are controlled by hydrodynamically effective radii according to Eqn.(7-3) which in turn depend on molar masses. The distribution of M can be calculated from the distribution of s_o.

Molar masses can also be obtained without any calibration. Frictional coefficients
ξ_s of sedimentation and ξ_d of diffusion are equal within the limits of experimental
error. Introduction of the Einstein–Sutherland equation (Eqn.(7-2)) into Eqn.(7-9)
delivers the **Svedberg equation** for $c_2 \rightarrow 0$ with $R = k_B N_A$:

$$(7\text{-}10) \quad M_2 = \frac{RT s_o}{D_o(1 - {}^*v_2 \rho_1)} = K_{sd}\, s_o D_o^{-1}$$

Molar mass determinations via Eqn.(7-10) are absolute since all quantities T, s_o,
D_o, $^*v_2$, and ρ_1 can be measured without any calibration. The molar mass average
nevertheless depends on the shape of molecules and their interactions with the
solvent. It is a complicated average for polymolecular solutes since it is the ratio of
two hydrodynamic quantities s_o and D_o that depend on molar masses according to D_o
$= K_D M^\delta$ and $s_o = K_s M^\gamma$, respectively. If s_o and D_o are both obtained as mass-averages,
a so-called **mixed molar mass average** is obtained:

$$(7-11) \quad \overline{M}_{s_wD_w} = \frac{K_{sD}K_s}{K_D} \frac{\sum_i w_i M_i^\gamma}{\sum_i w_i M_i^\delta} \quad ; \quad K_{sD}K_s/K_D = 1 \quad ; \quad \gamma - \delta = 1$$

The sum of the molar mass exponents must always equal unity according to the Π-theorem (left and right sides of Eqn.(7-11) must have the same physical units!). It can be shown that these exponents are related to the exponent α of the intrinsic viscosity–molar mass relationship via $\alpha = 2 - 3\,\gamma = -(1 + 3\,\delta)$.

The combination of mass-averages of sedimentation and diffusion coefficients delivers the number-average molar mass $\overline{M}_n$ for rigid rods since here $\delta = -1$ (Section 7.2.2.) and therefore $\gamma = 0$. For unperturbed coils ($\alpha = 1/2$), however, the mass-averages of s and D deliver a complicated average $\overline{M}_{s_wD_w} = \Sigma_j\, x_j(M_j)^{3/2}/\Sigma_j\, x_j(M_j)^{1/2}$.

If number-average sedimentation coefficients and number-average diffusion coefficients are combined, the number-average molar mass is always obtained, regardless of the molecular shape (i.e., regardless of the value of α). The combination of mass-average sedimentation coefficients and z-average diffusion coefficients, on the other hand, always leads to the mass-average molar mass for any α. More complicated molar mass averages are observed for other averages of s and D.

7.3.3. Sedimentation Equilibrium

Both sedimentation flux J_s and diffusion flux J_d add to the total flux $J = J_s + J_d$ during a sedimentation experiment. The sedimentation flux $J_s \equiv cv_s$ is defined as the product of polymer concentration c and molecular velocity $v_s = s\omega^2 r$. The diffusion flux is $J_d = - D(dc_2/dr)$ (Section 7.2.1.). The total flux J is zero at equilibrium, thus $s/D = (dc/dr)/(c\omega^2 r)$. Introduction of the Svedberg equation (Eqn.(7-10) for $c > 0$ (i.e., s instead of s_0; D instead of D_0) delivers a concentration-dependent apparent molar mass:

$$(7-12) \quad M_{2,app} = \frac{RT}{\omega^2(1 - {}^*v_2\rho_1)} \cdot \frac{dc/dr}{rc} \quad ; \quad 1/M_{2,app} = (1/\overline{M}_w) + Bc + \ldots$$

This equation can be evaluated in various ways. At small fill heights $\Delta r = r_b - r_m$ of the ultracentrifuge cell, the concentration loss between the meniscus r_m and at half of the fill height of the cell, $r = r_m + (\Delta r/2)$, practically equals the concentration gain between r and the bottom of the cell, r_b. The concentration c at distance r is then identical with the initial concentration c_0. One only has to measure the concentration gradient dc/dr at r.

The molar mass $M_{2,app}$ is a mass-average for $c \to 0$. Solving Eqn.(7-12) for dc/dr and summation of the variables delivers $\overline{(dc/dr)} = \omega^2 r(1 - {}^*v_2\rho_1)(RT)^{-1}(\sum_i c_i M_i)$. The sum of the right side of this equation also appears in the definition of the mass-average molar mass (Eqn.(2-6)) since $c = m/V$. This technique thus delivers the mass-average molar mass.

7.4. Viscosity of Solutions and Melts

7.4.1. Introduction

Rheology (G: *rheos* = current, stream, from *rhein* = flow; *logos* = speech, word, reason) tries to describe the temporal relationships between forces and deformations of bodies by so-called **constitutive equations.** Rheology was originally the science of flow of liquids and gases but it now also covers the time dependence of deformation of all matter including solids (Chapter 10). Polymers respond in a very complex way to deformation; no comprehensive constitutive equation has been found for their behavior. One rather describes their rheology by ideal constitutive equations that have been amended empirically by correction factors.

A body responds to a deformation by the formation of stresses and *vice versa.* Nine different stresses can be defined for a body (Fig. 7-4). The stresses σ_{11}, σ_{22} and σ_{33} are perpendicular ("normal") to areas and are thus called **normal stresses.** They are defined as positive for tension and as negative for compression. Usually only two types of deformation are considered: in tension and in shear (Fig. 7-5). Stresses are defined as forces per area attacked: areas $A_1 = HW$ for tensile stress and areas $A_2 = L_oW$ for shear stress. The difference $\sigma_{11}-\sigma_{22}$ is the first normal stress difference; the difference $\sigma_{22}-\sigma_{33}$, the second normal stress difference.

Strains are given by shifts ΔL in length per characteristic dimension: initial length L_o for tensile strain and distance H between boundaries for shear strain. Moduli are ratios of stress to strain in the limit of diminishing strains.

7.4.2. Viscometry

Viscometry characterizes the flow of liquids by their **viscosity**. Three types of viscosities can be distinguished:

- shear viscosity: shear rate = f(shear stress),
- extensional viscosity: extensional rate = f(tensile stress),
- bulk viscosity: deformation rate = f(hydrostatic pressure).

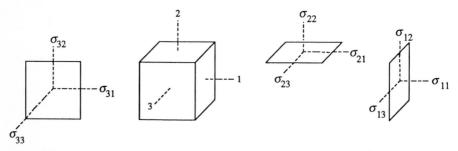

Fig. 7-4 Definition of stresses.

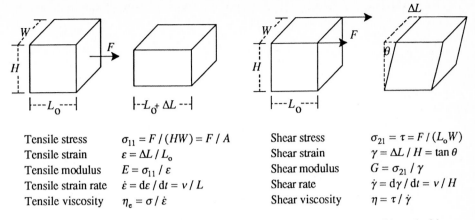

Tensile stress	$\sigma_{11} = F/(HW) = F/A$	Shear stress	$\sigma_{21} = \tau = F/(L_0W)$
Tensile strain	$\varepsilon = \Delta L/L_0$	Shear strain	$\gamma = \Delta L/H = \tan\theta$
Tensile modulus	$E = \sigma_{11}/\varepsilon$	Shear modulus	$G = \sigma_{21}/\gamma$
Tensile strain rate	$\dot{\varepsilon} = d\varepsilon/dt = v/L$	Shear rate	$\dot{\gamma} = d\gamma/dt = v/H$
Tensile viscosity	$\eta_e = \sigma/\dot{\varepsilon}$	Shear viscosity	$\eta = \tau/\dot{\gamma}$

Fig. 7-5 Tensile deformation and shear deformation by forces F and rates caused by velocities v.

Shear viscosities are most commonly investigated; they are thus simply called "viscosities" by most researchers. They are important for the molecular characterization of polymers via intrinsic viscosities of dilute solutions (Section 5.8.), for the application of paints, varnishes, adhesives, etc., as semidilute or concentrated solutions, and for the processing of polymer melts by extrusion, injection molding, or other fabrication methods. **Extensional viscosities** (elongational viscosities) affect the spinning of materials to fibers and the blowing of polymers to films (Section 7.4.6.). Very little is known about **bulk viscosities**.

Coil molecules behave very differently on shearing or extension by a flow field (Fig. 7-6). Shear flow causes coil molecules to undergo rotation in addition to that due to Brownian motion. Tensile flow stretches coil molecules.

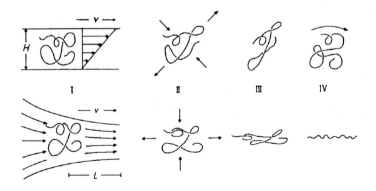

Fig. 7-6 Action of flow fields on coil molecules.
  Top: Shear flow. The shear gradient (I) generates principal stresses (II) which cause the molecule to expand in the direction of the main principal stress (III). The gradient forces the molecule to rotate (IV), allowing it to return to the macroconformation of a random coil. See also Fig. 7-7.
  Bottom: Tensile flow. The drawing action reduces the cross-sectional area of the specimen (I). Principal stresses (II) force the molecule to expand in the direction of the main principal stress (III). On further drawing, the molecule becomes completely stretched (IV).

A body with the three types i = 1, 2, and 3 of surface areas is by definition sheared in the 2-1 direction. This shear requires a **shear(ing) stress** $\sigma_{21} \equiv \tau$. The ratio of the first normal stress difference $\sigma_{11} - \sigma_{22}$ to the shear stress is the **elastic shear deformation** $\gamma = (\sigma_{11} - \sigma_{22})/\sigma_{21}$. The ratio of shear stress to elastic shear deformation is known as the **shear modulus**

$$(7\text{-}13) \quad G = \tau/\gamma = \sigma_{21}/\gamma = \sigma_{21}{}^2/(\sigma_{11} - \sigma_{22})$$

The **dynamic viscosity** η is defined as the ratio of shear stress σ_{21} to shear rate $\dot{\gamma}$ (sometimes called "absolute viscosity" but usually only "viscosity"); it thus equals the product of the shear modulus G and time t:

$$(7\text{-}14) \quad \eta = \tau/\dot{\gamma} = \sigma_{21}/\dot{\gamma} = G\gamma/\dot{\gamma} = Gt$$

The inverse of the dynamic viscosity is the **fluidity** $1/\eta$ and the ratio of dynamic viscosity η to density ρ the **kinematic viscosity** $v = \eta/\rho$.

Viscosities of melts and concentrated polymer solutions are very high. For example, air has a viscosiy of ca. 10^{-5} Pa s, water one of ca. 10^{-3} Pa s, and glycerol one of ca. 1 Pa s. Polymer melts possess viscosities of $(10^2\text{-}10^6)$ Pa s, pitch ca. 10^9 Pa s, and silicate glass ca. 10^{21} Pa s.

The large range of viscosities requires very different viscometers for various types of matter. Very large viscosities can be measured with *band viscometers* where an "infinitely long" plane band runs with a velocity v_{band} through the liquid which is contained between two "infinitely long" parallel plates. The distance between the plates is $2 R$. The liquid remains at rest near each surface of the plate but moves with the same velocity v_{band} as the band near each surface of the band. The velocity v of the liquid thus changes linearly with the distance y from the band from $v = v_{band}$ at $R = 0$ to $v = 0$ at $R = R$. The **velocity gradient** (or **shear rate**) $\dot{\gamma} = dv/dR$ is constant.

Such constant shear rates are highly desirable for the evaluation of rheological measurements. They can also be obtained with *rotational viscometers* where a rotor rotates in or around a stator (Fig. 7-7). Rotational viscometers have constant shear rates of ca. $(0.1\text{-}10^3)$ s^{-1} if the gap between rotor and stator is very narrow.

A cone rotates on a plate in *cone-and-plate viscometers*. For small angles between cone and plate, constant velocity gradients of $(10^{-3}\text{-}10^2)$ s^{-1} are obtained.

Capillary viscometers measure the flow time of a defined liquid volume. They consist of glass capillaries for dilute solutions and of glass or steel capillaries in a pressure chamber for melts and concentrated solutions. These viscometers generate hyperbolic velocity gradients (Fig. 7-7); they are usually $(1\text{-}10^5)$ s^{-1}.

Plastics are characterized industrially by their **melt flow**. The **melt flow index** *MFI* measures the mass of polymer in grams that is extruded at constant temperature T in 10 minutes from a standard plastometer under a standard load m. Division of *MFI* by the density of the polymer melt delivers the **melt volume index** *MVI*. Both *MFI* and *MVI* measure (shear-dependent) fluidities; they are lower, the higher the molar mass of the polymer.

The rheological behavior of elastomers (and sometimes also polymer melts) is characterized industrially by their **Mooney viscosities**. A polymer is deformed with constant rate at constant temperature in a standardized cone-and-plate viscometer. After a certain time, the elastic recovery is measured. Mooney viscosities thus determine elasticities, not viscosities.

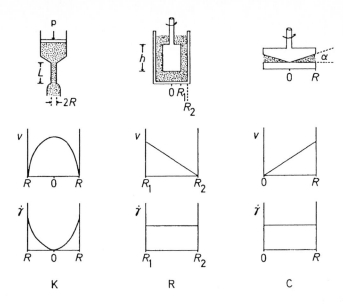

Fig. 7-7   Three types of viscometers and the variation of rates v and velocity gradients $\dot{\gamma} = dv/dR$ with the distance R, R_1 and R_2, respectively. K = Capillary viscometer, R = rotational viscometer (shear), C = cone-and-plate viscometer. The shear flow is a drag flow in K and a pressure flow in R and C.

7.4.3.   Newtonian   Viscosities

The viscosity $\eta = \sigma_{21}/\dot{\gamma}$ (Eqn.(7-14)) of a Newtonian liquid is by definition independent of the shear rate $\dot{\gamma}$ (**Newton's law**). This **Newtonian viscosity** $\eta_0 = G_0(\gamma/\dot{\gamma})$ $\neq f(\dot{\gamma})$ is also called **zero-shear viscosity, viscosity at rest,** or **stationary viscosity.** The **shear modulus at rest,** G_0, is independent of the deformation. η_0 and G_0 are thus true material constants.

Newtonian viscosities of *dilute solutions* are mainly used to characterize polymers via their intrinsic viscosities (Section 5.8.) which can be considered as the first viscosimetric virial coefficients if Eqn.(5-33) is written as $\eta = \eta_1[1 + [\eta]c + k_H[\eta]^2c^2 + ...]$. Like osmotic virial coefficients (Eqn.(6-23)), they are a measure of the space requirements of polymer molecules (see Eqn.(6-30)), i.e., hydrodynamic volumes, which in turn are related to molar masses (Eqn.(5-47)). The molar mass dependence of intrinsic viscosities is usually expressed by the KMHS equation $[\eta] = K_\eta \overline{M}_\eta{}^\alpha$.

Newtonian viscosities of *concentrated solutions* of non-electrolyte polymers increase strongly at higher concentrations (Fig. 5-17). This behavior is again similar to that of osmotic pressures (see Fig. 6-10) and can also be explained by the overlap of coil molecules. At low concentrations, one can write Eqn.(5-33) as $(\eta/\eta_1)-1 \sim [\eta]c$. A plot of $\log_{10}[(\eta/\eta_1)-1]$ *vs.* $c[\eta]$ should thus exhibit a slope of unity (Fig. 7-8).

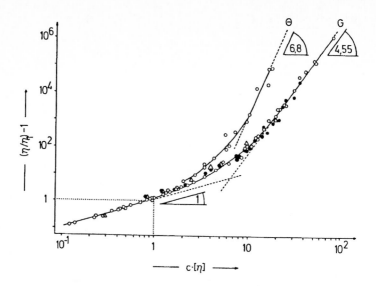

Fig. 7-8   Decadic logarithm of relative viscosity increments $(\eta/\eta_1) - 1 \equiv \eta_{sp}$ as a function of the decadic logarithm of $c[\eta]$ for polymers in good (G) and theta solvents (Θ).
  o Poly(styrene)s in trans-decalin (Θ) and toluene (G) at 25°C [4].
  ● cis-1,4-Poly(isoprene)s in toluene at 34°C (G) [5].
  ∆ Hyaluronates in water at 25°C (G) [6].

Coils start to overlap at a critical concentration $c_\eta^* = c[\eta]$ (Section 6.1.2.). At concentrations $c > c_\eta^*$, one observes $\eta_{sp} \sim (c[\eta])^q$ with q > 1 (Fig. 7-8). Since $[\eta]$ can be expressed by $[\eta] = K_\eta \overline{M}_\eta^{\,\alpha}$, one can also write $\eta_{sp} \sim c^q \overline{M}_\eta^{\,\alpha q}$. The relative viscosity increment of highly concentrated solutions is approximately $(\eta/\eta_1) - 1 \approx \eta/\eta_1$. Highly concentrated solutions also act similarly to melts. The molar mass dependence of Newtonian viscosity of melts can be expressed by $\eta_0 = K_v \overline{M}_w^{\,\varepsilon}$ (see below) where ε = 3.4 for the high molar mass regime. It follows that αq = 3.4. Since coil molecules in melts are unperturbed (Section 8.2.) and unperturbed coils exhibit an exponent α = 1/2 (see Section 5.8.3.), exponent q should become 6.8 for chain molecules in theta solvents (Fig. 7-8). The exponent α becomes 0.764 for coils in good solvents according to renormalization theory (Section 5.9.3.) and q adopts a value of q = 4.45 (experiment: 4.55). The q values for theta and good solvents are universal and not dependent on the specific polymer–solvent–temperature system.

 Melt viscosities also show a power dependence on molar masses similar to the KMHS equation for intrinsic viscosities:

$$(7\text{-}15)\quad \eta_0 = K_v \overline{M}_w^{\,\varepsilon}$$

albeit with two different molar mass regimes (Fig. 7-9). In the low molar mass regime, exponents ε are usually unity (but see also Fig. 7-16) whereas exponents ε adopt values of ca. 3.4 at higher molar masses (Fig. 7-9). Exponents of $\varepsilon \approx 3.4$ are thought to be due to reptations of chains (Section 7.5.1.).

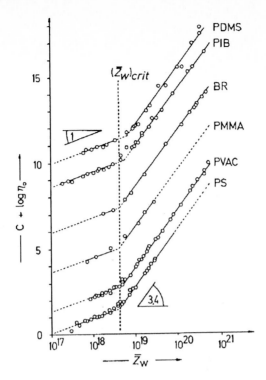

Fig. 7-9 Dependence of the decadic logarithms of melt viscosities η_o on the decadic logarithms of the characteristic parameter $\overline{Z}_w = [\langle s^2 \rangle / (\overline{M}_w v_2)]N_c$ where s = radius of gyration, v_2 = specific volume of polymer, and N_c = number of chain atoms. $\langle s^2 \rangle / \overline{M}_w$ is a constant for each polymer (Section 5.5.2) and N_c is proportional to the degree of polymerization and thus to the molar mass of polymers. Z is therefore proportional to the molar mass. A different numerical constant C was added to the $\log_{10} \eta_o$ values of each polymer for clarity. Physical units of η_o and $\overline{Z}_w$ are missing because they were not reported by the author [7].

BR = 1,4-cis-poly(butadiene), PDMS = poly(dimethylsiloxane), PIB = poly(isobutylene), PMMA = poly(methyl methacrylate), PS = poly(styrene), PVAC = poly(vinyl acetate).

The temperature dependence of Newtonian melt viscosities can be described by a Williams–Landel–Ferry (WLF) equation (Section 9.6.2.) if the shift factor a_T is expressed by the ratio of viscosity η at the temperature of the experiment to η_0 at a reference temperature. This reference temperature is usually chosen as $T_0 = T_{VT} + 50$ K with the Vicat temperature T_{VT} (Section 9.2.1.) as the "isoviscous" temperature. The resulting master curve indicates the change of viscosity with temperature (Fig. 7-10).

7.4.4. Non-Newtonian Viscosities

The shear stress σ_{21} and the dynamic viscosity η of non-Newtonian fluids change with shear rate (Fig. 7-11). **Bingham bodies (plastic bodies)** show a yield value; they require a minimum of shear stress $\sigma_{21,min}$ before the body begins to flow, i.e., they show **plasticity** (G: *plastein* = to form). Above $\sigma_{21,min}$, Bingham bodies behave as

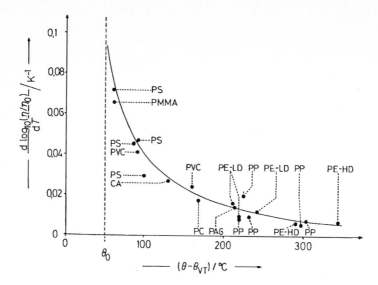

Fig. 7-10 Change of decadic logarithms of viscosity ratios η/η_0 with temperature T as a function of temperature differences $\theta - \theta_{VT}$. The solid line indicates the differentiated WLF equation: $d\log_{10}(\eta/\eta_0)/dT = -KK'/[K' + (T-T_0)]^2$. With permission by G.Thieme Publ. [8].

either Newtonian fluids (ideal Bingham bodies) or non-Newtonian fluids (pseudo-ideal Bingham bodies). An example of a Bingham body is tomato ketchup which needs a good whack before it flows from the bottle.

Most polymer melts behave as Newtonian fluids at small shear rates but then exhibit decreases in viscosity with increasing shear rate. This **shear-thinning** is sometimes called **pseudo-plasticity** because of the similarity to a Bingham body although there is no yield value. At very high shear rates, a second Newtonian regime is sometimes observed for concentrated solutions; there are indications, however, that such concentrated solutions are in reality dispersions caused by demixing or that the polymers have degraded. No second Newtonian regime has been discovered for melts.

Shear-thickening liquids are characterized by a stronger than proportional increase of shear stresses with increasing shear rates (Fig. 7-11). This **dilatancy** (L: *dilatare* = expand, enlarge) causes viscosities to rise with shear rates. Shear-thickening is observed for melts of ionomers and for ionically stabilized dispersions.

The viscosity of Newtonian, shear-thinning, and shear-thickening fluids is "immediately" attained on application of a velocity gradient. It also does not change with time as some other non-Newtonian fluids do. In contrast, the viscosity of **thixotropic materials** decreases with time at constant shear rate (G: *thixis* = movement; *tropos* = change) whereas that of **antithixotropic** (or **rheopectic**) materials increases (G: *rheos* = flow; *pektos* = coagulated, hardened).

The rheological behavior of matter can also be complicated by wall effects. On application of shear stresses, some dispersions and gels exude liquids which act as lubricants and cause a **plug flow**. An example is toothpaste. Other systems show turbulence which starts at lower Reynolds numbers than for Newtonian liquids.

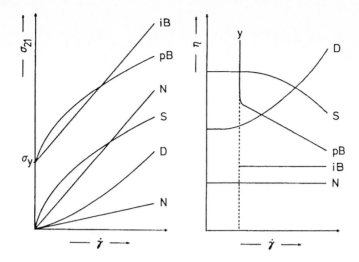

Fig. 7-11 Dependence of shear stress σ_{21} and dynamic shear viscosity η on the shear rate $\dot\gamma$ for Newtonian (N), shear-thickening (D), and shear-thinning (S) liquids and for ideal (iB) and pseudo-plastic (pB) Bingham bodies. σ_y = Yield value.

7.4.5. Flow Curves

Plots of $\sigma_{21} = f(\dot\gamma)$, $\dot\gamma = f(\sigma_{21})$, $\log_{10}\eta = f(\log_{10}\sigma_{21})$, $\eta = f(\dot\gamma)$, etc., are called **flow curves**. For shear-thinning fluids, these curves often show three regimes which are separated by more or less sharp transition regions: (I) Newtonian behavior at small shear rates, (II) shear-thinning at higher shear rates, and (sometimes) (III) a second Newtonian range for concentrated solutions at very high shear rates (very likely an artefact caused by turbulence and/or shear induced polymer degradation).

Very many empirical, semi-empirical and theoretical functions have been proposed for the non-Newtonian regime II. This regime is often described by the **Ost-wald-de Waele power law** which can be written for $\eta = f(\dot\gamma)$, etc., or $\dot\gamma = f(\sigma_{21})$ as:

$$(7\text{-}16) \quad \dot\gamma = K\sigma_{21}{}^{m}$$

with the empirical and system-specific constants K and m. The exponent m is called the **flow exponent** or **pseudo-plasticity index**. Eqn.(7-16) is sometimes given as a power law $\sigma_{21} = K'\dot\gamma^{n}$; the empirical constant K' is here known as the **consistency index** and the exponent n as the **power law index**.

Polymer melts that show strong shear-thinning are more easily processed than Newtonian liquids. The lower the viscosity, the smaller is the internal friction of the melt. Less energy has to be applied in order to maintain the processing temperature. Polymer melts are therefore processed at the highest shear rates that can be achieved by the chosen processing method (Chapter 14) and that do not adversely affect polymer properties through thermal and mechanical degradation of the polymers.

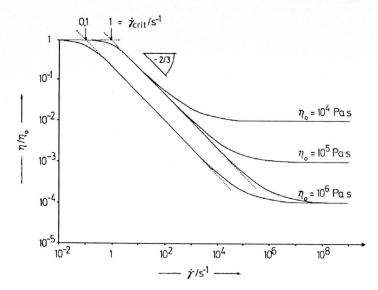

Fig. 7-12   Double logarithmic plot of the shear-rate dependence of normalized shear viscosities η/η_0 where η_0 = viscosity at rest. Calculations with the Carreau equation and $\eta_\infty = 10^2$ Pa s; $\eta_0 =$ 10^4 Pa s, 10^5 Pa s, or 10^6 Pa s; $\lambda = 1$ s ($\dot\gamma_{crit} = 1$ s^{-1}) or $\lambda = 10$ s ($\dot\gamma_{crit} = 0.1$ s^{-1}); $n = 1/3$ [9].

The **Carreau function** seems to cover all three regimes for shear-thinning liquids:

$$(7\text{-}17)\quad \eta = \eta_\infty + (\eta_0 - \eta_\infty)[1+(\lambda\dot\gamma)^2]^{(n-1)/2}$$

where η = viscosity, η_0 = first Newtonian viscosity, η_∞ = second Newtonian viscosity, $\dot\gamma$ = shear rate, and λ, n = system-specific constants (Fig. 7-12). This function is used by the CAMPUS® system (Section 10.1.).

The constant λ represents an inverse critical shear rate $\lambda = 1/\dot\gamma_{crit}$. Introducing the general constants ab instead of 2 and $1/b$ instead of $(n-1)/2$, one can also write Eqn.(7-17) for $\eta_\infty \to 0$ as:

$$(7\text{-}18)\quad \frac{\eta}{\eta_0} = \left[1+\left(\frac{\dot\gamma}{\dot\gamma_{crit}}\right)^{ab}\right]^{1/b}$$

In Vinogradov–Malkin diagrams, $\log_{10}(\eta/\eta_0)$ is plotted against $\log_{10}\eta_0\dot\gamma$ (Fig. 7-13). It is interesting that three very different systems show almost the same critical shear stress $(\sigma_{21})_{crit} \equiv \eta_0\dot\gamma_{crit} \approx 10^4$ Pa and practically the same exponents $a \approx -2/3$ and $b \approx -3$ (i.e., $n = 1/3$).

The temperature dependence of viscosities is described in the CAMPUS® system by either an Arrhenius type of equation, $\eta = A\dot\gamma^B \exp(C/T)$, or by a combined Arrhenius–WLF equation with five adjustable constants $K_1, K_2, K_3, K_4,$ and K_5:

$$(7\text{-}19)\quad \eta = \frac{K_1 a_T}{(1+K_2\dot\gamma a_T)^{K_3}};\quad \log_{10} a_T = \frac{8.86(K_4-K_5)}{101.6+K_4-K_5} - \frac{8.86(T-K_5)}{101.6+T-K_5}$$

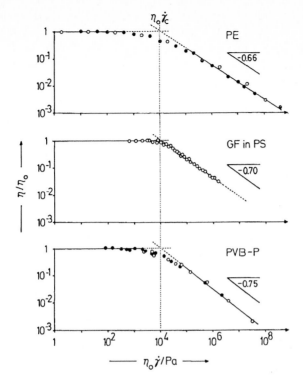

Fig. 7-13   Double logarithmic plot of normalized shear viscosities η/η_0 as a function of shear stresses $\sigma_{21} \equiv \eta_0 \dot{\gamma}$ for various polymers [10].
  Top: Poly(ethylene)s PE with different molar masses at 150°C [11];
  Center: Glass-fiber-filled poly(styrene) (GF in PS) [12];
  Bottom: Plasticized poly(vinyl butyral) at 125°C [13].

7.4.6. Extensional Viscosities

Long threads can be drawn from melts and concentrated solutions of polymers. The extensibility of these liquids allows the spinning of fibers from melts and solutions, the extrusion blowing of hollow bodies, and the stretching of films. Threads can also be drawn from honey and concentrated soap solutions because these materials form physical molecules by intermolecular association. These threads, however, are not stable because of large bond fluctuations.

Extensibilities are characterized by **extensional viscosities (elongational viscosities)** $\eta_e = \sigma_{11}/\dot{\varepsilon}$ which are defined as the ratio of tensile stress σ_{11} in the draw direction to the elongational rate, $\dot{\varepsilon} = d\varepsilon_H/dt$, as change of the Hencky strain ε_H with time. The Hencky strain $\varepsilon_H = \log_e(L/L_0)$ is calculated from length L and initial length L_0 (see also Section 10.2.2.). Contrary to shear viscosities, the type of deformation must always be specified for extensional viscosities (uniaxial, biaxial, etc.). Uniaxial extensional viscosities are important for fiber spinning whereas biaxial extensional viscosities control blow molding, film blowing, and vacuum forming.

Shear and extensional viscosities show different dependencies on deformation rates $\dot{q}$. Both types of viscosities are independent of the deformation rate at small deformation rates (Fig. 7-14). In this range, extensional viscosities η_e are three times (uniaxial) and six times (biaxial) as large as shear viscosities η_s. The extensional viscosity in the uniaxial case $\eta_e = 3\,\eta_s$ is also known as the **Trouton viscosity**.

Shear viscosities decrease with increasing shear rate for shear-thinning fluids whereas extensional viscosities may run through maxima with increasing elongational rate (Fig. 7-14).

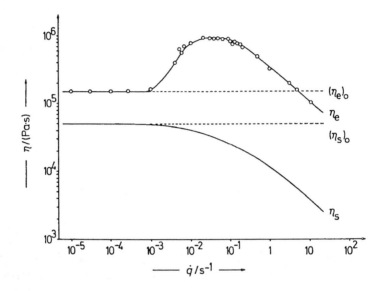

Fig. 7-14 Double logarithmic plot of the shear viscosity η_s as a function of the shear rate $\dot{q} = \dot{\gamma}$ and the uniaxial extensional viscosity as a function of the uniaxial elongational rate $\dot{q} = \dot{\varepsilon}$ for a branched poly(ethylene) at 150°C. With permission by Steinkopff Publ. [14].

7.5. Molecular Dynamics of Melts

7.5.1. Diffusion

Motions of polymer segments are so small in solid polymers that whole polymer chains cannot move at all. Only in melts is the mobility of chain molecules large enough to be measured experimentally. This **self-diffusion** of chains is caused by Brownian motion; it does not involve a net transport of mass.

In melts, flexible chain molecules are present as unperturbed coils that are filled with segments of other coils (Section 8.2.). The diffusion of segments in such an environment has to proceed via interchange of sites with other segments. This self-diffusion can be measured by NMR spectroscopy. The self-diffusion coefficient of a chain of N_{seg} segments with a frictional coefficient ξ_{seg} per segment is $D_2 = k_BT/(N_{seg}\xi_{seg})$ according to Eqn.(7-5). Self-diffusion coefficients are very small, for example, $D_2 \approx 10^{-18}$ cm^2/s for poly(styrene)s of molar mass $M \approx 10^6$ g/mol.

The mutual interpenetration of coil molecules (Section 5.7.2.) creates a physical network through which a chain (the test chain) has to diffuse like a snake through brush (Fig. 7-15). This **reptation** of the test chain proceeds in a kind of tube formed by segments of other chains according to the **Doi–Edwards theory**. The length L_{tub} = $N_{seg}L_{seg}$ of the tube is given by the number N_{seg} of segments of length L_{seg}.

Fig. 7-15   Reptation of a test chain (black) through a tube with "walls" - - - - formed by segments of other chains (white). The primitive chain is indicated by — — —.

The test chain needs a reptation time t_{rep} to find its way through the tube of length L_{tub}. Replacing the general diffusion law $D = \langle x^2 \rangle/(2\ t)$ by $D_2 = (L_{tub}^2)/(2\ t_{rep})$, solving for t_{rep}, and inserting $D_2 = k_BT/(N_{seg}\xi_{seg})$ and $L_{tub} = N_{seg}L_{seg}$ results in the reptation time $t_{rep} = N_{seg}^3L_{seg}^2f_{seg}/(2\ k_BT)$.

The center line of the tube is called the primitive chain; it is assumed to have the same radius of gyration as the test chain. The self-diffusion coefficient thus becomes $D_2 = \langle s^2 \rangle_o/(2\ t_{rep}) = k_BT\langle s^2 \rangle_o/(N_{seg}^3L_{seg}^2\xi_{seg})$. Since coils in melts are unperturbed, one can furthermore write $N_{seg}L_{seg}^2 = \langle r^2 \rangle_o = 6\ \langle s^2 \rangle_o = 6\ K_{s,o}^2M$ (Sections 5.3.2. and 5.3.3.). The self-diffusion coefficient of flexible linear chain molecules in melts should thus decrease with the square of the molar mass (Fig. 7-16):

$$(7\text{-}20) \quad D_2 = \frac{k_BT\langle s^2 \rangle_o}{N_{seg}^3 L_{seg}^2 \xi_{seg}} = \frac{k_BT}{6\,\xi_{seg}} M_{seg}^2 \cdot \frac{1}{M^2} = K_{rep}M^{-2}$$

7.5.2. Viscosity

Viscosities are much more difficult to calculate theoretically than self-diffusion coefficients. In diffusion, one considers the motion of a particle in a resting liquid whereas in viscosity, one has to calculate the motion of a particle in a moving liquid. In melts, this liquid is composed of segments of other polymer molecules. The segments try to move in unison if a shear stress is applied. They are prevented from doing so if the chains are long because molecules are entangled (Section 5.7.2.). The entanglements behave like physical cross-links: the viscosity of melts rises dramatically and the melts behave more like rubbers than liquids.

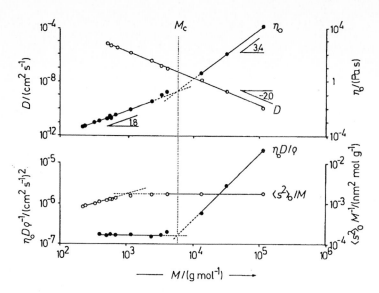

Fig. 7-16 Dependence of self-diffusion coefficients D [15], melt viscosities η_o at rest [15], reduced unperturbed radii of gyration, $\langle s^2 \rangle_o/M$, calculated by the RIS method [16], and the quantity $\eta_o D/\rho$ on the molar mass of alkanes and narrowly distributed linear poly(ethylene)s at 175°C [17]. M_c = critical molar mass between entanglements; ρ = density of the melt.

The different viscosimetric response of nonentangled and entangled chains leads to two regimes of the molar mass dependence of Newtonian viscosities (Fig. 7-9): the Rouse regime for low molar masses and the reptation regime for high molar masses (see below). Melt viscosities of alkanes $H(CH_2)_nH$ with $n < 71$ ($\overline{M}_w < 1000$ g/mol) show $\eta_o \sim \overline{M}_w^{1.8}$ while those of poly(ethylene)s with $n > 280$ ($\overline{M}_w > 4000$ g/mol) increase with $\overline{M}_w^{3.4}$ (Fig. 7-16). The critical molar mass for the onset of entanglements is given by the intersection of these two functions and found to be $M_c \approx 3700$ g/mol. No change of slope is found for the molar mass dependence of self-diffusion coefficients because the entangled chains behave like static obstacles for the very slowly self-diffusing chains.

Critical molar masses for entanglements are not only observed for Newtonian viscosities but also for other dynamic phenomena, for example, the frequency dependence of shear moduli (Section 10.5.7.). Molar masses M_c from melt viscosities are more than twice as large as critical molar masses M_e from shear moduli (Table 7-1).

Poly(ethylene) chains of low molar mass ($M_c < 1000$ g/mol) are too short for entanglements. If the chains are not very short, however, they can still form unperturbed coils since a poly(ethylene) molecule $H(CH_2)_nH$ of molar mass 1404.7 g/mol possesses $N_c = 100$ chain units and thus sufficient segments for the formation of unperturbed coils (Appendix A 5-1). The ratio $\langle s^2 \rangle_o/M$ thus becomes independent of M for $1000 < M/(\text{g mol}^{-1}) < 3700$ g/mol (Fig. 7-14).

In this molar mass range, chains can be modeled as hydrodynamically equivalent spheres with radii of $R_d = \langle s^2 \rangle_o^{1/2}$. Their self-diffusion through the melt as "solvent" with a viscosity of η_o can be described by Eqn.(7-4): $D = RT/(6 \pi N_A \eta_o \langle s^2 \rangle_o^{1/2})$.

Table 7-1 Average molar masses M_u of monomeric units, critical molar masses M_c from the molar mass dependence of Newtonian melt viscosities, entanglement molar masses M_e from plateau moduli (Section 10.5.7.), and critical chain unit numbers $N_c = M_c/M_u$ from melt viscosities at rest.

Polymer	$\dfrac{M_u}{\text{g mol}^{-1}}$	$\dfrac{T}{°C}$	$\dfrac{M_c}{\text{g mol}^{-1}}$	$\dfrac{M_e}{\text{g mol}^{-1}}$	$\dfrac{M_c}{M_e}$	N_c
Poly(ethylene)	28	140	-	940	-	-
Poly(oxyethylene)	44	140	-	2 280	-	-
		25	5 800	-	-	-
Poly(isobutylene)	54	25	15 200	5 690	2.67	203
Poly(dimethylsiloxane)	74	25	24 500	9 610	2.55	662
		140	-	12 300	-	-
Poly(α-methyl styrene)	118	100	28 000	12 800	2.19	237
Poly(styrene), atactic	104	140	-	13 300	-	-
		190	35 000	13 600	2.57	673
		270	-	14 000		

Multiplication of the right side of this equation by $\langle s^2 \rangle_o^{1/2}/\langle s^2 \rangle_o^{1/2}$ and introduction of the molecule volume $V_{mol} = 4\,\pi\,\langle s^2 \rangle_o^{3/2}/3$, the density $\rho = m_{mol}/V_{mol}$ of the melt, and the molar mass $M = m_{mol}N_A$ of the polymer delivers:

$$(7\text{-}21) \qquad \frac{D\eta_o}{\rho} = \frac{2}{9} RT \frac{\langle s^2 \rangle_o}{M}$$

The ratio $\eta_o D/\rho$ should be independent of the molar mass as long as entanglements are absent (Fig. 7-14). All terms of this equation can be determined experimentally; for poly(ethylene), the left side of Eqn.(7-21), $\eta_o D/\rho = (1.68\pm0.06)\cdot10^{-7}$ $(\text{cm}^2\ \text{s}^{-1})^2$, agrees excellently with the right side, $(2/9)\ RT\langle s^2 \rangle_o/M = 1.67\cdot10^{-7}$ $(\text{cm}^2\ \text{s}^{-1})^2$.

Eqn.(7-21) was obtained by a quasistatic derivation. It agrees with the result of hydrodynamic theories except for a numerical factor of $2/9 \approx 0.222$ instead of $1/6 \approx 0.167$ which is caused by a factor of 4/3 due to the assumption of equivalent spheres.

Hydrodynamic theories have been developed for both the low molar mass regime at $M < M_c$ (Rouse regime, see Section 7.2.3.) and the high molar mass regime at $M > M_c$ (reptation regime, see Section 7.5.1.). Theories for both regimes assume that hydrodynamic properties P_h can always be described by a local quantity f_h and a global quantity F_h. The local quantity is the frictional coefficient ξ_{seg} of a segment. The global quantity depends on the macroconformation. The general expressions are $D = k_B T F_d/\xi_{seg}$ for self-diffusion and $\eta_o = \xi_{seg}F_\eta$ for zero-shear viscosity (Table 7-2).

The global factors F_d and F_η differ for Rouse and reptation regimes. For *self-diffusion*, the Rouse regime predicts $F_d = 1/N_{seg}$ (Eqn.(7-5)) and thus $D \sim M^{-1}$ whereas reptation theory leads to $F_d = 1/(6\ N_{seg}^2)$, using Eqn.(7-20) and $M = N_{seg}M_{seg}$, and thus $D \sim M^{-2}$. Experimentally, an exponent of -2 is found for the whole molar mass range of poly(ethylene) (Fig. 7-14) and other polymers which agrees with the prediction of reptation theory.

For *Newtonian melt viscosities*, Rouse theory predicts the global factor to be $F_\eta = (1/6)\, \rho N_A(\langle s^2\rangle_o/M)N_{\text{seg}}$, i.e., $\eta_o \sim M^{+1}$. An exponent of +1 is indeed found for most polymers (see Fig. 7-9). The deviating exponent of +1.8 for poly(ethylene) melts is caused by a strong dependence of melt densities on molar masses in the low molar mass range (cf. Fig. 7-16).

Table 7-2   Self-diffusion coefficients D and viscosities at rest, η_o, in the Rouse and reptation regimes of melts. $v_o = \eta_o/\rho$ = kinematic viscosity; ρ = density of the melt. See text.

	Self-diffusion	Viscosities at rest	$D\eta_o/\rho$
General	$D = \dfrac{k_B T}{\xi_{\text{seg}}}\cdot F_d$	$\eta_o = \xi_{\text{seg}}\cdot F_\eta$	Dv_o
Rouse	$F_d = \dfrac{1}{N_{\text{seg}}}$	$F_\eta = \dfrac{1}{6}\rho N_A \dfrac{\langle s^2\rangle_o}{M}N_{\text{seg}}$	$Dv_o = \dfrac{1}{6}RT\dfrac{\langle s^2\rangle_o}{M}$
Reptation	$F_d = \dfrac{1}{6\,N_{\text{seg}}^2}$	$F_\eta = 6\,N_A\rho\dfrac{M_{\text{seg}}}{M_c}\left(\dfrac{\langle s^2\rangle_o}{M}\right)N_{\text{seg}}^3$	$Dv_o = RT\dfrac{M_{\text{seg}}}{M_c}\left(\dfrac{\langle s^2\rangle_o}{M}\right)N_{\text{seg}}$

Reptation theory starts with Eqn.(7-14) which gives the melt viscosity $\eta_o = Gt$ as the product of shear modulus G and time t. The time t is identified by reptation theory as the reptation time t_{rep} which the test chain needs to wind itself out of the tube. This time is given by $t_{\text{rep}} = N_{\text{seg}}^3 L_{\text{seg}}^2 \xi_{\text{seg}}/(2\,k_B T)$ (Section 7.5.1). The shear modulus is taken from the dynamic entanglement theory as $G = (4/5)\,RT\rho/M_e$ (see Eqn.(10-33)), where M_e = critical molar mass for entanglements from shear moduli. Inserting $R = k_B N_A$, $L_{\text{seg}}^2 = 6\,\langle s^2\rangle_o/N_{\text{seg}}$, and $N_{\text{seg}} = M/M_{\text{seg}}$, one obtains by setting $M_e \equiv (10/4)\,M_c$ (Table 7-1):

$$(7\text{-}22)\quad \eta_o = 6\,N_A\rho(M_{\text{seg}}/M_c)(\langle s^2\rangle_o/M)\xi_{\text{seg}}N_{\text{seg}}^3\;;\quad \eta_o \sim M^3$$

Newtonian melt viscosities are thus predicted to be proportional to the third power of molar masses ($\eta_o \sim M^3$) since $N_{\text{seg}} \sim M$ and $\langle s^2\rangle_o/M = const$.

Experimentally, a molar mass dependence of $\eta_o \sim M^{3.4}$ is obtained for all polymers with $M > M_c$. The reason for the difference between experiment (exponent ca. 3.4) and theory (exponent 3.0) is unclear. It may be due either to an additional effect that was not considered by theory (such as a "breathing" of the tube) or an experimental shortcoming (such as an insufficiently high molar mass range).

Reptation theory thus predicts that tube segments exert a resistance to the moving molecule: the viscosity is increased ($\eta_o \sim M^3$ instead of $\eta_o \sim M$) and the diffusion slowed ($D \sim M^{-2}$ instead of $D \sim M^{-1}$) as compared to the Rouse regime for nonentangled chains.

Appendix to Chapter 7

A 7-1: Hydrodynamically Effective Volume

The hydrodynamically effective volume V_h is the sum of the volume $V_2 = v_2 m_2$ of the dry macromolecule and the volume $V_s = v_s m_s$ of the solvating liquid where v_2 and v_s are the specific volumes and m_2 and m_s the masses of macromolecule 2 and solvent shell s. Introducing $m_2 N_A = M_2$ and $\Gamma_s \equiv m_s/m_2$, one obtains $V_h = V_2 + V_s = v_2 m_2 + v_s m_s = M_2(v_2 + \Gamma_s v_s)/N_A$.

The specific volume v_s of the liquid in the solvent shell of the macromolecule differs from the specific volume v_1 of the pure liquid 1 because of the interaction of the solvating liquid with the polymer. It is calculated as follows.

The total volume $V = V_2 + V_s + V_1$ of the solution is comprised of the volume V_2 of the dry macromolecules, the volume V_s of the bound solvent molecules, and the volume V_1 of the free solvent molecules: $V = m_2 v_2 + m_s v_s + (m_1 - m_s)v_1 = m_2 v_2 + m_1 v_1 + \Gamma_s m_2(v_s - v_1)$ where m_1 = total mass of all solvent molecules. Many more solvent molecules are present in dilute solutions than are needed for the solvation of the macromolecules. The degree of solvation, Γ_s, will thus be independent of the concentration. Differentiation of V with respect to the mass m_2 delivers the partial specific volume of the solute: $*v_2 = (\partial V/\partial m_2)_{p,T,m_1} = v_2 + \Gamma_s(v_s - v_1)$. The specific volume v_1 of the free solvent is by definition the inverse of the density ρ_1. Introduction of $\Gamma_s v_s = *v_2 - v_2 + (\Gamma_s/\rho_1)$ into the expression for the hydrodynamically effective volume V_h delivers $V_h = M_2[*v_2 + (\Gamma_s/\rho_1)]/N_A$.

Literature

7.1. INTRODUCTION
H.-G. Elias, Makromoleküle (in German), Hüthig and Wepf, Basel, Vol. I (1990): Chapter 23 (Diffusion and Permeation), Chapter 24 (Viscosity)
P.-G. de Gennes, Scaling Concepts in Polymer Physics, Cornell University Press, Ithaca (NY) 1979
M.Doi, S.F.Edwards, The Theory of Polymer Dynamics, Oxford University Press, Oxford 1987
R.B.Bird, R.C.Armstrong, O.Hassager, Dynamics of Polymeric Liquids, Vol. 1; R.B.Bird, C.F.Curtiss, R.C.Armstrong, O.Hassager, ibid., Vol. 2, Wiley, New York, 2nd ed. 1987
K.F.Freed, Renormalization Group Theory of Macromolecules, Wiley, New York 1987
H.Fujita, Polymer Solutions, Elsevier, Amsterdam 1990
P.-G.de Gennes, Introduction to Polymer Dynamics, Cambridge University Press, Cambridge 1990

7.2. DIFFUSION IN SOLUTION and 7.3. SEDIMENTATION IN SOLUTION
H.Fujita, Foundations of Ultracentrifugal Analysis, Wiley, New York 1975
V.N.Tsvetkov, Rigid-chain Polymers: Hydrodynamic and Optical Properties in Solution, Consultants Bureau, New York 1989
K.S.Schmitz, An Introduction to Dynamic Light Scattering by Macromolecules, Academic Press, San Diego (CA) 1990

W.Brown, ed., Dynamic Light Scattering; The Method and Some Applications, Clarendon Press, Oxford 1993

7.4. VISCOSITY OF SOLUTIONS AND MELTS

S.Middleman, The Flow of High Polymers, Interscience, New York 1968
J.A.Brydson, Flow Properties of Polymer Melts, Iliffe Books, London 1970
L.E.Nielsen, Polymer Rheology, Dekker, New York 1977
R.S.Lenk, Polymer Rheology, Appl.Sci.Publ., Barking, Essex 1978
G.Astarita, ed., Rheology, Plenum, New York 1980, 3 vols.
G.V.Vinogradov, A.Ya.Malkin, Rheology of Polymers, Mir Publ., Moscow; Springer, Berlin 1980
R.W.Whorlow, Rheological Techniques, E.Horwood, Chichester (UK) 1980
F.N.Cogswell, Polymer Melt Rheology, Wiley, New York 1981
M.Bohdanecky, J.Kovar, Viscosity of Polymer Solutions, Elsevier, Amsterdam 1982
J.Meissner, Alte und neue Wege in der Rheometrie der Polymer-Schmelzen, Chimia **38** (1984) 35, 65
J.Ferguson, N.E.Hudson, Extensional Flow of Polymers, in R.A.Pethrick, ed., Polymer Yearbook **2**, Harwood Academic Publ., Chur (Switzerland) 1985, p. 155
W.-M.Kulicke, Fliessverhalten von Stoffen und Stoffgemischen, Hüthig and Wepf, Basel 1986
R.Larsen, Constitutive Equations for Polymer Melts and Solutions, Butterworths, Stoneham (MA) 1988
S.W.Churchill, Viscous Flows: The Practical Use of Theory, Butterworths, Stoneham (MA) 1988

7.5. MOLECULAR DYNAMICS OF MELTS

W.W.Graessley, The Entanglement Concept in Polymer Rheology, Adv.Polym.Sci. **16** (1974) 1
J.Klein, The Self-Diffusion of Polymers, Contemp.Phys. **20** (1979) 611
R.T.Bailey, A.M.North, R.A.Pethrick, Molecular Motion in High Polymers, Clarendon Press, New York 1981
W.W.Graessley, Entangled Linear, Branched and Network Polymer Systems - Molecular Theories, Adv.Polym.Sci. **47** (1982) 67
M.Tirrell, Polymer Self-Diffusion in Entangled Systems, Rubber Chem.Techn. **57** (1984) 523

References

[1]   Data in C.Tanford, Physical Chemistry of Macromolecules, Wiley, New York 1961
[2]   H.-J.Cantow, Makromol.Chem. **30** (1959) 169
[3]   S.Bantle, M.Schmidt, W.Burchard, Macromolecules **15** (1982) 1604
[4]   W.-M.Kulicke, R.Kniewske, Rheol.Acta **23** (1984) 75, Fig. 5
[5]   D.S.Pearson, A.Mera, W.E.Rochefort, ACS Polymer Preprints **22** (1981) 102, Table 1
[6]   E.R.Morris, A.N.Cutler, S.B.Ross-Murphy, D.A.Rees, J.Price, Carbohydrate Polym. **1** (1981) 5
[7]   T.G.Fox, J.Polym.Sci. C **9** (1965) 35, Fig. 2
[8]   G.Menges, in H.Batzer, ed., Polymere Werkstoffe, G.Thieme Publ., Stuttgart, Vol. 2 (1984), Fig. 3.22 (p. 86)
[9]   H.-G.Elias, An Introduction to Plastics, VCH, Weinheim 1993, Fig. 6-5
[10] H.-G.Elias, An Introduction to Plastics, VCH, Weinheim 1993, Fig. 6-7
[11] BASF, Kunststoff-Physik im Gespräch, BASF, Ludwigshafen, 2nd ed. 1968, p. 111
[12] J.L.White, Plastics Compounding (Jan-Feb 1982) 45, Fig. 7
[13] W.Philippoff, F.H.Gaskins, J.Polym.Sci. **21** (1956) 205, Fig. 7
[14] H.M.Laun, H.Münstedt, Rheol.Acta **17** (1978) 415, Fig. 5
[15] D.S.Pearson, G. Ver Strate, E. von Meerwall, F.C.Schilling, Macromolecules **20** (1987) 1133
[16] A.Tonelli, quoted in [15]
[17] H.-G.Elias, An Introduction to Plastics, VCH, Weinheim 1993, Fig. 6-2

8   Polymer   Assemblies

8.1.   Introduction

Pure polymers are assemblies of macromolecules that may be in solid or fluid states. Matter is said to be in the *solid state* if small deformations are completely reversible. Such matter behaves elastically on deformation, i.e., it returns immediately and completely to the initial state if the load is removed (G: *elastos*, *elatos* = beaten; from *elaunein*: to drive) (Chapter 10). An example is steel. *Fluid matter* (gases, liquids) deforms completely and irreversibly; it exhibits a viscous behavior (L: *viscum* = birdlime from mistletoe berries). An example is water.

Low molar mass materials often show ideal elastic *or* ideal viscous behavior. Polymers, on the other hand, usually exhibit elastic *and* viscous behavior at the same time; they are viscoelastic materials. Whether elasticity or viscosity dominates depends on the type of polymer assembly, temperature and pressure.

Assemblies of polymers may exist in the solid state in two ideal types of assemblies. In ideal polymer **crystals,** macromolecules and/or their segments are completely ordered (L: *crystallum*, from G: *krustallos* = ice) (Section 8.3.). The long-range crystalline order is destroyed if a crystalline polymer is heated above its melting temperature (Section 9.4.). The resulting melt is a fluid (although it may not flow visibly due to a very high viscosity, see Section 7.4.2. ff.) and, in the ideal case, completely disordered with respect to the arrangement of polymer segments and molecules (Section 8.2.).

Polymer molecules and segments may also be completely disordered in the solid state; they are then said to be **amorphous** (G: *a-* = without; *morphe* = shape) (Section 8.2.). Such amorphous materials resemble silicate glass. On heating, the glass-like structure of an amorphous material is removed at a certain temperature, the glass "transition" temperature (Section 9.4.). Shortly above the glass transition temperature, high molar mass amorphous polymers resemble chemically cross-linked rubbers (although they are not cross-linked) whereas low molar mass polymers behave more like liquids. The fluid state of matter is often called a **melt**, regardless of whether it was produced by heating a crystalline polymer above its melting temperature or by heating an amorphous polymer above its glass temperature.

Crystalline and amorphous arrangements are ideal structures and their behavior as solids or fluids constitutes ideal states. There are also arrangements of polymer assemblies that show order similar to crystals and, at the same time, fluidity like liquids. These materials are "in the middle" between crystals with long-range order and liquids without any long-range order; they are therefore called **mesomorphous** (G: *meso* = middle; *morphe* = shape) (Section 8.4.). Their most prominent representatives are liquid-crystalline polymers which show one-dimensional or two-dimensional "crystalline" order yet flow like liquids in their "melts" or solutions. Other mesomorphic materials comprise block copolymers (Section 8.5.) and ionomers (Section 8.6.).

8.2. Melts and Amorphous States

X-Ray measurements of polymer melts indicate the absence of long-range order. Small angle neutron scattering, on the other hand, shows that the radius of gyration of linear polymer molecules in melts is identical with that of polymer coils in the unperturbed state (Fig. 5-16). Since the segment density of isolated coils decreases with increasing molar mass (Fig. 5-8) but the macroscopic density of melts of true polymers ($X > 100$) does not, it follows that polymer molecules must overlap in melts (see also Sections 5.7.2., 6.3.5. and 7.4.3.).

Segments of polymer molecules are surrounded in melts by segments of the same type. A segment cannot distinguish, however, whether an adjacent segment is part of the same or another molecule. Polymer chains in melts thus exhibit the same reduced radii of gyration, $[\langle s^2 \rangle_w / \overline{M}_w]^{1/2} \neq f(M) = const.$, as in theta solvents (Table 8-1).

Table 8-1   Reduced radius of gyration, $[\langle s^2 \rangle_w / \overline{M}_w]^{1/2}$, of polymers in different states of matter according to static light scattering (theta solvents) and small angle neutron scattering (melts, glassy states, semicrystalline polymers).

Polymer	$[\langle s^2 \rangle_w / \overline{M}_w]^{1/2} / (\text{nm mol}^{1/2} \text{ g}^{-1/2})$			
	Theta solvent	Melt	Glass	Semicrystalline
Poly(styrene), atactic	0.0275	0.0280	0.0278	-
Poly(methyl methacrylate), isotactic	0.024	-	0.029	-
Poly(methyl methacrylate), syndiotactic	0.028	-	0.030	-
Poly(methyl methacrylate), atactic	0.030	-	0.031	-
Poly(isobutylene)	0.030	0.0305	0.031	-
Poly(propylene), isotactic	0.034	0.035	-	0.036
Poly(ethylene)	0.045	0.045_5	-	0.045
Poly(oxyethylene)	-	0.042	-	0.052
Poly(dimethyl siloxane)	0.025	0.027	-	-

Physical structures of polymers are frozen-in if melts are quenched below their glass temperatures. Glassy polymers thus exhibit the same unperturbed dimensions as coils in the theta state. Since the distribution of segments is completely at random in the unperturbed state, it follows that neither melts nor glasses possess long-range order. An absence of long-range order does not exclude short-range order, however. The persistence of chains will cause short chain segments to pack parallel as has been shown for alkanes by X-ray measurements. This local order does not exceed 1 nm, however.

Viscosities rise from $(10^2\text{-}10^6)$ Pa s in melts to ca. 10^{12} Pa s in the glassy state which reduces the mobility of segments quite severely. Chains cannot pack as tightly as they would like since they have some persistence and segments are not infinitely thin. The polymer glass thus has some vacant sites; the density of amorphous polymers in the glassy state is smaller than the density of the melt. An example is poly(methyl methacrylate): $\rho = 1.19$ g/mL (glass) vs. $\rho = 1.22$ g/mL (melt).

Vacant sites are regions with the size of atoms. They generate in the glassy polymer a **free volume**. The volume fraction of the free volume can be calculated as $\phi_f = (v_g - v_m)/v_g$ from the specific volumes of the glass (v_g) and the melt (v_m). At the glass temperature, the fraction of free volume has been found as $\phi_f \approx 0.025$ for *all* polymers. It determines dynamic glass temperatures and many other properties.

8.3. Crystalline Polymers

8.3.1. Introduction

The meaning of the word "crystal" changed several times during the last century. In the mid 1800s, it denoted a material with plane surfaces that intersected each other at constant angles. At the end of the 1800's, a crystal was defined as a homogeneous, anisotropic, solid material. It is "homogeneous" because physical properties do not change on translation in the direction of the crystal axes, "anisotropic" because physical properties differ in various directions, and "solid" because it resists deformation.

In the early 1900s, crystals were redefined as materials with *three-dimensional order* in a three-dimensional **lattice** with atomic dimensions of lattice sites. Such lattice sites are occupied by carbon atoms in diamond and methylene groups in poly-(methylene) $\pm CH_2 \pm_{\overline{n}}$. Perfect lattices are called *ideal*.

Lattice sites may also be taken up by larger spheroidal entities, for example, spherical proteins or latex particles. Lattices with large, tightly packed spherical entities are called **superlattices**. Lattices with large spherical domains of polymer blocks that are separated by amorphous matrices are not considered superlattices but rather mesophases (Section 8.4.).

Three-dimensional lattices are composed of smaller units whose three-dimensional repetition (translation) generates the crystal. These units are called **unit cells**; they are the simplest parallelepipeds that can be given with lattice sites as corners (G: *para* = beside; *allelon* = of one another (from *allos* = other); *epipedos* = level (from *epi* = on, *pedon* = ground)). *Primitive unit cells* contain only lattice sites from corner sites; *centered unit cells* comprise also lattice sites which are not corner sites (Fig. 8-1).

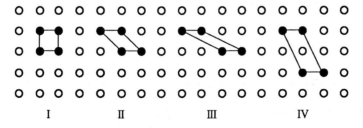

I II III IV

Fig. 8-1   Three primitive unit cells I-III and one centered unit cell IV in a two-dimensional point lattice. In three-dimensional lattices, centered unit cells may be body-centered (lattice point at the intersection of body diagonals) or face-centered (ditto for face diagonals).

All chain units must occupy crystallographic equivalent positions in ideal lattices of chain molecules. On crystallization, some chain units may not find their ideal positions, however, because of the high viscosity of the melt and the fact that chain units are not independent of each other but rather parts of a chain. The crystallized polymer may thus contain **lattice defects** or even only small **crystallites** besides non-crystalline regions. Such crystallized polymers are called **semi-crystalline**. Truly 100 % crystalline polymers are very rare.

Semi-crystalline polymers are not in thermodynamic equilibrium. According to the **phase rule** P + F = K + 2, only one phase (P = 1) can exist for a single component (K = 1) with two degrees of freedom (temperature and pressure; F = 2). Crystalline and non-crystalline regions must therefore be interconnected: any single macromolecule passes through both phases. The two phases of semi-crystalline polymers are therefore not separate entities; they cannot be separated by physical means.

One has thus to distinguish between crystallizability and crystallinity. **Crystallizability** denotes the maximum theoretical crystallinity; this thermodynamic quantity depends only on temperature and pressure. **Crystallinity** is affected by kinetics and thus crystallization conditions (nucleation, cooling time, etc.). It includes frozen-in non-equilibrium states and it is always lower than the crystallizability.

8.3.2. X-Ray Diffraction

X-Ray crystallography is the most important method for the determination of crystal structure and crystallinity. Lattices diffract rays if the distances between lattice points are comparable to the wavelength of the rays. Lattices with lattice distances of atomic size (ca. 0.1 nm) thus diffract X-rays (λ_o = 0.154 nm for Cu-K$_\alpha$) and electron beams (λ_o = 0.00123 nm for an acceleration of electrons by 10 000 eV).

The three-dimensional lattices of crystals diffract coherent rays from two-dimensional lattice planes. Two adjacent lattice planes G_1 and G_2 are at a distance d (Fig. 8-2). An approaching wave L hits the lattice site A of lattice plane G_1 at an angle θ (Bragg angle), a parallel wave the lattice point A_2 of the lattice plane G_2, etc. The phases of these two waves are shifted by $PA_2 + A_2Q = 2\ d \sin \theta$. They interfere constructively if they arrive simultaneously at the plane N–N_2. This requires the plane shift to be equal to or a multiple N of the wave length λ_o of the incident rays (**Bragg equation**):

(8-1) $N\lambda_o = 2\ d \sin \theta$

Most of the radiation travels linearly through the crystal and exits at the other side (primary beam). A smaller part interferes constructively with the lattice planes at the Bragg angle θ. This part is measured classically as a strong blackening of a photographic film; modern diffractometers use photoelectric counters. The position and intensity of the diffracted beams (the **reflections**) allows one to calculate the type and dimensions of the unit cells of the crystal.

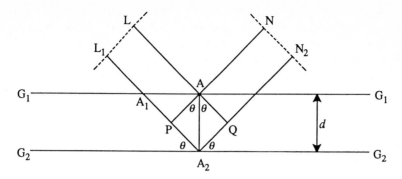

Fig. 8-2   Illustration of Bragg's law (see text).

Crystal structures of single crystals are determined by positioning crystals of ca. 1 mm diameter on a goniometer head in such a way that a known crystal axis (the reference axis) is perpendicular to the incident beam. In the **rotating crystal method** of Bragg, the crystal is then rotated which lets different crystal planes appear one after the other at the positions for maximum scattering intensity. The orientations of the lattice planes are fixed because of the fixed position of the reference axis. The resulting reflections appear as spots and not, as on so-called fiber diagrams, as arcs (planar films) or sickles (concentric films). The rotating crystal method is, for example, used for the determination of crystal structures of protein single crystals.

The uniaxial drawing of fibers and films of crystallized linear polymers leads to a preferential orientation of molecular axes in the draw direction. An incident beam perpendicular to the draw direction generates reflections which are more or less sharp depending on the degree of orientation (Fig. 8-3). For historic reasons, such pictures are called **fiber diagrams** although they can also be obtained from oriented films.

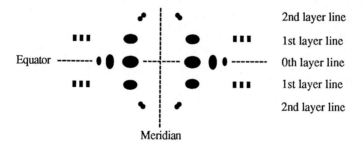

Fig. 8-3   Schematic representation of a fiber diagram of polymers which crystallize in the macro-conformation of a 3_1 helix. An example is the fiber diagram of a uniaxially drawn film of it-poly-(propylene) (see [1]). Insufficient orientation of crystallites causes the spots to degenerate to sickles. In powder patterns, non-oriented crystallites generate circles.

Reflections on the zeroth layer plane correspond to lattice planes that are parallel to the molecular axis (**equatorial reflections**). **Meridional reflections** are generated by lattice planes that are arranged perpendicular to the molecular axis on the plane which bisects the equatorial plane. Numbers and distances of reflections allow direct elucidation of the physical structure of helical macromolecules. In a 3_1 helix, each fourth, seventh ... chain unit is in the same position as the first one (Section 8.3.3.). Three layer lines can thus be expected for the 3_1 helix of isotactic poly(propylene), two each on both sides of the equator which gives the zeroth layer line (Fig. 8-3).

In single crystals, all lattice planes are oriented; one observes optimal X-ray diagrams with sharp reflections. Such single crystals of polymers can be obtained by topotactic polymerization of monomer crystals, e.g., diacetylenes $R-C\equiv C-C\equiv C-R$. Topotactic polymerizations of crystallized monomers are possible if the polymerizable groups are in the right positions relative to each other and if the resulting chains can be rearranged in a new crystal lattice.

Most crystalline polymers are however polycrystalline. Lattice layers are ordered in each crystallite but the crystallites themselves are not. Such microcrystalline materials are constantly rotated on irradiation in the **Debye–Scherrer powder method** which was originally used for crystal powders with particle diameters $d < 100\,\mu m$. A monochromatic X-ray beam thus finds sufficient lattice layers for all reflection positions which satisfy the Bragg condition. The many small crystallites with their multitude of orientations of layer lattices generate a system of coaxial radiation cones with a common tip in the center of the specimen. A vertical cut through this system of cones leads to a number of concentric circles (Fig. 8-4) or ellipses if concentric film strips are used instead of plane photographic plates.

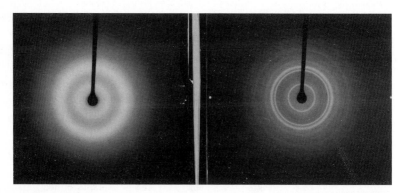

Fig. 8-4 Powder diagrams of undrawn, semi-crystalline isotactic poly(styrene) (left) and amorphous (non-crystalline) atactic poly(styrene) (right).

X-Ray diffractograms of semi-crystalline polymers also show weak rings and a background scattering besides the strong crystalline reflections. Weak rings are called **halos**; they are caused by short-range ordering of segments. The background scattering of polymers is always relatively strong; it originates primarily from scattering by air and secondarily from thermal motions in crystallites as well as from the **Compton scattering**. The Compton effect is an incoherent scattering caused by the scattering of an electromagnetic wave by the electrons of an atom. It is independent of the physical state of matter.

8.3.3. Crystal Structures

Lattice constants and lattice angles often allow one to deduce the crystal structure of polymers. An example is poly(ethylene) (Fig. 8-5):

Poly(ethylene) $+CH_2\text{-}CH_2\,]_{\overline{n}}$ in the thermodynamically stable state crystallizes in the all-trans conformation (modification I). Each third, fifth ... methylene group thus has the same relative position in the chain direction as the first (Fig. 8-5). Since the C-C bond length is 0.154 nm and the C-C-C valence angle is 111.5°, a short periodicity with a lattice constant of $c = 0.2546$ nm is expected (Section 5.3.1., Table 8-3). Chains are side-by-side but run antiparallel because of chain folding (Section 8.3.5.). A short periodicity exists for the first, third, fifth, ... chain in the *a*-direction ($a = 0.742$ nm) and for the first, second, third, ... chain in the *b*-direction ($b = 0.495$ nm). This poly(ethylene) modification belongs to the orthorhombic crystal class since all angles are $\alpha = \beta = \gamma = 90°$ and all lattice constants $a \neq b \neq c$ differ from each other. Drawing may cause poly(ethylene) to adopt monoclinic or triclinic modifications (see Table 8-3).

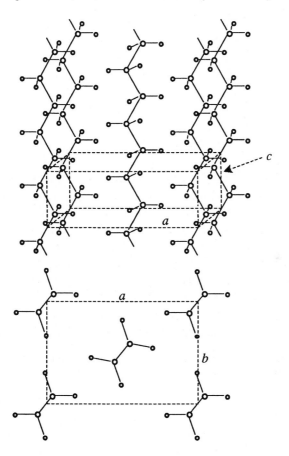

Fig. 8-5 Section with 5 chains of the orthorhombic crystal lattice of poly(ethylene) in side view (top) and as cross-section (bottom). O Carbon atoms, o hydrogen atoms. Lattice points are occupied by ethylene units -CH$_2$-CH$_2$-. The center chain runs antiparallel because of chain folding (see Section 8.3.5.).

With permission by the ACS Rubber Division [2a].

The parallelepipeds of unit cells are characterized by three lattice constants (axes) a, b and c and three planar angles α (between b and c), β (between a and c), and γ (between a and b) which leads to seven crystal systems (Table 8-2). The c direction usually indicates the direction of the polymer chain. This direction is characterized by relatively short chemical bonds whereas the lattice constants a and b are caused by relatively long physical bonds. Because of the great differences in bond lengths in c vs. a and b directions, no cubic lattice exists for chain molecules.

Table 8-2 Crystal Systems

Name	Axes	Angles	Symbol
Cubic	$a = b = c$	$\alpha = \beta = \gamma = 90°$	CUB
Tetragonal	$a = b \neq c$	$\alpha = \beta = \gamma = 90°$	TET
Hexagonal	$a = b \neq c$	$\alpha = \beta = 90°;\ \gamma = 120°$	HEX
Trigonal	$a = b = c$	$\alpha = \beta = \gamma \neq 90°$	TRG
Orthorhombic	$a \neq b \neq c$	$\alpha = \beta = \gamma = 90°$	RHO
Monoclinic	$a \neq b \neq c$	$\alpha = \gamma = 90° \neq \beta$	MON
Triclinic	$a \neq b \neq c$	$\alpha \neq \beta \neq \gamma \neq \alpha$	TRK

Lattice constants c are practically independent of temperature because neither lengths of chemical bonds nor valence angles change much with temperature. Lattice constants a and b increase slightly with temperature due to increasing vibrations.

The assignment of X-ray reflections to lattice constants and angles is often not unequivocal. An example is the modification II of isotactic poly(propylene) which has been classified as trigonal, hexagonal or orthorhombic (Table 8-3).

The existence of various crystal modifications of molecules with the same constitution and configuration is called **polymorphism**. Crystal modifications may be caused by different packing of chains and/or various macroconformations. They are generated by differences in crystallization conditions, for example, various crystallization temperatures, cooling rates, initial states, or the presence of nucleating agents.

The monoclinic modification I of it-poly(propylene) has equal proportions of left-handed and right-handed helices but the hexagonal modification II only either left-handed or right-handed ones. Chains possess the same macroconformation in both modifications but are packed differently. The three modifications of poly(1-butene) differ in the helix type, however, i.e., in their macroconformations.

Isomorphism is the phenomenon that different monomeric units can replace each other in the same crystal lattice. It is found for copolymers whose homopolymers possess analogous crystal modifications, similar lattice constants, and the same helix types.

Modification II of it-poly(propylene) and modification I of it-poly(1-butene) have identical crystal systems (orthorhombic), equal numbers of monomeric units per unit cell (18), the same lattice constant c (0.65 nm), and the same helix type (3_1). Copolymers of propylene and 1-butene thus show isomorphism. Such crystalline copolymers from two or more olefinic monomers are sometimes called **polyallomers**.

Table 8-3 Crystal structures of some polymers and their modifications (called α, β, or γ, or I, II or III). N_u = Number of monomeric units per unit cell; a, b, c = lattice constants; α, β, γ = angles of unit cell; * fiber axis (unless c).
PA 6 = Poly(ε-caprolactam), PA 66 = poly(hexamethylene adipamide), PB = poly(1-butene), PE = poly(ethylene), PEOX = poly(oxyethylene), PG = poly(glycine), PIB = poly(isobutylene), P3MB = poly(3-methyl-1-butene), POM = poly(oxymethylene), PP = poly(propylene), PPOX = poly(propylene oxide), PS = poly(styrene), PTFE = poly(tetrafluoroethylene), PVC = poly(vinyl chloride).

Polymer	Mod.	N_u	$\dfrac{a}{nm}$	$\dfrac{b}{nm}$	$\dfrac{c}{nm}$	$\dfrac{\alpha}{\circ}$	$\dfrac{\beta}{\circ}$	$\dfrac{\gamma}{\circ}$	Helix	Crystal system
PE	I	2	0.742	0.495	0.254	90	90	90	1_1	orthorhombic
	II	2	0.809	0.253*	0.479	90	90	107.9	1_1	monoclinic
PVC, st		4	1.040	0.530	0.510	90	90	90	2_1	orthorhombic
PTFE	I	15	0.566	0.566	1.950	90	87	120	15_7	hexagonal
	II	13	0.952	0.559	1.706	88	90	92	13_6	triclinic
PP, st		8	1.450	0.560	0.740	90	90	90	2_1	orthorhombic
it	I (α)	12	0.665	2.09	0.650	90	99.5	90	3_1	monoclinic
it	II (β)	18	1.908	1.101	0.649	90	90	90	3_1	orthorhombic
it	III (γ)	12	0.638	0.638	0.633	89	100	99	3_1	triclinic
PIB		16	0.688	1.191	1.860	90	90	90	8_3	orthorhombic
PB, it	I	18	1.770	1.770	0.651	90	90	90	3_1	orthorhombic
	II	44	1.485	1.485	2.060	90	90	90	11_3	tetragonal
	III		1.238	0.892	0.745	90	90	90	4_1	orthorhombic
PS, it		18	2.19	2.19	0.665	90	90	120	3_1	trigonal
P3MB		8	0.955	1.708	0.684	90	90	116.5	4_1	monoclinic
POM	I	9	0.446	0.446	1.730	90	90	90	9_5	trigonal
PEOX	I	28	0.803	1.209	1.948	90	125.4	90	3_1	monoclinic
PPOX, it		4	1.052	0.468	0.710	90	90	90	2_1	orthorhombic
PG	I	2	0.477	0.477	0.70	90	90	66	2_1	monoclinic
	II	3	0.48	0.48	0.93	90	90	120	3_1	hexagonal
PA 6	α	8	0.960	1.718*	0.805	90	68.6	90	2_1	monoclinic
	β	1	0.48	0.48	0.86	90	90	120	1_1	hexagonal
PA 66	I	1	0.49	0.54	1.73	48	77	63	1_1	triclinic

8.3.4. Macroconformation and Packing

Many chains adopt macroconformations of helices in crystals. A helix is characterized by the symbol $aA*B/N$, where a = type of repetition along the longitudinal axis, A = helix class (= number of skeletal chain atoms contained within the helix residue; the motif), B = integral number of conformational repeating units per N turns, N = number of turns needed to return to the original position, and *, / = separators. Repetitions may be translations t or screw repetitions s. a and/or A are often omitted and the helix structure is simply described by B_N. Chains in all-trans conformation can also be described by helix symbols.

Poly(ethylene) $+CH_2CH_2\xrightarrow{}_{\overline{n}}$ crystallizes in the all-trans conformation. It possesses $A = 2$ carbon atoms per $B = 1$ conformational repeating unit; the original position is obtained again after $N = 1$ turns. Poly(ethylene) is thus a "1_1 helix" with the symbol t2*1/1. If this polymer is thought of as poly(methylene) $+CH_2\xrightarrow{}_{\overline{n}}$, then it is a "$2_1$ helix" with the symbol t1*2/1.

Isotactic poly(propylene) with the chain $+CH_2CH(CH_3)+_{\overline{n}}$ forms a 3_1 helix (Fig. 8-6) with A = 2 chain units per configurational repeating unit; the symbol is thus s2*3/1. The helical syndiotactic poly(propylene) gets the symbol s4*2/1 because it possesses A = 4 chain units (2 monomeric units) per configurational repeating unit (see Fig. 2-14).

Macroconformations of linear chains in the crystalline state can be estimated as follows from constitution and configuration. For *apolar polymers*, practically only space requirements of ligands have to be considered, i.e., repulsion. The distance between the centers of non-bonded hydrogen atoms in poly(ethylene) $+CH_2-CH_2+_{\overline{n}}$ is 0.2546 nm. It is thus greater than the sum of the van der Waals radii of two hydrogen atoms (0.24 nm) (see Section 2.4.1. for the definition of "non-bonded"). The macroconformation of poly(ethylene) is not sterically hindered by substituents and poly(ethylene) possesses an all-trans conformation $(T)_n$ in the lowest energy state, i.e., a zigzag chain.

The shortest distance between non-bonded fluorine atoms of poly(tetrafluoroethylene) $+CF_2-CF_2+_{\overline{n}}$ in the hypothetical all-trans conformation (0.2564 nm) is smaller than the sum of the van der Waals radii of two fluorine atoms (0.28 nm). At temperatures below 19°C, -CF_2- units thus depart slightly from the ideal all-trans conformation by increasing the torsional angle to 16° from 0°; poly(tetrafluoroethylene) adopts the conformation of a 13_6 helix ($s1*13/6$) (modification II; macroconformation is still $(T)_n$). Increasing vibration of fluorine atoms at temperatures greater than 19°C causes the helix to expand perpendicular to the chain axis and the helix is now 15_7 (modification I; still $(T)_n$) (Table 8-3).

The much larger methyl groups of isotactic poly(propylene) PP force every other chain bond into a gauche conformation. The resulting macroconformation $(TG)_n$ of

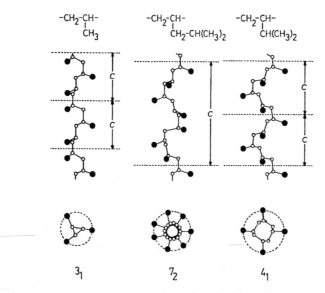

Fig. 8-6 Helix types of some isotactic poly(α-olefin)s $+CH_2-CHR+_{\overline{n}}$ in side view (top) and as cross-section (bottom): 3_1 helix of poly(propylene), 7_2 helix of poly(4-methyl-1-pentene), and 4_1 helix of poly(3-methyl-1-butene). ○ Chain atoms, ● substituents; hydrogen atoms are not shown. c = Short periodicity. With permission by the ACS Rubber Division [2b].

crystalline it-PP leads to a 3_1 helix with three monomeric units $-CH_2-CH(CH_3)-$ per 1 turn ($s2*3/1$) (Fig. 8-6). Crystals of isotactic poly(propylene) consist of equal amounts of $(TG^+)_n$ and $(TG^-)_n$ chains since these two macroconformations are energetically equal. The energetic equivalency of G^+ and G^- applies to all non-chiral isotactic polymers, even in solution (cf. Fig. 5-4).

Still larger substitutents expand the helix further to the 7_2 helix of it-poly(4-methyl-1-pentene) P4MP and the 4_1 helix of it-poly(3-methyl-1-butene) P3MB. All these helices still have the macroconformation $(TG)_n$ but the ideal torsional angles of $0°$ (T) and $120°$ (G) in it-PP now become $-13°/110°$ (P4MP) and $-24°/96°$ (P3MB).

Modification II of poly(glycine) $+NH-CH_2-CO+_{\overline{n}}$ (PG) also forms a 3_1 helix but this helix is deformed because of intramolecular hydrogen bonds between amide groups. Modification I of the same polymer is in the pleated sheet conformation (Fig. 4-4) with "intermolecular" hydrogen bonds. Such pleated sheet structures are also found for polyamides 6 and 66 (Fig. 8-7).

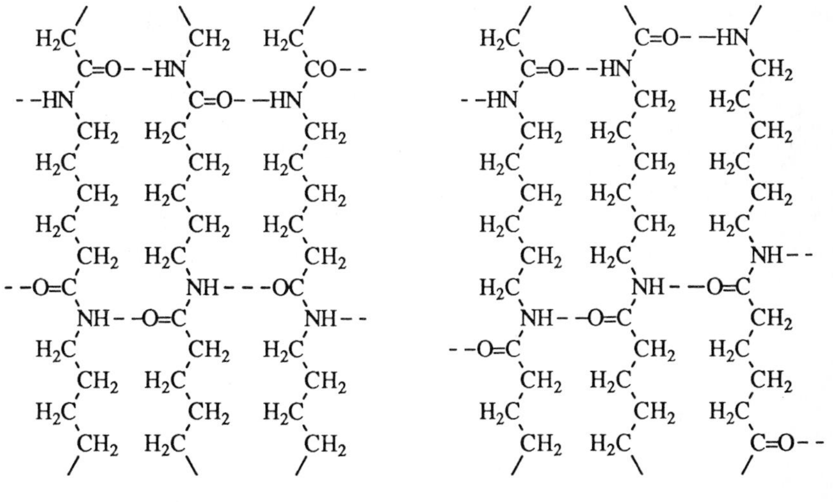

Fig. 8-7 Pleated sheet structures of polyamides 6 (left) and 66 (right). See also Fig. 4-4.

For *polar polymers*, one has to consider **gauche effects** by unpaired electron pairs (Section 2.4.2.) as well as intermolecular and intramolecular attraction between non-bonded atoms. Because of the gauche effect, poly(oxymethylene) POM crystallizes in the all-gauche macroconformation $(G)_n$ and poly(oxyethylene) in $(TTG)_n$. The repulsion of adjacent methyl groups in it-poly(oxypropylene) PPOX reduces bond orientations; PPOX crystallizes in the all-trans macroconformation $(T)_n$.

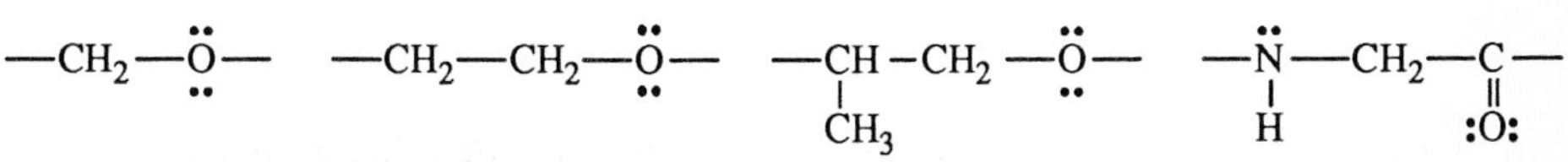

POM	PEOX	PPOX	PG
$(G)_n$	$(TTG)_n$	$(T)_n$	$(TTG)_n$

The packing of chains in crystals is described by the cross-sectional area $A_m = V/(N_c c)$ of chains, where N_c = number of chains per unit cell and c = lattice constant. The volume $V = a \times b \times c$ of unit cells has to be calculated from the vectors a, b and c and not from the lattice constants a, b and c since lattices are not always rectangular (see Fig. 8-1 and Table 8-3).

The packing of chains in crystals affects the density of polymers and thus the temperatures, enthalpies and entropies of melting. For example, the densities of the three modifications of poly(1-butene) are $\rho/(g\ mL^{-1}) = 0.95$ (I), 0.91 (II) and 0.90 (III) and the melting temperatures $T_M/^\circ C$ correspondingly 142 (I), 130 (II) and 108 (III).

8.3.5. Chain Folding

The strong reflections by lattice sites are observed at relatively large angles according to Bragg's law. They provide small lattice constants and are thus called **short periodicities**. Besides these periodicities from unit cells (X-ray wide angle scattering) one also observes for chain molecules **long periodicities** at small angles (SAXS).

For low molar mass alkanes $H(CH_2)_n H$ with $n < 75$, long periodicities d equal conventional contour lengths r_{cont} of alkanes (Fig. 8-8). The long periodicity comprises whole molecules; i.e., chain axes must be perpendicular to the base plane.

Long periodicities of alkanes with $n > 75$ are independent of chain lengths. Since conventional contour lengths continue to increase linearly with increasing n, chains must fold back in crystals. **Chain folding** is also observed for aliphatic polyurethanes (Fig. 8-8), except that long periodicities and contour lengths are only proportional to each other since the molecular axes are inclined to the base plane.

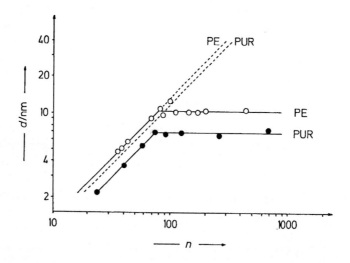

Fig. 8-8 Double logarithmic plot of long periodicities d as function of the number n of units for polyurethanes (PUR) $HO(CH_2)_2O(CH_2)_2O + OCNH(CH_2)_6NHCOO(CH_2)_2O(CH_2)_2O +_n H$ and alkanes (PE) $H(CH_2)_n H$. - - - Calculated for an all-trans macroconformation. Measurements on PE according to various authors, on PUR according to [3].

The presence of **chain folds** in chain molecules was first inferred in 1938 from the electron diffraction pattern of very thin films of 1,4-trans-poly(isoprene) (gutta percha). However, the phenomenon received widespread attention only after three research groups observed independently of each other in 1957 that the cooling of very dilute solutions of poly(ethylene) delivered very thin rhombohedral platelets (Fig. 8-9). At constant undercooling, the thicknesses of these platelets are constant, independent of the solvent, and identical with the long periodicity ($d = 10.5$ nm in Fig. 8-8). The size of the platelets indicates that one platelet contains many poly(ethylene) molecules. According to electron diffraction, such platelets are indeed **single crystals**.

Such crystallites with folded chain molecules are also called **fold micelles** in analogy to **fringed micelles** (see below). Fold micelles are not 100 % crystalline but contain non-crystalline ("amorphous") regions. The proportion of non-crystalline regions increases with increasing molar mass. Melt-crystallized poly(ethylene) was 94 % crystalline at $M = 10^4$ g/mol but only 65 % crystalline at $M = 3 \cdot 10^6$ g/mol.

Fold micelles are not only obtained as thin platelets from dilute solutions but also as **lamellae** of stacked platelets from crystallized melts. The upper layers of poly(ethylene) lamellae can be oxidized by fuming nitric acid. The remaining platelets are 100 % crystalline. Lamellae must be comprised of highly crystalline interiors and less ordered cover layers (Fig. 8-10).

Folds are not very sharp; they consist of ca. 6-7 chain bonds in gauche conformation in poly(ethylene). The re-entry of chains into the lamella can be in adjacent sites with sharp folds or loose loops, to far distant stems of the same lamella (switchboard), or via **interlamellar bridges** (**crystal bridges, tie molecules**) to other lamellae.

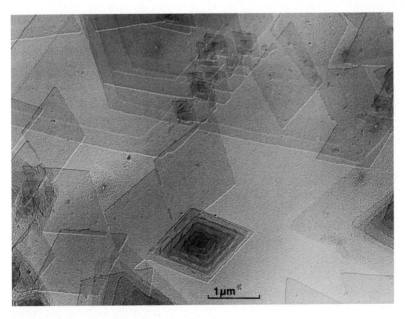

Fig. 8-9 Electron microscopic picture of fold micelles of poly(ethylene). Bottom center: screw dislocation. With permission by the authors [4].

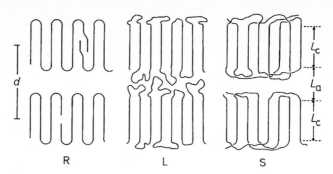

Fig. 8-10   Chain folding in fold micelles. L_c = Thickness of crystalline lamella; L_a = thickness of "amorphous" cover layers; d = X-ray long periodicity.
R = Regular lamellae with sharp folds and crystal defects by chain ends and dislocations.
L = Loose loops with adjacent re-entry.
S = Switch-board with distant re-entries (leads to an unacceptable density ot the lamellar surface).

Crystallinities of melt-crystallized lamellae decrease with increasing molar masses. This effect is caused by the kinetic difficulty of forming micelles with regular folds from melts with entangled, unperturbed coil molecules and the more ready incorporation of a single chain in two or more lamellae ("L"). Quenching of melts generates lamellae with random entries or exits of chains. Since the melt solidifies rapidly, many high molar mass chains cannot fold back. They rather run through several lamellae and form a kind of fringed micelle (Fig. 8-11).

A melt-crystallized chain molecule is thus comprised of N_{lam} rod-like stems of length L_{lam} that are interconnected by short, coil-like sections. Since chains enter lamellae at random and lamellae are oriented at random, too, chains adopt the macro-conformation of a random coil with a radius of gyration $s_o = N_{lam}L_{lam}^2/6$. This situation is analogous to the one in concentrated solutions (Section 5.7.2.): instead of blobs with coiled segments one now has lamellae of folded segments.

Radii of gyration of chain molecules in fold micelle lamellae not only show the root dependency on molar masses like melts but also identical reduced mean-square radii of gyration, $[\langle s^2 \rangle_w / \overline{M}_w]^{1/2}$ (Table 8-1). However, at higher degrees of crystallinity, reduced radii of gyration of polymer molecules are greater in fold micelles than in melts. Such higher degrees of crystallinity are obtained by the crystallization of very flexible chain molecules (such as poly(oxyethylene) with very low rotational barriers around the CH_2-O bond (see Table 8-1) or by crystallization under pressure to so-called chain-extended crystals.

8.3.6. Morphology

Crystalline polymers can thus have various degrees of crystallinity and different morphologies depending on the cooling conditions for melts or solutions. Degrees of crystallinity are usually calculated using a **two-phase model** which assumes (incorrectly, see Section 8.3.1.) that perfect crystalline domains exist besides totally dis-

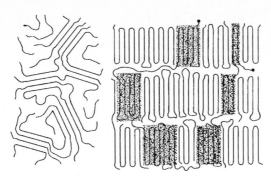

Fig. 8-11   Left: Historic representation of the structure of fringed micelles. Right: "Random coils" composed of chain folds and short, coiled segments in fold micelles.

ordered regions. **One-phase models** interpret the same experimental data as caused by lattice defects (end groups, dislocations, kinks, flaws, etc.) in an otherwise ideal crystal lattice. A degree of crystallinity of 83 % (from density) of poly(ethylene) according to the 2-phase model corresponds to a proportion of 2.9 % of lattice defects in the 1-phase model.

The **degree of crystallinity** of a polymer is not an absolute quantity since the border between "crystalline" and "amorphous" regions is not sharp (see also Fig. 8-10). Different experimental methods measure different degrees of order and thus different "average" crystallinities (Table 8-4). Degrees of crystallinity can be furthermore calculated as mass fractions w_c or volume fractions ϕ_c. They can be interconverted by $w_c = \phi_c \rho_c / \rho$ with the densities of the specimen (ρ) and a 100 % crystalline polymer (ρ_c).

The density crystallinity $w_{c,d} = [\rho_c(\rho_P - \rho_a)]/[\rho_P(\rho_c - \rho_a)]$ of a specimen with density ρ_P is calculated from the densities ρ_a of the completely amorphous and ρ_c of the ideal crystalline polymer. The X-ray crystallinity $w_{c,x} = I_c/(I_c + K_a I_a)$ results from the integrated intensities I_a and I_c where K_a is a calibration constant. Infrared crystallinities $w_{c,i} = (a_c \rho L) \log_{10}(I_0/I)$ of a specimen with thickness L are obtained from the intensities of the incident (I_0) and transmitted (I) beam at the frequency of the absorption band and the absorptivity a_c of the crystalline part. Calorimetric measurements deliver $w_{c,h} = \Delta h_M / \Delta h_{M,c}$ from the corresponding specific melt enthalpies. Crystallinities have also been calculated from the degradation rates of hydrolyzable polymers since chains in amorphous regions are hydrolyzed much faster than those in crystalline domains.

Densities ρ_P of solid polymers are usually determined with the help of a **density gradient column**. Two liquids A and B with densities $\rho_A < \rho_P$ and $\rho_B > \rho_P$ are mixed in such a way that the density ρ of the blend increases continuously from the meniscus to the bottom of the tube. The insoluble and unswellable polymer moves to that part of the tube where the density ρ of the liquid mixture equals the density ρ_P of the polymer.

Table 8-4   Degrees of crystallinity (as mass fractions in percent).

Method	Poly(ethylene)	Celluloses		Poly(ethylene terephthalate)	
		Cotton	Rayon	undrawn	drawn
Infrared spectroscopy	67	62	42	61	59
X-ray diffraction	–	70	38	29	2
Density	67	60	25	20	20

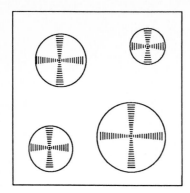

Fig. 8-12 Schematic representation of spherulites as observed for melt-crystallized it-poly(propy-lene) under the polarizing microscope (left) and the phase contrast microscope (right) [5]. The Maltese cross (left) arises from the polarizer and analyzer axes, i.e., where the vibration plane is parallel to one of them and, hence, blocked by the other. Four positions with extinction are obtained because spherulites behave like crystals with radial optical symmetry.

Crystallized polymer melts sometimes exhibit spherical, polycrystalline domains, so-called **spherulites** (Fig. 8-12). Spherulites with diameters greater than the wavelength of light show radial symmetry under the phase contrast microscope. By definition, positive spherulites have the highest refractive index in the radial direction and negative spherulites in the tangential direction. Spherulites make films opaque if spherulite diameters are greater than one-half of the wavelength of light and if spherulites are inhomogeneous with respect to density or refractive index. They develop if crystallization rates are equal in all spatial directions. Otherwise, snowflake-like or tree-like entities result, so-called **dendritic crystals** or **dendrites** (G: *dendron* = tree).

Chains are approximately tangential in all spherulites. For example, poly(vinylidene chloride) $\{CH_2\text{-}CCl_2\}_{\overline{n}}$ forms positive spherulites. Since the refractive index is smaller in chain direction than perpendicular to it, chain axes must be tangential to the spherulite radius. Poly(ethylene), on the other hand, exhibits negative spherulites. Here, the refractive index in chain direction is larger than the one perpendicular to it. The chain axes of poly(ethylene) molecules are therefore also tangential to the spherulite radius.

Strongly stirred dilute polymer solutions deliver **shish-kebab** structures on crystallization (Turkish: *sis* = skewer; *kebab* = roast meat) (Fig. 8-13). Strong stirring forces molecular axes to become parallel and form fibrils. The shear gradient is strongly reduced between fibrils so that fibrils can act as nuclei for the epitaxial overgrowth of the remaining macromolecules. This overgrowth results in lamellae that are perpendicular to fibril axes; axes of folded molecules in lamellae are thus parallel to fibrils.

8.3.7. Orientation

Chain segments and crystallites may orient themselves in the draw direction on injection molding or extrusion of plastics and on drawing of fibers and films. The **degree of orientation** is often difficult to measure and the distribution of orientation

Fig. 8-13   Left: Electron micrograph of shish-kebab structures of poly(ethylene) crystallized from strongly stirred 5 % solutions in toluene. Right: Schematic representation of chain folding in these structures. With permission by Steinkopff Publ. [6].

is usually impossible to measure. Often, the **degree of drawing** $\lambda = L/L_o$ is taken as a measure of the degree of orientation, f_{or}, of segments, molecules, and crystallites. It is however not a good measure of f_{or} since it only gauges the length of the specimen after (L) and before (L_o) the drawing which may be due to viscous flow.

Orientation can be measured by the velocity of sound, infrared dichroism, X-ray wide angle measurements, or optical birefringence. *X-Ray wide angle measurements* perpendicular to the draw direction of semi-crystalline polymers show that concentric circles (Fig. 8-4) first degenerate to arcs and then to point-like reflections on drawing. The length of an arc is therefore a measure of the degree of orientation of the crystallite (more exactly: the lattice layer).

An orientation factor f exists for each of the three spatial coordinates. The **Hermans orientation factor** $f_{orient} = (1/2)[3 \langle \cos^2\beta \rangle - 1]$ (**uniaxial order parameter**) is calculated from the orientation angle β which is defined as the angle between the draw direction and the main optic axis. The order parameter becomes unity for a complete orientation in the chain direction ($\beta = 0°$), $-1/2$ for a complete orientation perpendicular to the chain direction ($\beta = 90°$), and zero for a random orientation.

Orientations in transparent and translucent materials can also be characterized by **optical birefringence**. A transparent material exhibits refractive indices n_x, n_y and n_z along the three main axes x, y, and z. At least two of these refractive indices are different for optically anisotropic materials. The difference between any two of these refractive indices is the **birefringence** Δn.

Monomeric units are always optically anisotropic but unoriented, amorphous polymers are not because their monomeric units are randomly oriented in space. These polymers become birefringent if chains are oriented or under mechanical stress. Such stressed regions can be detected by **stress birefringence**. In general, this requires polarized light. For poly(styrene) with its strongly anisotropic phenyl groups, unpolarized light suffices, however.

8.4. Liquid Crystals

8.4.1. Introduction

Mesophases and mesomorphic substances, respectively, possess microscopic structures whose order is between that of crystals with long-range three-dimensional order and liquids and amorphous solids without any long-range order (G: *mesos* = middle; *morphe* = shape). Mesophases comprise liquid crystals, plastic crystals, and conformationally disordered crystals. Since liquid crystals were discovered first and are much more prevalent than plastic or conformationally disordered crystals, "mesophase" is often used as a synonym for "liquid crystal".

Liquid crystals LC show order like crystals but flow like liquids which is caused by anisotropic entities: molecules of low molar mass compounds and segments of liquid-crystalline polymers LCP. These anisotropic entities are called **mesogens**.

Some liquid crystals are **thermotropic**; they are liquid-crystalline in the temperature range between melts and solids (crystals or LC glasses). Other liquid crystals are **lyotropic**; they form LCs or LCPs in concentrated solutions.

Special names are usually used for phase transitions such as melting temperature for the transition crystal → melt, glass temperature for glass → melt, and boiling temperature for melt → gas. No special names exist however for thermal transitions of crystals or LC glasses to LCs or LCPs or for transitions between the various types of liquid crystals, except for the transition liquid crystal → melt. This **isotropization temperature** is also called the **clearing temperature** because LCs and LCPs are turbid whereas melts are clear. The turbidity of LCs and LCPs is caused by anisotropic structures with dimensions similar to the wavelength of light. Melts are isotropic.

8.4.2. Mesogens and Mesophases

LCs and LCPs are classified according to the shape of mesogens and the structure and appearance of mesophases. Mesogens are either **rod-like** (**calamitic**; G = *kalamos* = reed) or **disc-like** (**discotic**; L = *discus*, G = *diskos* = circular plate, from *dikein* = to throw):

calamitic   discotic

X may be -O- or -COO-, Y may be -COO-, *p*-C₆H₄-, -CH=CH-, -N=N-, and R, e.g., -CO(CH₂)ₙCH₃. Mesogens form three classes of mesophases:

– **Smectic** mesophases exhibit fan-like structures under the polarizing microscope. They are generated by calamitic mesogens that form two-dimensional layers which feel like soap for low molar mass LCs (G: *smegma* = soap). Smectic mesophases A consist of layers with parallel rod-like mesogens whose longitudinal axes are perpendicular to the layer planes. Molecular axes are inclined to layer planes in smectic mesophases C, etc. At least eight different types of smectic mesophases are known.

– **Nematic mesophases** are most common. They appear as thread-like *schlieren* (streaks) under the polarizing microscope (G: *nema* = thread). Nematic mesophases are only one-dimensionally ordered (Fig. 8-14). Mesogen axes are parallel for calamitic LCs and LCPs and parallel to disc planes for discotic ones.

– **Cholesteric mesophases** are only obtained from chiral mesogens. The mesogens are nematically ordered but are forced by chirality centers to adopt a screw sense.

In polymers, mesogens may be either in main chains or in side chains (Fig. 8-14). **Liquid-crystalline main chain polymers** MCLCP are usually synthesized by polycondensation, less often by polyaddition. Nematic structures are easy to produce since only large mesogens are required or short mesogens that are connected by *short* flexible spacers. Smectic structures are usually obtained if rigid rod-like mesogens are present in periodic sequences or if mesogens of equal length are connected by *large* flexible spacers.

Liquid-crystalline side-chain polymers SCLCP contain mesogens in side chains that are bound to non-mesogenic main chains via flexible spacers. SCLCPs result from the polymerization of suitable macromonomers or the reaction of mesogenic oligomers and flexible chains. The best known SCLCPs have flexible backbones of polysiloxane or poly(methylmethacrylate) chains.

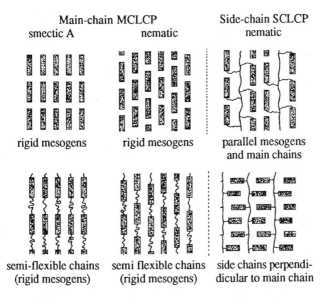

Fig. 8-14 Arrangement of mesogens in liquid-crystalline polymers (schematic). The mesogens of the top left and top center arrangements do not have the same scale as the others [7].

LCs and LCPs consist macroscopically of domains with diameters in the range of micrometers. Within each domain, mesogens are oriented relative to a preferential axis. The orientation is described by the Hermans orientation factor f_{orient} (Section 8.3.7.). This orientation factor is $0.85 < f_{orient} < 0.95$ for smectic LCPs but only $0.4 < f_{orient} < 0.65$ for nematic ones. The preferential axes of the domains themselves are oriented at random, however.

8.4.3. Thermotropic Liquid Crystals

Molecular axes of rod-like molecules (or rod-like segments) with length L and diameter d are parallel in mesophases. Rods with small aspect ratios $\Lambda = L/d$ resemble spheres with no preferential axis. There must be therefore a critical aspect ratio Λ_{crit} above which the simple geometric anisotropy of rods is sufficient to stabilize mesophases. This critical value has been calculated as $\Lambda_{crit} = 6.42$ by the lattice model. For $\Lambda < 6.4$, repulsion is insufficient for stabilization. These small mesogens must exert additional, orientation-dependent attraction forces in order to generate mesophases.

Attractive forces are almost always present in LCPs, for example, as π–π interactions. The ordered states of rigid mesogens with $\Lambda_{crit} > 6.4$ are so stable that thermotropic polymers begin to decompose at temperatures below the melting temperature. Examples are poly(p-benzamide) PBA and poly(p-hydroxybenzoic acid) PHB:

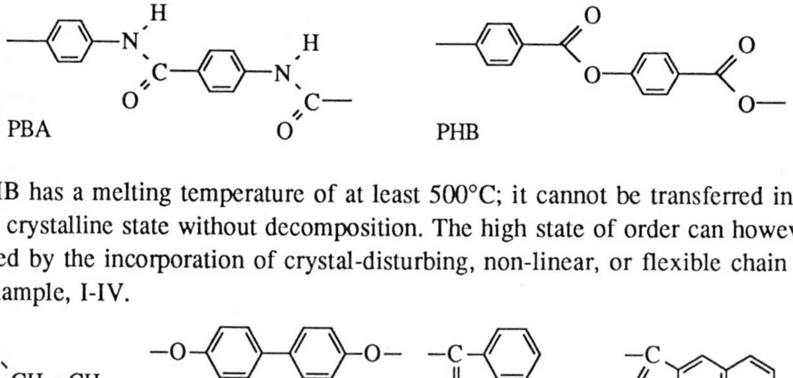

PHB has a melting temperature of at least 500°C; it cannot be transferred into the liquid crystalline state without decomposition. The high state of order can however be reduced by the incorporation of crystal-disturbing, non-linear, or flexible chain units, for example, I-IV.

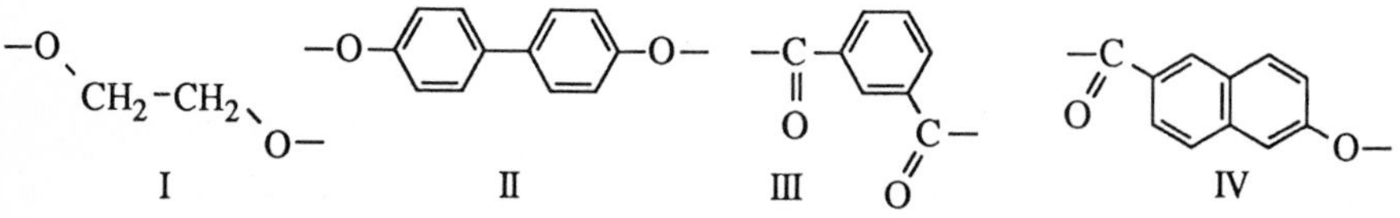

Mesogens become oriented on shearing; thermotropic nematic mesophases thus have considerably smaller viscosities than isotropic melts. Processing of these materials requires less energy. The orientation of mesogens is maintained if the mesophase structure is frozen-in by cooling below the glass temperature to LCP glasses. Mechanical properties of these materials are improved in the orientation direction. Such **self-reinforcing plastics** from p-hydroxybenzoic acid units -O-C_6H_5-CO- and II + III units or from -O-C_6H_5-CO- units + IV units compete however with the less expensive glass-fiber-reinforced saturated polyesters (Chapter 14).

8.4.4. Lyotropic Liquid Crystals

Solutions of mesogenic compounds separate into a higher concentrated liquid-crystalline phase and a less concentrated isotropic phase at concentrations above a critical concentration. The larger the axial ratio of mesogens, the lower is the critical volume fraction for phase separations (Fig. 8-15). Critical volume fractions can be calculated from second virial coefficients, with lattice theories, or via attraction forces.

Second virial coefficients are controlled by excluded volumes (Eqn.(6-30)) and thus by the aspect ratios of calamitic mesogens (Section 5.6.3.). The Onsager theory describes correctly the critical concentrations for phase separations of dilute solutions of very long, rigid rods with dominating repulsive forces (e.g., imogolite). It fails for the higher concentrations at which solutions of LCPs with flexible spacers and weak attraction forces separate into phases. The reason is that second virial coefficients describe only the mutual orientation of *two* molecules (Section 6.3.1.) but not the mutual orientation of *many* mesogens which is important for the phase separation of mesophases.

Lattice theories place calamitic mesogens on two-dimensional lattices where they reside at an angle to the preferential direction of the longitudinal axes of mesogens in a nematic mesophase. In the simplest case, nematic phases are only generated by the space requirements and arrangements of rods (entropic effects) and not by attractive

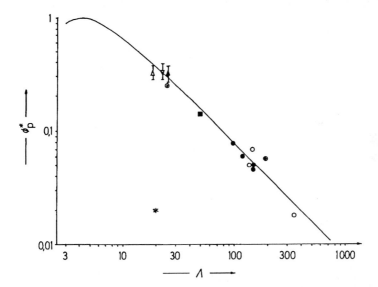

Fig. 8-15   Volume fractions $\phi_p{}^*$ for the phase separation isotropic $\rightleftarrows$ nematic as a function of the axial ratio Λ of mesogens (double-logarithmic plot). Solid line calculated with Eqn.(8-2).

Helices or worm-like chains: ● Poly(γ-benzyl-L-glutamate) with various molar masses in *m*-cresol, ○ ditto in mixtures of *N,N*-dimethylformamide and methanol; ⊕ poly(γ-phenyltrimethylene-L-glutamate); ⊙ deoxyribonucleic acids in 1-2 mol/L aqueous NaCl solution. The deviation of the datum for the tobacco mosaic virus (∗) is probably caused by charge effects.

Semi-flexible molecules: ■ Poly(*p*-benzamide) in sulfuric acid; ∇▵▲ cellulose acetates and cellulose ethers in various solvents.

forces (enthalpic effects). The Flory theory predicts that the critical volume fraction ϕ_p^* for the onset of phase separation depends only on the aspect ratio $\Lambda = L/d$; the dependence can be approximated by (Fig. 8-15):

(8-2) $\phi_p^* = 8\,(1 - 2\,\Lambda^{-1})/\Lambda$

The aspect ratio is obtained from diameters d of mesogens and from lengths $L = L_s$ of rods and Kuhn lengths $L = L_K$ of worm-like or semi-flexible chains, respectively.

The formation of lyotropic nematic mesophases is utilized for the generation of "self-orienting" fibers. These fibers are spun from lyotropic solutions at concentrations $\phi > \phi_p^*$; solvents are then removed by precipitation baths. Films are cast from isotropic solutions at $\phi < \phi_p^*$, however. After removal of the solvent, films are annealed at temperatures above the glass temperature but below the transition temperature nematic glass $\rightleftarrows$ isotropic melt. Industrially used fibers comprise poly(p-phenylene terephthalamide) PPTA (Kevlar®), its copolymer with 2,4'-diaminodiphenyleneoxide and terephthalic acid (Technora®), poly(p-phenylenebenzbisoxazol) PPBO, and poly(p-phenylenebenzbisthiazol) PPBT.

Polymolecular LCPs fractionate on phase separation. Higher molar masses assemble almost exclusively in the nematic phase which is always more concentrated than the isotropic phase. Polymer coils are nearly excluded from nematic phases since the Gibbs energies of nematic phases increase strongly if coils are present.

Evaporation or coagulation of dilute isotropic solutions of "rod-like" semi-flexible molecules with calamitic mesogens and coil-like flexible molecules do not lead to *molecular composites* where calamitic segments self-reinforce amorphous matrices consisting of polymer coils. With increasing concentration, solutions pass through the critical concentration for phase separation and rods and coils separate into different phases. Domains of mesophases will reside in matrices composed of random coils but there will be no *molecular* mixing of LCP and random coil molecules.

8.5. Block Copolymers

Two constitutionally different homopolymers A_p and B_q are usually incompatible; their mixtures demix and two separate phases are formed (Section 6.2.3.). The separation into macrophases can be prevented if small proportions of diblock copolymers $A'_m\text{-}B'_n$ are added. Blocks A'_m should be miscible with homopolymer A_p and blocks

B'_n with homopolymer B_q but it is not required that units A and A' have the same constitution; the same is true for B and B'. The blocks will move into the respective domains of A_p and B_q polymers and act as anchors (Fig. 8-16 C); diblock copolymers are **compatibilizers** for the blend of two homopolymers.

Blocks of pure diblock copolymers A_m–B_n, triblock copolymers $A_{m/2}$–B_n–$A_{m/2}$, etc., also try to demix. A complete demixing into two phases is however not possible since different blocks are coupled to each other. Constitutionally identical blocks can only aggregate and form domains in the matrix of other blocks. These **microphases** are special cases of thermotropic mesophases.

In the simplest case, blocks A_m and B_n cannot crystallize. Triblock copolymers $A_{m/2}$–B_n–$A_{m/2}$ will furthermore behave like diblock copolymers A_m–B_n since a block $A_{m/2}$ of a triblock copolymer corresponds to half a block of a diblock copolymer. Each block A_m (or $A_{m/2}$) tries to form a random coil; identical blocks will aggregate. If the space requirement of A_m blocks matches that of the B_n blocks, A_m-blocks will form layers where the A_m blocks of diblock copolymers A_m–B_n face other A_m blocks; the same is true for B_n blocks. This **microphase separation** results in the formation of lamellae in which layers of A_m blocks alternate with layers of B_n blocks (Fig. 8-16 L).

A_m blocks can no longer be packed into planar layers if the space required by B_n blocks in diblock copolymers A_m–B_n is much greater than that of A_m blocks ($n >> m$). A packing in lamellae would either violate the requirement of a densest packing of segments or lead to a strong deviation from the structure of unperturbed coils. These two possibilities are thermodynamically unfavorable. The small A_m blocks thus form spherical domains in a continuous matrix of B_n blocks.

The A_m domains are not connected by chains in diblock copolymers A_m–B_n. They are bound to each other via B_n segments in triblock copolymers $A_{m/2}$–B_n–$A_{m/2}$, however, since the two $A_{m/2}$ blocks of an $A_{m/2}$–B_n–$A_{m/2}$ molecule will reside in two different domains (Fig. 8-16 S). The A_m domains thus act as physical cross-links.

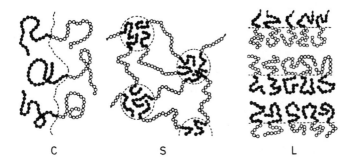

C   S   L

Fig. 8-16   Arrangement of A–blocks with A–units ● and B–blocks with B–units ○ in diblock copolymers A_m–B_n (C, L) and in triblock copolymers $A_{m/2}$–B_n–$A_{m/2}$ (S) [8]. Note that m and n in this figure refer to the respective space requirements and not to the amounts of monomeric units.
C: Compatibilizer at the phase boundary - - - - between A–polymers and B–polymers.
S: Spherical A–domains in a continuous B–matrix ($m << n$).
L: Lamellae with A–layers and B–layers ($m = n$).

If the space required by A_m blocks relative to B_n blocks is between that for lamellae ($m = n$) and that for spherical domains ($m \ll n$), cylindrical domains of A_m blocks will form in a continuous matrix of B_n blocks since one can imagine cylinders as one-dimensionally shrunk lamellae or one-dimensionally stretched spheres.

The formation of spherical poly(styrene) domains PS (glass temperature of domains ca. 80°C) in a continuous matrix of poly(butadiene) segments BR (glass temperature ca. −10°C) is utilized for **thermoplastic elastomers** from triblock copolymers $S_{m/2}$–Bu_n–$S_{m/2}$. The "hard" PS domains act at temperatures $T < 65°C$ as physical cross-links for the "soft" BR matrix. At room temperature, the polymer behaves like an elastomer (Chapter 12). Heating the triblock copolymer above 80°C disbands the reversible physical cross-links and the polymer can be processed like a thermoplastic (Chapter 14). Similar thermoplastic elastomers are based on other triblock copolymers, for example, with center blocks composed of isoprene units.

Thermoplastic elastomers are also formed from multiblock copolymers with "hard" blocks (high glass temperature) and "soft" blocks (glass temperature below room temperature). Examples are thermoplastic polyesters with "hard" segments of ethylene terephthalate units $-(OCH_2CH_2O\text{-}OC(p\text{-}C_6H_4)CO)_m\text{-}$ and "soft" segments of tetrahydrofuran units $-(OCH_2CH_2CH_2CH_2)_p\text{-}$ and some polyurethanes with "hard" aromatic segments and "soft" polyether segments.

The property spectrum of a thermoplastic elastomer can, in principle, be obtained from any multiphase polymer which has a "hard" microphase dispersed in and connected to a "soft" continuous phase. Examples are not only graft copolymers of, e.g., isobutylene on poly(ethylene) or of vinyl chloride on poly(ethylene-*co*-vinyl acetate) but also blends of isotactic poly(propylene) with semicrystalline ethylene–propylene rubbers. These thermoplastic olefin elastomers "cross-link" via crystallization.

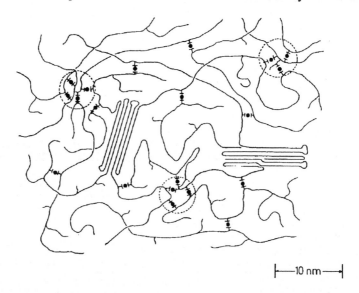

|—10 nm—|

Fig. 8-17 Schematic representation of a semicrystalline ionomer with crystalline lamellae, triple ions ~COO⁻/Mt²⁺/⁻OOC~ (-l●l-), and ion clusters (broken circles). The scale is approximate.

8.6. Ionomers

Ionomers are water-soluble copolymers with a high proportion of hydrophobic monomeric units and a small proportion of monomeric units with ionic groups. The best known example are polymers from the copolymerization of ethylene with less than 10 mol-% methacrylic acid where the acid has been converted into sodium or zinc salts. The ionic groups form intermolecular and intramolecular ion associations because of the high coordination numbers of Na and Zn (the valency does not matter). The resulting ionic domains are smaller than the spherical domains of triblock copolymers (Section 8.5.) but act similarly to physical cross-links in the continuous matrix of hydrophobic segments (Fig. 8-17)..

Literature

8.1. GENERAL
L.J.Mathis, ed., Solid State NMR of Polymers, Plenum Press, New York 1991
S.J.Spells, ed., Characterization of Solid Polymers: New Techniques and Developments, Chapman
    and Hall, London 1994
A.Teramoto, M.Kobayashi, T.Norisuye, eds., Ordering in Macromolecular Systems, Springer,
    Berlin 1994

8.2. MELTS and AMORPHOUS STATES
R.N.Haward, The Physics of the Glassy State, Interscience, New York 1973
G.Allen, S.E.B.Petrie, eds., Physical Structure of the Amorphous State, New York 1977
R.Zallen, The Physics of Amorphous Solids, Wiley, New York 1983
S.E.Keinath, R.L.Miller, J.K.Rieke, eds., Order in the Amorphous "State" of Polymers,
    Plenum, New York 1987

8.3. CRYSTALLINE STATE
A.Guinier, G.Fournet, Small Angle Scattering of X-Rays, W.Cey, New York 1955
L.E.Alexander, X-Ray Diffraction Methods in Polymer Science, Wiley, New York 1969
B.Wunderlich, Macromolecular Physics, Academic Press, New York 1973 ff. (3 vols.)
I.M.Ward, ed., Structure and Properties of Oriented Polymers, Halsted Press, New York 1975
H.Tadokoro, Structure of Crystalline Polymers, Wiley, New York 1979
D.C.Bassett, Principles of Polymer Morphology, Cambridge Univ.Press, Cambridge 1981
F.A.Bovey, Chain Structure and Conformation of Macromolecules, Academic Press, New York
    1982
O.Glatter, O.Kratky, Small Angle X-Ray Scattering, Academic Press, New York 1982
I.H.Hall, ed., Structure of Crystalline Polymers, Elsevier Appl.Sci.Publ., London 1984
D.A.Hemsley, The Light Microscopy of Synthetic Polymers, Oxford Univ.Press, New York 1985
R.A.Komoroski, ed., High Resolution NMR Spectroscopy of Synthetic Polymers in Bulk,
    VCH, Weinheim 1986
L.C.Sawyer, D.T.Grubb, Polymer Microscopy, Chapman and Hall, New York 1987
A.E.Woodward, Atlas of Polymer Morphology, Hanser/Oxford Univ.Publ., Munich/Oxford 1988
P.Corradini, G.Guerra, Polymorphism in Polymers, Adv.Polym.Sci. **100** (1992) 183
G.Rhodes, Crystallography Made Crystal Clear. A Guide for Users of Macromolecular Models,
    Academic press, New York 1993
A.E.Woodward, Understanding Polymer Morphology, Hanser, Munich 1995

8.4. LIQUID CRYSTALS

P.G.de Gennes, The Physics of Liquid Crystals, Clarendon Press, Oxford 1974

D.Demus, L.Richter, Textures of Liquid Crystals, Verlag Chemie, Weinheim 1978

N.A.Platé, V.P.Shibaev, Comb-Shaped Polymers and Liquid Crystals, Khimia, Moscow 1980 (in Russian); Plenum, New York 1987 (in English)

A.Ciferri, W.R.Krigbaum, R.B.Meyer, ed., Polymer Liquid Crystal, Academic Press, New York 1982

P.J.Flory, Molecular Theory of Liquid Crystals, Adv.Polym.Sci. **59** (1984) 1

N.March, M.Tosi, ed., Polymers, Liquid Crystals, and Low-Dimensional Solids, Plenum, New York 1984

G.W.Gray, ed., Thermotropic Liquid Crystals, Wiley, New York 1987

A.E.Zachariades, R.S.Porter, ed., Structure and Properties of Oriented Thermotropic Liquid Crystalline Polymers in the Solid State, Dekker, New York 1988

C.B.McArdle, Side Chain Liquid Crystals, Blackie and Sons, Glasgow (UK) 1988

A.A.Collyer, ed., Liquid Crystal Polymers: From Structure to Applications, Chapman and Hall, New York 1992

A.M.Donald, A.H.Windle, Liquid Crystalline Polymers, Cambridge Univ. Press, Cambridge, UK, and New York, 1993

N.A.Platé, ed., Liquid Crystal Polymers, Plenum, New York 1993

A.Keller, M.Warner, A.H.Widle, eds., Self-Order and Form in Polymeric Molecules, Chapman and Hall, London 1995

8.5. BLOCK COPOLYMERS

J.A.Manson, L.H.Sperling, Polymer Blends and Composites, Plenum, New York 1976

A.Nohay, J.E.McGrath, Block Copolymers: Overview and Critical Survey, Academic Press, New York 1976

B.R.M.Gallot, Preparation and Study of Block Copolymers with Ordered Structures, Adv.Polym.Sci. **29** (1978) 87

I.Goodman, ed., Developments in Block Copolymers **1** (1982) ff., Appl.Sci.Publ., Barking

D.J.Meier, ed., Block Copolymers. Science and Technology, Harwood Academic Publ., New York 1983

M.J.Folkes, ed., Processing, Structure and Properties of Block Copolymers, Elsevier, New York 1985

N.R.Legge, G.Holden, H.E.Schroeder, ed., Thermoplastic Elastomers, Hanser, Munich 1987

8.6. IONOMERS

M.Pineri, A.Eisenberg, ed., Structure and Properties of Ionomers, NATO ASI Series, Reidel, Dordrecht 1987

References

[1] R.J.Samuels, Structured Polymer Properties, Wiley, New York 1974, Fig. 2-4

[2] G.Natta, P.Corradini, I.W.Bassi, Rubber Chem.Technol. **33** (1960) 703, Fig. 10 [2a] and 13 [2b]; both modified

[3] W.Kern, J.Davidovits, K.J.Rauterkus, G.F.Schmidt, Makromol.Chem. **43** (1961) 106, Table 4

[4] A.J.Pennings, A.M.Kiel, private communication

[5] R.J.Samuels, Structured Polymer Properties, Wiley, New York 1974, Figs. 2-31 and 3-22

[6] A.J.Pennings, J.M.M.A. van der Mark, A.M.Kiel, Kolloid Z.Z.Polym. **237** (1970) 336, Fig. 2 (electron microscopy) and 14 (drawing)

[7] H.-G. Elias, An Introduction to Plastics, VCH, Weinheim 1992, Fig. 4-4

[8] H.-G. Elias, An Introduction to Plastics, VCH, Weinheim 1992, Fig. 4-7

9. Transitions and Relaxations

9.1. Introduction

Melting temperatures of most low molar mass substances are very easy to detect since solid, crystalline materials fuse together to become melts with low viscosities at these temperatures. The transition mesophase $\rightleftarrows$ isotropic liquid of low molar mass LCs is also easy to see because it involves a clearing of the liquid mesophase. Since polymer melts are turbid and flow only slowly under their own weight because of their high viscosities, changes of viscosities at the melting temperature and other thermal transitions are hard to see with the naked eye. These and other polymer transitions are therefore usually explored with thermal, mechanical or electrical methods.

9.2. Thermal Properties

9.2.1. States and Methods

The thermodynamic state of matter is described by the Gibbs energy G and its partial first derivatives with respect to temperature T and pressure p, i.e., by entropy S, volume V and enthalpy H, resp.:

(9-1) $\quad S = -(\partial G/\partial T)_p$

(9-2) $\quad V = (\partial G/\partial p)_T$

(9-3) $\quad H = G + TS = G - T(\partial G/\partial T)_p$

Partial second derivatives of G are the isobaric heat capacity C_p (formerly called "specific heat"), the cubic expansion coefficient β, and the isothermal cubic compressibility κ:

(9-4) $\quad C_p = (\partial H/\partial T)_p = T(\partial S/\partial T)_p = -T(\partial^2 G/\partial T^2)_p$

(9-5) $\quad \beta = (1/V)(\partial V/\partial T)_p$

(9-6) $\quad \kappa = -(1/V)(\partial V/\partial p)_T$

IUPAC recommends the symbols α, α_V or γ for the cubic expansion coefficient and the symbol α_l for the linear expansion coefficient. The polymer literature uses mostly α for the cubic expansion coefficient of solid and liquid polymers. Since α is also the symbol for the *linear* expansion coefficient of coils, this book will use α for the linear expansion coefficient and β for the cubic one in order to simplify the notation.

The isochoric heat capacity C_V (heat capacity at constant volume; G: *chora* = room) is an important theoretical quantity. It is calculated from the experimental isobaric heat capacity via $C_V = C_p - (TV\beta^2/\kappa)$. Molar heat capacities $C_{V,m} = C_V/n$ and specific heat capacities C_V/m are obtained by dividing heat capacities C_V by amount of substance n and mass m, respectively (analog of $C_{p,m}$ and c_p).

Thermodynamic quantities can be obtained by classical methods such as calorimetry. Much more important for polymers is **thermal analysis** which determines the temperature dependence of a physical property with the help of a controlled temperature program (Fig. 9-1).

Heat capacities are usually determined by (power-compensated) **differential scanning calorimetry** (DSC). Specimen and reference material are heated in such a way that both are always at the *same temperature* ($\Delta T = 0$). DSC measures the temperature dependence of the electric power P that is necessary to compensate caloric effects and thus the temperature dependence of $d\Delta Q/dt$, the change of added heat ΔQ with time (Fig. 9-1). Positive signals indicate exothermic effects (crystallization, exothermic chemical reactions) and negative signals endothermic ones (solid–solid transitions, crystal melting, endothermic chemical reactions). The added heat ΔQ allows to calculate the heat difference between specimen and reference, and, if the heat capacity of the reference is known, also heat, enthalpy, and heat capacity of the specimen.

Specimen and reference are heated at a *constant rate* in **differential thermal analysis** (DTA). At the melting temperature, a crystalline specimen will take up heat until it is melted. The temperature of the specimen stays constant during the melting process whereas the temperature of the reference continues to increase. The observed temperature difference ΔT between specimen and reference is thus proportional to the heat flux Q_p.

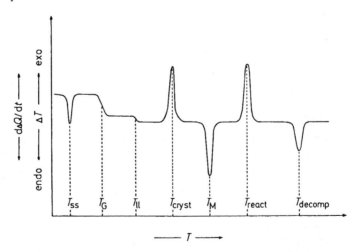

Fig. 9-1 Idealized thermogram of a semi-crystalline polymer with solid–solid transition at T_{ss}, glass temperature T_G (transformation of a glass into a melt), liquid–liquid transition at T_{ll} (controversial), crystallization at T_{cryst}, crystalline melting temperature T_M, exothermic reaction at T_{react}, and endothermic polymer decomposition at T_{decomp} (schematic). Measurements by DSC {$d\Delta Q/dt = f(T)$} and DTA {$\Delta T = f(T)$}.

Thermomechanical analysis (TMA) measures the deformation of a specimen by a load. **Dynamic mechanical analysis** (DMA) determines either the temperature dependence of shear moduli by forced oscillations or by damping of free oscillations. Several thermoanalytical methods measure the decomposition of polymers on heating: **gas evolution analysis** (EGA) by determination of evolved gases and **thermogravimetry** (TGA) by the loss of mass of the specimen.

Industrial thermoanalytical methods measure **heat distortion temperatures,** i.e., the temperatures at which a specimen is deformed by ΔL under controlled conditions such as standardized load (in MPa) and heating rate (in K/h). The deformation is either the bending of the specimen by a controlled change of length, ΔL, (Martens temperature, heat distortion temperature) or the penetration of a needle to a depth of ΔL (Vicat temperature). The resulting **Martens temperatures, heat distortion temperatures** and **Vicat temperatures** depend differently on melting or glass temperatures since the results are also affected by the elasticity of the specimen. Vicat temperatures and heat distortion temperatures are furthermore influenced by surface hardnesses.

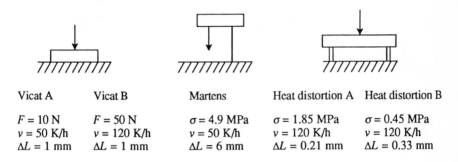

Vicat A	Vicat B	Martens	Heat distortion A	Heat distortion B
$F = 10$ N	$F = 50$ N	$\sigma = 4.9$ MPa	$\sigma = 1.85$ MPa	$\sigma = 0.45$ MPa
$v = 50$ K/h	$v = 120$ K/h	$v = 50$ K/h	$v = 120$ K/h	$v = 120$ K/h
$\Delta L = 1$ mm	$\Delta L = 1$ mm	$\Delta L = 6$ mm	$\Delta L = 0.21$ mm	$\Delta L = 0.33$ mm

9.2.2. Thermal Expansion

Atoms, segments and molecules increase their vibration on heating. More space is required and the specimen expands. The expansion is measured by **thermodilatometry** (L: *dilatare* = to enlarge, extend) and characterized by the **cubic expansion coefficient** $\beta = (1/V)(\partial V/\partial T)_p$ or the **linear expansion coefficient** $\alpha = (1/L)(\partial L/\partial T)_p$. Linear expansion coefficients of polymers are similar to those of liquids and may adopt values up to ca. $300 \cdot 10^{-6}$ K^{-1} (Table 9-1).

Expansion coefficients are controlled by changes of distances between atoms with temperature. These changes are small for covalent bonds and large for dispersion forces: the expansion coefficient of quartz is thus small and that of carbon disulfide is large (Table 9-1). Expansion coefficients of polymers are between these two extremes because polymers have covalent bonds in the chain direction and dipole–dipole interactions and/or dispersion forces in the two other directions.

Oriented polymers are therefore anisotropic; they expand differently in the three spatial directions. Their expansion coefficients are negative in the direction of chain axes because chains contract with increasing temperature due to the increasing ampli-

Table 9-1 Effect of bond forces in the three spatial directions on linear expansion coefficients $\alpha = \beta/3$, isobaric specific heat capacities c_p, and heat conductivities λ of isotropic materials at 25°C.
  Bonds in the three spatial directions: c = Covalent bonds, m = metallic bonds, h = hydrogen bonds, p = dipole–dipole interactions, d = dispersion forces.

Material		Bonds	$\dfrac{10^6\alpha}{K^{-1}}$	$\dfrac{c_p}{J\,K^{-1}g^{-1}}$	$\dfrac{\lambda}{W\,m^{-1}K^{-1}}$
Quartz	$(SiO_2)_n$	c c c	1	0.72	10.5
Iron	[Fe]	m m m	12	0.54	58
Water	$\{H_2O\}$	h h h	70	4.2	
Polyamide 6	$+NH–(CH_2)–CO\,\}_{\overline{n}}$	c h d	60	1.6	0.31
Poly(styrene), at	$+CH_2–CH(C_6H_5)\,\}_{\overline{n}}$	c p p	70	1.3	0.16
Poly(vinyl chloride), at	$+CH_2–CHCl\,\}_{\overline{n}}$	c p p	80	1.2	0.18
Poly(ethylene), amorphous	$+CH_2–CH_2\,\}_{\overline{n}}$	c d d	287	2.1	0.35
Carbon disulfide	CS_2	d d d	380		

tude of thermal motions perpendicular to chain axes. Cubic expansion coefficients β of anisotropic materials can therefore not be converted by $\beta = 3\,\alpha$ into true linear expansion coefficients α because $\alpha_x \neq \alpha_y \neq \alpha_z$. Polymers are isotropic if chain axes are distributed at random, for example, in amorphous or in non-oriented crystalline polymers.

9.2.3. Heat Capacities

Specific heat capacities are independent of crystallinity at very low temperatures (Fig. 9-2). At 0 K, amorphous and semi-crystalline materials do have a residual entropy $S_{0,a}$, however, whereas $S_{0,c} = 0$ for ideal crystals.

At low temperatures, isobaric specific heat capacities c_p increase practically linearly with temperature because atoms vibrate more and more around their positions at rest. Additional new vibrations and rotations around chain bonds cause c_p to increase strongly at the glass temperature T_G. Since such motions start below the glass temperature, chain segments of some crystallizable polymers gain enough mobility to move into ordered positions; these polymers **recrystallize** below T_G.

Heat is required to melt crystals and c_p thus runs through a maximum at the melting point T_M. The true thermodynamic melting temperature $T_{M,0}$ is the highest temperature of the melting range because it is here where the largest and most perfect crystals melt (Fig. 9-2; see also Fig. 9-7).

Heat capacities of spherical gas atoms can have a maximum value of $C_{V,m} = 3\,R = 3 \cdot 8.314$ J K^{-1} mol^{-1} according to the energy distribution. In polymers, some degrees of freedom are always frozen-in; experimental heat capacities are therefore lower than $C_{V,m} = 3\,R$. For example, PPE $+O\text{-}p\text{-}C_6H_2(CH_3)_2\}_{\overline{n}}$ with the monomeric unit C_8H_8O exhibits $c_p = 1.22$ J K^{-1} g^{-1} and thus $C_{p,m} = 146.4$ J K^{-1} (mol mer)$^{-1}$ and $C_{p,m} = 8.61$ J K^{-1} (mol atom)$^{-1} \approx 1\,R = 8.314$ J K^{-1} mol^{-1} at 25°C.

Specific heat capacities of plastics vary between $c_p = 0.85$ J K^{-1} g^{-1} [poly(vinyl chloride)] and 2.7 J K^{-1} g^{-1} [high density poly(ethylene)]. Since specific heat capacities of mineral fillers are usually ca. 0.9 J K^{-1} g^{-1}, most filled plastics use less energy for processing than unfilled ones (Chapter 14).

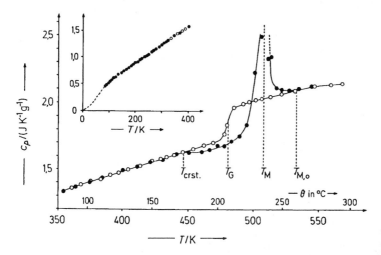

Fig. 9-2   Temperature dependence of isobaric specific heat capacities of semi-crystalline (●) and amorphous (○) poly[oxy-(2,6-dimethyl)-1,4-phenylene] PPE. T_{crst} = Start of recrystallization, T_G = glass temperature, T_M = conventional melting temperature, $T_{M,0}$ = melting temperature of the largest and most perfect crystals. With permission by General Electric Co. [1].

9.2.4. Heat Conductivity

Heat is conducted in metals by electrons with a radius of ca. 10^{-5} nm but in conventional (not electrically conducting) polymers by much larger phonons (elastic waves) with a free wavelength of ca. 0.7 nm. Heat conductivities of polymers are therefore much lower than those of metals (Table 9-1).

Heat conductivities λ of natural rubber and other *amorphous polymers* increase slightly with temperature T at low temperatures (Fig. 9-3). This slight increase of λ with T must come from an increase of specific heat capacity c_p with T since the free path length of phonons is almost independent of temperature for liquids and amorphous glasses. Heat conductivities are practically identical just below and just above the glass temperature since segments are packed with similar densities in both states. Increasing temperatures cause stronger thermal motions: segments are packed less densely and λ decreases.

Heat conductivities of *crystalline polymers* are very low near absolute zero, then increase with temperature and run through a maximum (not shown in Fig. 9-3). For poly(ethylene), this maximum is $\lambda \approx 100 \cdot 10^{-4}$ J cm^{-1} s^{-1} K^{-1} at $-170°C$. Heat conductivities begin to decrease long before the melting process starts and drop dramatically near the melting temperature T_M (Fig. 9-3).

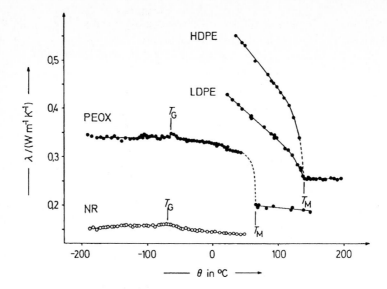

Fig. 9-3 Heat conductivities of natural rubber NR, poly(oxyethylene) PEOX, and poly(ethylene)s
with high density (HDPE) and low density (LDPE). ○ Amorphous polymers, ● semi-crystalline
polymers. T_G = Glass temperature, T_M = melting temperature. After a compilation in [2].

9.3. Crystallization

9.3.1. Nucleation

The crystallization of polymers is controlled by the macroconformations of
macromolecules. *Spheres* arrange themselves in superlattices, e.g., spherical enzymes
or latex particles (Section 8.3.1.). Rigid molecules with high aspect ratios form parallel *rods*. Flexible molecules fold to microlamellae (Section 8.3.5.) and spherulites
(Section 8.3.6.), depending on crystallization conditions.

Crystallizations are initiated by nuclei with concentrations between ca. 1 nucleus
per cm³ [poly(oxyethylene)] and 10^{12} nuclei per cm³ [poly(ethylene)]. In the very
rare **homogeneous nucleation**, thermal motion causes molecules and segments of the
crystallizing polymer to cluster spontaneously and to form unstable embryos which
develop into stable nuclei upon further growth. The nucleation is *sporadic* since nuclei are formed one after the other. It is also "primary", i.e., three-dimensional
because surfaces of nuclei are increased by the addition of molecule segments in all
three spatial directions (Fig. 9-4).

Heterogeneous nucleations are *athermal*; they involve extraneous nuclei with diameters of at least (2-10) nm. Nuclei may be dust particles, walls of containers, or deliberately added nucleating agents. They may even consist of residual nuclei of the

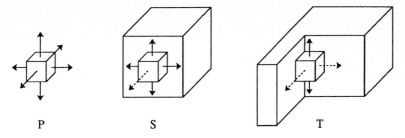

Fig. 9-4 Primary (P), secondary (S) and tertiary (T) nucleation. The surface of the nucleus is enlarged in direction → but not in direction - ->.

polymer. Melting of polymers with broad melting ranges may leave some higher melting crystallites intact and it is these crystallites that may act as nuclei on subsequent cooling and crystallization. Residual nuclei are also responsible for the "memory effect" of polymer melts. Spherulites appear on cooling of melts at the same spots they occupied before the melting since residual nuclei were unable to diffuse away because of high melt viscosities.

Chain segments add to surfaces of polymer nuclei in secondary nucleations and most likely to corners and furrows of nucleating agents in tertiary nucleations. Industrial nucleating agents are alkaline earth metal salts of organic acids for poly-(olefin)s, sulfates of bivalent metals for aromatic polyesters, and titanium dioxide or carbon black for polyamides.

Secondary nucleation and supercooling of the melt control chain folding and lamella heights (Fig. 9-5). If a chain segment of variable length L_c is added to a nucleus, the crystallite surface is enlarged by the contribution $2\,L_cL_d$ from the two side planes and the contribution $2\,L_dL_b$ from the two ebd planes. The gain of Gibbs

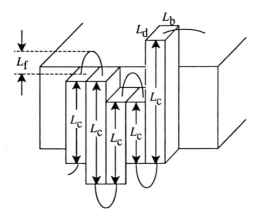

Fig. 9-5 Schematic representation of the addition of an infinitely long chain molecule to the side plane of a lamella of folded chain molecules. The extended chain segments of the added chain have various lengths L_c; the cross-sectional area of the added chain is L_dL_b. Gibbs surface energies are σ_s for each side plane L_cL_d and $\sigma_f = L_f\Delta H_f$ for each end plane L_dL_b where L_f = fold length and ΔH_f = surface enthalpy per volume.

energy by creation of new surfaces is counteracted by a loss of Gibbs energy ΔG_{cryst} per unit volume. One obtains for the first segment on the surface:

$$(9\text{-}7) \qquad \Delta G_i = 2\, L_b L_d \sigma_f + 2\, L_c L_d \sigma_s - L_c L_b L_d \Delta G_{cryst}$$

Differentiating Eqn.(9-7) with respect to L_b and equating the result with zero delivers the critical (minimal) height $L_{c,o} = 2\,\sigma_e/\Delta G_{cryst}$ at which the Gibbs energy of crystallization just balances the formation of an end surface, i.e., the addition of the first segment.

Since the change of the Gibbs energy is zero for such an addition, a nucleus of this size can never become stable. For the nucleus to grow to a stable crystal, Gibbs energies have to be slightly negative and fold heights thus slightly larger than $L_{c,o}$. This additional length ΔL will be ignored.

Since the Gibbs energy of crystallization per unit volume of an extended chain is given by $\Delta G_{cryst} = \Delta H_{M,o} - T_{cryst}\Delta S_{M,o}$ and a crystal composed of such chains has a melting temperature of $T_{M,o} = \Delta H_{M,o}/\Delta S_{M,o}$, one obtains:

$$(9\text{-}8) \qquad L_{c,o} = \frac{2\,\sigma_e T_{M,o}}{\Delta H_{M,o}(T_{M,o} - T_{cryst})} \qquad \text{(plus } \Delta L)$$

The critical theoretical lamella height thus decreases with increasing supercooling $(T_{M,o} - T_{cryst})$ which is confirmed by experiment.

9.3.2. Crystallization Rate

Embryons require a critical size before they become stable nuclei and then crystallites. At the melting temperature T_M, crystallites are dissolved and the crystallization rate is thus zero. Nuclei and crystallites can also not grow at temperatures below the glass temperature T_G since the high viscosity prevents the diffusion of chain segments to crystallites. The crystallization rate must therefore run through a maximum with increasing temperature. This maximum is found experimentally at $T_{cryst,max} \approx (0.80\text{-}0.87)\, T_{M,o}$ (in K) where $T_{M,o}$ = melting temperature of perfect crystals.

Crystallization can be subdivided into a primary and a secondary stage. At the end of primary crystallization, the whole volume of the vessel is macroscopically completely filled with crystalline entities, e.g., spherulites (see Fig. 8-12). Although the specimen appears to be totally crystallized, it is not totally crystalline since the spherulites contain non-crystallized material. The crystallinity is thus lower than the crystallizability (Section 8.3.1.). The polymer may then continue to crystallize slowly during a secondary stage. This after-crystallization may lead to shrinking and warping of freshly injection molded plastics.

Primary crystallization is described by the **Avrami equation**. In a volume V, N nuclei develop into N crystalline entities with volumes V_i each, e.g., spherulites. The probability is $p_i = 1 - (V_i/V)$ that a crystallizable unit (segment, molecule, etc.) is not in a certain entity i. The total probability that the unit is not at all in *any* entity is thus given by the product of single probabilities $p = \Pi_i [1 - (V_i/V)]$ (with $1 \leq i \leq N$) and thus $\log_e p = \Sigma_i \log_e [1 - (V_i/V)]$.

The logarithm can be developed into a series, $\log_e (1 - y) = - y - y^2/2 -...$, because the volume V_i is much smaller than the total volume ($V_i/V = y \ll 1$). Higher terms are neglected and one obtains $\log_e p = - \Sigma_i (V_i/V) = - (1/V) \Sigma_i V_i$ and $p = \exp[- (1/V) \Sigma_i V_i]$. Introduction of the number-average of volumes of entities, $\overline{V}_n = (\Sigma_i V_i)/N$, and their number concentration $C = N/V$ results in $p = \exp[- C \overline{V}_n]$.

The probability p equals the volume fraction $\phi_F = V_F/V_0$ of the non-crystallized fraction in the partially crystallized melt. The melt has a volume V_0 before the crystallization, V during crystallization, and V_∞ at the end of the primary crystallization. The density $\rho_0 = m_0/V_0$ of the melt before the crystallization equals the density $\rho_F = m_F/V_F$ of the non-crystallized liquid during crystallization. Thus, $\phi_F = V_F/V_0 = m_F/m_0$ and therefore also $m_F/m_0 = \exp[- C \overline{V}_n]$.

The volume $V = V_F + V_S$ of the melt at the time t is given by the volume $V_F = m_F/\rho_F$ of the liquid and the volume $V_S = m_S/\rho_S$ of the crystallized entities. The mass of these entities is $m_S = 0$ at $t = 0$, $m_S = m_0 - m_F$ at time t, and $m_S = m_0$ at $t \to \infty$. Using $\rho_F = m_0/V_0$ and $\rho_S = m_0/V_\infty$ one obtains $V = V_\infty + (m_F/m_0)(V_0 - V_\infty)$ and therefore also $(V - V_\infty)/(V_0 - V_\infty) = \exp[- C \overline{V}_n]$. The number-average $\overline{V}_n$ of volumes of entities is obtained as follows:

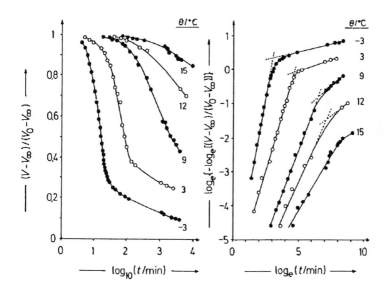

Fig. 9-6   Crystallization kinetics of a poly(butadiene) with 80 % 1,4-trans units at various temperatures [3].
Left: Time dependence of relative contractions (time scale is logarithmic for clarity).
Right: Avrami plot according to Eqn.(9-10).
The intercept of the two linear parts of the curve indicates the transition from primary to secondary stage. Avrami exponents of poly(butadiene) decrease from $z = 2.20$ at $-3°C$ to $z = 0.66$ at $15°C$.

All nuclei are generated at the same time in *simultaneous nucleation*; the number concentration of nuclei does not change with time. The nuclei occupy identical volumes at a certain time but these volumes increase with time. Rods extend in the longitudinal direction but maintain their cross-sectional areas A, radii R of discs grow but heights H remain constant, and spheres with radii R expand radially:

$$\text{Rods:} \quad \bar{V}_n = AL \quad ; \quad \text{with } L = k_1 t \quad \rightarrow \quad \bar{V}_n = Ak_1 t \quad = K_1 t$$
$$\text{Discs:} \quad \bar{V}_n = \pi H R^2 \quad ; \quad \text{with } R = k_2 t \quad \rightarrow \quad \bar{V}_n = \pi H k_2^2 t^2 \quad = K_2 t^2$$
$$\text{Spheres:} \quad \bar{V}_n = (4\,\pi/3)R^3 \quad ; \quad \text{with } R = k_3 t \quad \rightarrow \quad \bar{V}_n = (4\,\pi/3)k_3^3 t^3 \quad = K_3 t^3$$

The general case is characterized by the **Avrami equation**:

(9-9) $(V - V_\infty)/(V_0 - V_\infty) = \exp[-CK_j\, t^z] = \exp[-Kt^z]$

where the exponent z of the stretched exponential adopts values of 1, 2 or 3. Taking double logarithms lets one linearize the equation as:

(9-10) $\log_e\{-\log_e[(V - V_\infty)/(V_0 - V_\infty)]\} = \log_e K + z \log_e t$

Concentrations C of nuclei increase with time during *sporadic nucleations*, e.g., as $C = k_c t$. New nuclei are formed both outside and inside existing entities. This double occupation of the same space does not change the result since the newly generated entities in the interior do not affect the fraction of free space.

With $C = k_c t$, one obtains for spheres an average volume of $\bar{V}_n = K_3' t^3$, similar to simultaneous nucleation. Because of $\exp(-C\bar{V}_n) = \exp(-k_c t K_3' t^3)$, the Avrami exponent increases however to $z = 4$. Other theoretical Avrami exponents are obtained for diffusion-controlled nucleations (Table 9-2). Experiments may also deliver fractional exponents. These exponents are difficult to interpret since constants z and K depend on experimental methods. For example, dilatometry measures the growth of spherulites whereas calorimetry also determines that of lamellae in spherulites.

Crystallization rates vary widely with the constitution and configuration of polymers. At at crystallization temperature of 30 K below the melting temperature, the linear crystallization rate of poly(ethylene) is ca. 5000 μm/min but that of poly(vinyl chloride) only ca. 0.01 μm/min.

Table 9-2 Theoretical Avrami exponents z

Growth	Simultaneous nucleation		Sporadic nucleation
	Without diffusion control	With diffusion control	No diffusion control
One-dimensional	2	3/2	$2 \geq z \geq 1$
Two-dimensional	3	4/2	$3 \geq z \geq 2$
Three-dimensional	4	5/2	$4 \geq z \geq 3$
Sheaf-like	≥ 6		$7 \geq z \geq 5$

9.4. Melting Temperature

9.4.1. Phenomena

Increasing vibration of atoms on heating causes crystal lattices of linear macro-molecules to expand perpendicular to chain axes. For example, the lattice constant b of poly(ethylene) enlarges by ca. 7 % between $-196°C$ and $+138°C$. Monomeric units are more and more dynamically disordered around their ideal positions at rest; even crystal defects may occur. Disorder ia especially great at surfaces, edges and corners of crystallites (see Section 8.3.5.) and it is here that the melting process starts. Nuclei are not required. The number of chain units involved in the melting process has been estimated as 60 to 160 from the ratio of molar activation energy to molar melting enthalpy (both per mole chain unit).

Crystals of low molar mass compounds are relatively perfect. For example, crystals of $C_{44}H_{90}$ melt at $\theta = 86.4°C$ within a temperature interval of $\Delta T = 0.25$ K (Fig. 9-7). The larger chains of $C_{94}H_{190}$ cannot crystallize that perfectly; due to defects, some segments are therefore somewhat more mobile in the lattice. As a consequence, segments are constantly redistributed between crystalline and non-crystalline regions on heating and the melting of $C_{94}H_{190}$ starts at ca. 110°C and finishes at ca. 114.6°C ($\Delta T = 3.6$ K). The imperfect crystal structure produces a **melting range**. The largest and most perfect crystals melt at the high-temperature end of this range. For low molar

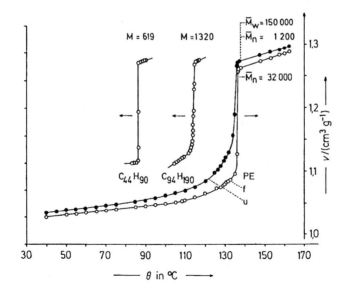

Fig. 9-7   Temperature dependence of specific volumes v of two chemically pure paraffins $C_{44}H_{90}$ and $C_{94}H_{190}$ [4], a broadly distributed, unfractionated (u) poly(ethylene) PE and a fraction (f) there-of [5]. The scales for $C_{44}H_{90}$ and $C_{94}H_{190}$ were not given by the authors. Poly(ethylene)s were crystallized for 40 days at a temperature of 131.5°C, just below the melting temperatures.

mass materials, this transition perfect crystal $\rightleftarrows$ melt is relatively sharp (Fig. 9-7); it constitutes the **thermodynamic melting temperature** of the *specimen*.

Chain folds, end groups, and branch points generate additional defects (Fig. 8-10). Polymers thus have broader melting regions than analog oligomers, especially if molar mass distributions are broad (Fig. 9-7). The jumps of specific volumes v or enthalpies H at T_M degenerate to S-shaped curves and the sharp signals for the first derivatives $(\partial V/\partial T)_p = \beta V$ and $(\partial H/\partial T)_p = C_p$, broaden to become bell curves (cf. Figs. 9-2 and 9-10). The upper end of the melting range is no longer sharp and the middle of the melting range is therefore usually taken as the melting temperature T_M. The melting temperature T_M of the specimen is usually smaller than the thermodynamic melting temperature $T_{m,o}$ but it may also be larger if crystals are overheated.

9.4.2. Effect of Molar Mass

Melting temperatures increase with increasing degree of polymerization to a limiting value $T_{M,\infty}/K = \theta_{M,\infty}/°C + 273.15$ (Fig. 9-8). The dependence can be described by $(1/T_M) = (1/T_{m,\infty}) + (K_M/\bar{X}_n)$ where K_M is often thought to be due to the effect of end groups. This equation was first derived for linear polymers with a Schulz–Zimm distribution where $K_M = 2\,R/\Delta H_{M,u,m}$ and $\Delta H_{M,u,m}$ = molar melting enthalpy of a monomeric unit. It also applies to cyclic macromolecules (no end groups) as long as the macroconformations of polymer molecules are self-similar (Fig. 9-8).

Cycloalkanes with $N < 60$ have different ring conformations; they are not self-similar. At $N > 60$, they form fold micelles that are self-similar. At $N > 160$, they have the same melting temperature as alkanes which start folding at $120 < N < 160$.

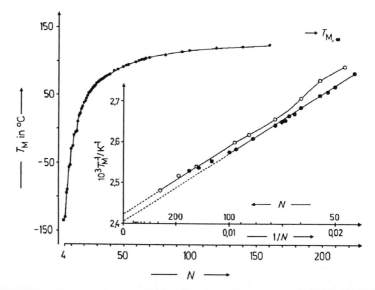

Fig. 9-8 Melting temperatures T_M of alkanes $H(CH_2)_N H$ as a function of the number N of methylene groups per molecule. Insert: $1/T_M = f(1/N)$ for alkanes ($\bullet$) and cycloalkanes ($\circ$).

9.4.3. Effect of Crystallite Size

The thermodynamic **melting temperature (fusion temperature)** of crystals from infinitely long ($N_u \rightarrow \infty$) extended chains (i.e., infinitely thick lamellae) is given by the melting enthalpies and melting entropies of the N_u monomeric units: $T_{M,o} = N_u \Delta H_{M,u}/N_u \Delta S_{M,u} = \Delta H_{M,u}/\Delta S_{M,u}$.

Folded chains have a lower melting temperature $T_M = \Delta H_M/\Delta S_M$ since the melting enthalpy of a segment, $\Delta H_M = N_u \Delta H_{m,u} - 2 \Delta H_\sigma = N_u \Delta H_{m,u} - 2 (\sigma_e/L_f)$, is lowered by the interfacial enthalpies of the folds. Combining these equations with the expressions for the melting temperature T_M of lamellae and $T_{M,o}$ of extended chains and neglecting differences in entropies ($\Delta S_M \approx \Delta S_{M,o}$) gives:

$$(9\text{-}11) \quad T_M = T_{M,o}\left(1 - \frac{2\sigma_e}{L_f \Delta H_{M,u}} \cdot \frac{1}{N_u}\right) = T_{M,o}\left(1 - \frac{2\sigma_e L_u}{\Delta H_{M,u} L_f} \cdot \frac{1}{L_c}\right)$$

Extrapolation of melting temperatures T_M to zero inverse lamella thickness ($1/L_c \rightarrow 0$) delivers the melting temperature $T_{M,o}$ of infinitely thick lamellae $L_c = N_u L_u \rightarrow \infty$.

Lamella heights can only be measured with great effort. A simpler means for the determination of thermodynamic melting temperatures $T_{M,o}$ is provided by the **Hoffman–Weeks equation**. Inserting the expression for $2\sigma_e$ from Eqn.(9-8) and setting $\gamma \equiv L_f/L_{c,o}$, one obtains an equation for the dependence of the melting temperatures T_M of lamellae on the crystallization temperature T_{cryst}:

$$(9\text{-}12) \quad T_M = (1 - \gamma^{-1})T_{M,o} + \gamma^{-1} T_{cryst}$$

Experiments indicate $\gamma \approx 2$ for many polymers. The fold length $L_f = \gamma L_{c,o}$ is thus ca. twice as large as the critical lamella height $L_{c,o}$.

9.4.4. Effect of Constitution

Melting temperatures of high molar mass polymers are strongly affected by the constitution of polymers. They decrease with increasing number i of methylene groups in repeating units $-NH(CH_2)_i CO-$ of aliphatic polyamides (Fig. 9-9). Melting temperatures of polyoxides with monomeric units $-O(CH_2)_i-$, aliphatic polyesters with $-OC(CH_2)_i COO(CH_2)_3 O-$, and isotactic poly($\alpha$-olefin)s with $-CH_2 CH\{(CH_2)_i H\}$ first decrease with increasing i but then raise again until they become identical with the melting temperature of poly(ethylene) for $i \rightarrow \infty$. The reason is as follows.

Aliphatic polyamides have much higher melting temperatures than aliphatic polyesters and polyethers which is usually taken as an effect of cohesion energies: 35.6 kJ/mol for amide groups vs. 12.1 kJ/mol for ester groups, 4.2 kJ/mol for oxygen groups and 2.9 kJ/mol for methylene groups. However, cohesion energies measure the intermolecular forces that are needed for the transition liquid $\rightleftarrows$ gas whereas melting is concerned with the transition crystal $\rightleftarrows$ melt. Since most hydrogen bonds are still present in polyamide melts according to infrared measurements, differences in cohesion energies of crystals and melts cannot be important. The melting enthalpy per mole chain unit averages only 3.7 kJ/mol for the seven chain units of $-NH(CH_2)_5CO-$ which is even slightly smaller than that of $-CH_2-$ with ca. 4.1 kJ/mol.

The important factor is not the cohesion energy but the packing of chains in crystals and the flexibility of segments. Aliphatic polyesters and polyoxides with small numbers i of methylene groups in repeating units are tightly packed in helical conformations whereas they adopt extended conformations for higher i. The macroconformations of these polymers are not self-similar at small and large i which explains the minimum in the function $T_M = f(i)$. Segments of these chains are also much more flexible than methylene segments because rotational barriers around bonds marked by $-$ are much smaller for $\sim CH_2-COOCH_2\sim$ (ca. 2.1 kJ/mol bond), $\sim O-CH_2\sim$ (ca. 5.0 kJ/mol) and $\sim CH_2-OOCCH_2\sim$ (ca. 5.0 kJ/mol) than for $\sim CH_2-CH_2\sim$ (12.3 kJ/mol). Melting temperatures of polyesters and polyoxides are thus lower than that of poly(methylene) (except for $i = 1$); they approach with increasing i (for $i > 4$) the melting temperature of poly(methylene).

Effects of packing are also responsible for the melting behavior of isotactic poly(α-olefin)s $+CH_2CH\{(CH_2)_iH\}+_{\overline{n}}$. The packing of short side chains becomes less tight with increasing i and the melting temperature decreases (Fig. 9-9). Longer side chains can however easily pack parallel due to **side-chain crystallization** and the melting temperature increases.

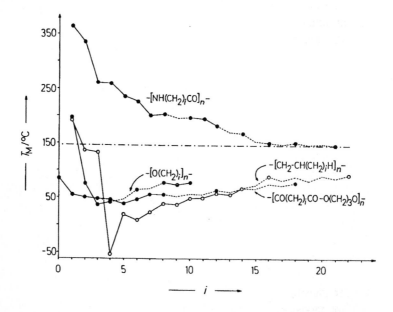

Fig. 9-9   Melting temperatures as function of the number i of methylene groups per repeating unit of high molar mass ($n \rightarrow \infty$) polyamides $+NH(CH_2)_iCO+_{\overline{n}}$, polyoxides $+O(CH_2)_i+_{\overline{n}}$ and polyesters $+OC(CH_2)_iCOO(CH_2)_3O+_{\overline{n}}$ and the number i of methylene groups per substituent in it-poly(α-olefin)s $+CH_2-CH\{(CH_2)_iH\}+_{\overline{n}}$. - - - Melting temperature of poly(ethylene) [6].

Melting temperatures are lowered if the regular packing of crystallizable macromolecules is disturbed by constitutional or configurational mistakes. They thus usually decrease with lower tacticities and with increasing proportions of irregular short-chain branches. Correspondingly, the melting temperatures of random bipolymers from two non-isomorphic monomers show a minimum with the composition whereas those of random bipolymers from isomorphic units increase steadily with the proportion of the higher melting comonomer.

9.5. Thermal Transitions and Relaxations

Melting is but one of many possible thermal transformations of polymers (Fig. 9-1). In **thermal transitions**, compounds are in thermal equilibrium below and above the transition temperature. An example is the melting temperature where crystallites are in thermal equilibrium with the melt. Polymers may also be present in non-equilibrium states that relax at a certain temperature. **Thermal relaxations** are thus kinetic phenomena that depend on the time scale, for example, on the frequency (a reciprocal time!). The best known example is the glass temperature (Section 9.6.).

Thermal transitions are subdivided into those of first, second ... nth order. Classic first order transitions are crystal $\rightleftarrows$ liquid (melting), liquid $\rightleftarrows$ gas (boiling), smectic $\rightleftarrows$ nematic (two-dimensionally ordered LCs $\rightleftarrows$ one-dimensionally ordered LCs), and nematic $\rightleftarrows$ isotropic liquid (clearing temperature).

At the melting temperature, heat has to be added until all ordered monomeric units in crystallites have been transformed into disordered units in the melt. Enthalpy H, volume V, and entropy S all jump to higher values at the melting temperature (Fig. 9-10). The first derivatives of H and V with respect to temperature (C_p, α; C_V) and pressure (κ) show corresponding infinitely high signals in the ideal case, the melting of infinitely large, perfect crystals. For imperfect crystals, discontinuities (H, V, S) degenerate to S-shaped curves and sharp signals (C_V, α, κ) to bell curves (Fig. 9-7).

A transition of nth order is defined as that transformation where the nth derivative of the Gibbs energy shows a discontinuity. An ideal first order thermodynamic transition thus has discontinuities in H, S and V at the transition temperature (e.g., T_M). An ideal second order thermodynamic transition shows discontinuities in α, C_V and κ at the transition temperature T_{tr}. Examples of true second order transitions are the lambda transition of liquid helium at 2.2 K, the rotational transformations of crystalline ammonium salts, and the disappearance of ferromagnetism at the Curie point. Some transformations of smectic LCs into other smectic LCs are also suspected to be true second order transitions.

This formal thermodynamic classification of thermal transitions corresponds to the phase behavior. All first order transitions exhibit two phases at their transition temperatures (crystal $\rightleftarrows$ melt $\rightleftarrows$ gas; nematic LC $\rightleftarrows$ isotropic liquid; etc.); all true second order transitions happen in a single phase.

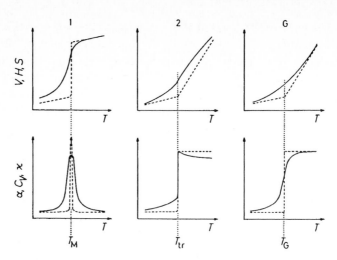

Fig. 9-10 First order (1) and second order (2) transitions and glass transformations (G):
First order: – – – Melting of an infinitely large, perfect crystal; ------ melting of a crystal with defects. See also Figs. 9-7 and 9-2.
Second order: ------ Transition with dominating intermolecular cooperative effects; - - - transition with exclusively intermolecular-cooperative effects. See text and Fig. 9-11.
Glass transformation: ----- Conventional experiment; - - - infinitely slow measurement. See Figs. 9-2 and 9-12.

This classification does not correlate with molecular processes, however.

In crystalline polymers, polymorphism can be generated by conformational iso-merism (Section 8.3.3.). If conformational isomers are neither iso-energetic nor ki-netically hindered, a conformer will be converted at a temperature $T_{tr} < T_M$ into the other conformer, for example, $(TG)_{3i} \rightleftarrows (TTG)_{2i}$. Such a conversion of sequences of conformers proceeds by rotation around chain bonds. The rotation is easy for loosely packed chains. In tightly packed crystals, it is hindered by neighboring chains. Rota-tions progress intramolecular-cooperatively between local conformations of the same chain in very loosely packed crystals but must involve intermolecular-cooperative effects of many other chains in tightly packed crystals. At a given packing density, transformations are easier if many crystal defects are present. One can thus distin-guish chains without an adjacent defect (O) and chains with an adjacent defect (D). An energy difference $\Delta E = (E_{DO} - E_{OO}) - (E_{DD} - E_{DO})$ per mole chain must thus exist for the various pairs of chains OO, DO, and OO. This energy difference mea-sures the tendency of defects to agglomerate near chains which favors intermolecular-cooperative movements. Intramolecular rotational transformations dominate at small ΔE whereas intermolecular ones are prevalent at large ΔE.

Model calculations of the temperature dependence of isochoric molar heat capaci-ties C_V showed that an increase of ΔE first leads to the so-called **Schottky anomaly**, followed by a diffuse and a sharp first order transition, a lambda transition (often in-terpreted as a second order transition), and finally a true second order transition (Fig. 9-11). Thus, rotational transformations appear as either first order or second order transitions, depending whether intra- or intermolecular-cooperative effects dominate.

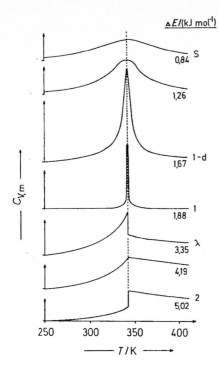

$\Delta E/(\text{kJ mol}^{-1})$

0,84 S

1,26

1,67 1-d

1,88 1

3,35 λ

4,19

5,02 2

250   300   350   400

—— T/K ——

Fig. 9-11 Model calculations of isochoric molar heat capacities for rotational isomerism in crystals in a system with various agglomerization energies ΔE of defects per chain (for details see original literature [7]). Increasing agglomerization energy first leads to a Schottky defect (S), then a diffuse first–order transition (1-d), a sharp first–order transition (1), a lambda transition (λ), and finally a second order transition (2). y axes are not to scale. With permission by Springer-Verlag [7].

9.6. Glass Temperatures

9.6.1. Static Glass Temperature

Amorphous substances convert at the **glass temperature** T_G from a glassy state to a "liquid" state, i.e., into a melt (low molar mass compounds) or a rubbery state (high molar mass chains). This liquid state vitrifies at a temperature T_E on cooling (L: *vitrum* = glass). T_G and T_E are affected by the rates of heating and cooling, resp. (Fig. 9-12). In general, they are not very different so that it suffices to discuss only T_G.

The transformation glass $\rightleftarrows$ (rubber-like) liquid resembles phenomenologically a thermodynamic second order transition since a more or less pronounced "jump" is observed at T_G for heat capacities or expansion coefficients (cf. Figs. 9-2 and 9-12 with Fig. 9-11). The transformation is not a true thermodynamic transition, however, because of the strong effect of heating/cooling rates on T_G. Because of these kinetic effects, the glass transformation is rather a relaxation; it should not be called a "glass *transition*" or a "second–order transition temperature".

The curves for the temperature dependence of C_V, C_p, c_p and β look similar for glass transformations and true thermodynamic second-order transitions but they are not one-phase transitions like the latter. For glass transformations, they are rather caused by strong intermolecular-cooperative movements of chain segments. A rapid cooling of a polymeric liquid prevents monomeric units from finding their equilibrium positions. The frozen-in structure of the liquid thus contains "defects" of atomic size, the **free volume**. These defects agglomerate similar to crystal defects (Section 9.5.) if glasses are heated. The resulting larger free volumes allow intermolecular-cooperative movements of chain segments in which ca. 20-60 chain atoms participate. These segment sizes can be deduced from the ratio of activation energy for the glass transformation to the melt enthalpy of semi-crystalline polymers and from the dependence of glass temperatures of amorphous polymers on the lengths of segments between cross-links.

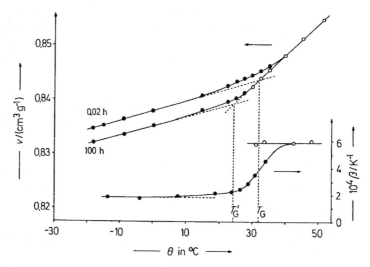

Fig. 9-12   Temperature dependence of specific volumes v and cubic expansion coefficients β of a poly(vinyl acetate) ($\overline{M}_\eta = 60\ 000$ g/mol) on fast cooling (0.02 h) and slow cooling (100 h). o Liquid state, ● glassy state. With permission by Wiley-Interscience [8].

9.6.2. Dynamic Glass Temperatures

Chain segments move with a certain frequency at temperatures $T > T_G$. The frequency v used in measurements and the deformation time $t_d = 1/v$ of the specimen must therefore affect the numerical value of the glass temperature.

One distinguishes correspondingly between static and dynamic methods. **Static glass temperatures** are delivered by calorimetry, differential scanning calorimetry, differential thermal analysis, and thermodilatometry and **dynamic glass temperatures** by broad-line nuclear magnetic resonance, mechanical and dielectric loss, resilience, and heat deformation temperatures (Section 9.2.1.). At static glass temperatures, slowly deformed bodies convert from brittle behavior into tough behavior (see Section 10.2.4.) whereas dynamic glass temperatures are important for short–term deformations such as impacts.

Poly(methyl methacrylate) has a static glass temperature of 110°C by thermodilatometry ($t_d \approx 10^4$ s) and dynamic glass temperatures of 120°C by penetrometry ($t_d \approx 10^2$ s) and 160°C by falling ball rebound ($t_d \approx 10^{-6}$ s). At 140°C, it behaves either as a glass (resilience) or as a rubber (penetrometry).

Static and dynamic glass temperatures can be interconverted by the **Williams–Landel–Ferry** equation (**WLF equation**). The glass transformation is assumed to be a relaxation process similar to viscosity; both processes depend on free volumes V_f. The **Doolittle equation** $\log_e \eta = \log_e A + B(V - V_f)/V_f$ relates viscosities to the total volume V and the free volume V_f per total mass. The free volume fractions are $\phi_f \equiv V_f/V$ for a temperature T and $\phi_{f,o} \equiv V_{f,o}/V_o$ for the reference temperature T_o. Temperatures shift viscosities which can be described by a shift factor $a_T = (\eta T_o \rho_o)/(\eta_o T \rho)$ where the densities ρ at temperature T and ρ_o at temperature T_o correct for the thermal expansion. The shift factor a_T corresponds to the ratio t/t_o of relaxation times at temperatures T and T_o.

Introduction of the Doolittle equations for T and T_o into the shift factor results in:

$$(9\text{-}13) \quad \log_{10} a_T = \frac{B}{2.303}\left(\frac{1}{\phi_f} - \frac{1}{\phi_{f,o}}\right) + \log_{10}\left(\frac{T_o \rho_o}{T \rho}\right) \approx \frac{B}{2.303}\left(\frac{1}{\phi_f} - \frac{1}{\phi_{f,o}}\right)$$

It is further assumed that the free volume fraction $\phi_f = V_f/V$ increases *linearly* with temperature according to $\phi_f = \phi_{f,o} + \beta_f (T - T_o)$. The expansion factor β_f approximates the true cubic expansion factor $\beta = (1/V)(dV/dT)$ for the *exponential* increase of volume with temperature. Because of this approximation, the WLF equation is restricted to a temperature range of $T_o < T < (T_o + 100\ \text{K})$.

Introduction of the expansion factor $\beta_f = (\phi_f - \phi_{f,o})/(T - T_o)$ into Eqn.(9-13) delivers the **WLF equation**:

$$(9\text{-}14) \quad \log_{10} a_T = \frac{-[B/(2.303\,\phi_{f,o})][T - T_o]}{[\phi_{f,o}/\beta_f] + [T - T_o]} = \frac{-K[T - T_o]}{K' + [T - T_o]} = \log_{10} t - \log_{10} t_o$$

Eqn.(9-14) applies to all relaxation processes. The adjustable parameters K, K' and $\phi_{f,o}$ are often assumed to be *universal* parameters, for example, first as $K = 17.44$, $K' = 51.6$ K, and $\phi_{f,o} = 0.025$ for $T_o = T_G$ and later as $K = 8.86$ K and $K' = 101.6$ K for $T_o = T_G + 50$ K. For more accurate calculations, different values of K, K' and $\phi_{f,o}$ should be used for *individual* polymers.

9.6.3. Effects of Constitution

The assumption of a free volume fraction explains many effects on glass temperatures, for example, that of molar mass. An end group E should generate a greater free volume than a monomeric unit. The excess free volume ΔV_E per chain end should be constant at T_G and independent of the number-average of molar mass, $\overline{M}_n$, for *linear chains*. Since the number concentration of end groups in melts of density ρ is given by $C_E = 2\rho N_A/\overline{M}_n$, the product $C_E\Delta V_E$ must indicate the change of free volume, $\beta_f(T_{G,\infty} - T_G)$, that is caused by end groups; T_G and $T_{G,\infty}$ are the glass temperatures for $\overline{M}_n$ and $\overline{M}_n \to \infty$, respectively.

The expansion coefficient β_f of the free volume at the glass temperature can be approximated by the difference $\beta_f \approx (\beta_L - \beta_G)$ of the cubic expansion coefficients of the melt (β_L) and the glass (β_G) near the glass temperature. The combination of $\beta_f = (\beta_L - \beta_G)$, $\beta_f(T_{G,\infty} - T_G) = C_E\Delta V_E$, and $C_E = 2\rho N_A/\overline{M}_n$ delivers:

$$(9\text{-}15) \quad T_G = T_{G,\infty} - \frac{2\rho N_A \Delta V_E}{\beta_L - \beta_G} \cdot \frac{1}{\overline{M}_n} = T_{G,\infty} - K \cdot \frac{1}{\overline{M}_n}$$

The glass temperature of linear polymers thus decreases with the inverse number-average molar mass (Fig. 9-13). The excess volume of poly(dimethyl siloxane) PDMS is calculated as $\Delta V_E = 2.16 \cdot 10^{-3}$ nm^3 using $\rho \approx 0.97$ g/cm^3, $\beta_L = 8.5 \cdot 10^{-4}$ K^{-1} and $\beta_G = 4.5 \cdot 10^{-4}$ K^{-1} whereas that of linear poly(styrene) PS is $\Delta V_E = 38.8 \cdot 10^{-3}$ nm^3 using $\rho \approx 1.04$ g/cm^3, $\beta_L = 6.0 \cdot 10^{-4}$ K^{-1} and $\beta_G = 2.6 \cdot 10^{-4}$ K^{-1}. The linear dimensions of end groups exceed those of center groups by 0.13 nm (PDMS) and 0.34 nm (PS).

Cyclic polymers possess no end groups and thus no excess free volume. Their glass temperatures should thus be independent of the molar mass. Instead, they increase linearly with increasing reciprocal molar mass whereas those of linear polymers decrease (Fig. 9-13). Small rings should however adopt less macroconformations than larger ones. Their lower flexibility results in higher glass temperatures. Glass temperatures are therefore not only controlled by the free volume (i.e., the translational entropy) but also by other entropic effects, hence, "thermodynamic" behavior.

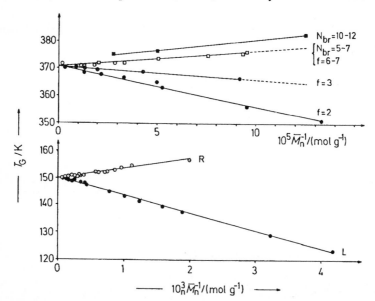

Fig. 9-13 Dependence of glass temperatures on inverse number-averages of molar masses.
    Bottom: Linear (L) and cyclic (R) poly(dimethyl siloxane)s PDMS [9].
    Top: Poly(styrene)s PS, either linear ($f = 2$) or star-like with $f = 3$ or 6-7 arms per molecule or branched via divinyl benzene units with $N_{br} = 5$-7 or 10-12 branches per molecule [10].

Chain segments of *star-like macromolecules* adopt less conformations near the core than farther away from it (Section 5.5.3.). The total flexibility of the molecule is reduced and the glass temperature is higher than that of linear chains with the same molar mass (Fig. 9-13). For star molecules with $f = 3$ arms per molecule, end group effects dominate and T_G decreases with increasing $1/\overline{M}_n$. For stars with $f = 6$-7, however, the reduction of the number of possible conformations is more important and T_G increases with increasing $1/\overline{M}_n$. The same effects increase the glass temperatures of randomly branched polymers and that of amorphous regions in semi-crystalline polymers. Glass temperatures cease to exist for strongly cross-linked polymers because segments are no longer mobile.

9.6.4. Plasticization

Glass temperatures of random bipolymers usually decrease with increasing proportion of comonomeric units 2 if $T_{G,1} > T_{G,2}$ (Fig. 9-14). Chains of such bipolymers pack less well than chains of the corresponding homopolymer 1. The increased flexibility of chains lets segments disperse the force of impacts more easily. Polymers become less brittle by this **internal plasticization** of chains.

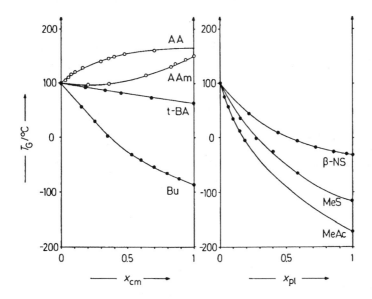

Fig. 9-14   Glass temperatures of poly(styrene)s as function of the mole fraction of comonomers (x_{cm}) or added plasticizer (x_{pl}) [11].
    Left: Internal plasticization of poly(styrene) by copolymerization with butadiene Bu or *t*-butyl acrylate t-BA and "hardening" of poly(styrene) by copolymerization with acrylic acid AA or acrylamide AAm (= internal plasticization of poly(acrylic acid) and poly(acrylamide) by copolymerization with styrene).
    Right: External plasticization of poly(styrene) by added β-naphthyl salicylate β-NS, methyl salicylate MeS, or methyl acetate MeAc.

Internal plasticizations can be interpreted as effects of either free volume or entropy. The **Gordon–Taylor equation** for the dependence of glass temperatures of copolymers on mass fractions w_1 and w_2 of monomeric units is obtained if volumes are additive and specific volumes of copolymers are mass-averages of the specific volumes of the corresponding homopolymers for both the glassy state $(T < T_G)$ and the rubbery state $(T > T_G)$:

$$(9\text{-}16) \quad T_G = \frac{K w_1 T_{G,1} + w_2 T_{G,2}}{w_1 + K w_2}; \quad K = \frac{\alpha_{R,2} - \alpha_{G,2}}{\alpha_{R,1} - \alpha_{G,1}}$$

K becomes unity if the differences of linear expansion coefficients are the same in the rubbery and glassy state for each of the two homopolymers. In this case, the glass temperature of copolymers varies linearly with the mass fraction of one of the comonomers according to $T_G = w_1 T_{G,1} + w_2 T_{G,2}$.

If glass transformations are treated as thermodynamic second order transitions and if the entropy change of the system is assumed to be that of a regular solution, then the **Couchman equation** is obtained:

$$(9\text{-}17) \quad \log_e T_G = \frac{K' w_1 \log_e T_{G,1} + w_2 \log_e T_{G,2}}{K' w_1 + w_2}; \quad K' = \frac{\Delta c_{p,1}}{\Delta c_{p,2}}$$

The Couchman equation reduces to the **Pochan equation** $\log_e T_G = w_1 \log_e T_{G,1} + w_2 \log_e T_{G,2}$ if the difference between the specific heat capacities in the glassy and the rubbery state is the same for both homopolymers, i.e., if $\Delta c_{p,1} = \Delta c_{p,2}$.

The logarithms of the temperatures in Eqn.(9-17) can be replaced by the temperatures themselves [$\log_e(1 + y) \approx y + ...$] if the two glass temperatures $T_{G,1}$ and $T_{G,2}$ do not differ very much. If furthermore $K' \approx T_{G,2}/T_{G,1}$, Eqn.(9-17) converts into the **Fox equation** $1/T_G = (w_1/T_{G,1}) + (w_2/T_{G,2})$.

An **external plasticization** is obtained by dissolved gases or liquids in polymers (Fig. 9-14). External plasticizers are the more effective, the less good they are as solvents. Polymers in bad solvents may however demix; low molar mass plasticizers then exude to the surface (Section 11.6.). This transport is prevented by **polymer(ic) plasticizers** with molecular weights of several thousand. Plasticizers are mainly used for poly(vinyl chloride).

It is often not realized that many polymeric materials are plasticized. Paper is plasticized by water; on heating above 110°C, it becomes brittle since the glass temperature of cellulose is ca. 225°C. The plasticizing of cellulose by water is utilized for the steam-ironing of cotton fabrics. Synthetic quartz usually contains ca. 0.1 % water; this plasticized quartz can be deformed without fracture at temperatures of 400°C whereas dry quartz will break. Glassy polymers are also plasticized by gases with high critical temperatures; an example is CO_2 which lowers the glass temperature of bisphenol A polycarbonate by 9 K at 6.8 atm. Gases with lower critical temperatures require far higher pressures of several thousand atmospheres for plasticization effects.

9.7. Diffusion and Permeation

9.7.1. Permeation of Gases

The transport of gases and liquids into, through or from polymers is called **permeation**. Permeations are sometimes desired, for example, the entry of colorants into fibers and fabrics, the permeation of drugs through the skin, the transport of water in membrane desalination processes or the flow of oxygen through membranes in artificial lungs. In most cases, permeations are unwanted, however; for example, if air escapes through rubber tires, if the carbon dioxide of soft drinks permeates through the walls of plastic bottles, ifoxygen enters foodstuffs through packaging films, or if plasticizers exude from plastics.

Permeations are caused by a difference in the chemical potentials of the permeant in the two phases. If a gas in a bottle has a pressure p_i and the outer pressure is $p_a <$ p_i, then the pressure difference $\Delta p = p_i - p_a$ will lead to a permeation of the gas from the bottle through the walls of thickness ΔL to the outer side. After some time, a steady state is reached with a concentration difference Δw between the two surfaces. This concentration difference is proportional to the pressure difference according to **Henry's law** $\Delta w = S \Delta p$ for permanent gases.

The **solubility coefficient** S of gases has the physical unit of an inverse pressure. For liquids, Δp has to be replaced by the concentration difference Δc so that $\Delta w = S \Delta c$. The solubility coefficient S has again a physical unit of unity if Δw and Δc are measured in the same physical units. Depending on the units for Δc or Δw, S may be a mass fraction, volume fraction, etc.

The steady state is characterized by a constant diffusion flux. Application of **Fick's first law** (Section 7.2.1.) for differences ($\partial r \rightarrow \Delta L$; $\partial c \rightarrow \Delta w$; $\delta m \rightarrow \Delta m$; $\delta t \rightarrow \Delta t$) delivers $-J_d = -\Delta m/(A\Delta t) = D\Delta w/\Delta L$ and, after Henry's law for gases is introduced:

$$(9\text{-}18) \quad -J_d = -\Delta m/(A\Delta t) = D\Delta w/\Delta L = DS(\Delta p/\Delta L) \equiv P(\Delta p/\Delta L)$$

The product $P = DS$ is the **permeation coefficient**. It has the physical unit length^2 time^{-1} if the solubility coefficient has a physical unit of unity. The physical unit is length^2 time^{-1} pressure^{-1} if S is measured as an inverse pressure. Industry usually uses "practical units" which indicate the measuring conditions. If, for example, the mass Δm of permeated gases is measured as volume ΔV in cm^3 that has permeated in the time $\Delta t = 24$ h through a film with area A in m^2 and thickness ΔL in mm under the action of a pressure Δp in atm, then the physical unit of the permeation coefficient P is (cm^3 mm)/(24 h m^2 atm).

Permeations can occur either by a diffusion through pores or by a molecular transport through the polymer itself. For example, nitrogen gas diffuses through pores in parchment paper but is molecularly transported through polymer coils in poly(ethylene) films. Pores are channels whose diameters exceed the diameter of the permeant molecules by several times. Molecular transport proceeds by interchange of sites between gas molecules and polymer segments.

The two types of transport can thus be distinguished by their temperature dependence. Gases flow "freely" through pores. Since the interaction of gas molecules and pore walls is not much altered by temperature, changes of solubilities of gases with temperature are negligible. The permeation is just affected by gas diffusion. Diffusion coefficients of gases are inversely proportional to viscosities and viscosities of gases increase with temperature. The permeation coefficient $P = DS$ of gases in pore membranes thus decreases with increasing temperature (Table 9-3).

The solubility coefficient S of molecularly dissolved gases decreases with temperature. The diffusion of gases proceeds by interchange of sites with polymer segments. This interchange is more frequent at higher temperatures because polymer segments become more flexible; the diffusion coefficients increase. In a pure "solubility membrane", the effect of temperature on diffusion is much stronger than that on solubility and the permeability coefficient $P = DS$ increases with temperature (Table 9-3), Most polymer membranes do contain small pores, however, and P may either increase or decrease with temperature, depending on the relative magnitude of the two effects.

Permeation coefficients of gases through polymers vary widely; the permeation coefficient of oxygen in silicone rubber is, for example, ten million times greater than that of oxygen in poly(acrylonitrile). Polymers with especially low permeation coefficients are called **barrier polymers**. Such polymers usually have small solubility coefficients S. Permeation coefficients may also be lowered by **tortuosity factors** such as bulky monomeric units, orientation of polymer segments, crystalline regions, and filler particles because they increase the path the gas molecule has to travel.

9.7.2. Permeation of Liquids

The time dependence of the permeating mass m of non-dissolving liquids into polymers is proportional to the cross-sectional area A, the permeated mass m_∞ at infinite time, and a system-dependent constant K:

(9-19) $m = KAm_\infty t^n$

Exponents n depend on the ratio of the relaxation rate of polymer segments to the diffusion rate of the liquid, the **diffusion Deborah number** *De* (see also Section 10.5.3.). In the so-called *Case I*, relaxation times of liquids are much smaller than mobilities of the molecules of the permeating liquid ($De < 0.1$). The permeants thus produce "immediate" conformation changes of the polymer and permeant and polymer behave as viscous liquids. They follow Fick's diffusion laws and the exponent n becomes 1/2.

Case II is just the opposite: the relaxation time of segments is far greater than the mobility of permeant molecules. The polymer behaves like an elastic body for the permeant molecules. The permeating mass is directly proportional to time (n = 1). This type of permeation is characterized by a sharp boundary between the glassy core in the interior and the swollen zone which advances with constant speed.

Table 9-3 Permeation coefficients *P* of nitrogen in poly(ethylene) and parchment paper (= cellulose paper treated with sulfuric acid)

| | $10^{14}\ P/(\mathrm{cm}^2\ \mathrm{s}^{-1}\ \mathrm{Pa}^{-1})$ at $T =$ | | | |
	0°C	30°C	50°C	70°C
Poly(ethylene)	25	210	740	2200
Parchment paper	1120	940	930	840

The range $0.1 < De < 10$ is the range of anomal (viscoelastic) diffusion. Motions of permeant molecules and conformational changes of polymer segments happen almost simultaneously and the exponent becomes $1/2 < n < 1$.

Literature

9.2. THERMAL PROPERTIES
B.Wunderlich, H.Baur, Heat Capacities of Linear High Polymers, Adv.Polym.Sci. **7** (1970) 151
W.Knappe, Wärmeleitung in Polymeren, Adv.Polym.Sci. **7** (1971) 477
R.N.Haward, The Physics of the Glassy State, Interscience, New York 1973
E.A.Turi, ed., Thermal Characterization of Polymeric Materials, Academic Press, New York, 2nd ed. 1982
V.B.F.Mathot, ed., Calorimetry and Thermal Analysis of Polymers, Hanser, Munich 1994

9.3. CRYSTALLIZATION
L.Mandelkern, Crystallization of Polymers, McGraw-Hill, New York 1964
A.Sharples, Introduction to Polymer Crystallisation, Arnold, London 1966
B.Wunderlich, Macromolecular Physics, Vol. **2**, Crystal Nucleation, Growth, Annealing, Academic Press, New York 1976
K.Armitstead, G.Goldbeck-Wood, Polymer Crystallization Theories, Adv.Polym.Sci. **100** (1992) 219

9.4. THERMAL TRANSFORMATIONS
G.M.Bartenev, Yu.V.Zenlenev, eds., Relaxation Phenomena in Polymers, Halsted, New York 1974
B.Wunderlich, Macromolecular Physics, Vol. **3**, Crystal Melting, Academic Press, New York 1980
E.-J.Donth, Glasübergang, Akademie-Verlag, Berlin 1981
R.T.Bailey, A.M.North, R.A.Pethrick, Molecular Motion in High Polymers, Clarendon Press, Oxford 1981
B.Wunderlich, S.Grebowicz, Thermotropic Mesophases and Mesophase Transitions of Linear, Flexible Macromolecules, Adv.Polym.Sci. **60/61** (1984) 1

9.5. DIFFUSION AND PERMEATION
J.Crank, G.S.Park, eds., Diffusion in Polymers, Academic Press 1968
H.B.Hopfenberg, ed., Permeability of Plastic Films and Coatings, Plenum Press, New York 1974
T.R.Crompton, Additive Migration from Plastics into Food, Pergamon Press, Oxford 1979
J.Comyn, ed., Polymer Permeability, Elsevier Appl.Sci.Publ., London 1985
P.Tyle, ed., Drug Delivery Devices, Dekker, New York 1988

References

[1]   F.E.Karasz, H.E.Bait, J.M.O'Reilly, General Electric Report 68-C-001 (1968)

[2]   After many authors according to a compilation by W.Knappe, Adv.Polym.Sci. **7** (1971)
      477, Figs. 2, 18 and 19

[3]   R.H.Doremus, B.W.Roberts, D.Turnbull, eds., Growth and Perfection of Crystals, Wiley,
      New York 1958

[4]   S.H.Kim, L.Mandelkern; reported by L.Mandelkern in G.Allen, J.C.Bevington, eds., Com-
      prehensive Polymer Science **2** (1989) 363, Fig. 4; Pergamon Press, Oxford

[5]   R.Chiang, P.J.Flory, J.Am.Chem.Soc. **83** (1961) 2857, Fig. 2

[6]   H.-G.Elias, An Introduction to Plastics, VCH, Weinheim 1993, Fig. 5-7

[7]   B.Wunderlich, M.Möller, J.Grebowicz, H.Baur, Adv.Polym.Sci. **87** (1988) 1, Figs. 2.7-2.9

[8]   A.J.Kovacs, J.Polym.Sci. **30** (1958) 131, Fig. 5

[9]   S.J.Clarson, K.Dodgson, J.A.Semlyen, Polymer **26** (1985) 930

[10] F.Rietsch, D.Daveloose, D.Froelich, Polymer **17** (1976) 859

[11] H.-G.Elias, An Introduction to Plastics, VCH, Weinheim 1993, Fig. 13-6

10. Solid State Properties

10.1. Introduction

The previous chapters were mainly concerned with the molecular and super-molecular properties of **chemical substances** and their assemblies. These properties are generated by molecules and their assemblies; they are specific for a compound with defined constitution, configuration and macroconformation. Several of these properties are independent of the degree of polymerization; some global properties do not even vary with the chemical structure of polymers.

The situation is quite different for polymers used in plastics, elastomers and fibers. In general, these materials are not polymers *per se* but **polymer systems** composed of polymers and additives (fillers, plasticizers, antioxidants, etc.) (Chapter 11).

Mechanical, electrical and optical properties of solid polymers depend however not only on the chemical and physical structure of the components of such polymer systems but also on test conditions and the preparation, shape and dimensions of specimens. Many mechanical, electrical and optical properties are furthermore not equilibrium properties; they are thus affected by the kind and rate of measurements. Many testing procedures are also supposed to simulate the behavior of polymers under realistic use conditions; they thus measure a combination of several physical processes or properties. Physical properties of solid polymers are therefore very often not the properties of chemical compounds but properties of **test specimens.**

Comparable values of physical properties can only be obtained by standardization of dimensions and shapes of test specimens as well as testing conditions (old French: *estandart* = flag marking a place for rallying). Such **standards** are proposed by the *International Organization for Standardization* (ISO) and by national organizations such as the *American Society for Testing and Materials* (ASTM), the *British Standards Institution* (BSI), and the German *Deutsches Institut für Normung* (DIN).

Unfortunately, standards are different from country to country. They often do not agree with ISO standards which is especially true for ASTM standards. Many standards also let the experimentalist choose between different specimen preparations and testing conditions. This problem has been remedied to some extent by the CAMPUS® system which is based on ISO standards. CAMPUS® is an acronym for **C**omputer **A**ided **M**aterials **P**reselection by **U**niform **S**tandards, a system of ca. 50 different characteristic mechanical, rheological, electrical, optical, and thermal data that are obtained from standardized test specimens as well as the same testing conditions (temperature, deformation speeds, loads, etc.). The CAMPUS® system was developed in Germany and is now adhered to in Europe by more than 30 major European plastics producers and European subsidiaries of American companies; in 1996, it will be joined by Japanese companies. In the United Kingdom, a similar system named PLASCAM was developed by RAPRA (Rubber and Plastics Research Association).

10.2. Deformation

10.2.1. Introduction

Mechanical properties include the deformation of polymers or their surfaces, the resistance to deformation, and the ultimate failure, all with static or dynamic loads. Deformations may be reversible or irreversible. They can be generated by drawing, shearing, compression, pressurizing, bending, or torsion as well as by combinations thereof (Fig. 10-1). For comparison purposes, forces F causing these deformations are usually related to the surface areas A, i.e., they are reported as mechanical stresses $\sigma = F/A$. Since nine different stresses can be defined for a body (Fig. 7.4.), evaluations of mechanical properties can be very complex.

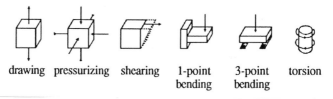

drawing   pressurizing   shearing   1-point   3-point   torsion
                                     bending   bending

Fig. 10-1   Some simple deformations.

10.2.2. Tensile Tests

The simplest and most often applied mechanical test method is tensile testing where a rectangular or dumbbell-type test specimen is placed between two clamps and then uniaxially drawn with constant speed. The (nominal) **tensile stress (engineering stress)** $\sigma_{11} = F/A_0$ is the ratio of force F to the initial area A_0; it is usually symbolized simply by σ. Tensile stresses are recorded as function of time t, **draw ratio (strain ratio)** $\lambda = L/L_0$, or **elongation (relative elongation, Cauchy elongation, linear strain, tensile strain, engineering strain)** $\varepsilon = \lambda - 1 = (L - L_0)/L_0 = \Delta L/L_0$. Drawing a test specimen to three times its original length ($L = 3\,L_0$) thus corresponds to $\lambda = 3$ and an elongation of 200 % ($\varepsilon = \lambda - 1 = 2$).

The tensile stress σ_{11} is directly proportional to the elongation ε at very small strains (**Hooke's law**):

(10-1) $\sigma_{11} = (F/A_0)(\Delta L/L_0) = E\varepsilon$; $E \equiv F/A_0$; $\Delta L/L_0 \equiv \varepsilon$

where A_0 = smallest cross-sectional area of the test specimen. Materials following Hooke's law are called **Hookean bodies**. The highest elongation at which Hooke's law still applies is called the **elastic limit**.

The **modulus of elasticity (tensile modulus, Young's modulus)** is the ratio of tensile stress to tensile strain at infinitesimal elongation. This asymptote is difficult to determine and the modulus is thus calculated from a secant, for example, from the origin of the coordinate system to the **proportionality limit**. This limit is not identical with the elastic limit but is in the same range of elongations. For plastics, the proportionality limit is defined as the elongation of 0.1 % that remains after removal of the load. In the CAMPUS® system, moduli refer to the secant of the stress–strain curve between the origin and elongations of 0.05 % to 0.25 %. So-called moduli of elastomers are given as the *stress* at 100 % or 300 % elongation; they are not moduli.

Polymers **yield** at elongations beyond the proportionality limit, i.e., they flow and deform irreversibly. On further drawing, thermoplastics often show **necking**, followed by a **telescope effect** (Fig. 10-2). Necking is not found for elastomers but is sometimes observed for fibers. It is caused by small differences in the cross-section A of the test specimen at which the tensile stress $\sigma = F/A$ is enlarged and the extensional viscosity is lowered. Necking starts near the clamps holding the test specimen because it is here that the stress concentration is highest since clamps exert pressure on the specimen. The flow of the segments causes internal friction which leads to a temperature increase at the neck. In turn, the viscosity is reduced further and additional flow is promoted. The heat developed does not *cause* necking, however, because telescope effects can also be observed under isothermal conditions. The neck travels along the specimen which maintains a constant cross-section. Tensile stresses do not increase further. This **stress softening** elongates the specimen without the usually necessary application of heat. The phenomenon is thus also called **cold flow**.

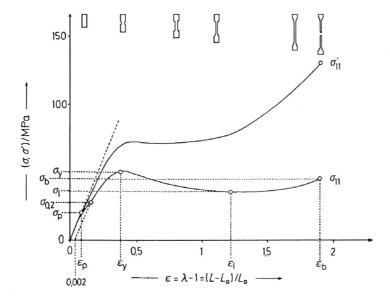

Fig. 10-2 Necking of a rectangular test specimen. Nominal tensile stress σ_{11} and true tensile stress σ_{11}' as a function of the nominal strain $\varepsilon = (L - L_o)/L_o$ for a thermoplastic polymer with a tensile modulus $E = 200$ MPa, a yield strength $\sigma_y = 50$ MPa at an elongation $\varepsilon_y = 0.38$ (38 %), and a fracture strength $\sigma_b = 45$ MPa at a fracture elongation $\varepsilon_b = 1.9$ (190 %) [1]. $\sigma_{0.2}$ is the off-set yield stress (proof stress) for an off-set elongation of 0.2 % ($\varepsilon = 0.002$). σ_l is the lower yield stress at the lower yield elongation ε_l.

Necking and subsequent stress-softening causes a maximum in the stress-strain curve. This maximum is called the (upper) yield point in polymer science. It is not the point where deviations from Hooke's law set in but rather the point where the change of stress with elongation is zero ($d\sigma/d\varepsilon = 0$).

The yield point is characterized by a **yield stress** σ_y at the elongation at yield, ε_y. For amorphous polymers, yield points are observed at temperatures $T < T_G$ (glass temperature); for semi-crystalline polymers, at $T_G < T < T_M$ (melting temperature).

The yield point is followed by strain-softening where the nominal stress σ_{11} either stays constant or (more commonly) decreases with increasing nominal strain (Fig. 10-2). Stress softening is often not observed if the stress is plotted as true stress $\sigma_{11}' = F/A$ instead of nominal stress. For *volume-invariant* elongations, the **true stress** σ_{11}' is greater than the nominal stress σ_{11} because the cross-sectional area decreases to A from A_o at constant force F when the volume $V = A_o L_o = AL$ remains constant:

(10-2) $\sigma_{11}' = F/A = (F/A_o)(L/L_o) = \sigma_{11}(L/L_o)$

Note that the volume does not remain constant if the drawing produces crazes, voids, additional crystallization, or orientation of segments (see below).

Nominal strains and true strains are also different. The total elongation of a body is $\varepsilon_{1+2} = (\Delta L_1 + \Delta L_2)/L_o$ if it is continuously drawn to $\Delta L = \Delta L_1 + \Delta L_2$. This elongation is however different from the one which results from two successive elongations $\varepsilon_1 = \Delta L_1/L_o$ and $\varepsilon_2 = \Delta L_2/(L_o + \Delta L_1)$ giving $\varepsilon_1 + \varepsilon_2 = (\Delta L_1/L_o) + (\Delta L_2/(L_o + \Delta L_1))$ and thus $\varepsilon_{1+2} \neq \varepsilon_1 + \varepsilon_2$. The Cauchy strain $\varepsilon = (L - L_o)/L_o$ can therefore not be the true elongation. By definition, the "**true strain**" (**Hencky strain**) is rather obtained by the summation of infinitesimal length changes per instantaneous length:

(10-3) $\varepsilon' = \int_{L_o}^{L'} (dL/L) = \log_e(L'/L_o) = \log_e(1+\varepsilon)$

Stress softening is generated by **crazing** and/or by **shear flow**. A shear flow leads to shifts in the mutual positions of chain segments. The volume of the test specimen remains constant. **Shear bands** are observed at angles of $(38-45)°$ to the stress direction if the shear flow in the test specimen is not homogeneous but localized (heterogeneous). In the latter case, segments are arranged at angles between shear bands and stress direction (Fig. 10-3). In thin films, shear zones are observed.

Eqns.(10-2) and (10-3) do not apply to non-homogeneous deformations. **Crazes** are up to 100 μm long and up to 10 μm wide. Their longitudinal axes are perpendicular to the stress direction. They are not cracks and therefore not voids. The interior of the crazes is rather filled with amorphous microfibrils of ca. (0.6-30) μm diameter which are bound on both sides to the matrix (Fig. 10-3). The microfibrils lie in the stress direction and are thus perpendicular to the longitudinal axes of the crazes.

Crazes scatter light since the refractive indices of matrix, microfibrils, and air differ and crazes are larger than the wavelength of visible light $[\lambda_o = (0.4-1.1) \mu m]$. The light scattering leads to a **stress whitening** of the specimen.

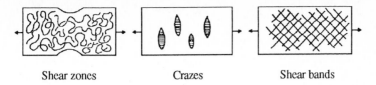

Fig. 10-3   Deformation mechanisms of polymers. Arrows indicate directions of applied stresses.

Continued drawing generates more and more crazes. The increasing number of microfibrils causes more resistance to deformation and the stress increases. One observes a **strain hardening**. The specimen finally breaks at an **elongation at break (fracture elongation)**, ε_b, with a **tensile strength at break (fracture strength)**, σ_b (Section 10.6.1.). The **tensile strength** is the highest tensile stress at any of the two points of **failure**, i.e., either at the upper yield point (as in Fig. 10-2) or at the fracture point (e.g., PE in Fig. 10-4). However, "tensile strength" usually refers exclusively to fracture strength.

10.2.3. Tensile Behavior

Polymers behave very differently on drawing (Fig. 10-4). They can be subdivided into six classes according to moduli and stress-strain behavior (Table 10-1).

Table 10-1   Classification of polymers according to their tensile behavior (see also Fig. 10-4).

Correct terms	Conventional terms	E	σ_b	ε_b	Examples
Rigid-brittle	hard-brittle	large	-	small	PS, PF
Rigid-strong	hard-strong	large	large	small	PMMA
Rigid-ductile	hard-tough	large	large	large	POM, PC
Soft-ductile	soft-tough	small	small	large	PE-LD
Soft-strong	soft-strong	small	small	small	PTFE
Soft-elastic	soft	small	large	large	SBS

The **rigidity** of a polymer is described by the *modulus of elasticity*. Polymers are defined as rigid if $E > 700$ MPa, semi-rigid if $700 \geq E/\text{MPa} \geq 70$, and soft if $E < 70$ MPa (ASTM). Rigid polymers should not be called "hard" because of a possible confusion with surface hardness (Section 10.7.).

The *stress–strain behavior* between elongation at yield and elongation at break controls the **deformability** of polymers. Non-yielding polymers do not flow and thus do not absorb energy; they are **brittle**. By definition, polymers are **brittle** if ε_b is smaller than 10 % (USA) or 20 % (Europe). Polymers with large stresses at yield are **strong** if the strain softening is small and **ductile** if the strain softening is large. Ductile polymers should not be called "tough" since toughness is the behavior on impact (Section 10.6.6.).

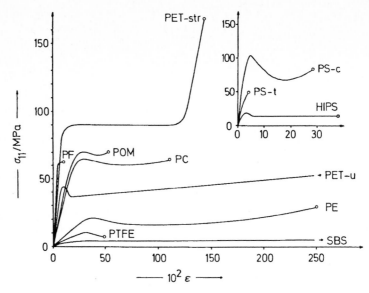

Fig. 10-4   Stress–strain behavior of some polymers at room temperature.
HIPS = Poly(styrene) with high impact strength;
PC   = Bisphenol A polycarbonate;
PE   = Poly(ethylene) with high density;
PET   = Poly(ethylene terephthalate), undrawn (u) or biaxially drawn (str) before the tensile test;
PF   = Cross-linked phenol–formaldehyde resin;
POM   = Poly(oxymethylene) (acetal resin);
PS   = Poly(styrene), deformation by compression (c) or uniaxial drawing (t);
PTFE = Poly(tetrafluoroethylene) (Teflon);
SBS   = Thermoplastic styrene–butadiene–styrene triblock copolymer. Note that σ_b can be as
        high as ca. 25 MPa (based on A_o) and ca. 200 MPa (based on final A)!

This classification of polymers applies only to standard test conditions. The general stress–strain behavior of polymers depends on the relative mobility of chain segments. A polymer that is rigid at low temperatures becomes soft near its glass temperature. All thermoplastics behave rigidly if temperatures are low and strain rates are high; they are all ductile if temperatures are high and drawing is slow (see also Fig. 10-12). The type of deformation is also important. Poly(styrene) is a rigid-brittle material on tensile testing because it forms crazes (Fig. 10-3). This is not possible on compression where it behaves as a rigid-ductile material (Fig. 10-4, insert).

10.2.4. Moduli and Poisson Ratios

Rigidities of polymers are characterized by tensile moduli E (see Eqn.(10-1)) or shear moduli G (see Eqn.(7-14)); the bulk modulus K is rarely used. These moduli are ratios of stresses [tensile stress σ_{11}, shear stress σ_{21}, pressure p (a negative stress)] to the corresponding strains [elongation ε, shear strain (shear deformation) γ_e, bulk strain (volume strain) $\Delta V/V_o$]. The three moduli are inverse **compliances** for *static deformations* of isotropic bodies (but not for dynamic ones):

Moduli				Compliances		

Moduli

Tensile modulus $E = \sigma_{11}/\varepsilon$

Shear modulus $G = \sigma_{21}/\gamma_e$

Bulk modulus $K = p/(-\Delta V/V_o)$

Compliances

Tensile compliance $D = 1/E$

Shear compliance $J = 1/G$

Bulk compliance $B = 1/K$

The term "modulus of elasticity" for E is not very accurate because G and K are also moduli of elasticity; E is better called "tensile modulus" or "Young's modulus". The often used term "torsion modulus" for shear modulus is etymologically incorrect since the Latin word *torquēre* means "to twist" (see also Fig. 10-1) and not "to shear". The bulk modulus is also called **compression modulus** and the bulk compliance also **compressibility**. The symbols K and B are sometimes interchanged in the literature.

The three simple moduli E, G and K of *Hookean bodies* are interrelated by the **Poisson ratio (lateral strain contraction)** $\mu = -(\Delta d/d_o)/(\Delta L/L_o)$ where d = diameter and L = length of test specimen:

$$(10\text{-}4) \quad E = 2\,G(1 + \mu) = 3\,K(1 - 2\,\mu)$$

Poisson ratios of isotropic Hookean bodies adopt values between 0 and 1/2. The lower limit of 0 is approached if transverse contractions are absent; examples are all ideal-elastic solid bodies. Steel is thus not a 100 % energy-elastic body (Table 10-2). Poisson ratios may exceed a value of 1/2 if bodies are anisotropic. Eqn.(10-4) does not apply to anisotropic bodies and/or viscoelastic materials. However, the inequalities $E/3 \leq G \leq E/2$ and $0 \leq K \leq E/3$ are always true.

The upper limit of 1/2 is obtained for volume-invariant materials such as water, gelatin gel and natural rubber (Table 10-2). However, the Poisson ratios of such materials are rather meaningless because they belong to non-Hookean materials: water is

Table 10-2 Densities ρ, Poisson ratios μ, tensile moduli E, shear moduli G, bulk moduli K, and "specific" moduli E/ρ at room temperature. $\|$ = Fiber direction. [a] Ultra-drawn.

Material	$\dfrac{\rho}{\text{g cm}^{-3}}$	μ	$\dfrac{G}{\text{GPa}}$	$\dfrac{K}{\text{GPa}}$	$\dfrac{E}{\text{GPa}}$	$\dfrac{E\rho^{-1}}{\text{GPa cm}^3\text{ g}^{-1}}$
Water (4°C)	1.000	0.50	0	2.04	≈ 0	≈ 0
Natural rubber, cross-linked	0.92	0.499	0.00035	2	0.001	≈ 0.0011
Gelatin gel (80 % water)	1.01	0.50			0.002	≈ 0.002
Poly(ethylene), low density	0.92	0.49	0.070	3.3	0.20	0.22
Polyamide 6.6	1.14	0.44	0.70	5.1	1.9	1.67
Poly(styrene)	1.05	0.38	1.2	5.0	3.4	3.23
Ice (-4°C)	0.917	0.33	3.7	10.0	9.9	10.8
Aluminum	2.72	0.34	27	75	72	26.5
Glass (E glass)	2.54	0.23	25	37	72	28.3
Steel (V2A)	7.86	0.28	80	160	211	26.8
Quartz	2.65	0.07	47	39	101	38.1
Poly(ethylene) fiber ($\|$) [a]	1.00				130	130
Poly(ethylene), theory for $\|$	1.00				340	340
Diamond, [110] direction	3.515	0.16		≈ 5000	1160	368
Alumina fibers ($\|$)	3.97	0	1000	667	2000	504

a viscous liquid, gelatin gel a thermoreversible gel, and cross-linked natural rubber an entropy-elastic body. Bulk moduli K and tensile moduli E (or shear moduli G) of these materials arise from completely different molecular mechanisms.

Moduli depend on mobilities of chain segments and thus on the testing temperature in relation to the glass temperature T_G. Five characteristic regions are observed for tensile and shear moduli of *amorphous linear (or branched) polymers* as a function of temperature (Fig. 10-5): glass-like at $T < T_G$, leathery at $T \approx T_G$, rubbery at $T > T_G$ (if the molar mass is high), then viscoelastic and finally viscous. The rubber-like behavior is caused by entanglements of chains that act as physical cross-links..

Semi-crystalline (linear or branched) polymers possess only small amorphous domains. Their moduli thus do not drop as strongly at T_G as they do for amorphous polymers. Crystalline domains act as large physical cross-links that restrict severely the mobility of segments in the crystallites. Moduli of semi-crystalline polymers are therefore much higher than those of amorphous polymers of the same constitution. Some crystalline domains start to melt if the melting temperature T_M is approached which causes moduli to decrease slowly with increasing temperature at $T < T_M$. All remaining crystalline domains melt at T_M and the modulus drops catastrophically. Both semi-crystalline and amorphous polymers behave as **thermoplastics** (Section 14.3.).

Chemically lightly *cross-linked polymers* remain rubbery at temperatures $T > T_G$ because their network chains cannot flow freely relative to each other due to cross-links; they are **elastomers** (Chapter 12). Chemically strongly cross-linked polymers have only short segments between cross-links. The segments are not mobile; these **thermosets** have no glass temperature (Section 14.4.)

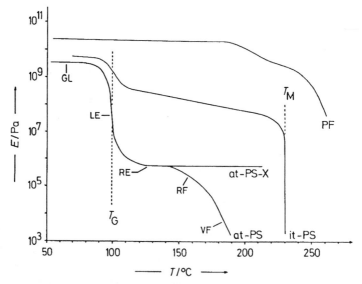

Fig. 10-5 Temperature dependence of tensile moduli of an amorphous atactic linear poly(styrene) (at-PS), its slightly chemically cross-linked product (at-PS-X), a semi-crystalline isotactic poly-(styrene) (it-PS), and a strongly cross-linked phenol–formaldehyde resin (PF).

GL = Glassy state, LE = leathery state, RE = rubbery state (elastomeric behavior), RF = rubbery flow, VF = viscous flow. T_G = Glass temperature, T_M = melting temperature.

10.3. Energy Elasticity

10.3.1. Introduction

Loads deform **ideal-elastic bodies** but the removal of loads allows the bodies to return "immediately" to the original conditions without permanent deformations. Examples of such reversible deformations are steel spheres hitting steel plates and moderately expanded rubber bands.

Steel spheres and rubber bands differ however in their behavior on deformation. Steel cools on drawing but rubber heats up. These differences are produced by different mechanisms of the response to the external stress. The deformation of steel causes iron atoms to move away from their positions at rest. The energy required for this work is taken from the system and steel cools. The deformation of rubber makes chain segments slip from each other; the internal friction produces heat.

Positions of iron atoms change on deformation of steel but the deviations from the positions at rest are affine. The enthalpy changes but not the entropy: steel is an energy-elastic body. Highly oriented polymer molecules, such as drawn fibers, are also predominantly energy-elastic. Their rigidity is usually characterized by the tensile modulus E.

The deformation of rubber creates however less probable macroconformations of network chains: rubber is an entropy-elastic body (Section 10.4.). Its elasticity is characterized by the shear modulus G. Most plastics behave not only elastic but are also viscous. Their rigidity is therefore also measured by the shear modulus G.

10.3.2. Theoretical and Experimental Moduli

With respect to moduli and Poisson ratios, conventional plastics often resemble liquids more than metals (Table 10-2). The chain segments of amorphous polymers are randomly positioned in space. Deformation forces are thus not evenly distributed on all chain segments but are concentrated on the few which happen to lie in the direction of the force. Orientation of segments increases tensile moduli; examples are the liquid crystalline polymers PPBT and PPTA (Table 10-3). Ultradrawn fibers also have very high tensile moduli in the longitudinal direction.

Even the high moduli of ultradrawn or liquid crystalline polymers are lower than those for fully aligned chains since some chain segments are still not completely oriented. Moduli of fully aligned chains can be determined experimentally by micromechanical deformation methods employing X-ray diffractometry, Raman spectroscopy or inelastic coherent neutron scattering. These experimental methods deliver **lattice moduli**. For example, loads σ shift Bragg reflections and thus lattice positions in X-ray spectroscopy. Since deformations ε are small, Hooke's law applies and lattice moduli can be calculated for the chain direction and perpendicular to it.

Table 10-3   Tensile moduli of polymers in longitudinal ($E_\parallel$) and transverse ($E_\perp$) directions and cross-sectional areas A_c of polymer chains. Theory = Calculated from bond lengths, valence angles, cross-sectional areas and force constants; Lattice = experimental from microdeformation experiments (X-ray, Raman, SANS); Tensile = experimental from tensile drawing.

Polymer	Confor-mation	A_c/nm^2	$E_\parallel$/GPa Theory	Lattice	Tensile	$E_\perp$/GPa Lattice
Poly(*p*-phenylenebenzbisthiazol) (PPBT)	trans		640		120	
Poly(ethylene) (PE)	trans	0.183	340	325	< 1	3.4
Poly(ethylene) (PE), ultradrawn fiber	trans	0.183	340	325	120	
Poly(*p*-phenylene terephthalamide) (PPTA)	trans	0.203	182	200	132	10
Poly(*p*-benzamide)	trans	0.198	238	182	77	
Poly(oxymethylene), orthorhombic (POM)	9_5	0.182	220	189	< 2	7.8
Poly(oxymethylene), trigonal		0.172	48	54	2	
it-Poly(propylene) (PP)	3_1	0.343	50	42	< 3	2.9
Poly(4-methyl-1-pentene) (P4MP)	7_2	0.864	6.7		1	2.9

The resulting lattice moduli are surprisingly high but not unreasonable as one can see from the following (rather crude) estimate:

Carbon atoms of diamond in the [110] direction can be considered as "chains" in the all-trans conformation. The "diamond chains" –C–C–C– are "naked"; their cross-sectional area is $A_c = 0.0488$ nm^2. Poly(ethylene) chains –CH$_2$–CH$_2$–CH$_2$–CH$_2$– have hydrogen as substituent and thus a larger cross-sectional area of $A_c = 0.182$ nm^2. Since the tensile modulus of diamond is $E_\parallel = 1160$ GPa in the [110] direction, poly(ethylene) chains should have a modulus of $E_\parallel = 1160 \cdot (0.0488/0.182)$ GPa = 311 GPa which agrees well with the theoretical values of $250 \leq E_\parallel/\text{GPa} \leq 340$ GPa.

In theory, experimental lattice moduli should agree with moduli calculated by various theoretical approaches: valence force field (Treloar's method), dynamic lattice theory (Born's method) and molecular mechanics. The dynamic lattice theory calculates physical properties of crystals from the thermal motions of chain atoms. It requires an exact knowledge of the potential function, force constants, lattice distances, etc. Molecular mechanics methods depend on minimizing the internal energy of the molecule by adding up all possible interatomic interactions. Both methods have been used relatively little.

Most theoretical tensile moduli have been calculated by the **valence force field** method (Treloar's method). This method assumes an isolated polymer chain in all-trans conformation. Such a chain contains N chain bonds of length b. Its length is thus $L_0 = r_{\text{cont}} = Nb \cdot \sin(\tau/2) = Nb \cdot \cos\beta$ where τ = bond angle between chain atoms and $\beta = \alpha/2 = (180° - \tau)/2$ = one-half of the complementary angle α to the bond angle τ (see Fig. 5-3).

The chain should be stretched but not bent (the following derivation can thus not be used for helical chains where the main deformation would be a change of the torsional angles). The applied force F increases the bond length b by Δb and the angle β by $\Delta\beta$. The chain is thus elongated by $\Delta L = \Delta[Nb \cdot \cos\beta] = N[\Delta b \cos\beta - b\Delta\beta \sin\beta]$. Δb and $\Delta\beta$ are calculated as follows.

Along the bond direction, the component of the force is $F \cos \beta$. The bond is thus extended by $\Delta b = (F \cos \beta)/K_b$ where the force constant K_b can be obtained by infrared or Raman spectroscopy.

The bond angle τ is enlarged by $\Delta\tau = M/K_\tau$ where M is the torque acting around each of the bond angles. The torque equals the moment of the applied force about the angular vertices, i.e., $M = (1/2) Fb \sin \beta$. Since the angle $\beta = 90° - (\tau/2)$, one obtains $\Delta\beta = -\Delta\tau/2 = -(Fb \sin \beta)/(4 K_\tau)$.

Inserting the expressions for Δb and $\Delta\beta$ into the equation for ΔL and introducing the simple Hooke equation $E_\| = (F/A_c)(L_o/\Delta L)$ and $L_o = Nb \cdot \cos \beta$ delivers:

$$(10\text{-}5) \quad E_\| = \frac{b \cos\beta}{A_c}\left[\frac{\cos^2 \beta}{K_b} + \frac{b^2 \sin^2 \beta}{4 K_\tau}\right]^{-1} = \frac{b \sin(\tau/2)}{A_c}\left[\frac{\sin^2(\tau/2)}{K_b} + \frac{b^2 \cos^2(\tau/2)}{4 K_\tau}\right]^{-1}$$

The longitudinal modulus $E_\|$ thus decreases with increasing cross-sectional area A_c of the chain for constitutionally and conformationally similar chains (Fig. 10-6). Examples are chains with -C-C- or similar bonds (e.g., -C-N- or -C-O-) that crystallize in the all-trans conformation (2_1) or almost all-trans conformation (9_5 helices).

Eqn.(10-5) was derived by assuming constant torsional angles which is true for the drawing in the chain direction of chains with all-trans conformations of chain atoms. The assumption cannot be valid for helical chains since their torsional angles are bound to change if chains with, e.g., gauche conformations are drawn. A change of torsional angles requires less energy than that of valence angles. Helical chains thus have smaller longitudinal moduli $E_\|$ than all-trans chains with the same cross-sectional area (Fig. 10-6).

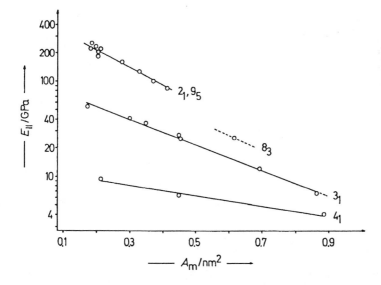

Fig. 10-6 Logarithm of the longitudinal lattice moduli $E_\|$ as function of the cross-sectional area $A_m = A_c$ of polymer chains in all-trans conformation (2_1) or as helices (9_5, 8_3, 3_1, 4_1).

Poly(ethylene) has the highest modulus $E_{||}$ of all chains with σ bonds between chain atoms (Table 10-2). Even higher moduli are obtained if the relatively easy to deform aliphatic C-C chain units are replaced by less easily deformable ones, for example, 1,4-phenylene rings or rigid benzbisthiazol units (Table 10-3).

Longitudinal moduli $E_{||}$ are always larger than transverse moduli $E_{\perp}$ since the former are controlled by strong covalent bonds and the latter by weaker dispersion or dipole forces (Table 10-3).

10.4. Entropy Elasticity

Elastomers are slightly cross-linked polymers with glass temperatures T_G below use temperatures (Chapter 12). They simultaneously have characteristics of solids, liquids and gases. Like solids, they show Hookean behavior at not too large deformations, i.e., no permanent deformations after removal of loads. Their moduli and expansion coefficients are however similar to those of liquids and not to those of solids. Like compressed gases, their stresses increase with increasing temperature T for $T > T_G$.

Elastomers with flexible chain segments exhibit a marked **entropy elasticity (rubber elasticity)** which makes them very useful for industry and commerce (Chapter 12). Elastomers with rigid chain segments are predominantly enthalpy elastic; they are at present only of academic interest.

Entropy-elastic bodies and enthalpy-elastic bodies differ characteristically:

	energy-elastic	*entropy-elastic*
Reversible deformation	small (ca. 0.1 %)	large (several 100 %)
Modulus of elasticity	large	small
Temperature change on drawing	cooling	warming
Length change on heating	expansion	contraction

10.4.1. Chemical Thermodynamics

Changes in the thermodynamic state of bodies are described quantitatively by the three basic equations of chemical thermodynamics which relate pressure p, temperature T and volume V to Helmholtz energy A, internal energy U, entropy S and isobaric heat capacity C_p. For entropy-elastic bodies, the volume change ∂V is replaced by the length change ∂L and the pressure p (a negative force) by the drawing force F:

(10-6) $p = -(\partial U/\partial V)_T + T(\partial p/\partial T)_V$; $F = (\partial U/\partial L)_T + T(\partial F/\partial T)_L$

(10-7) $(\partial A/\partial V)_T = (\partial U/\partial V)_T - T(\partial S/\partial V)_T$; $(\partial A/\partial L)_T = (\partial U/\partial L)_T - T(\partial S/\partial L)_T$

Combination of the two right-side equations delivers:

(10-8) $(\partial A/\partial L)_T + T(\partial S/\partial L)_T = F - T(\partial F/\partial T)_L$

Thermodynamically, $(\partial S/\partial V)_T = (\partial p/\partial T)_V$ is always valid and thus $(\partial S/\partial L)_T = (\partial F/\partial T)_L$. Insertion of the latter equation into Eqn.(10-8) leads to the **thermodynamic equation of state** of entropy-elastic bodies:

(10-9) $F = (\partial A/\partial T)_L$

Experiments have shown for (not too strongly) stretched elastomers that the force F is proportional to the temperature T. With $F = KT$ (K = constant) one gets $(\partial F/\partial T)_L = K$ and thus $F/T = (\partial F/\partial T)_L$. Introduction of this expression into Eqn.(10-6) delivers $(\partial U/\partial L)_T = 0$: the internal energy of an entropy-elastic body does not change on isothermal stretching. This behavior makes entropy-elastic bodies fundamentally different from enthalpy-elastic ones.

Heating to temperature T_2 from temperature T_1 changes the length of a body to L_2 from L_1. The change of internal energy is given by $\partial U = FdL + C_p dT = 0$ where C_p = heat capacity at constant pressure. Integration of the expression for ∂U from state 1 to state 2 leads to $F(L_2 - L_1) = - C_p(T_2 - T_1)$. Since $T_2 > T_1$, one gets $L_2 < L_1$: an entropy-elastic body contracts on heating whereas an energy-elastic body expands if the load (force) is constant.

A constant force F causes the stress to increase on heating: a rubber band clamped on both sides becomes more tense whereas a metal band slackens.

10.4.2. Structure of Networks

Entropy elasticity results from flexible chains that are loosely interconnected to **networks**. Chains can adopt many different spatial positions (Section 5.3.). They must be flexible, however, if *all* possible positions are to be accessed *rapidly*. Chains must also be interconnected so that they cannot flow away from each other under tension.

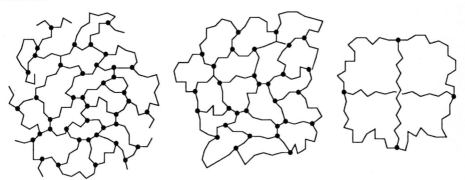

Fig. 10-7   Some networks with trifunctional junctions or branching points.
Left: different network chains and different mesh sizes, loose ends.
Center: different network chains but identical mesh sizes (14 segments each); no loose ends.
Right: perfect network with identical network chains (15 segments each) and identical mesh sizes, no loose ends.

A network consists of many **network chains** which extend from one **junction** (cross-link) to another (Fig. 10-7). In elastomers, a network chain consists of several hundred chain units. Junctions are usually trifunctional or tetrafunctional.

A *perfect* network is defined as a network where all junctions have the same functionality and all network chains consist of identical numbers of chain units. All meshes of the network thus have the same size. A perfect network furthermore has no loose ends, loops, knots, entanglements, and excluded volumes of network chains.

10.4.3. Statistical Thermodynamics

The interdependence of stress and strain of elastomers can be calculated by statistical thermodynamics. The spatial distribution of network chains can be treated like the spatial distribution of segments of linear chains if the network chains are sufficiently long (see Eqn.(5-18)). The "segment" is then the network chain ($X = 1$). Replacing radii of gyration by end-to-end distances, one obtains the Gauss distribution function of the end-to-end distances of network chains:

(10-10) $W(r) = [3/(2 \pi \langle r^2 \rangle_o)]^{3/2} \exp[- 3 r^2/(2 \langle r^2 \rangle_o)]$

The Gibbs energy of a network chain is $G'(r) = H - TS$. Insertion of $S = k_B \ln W(r)$ and Eqn.(10-10) leads to $G'(r) = H - k_BT \log_e[3/(2 \pi \langle r^2 \rangle_o)]^{3/2} + k_BT[3 r^2/(2 \langle r^2 \rangle_o)]$. The first and the second term of the right-hand side of this equation are combined to a temperature-dependent constant $C'(T)$. The Gibbs energy $G(r)$ of an ensemble of N_c network chains with an average mean-square end-to-end distance $\langle r^2 \rangle = r^2$ of *all* network chains is thus:

(10-11) $G(r) = C(T) + N_c k_BT[3 \langle r^2 \rangle/(2 \langle r^2 \rangle_o)]$

Stretching of the polymer changes the Gibbs energy in two ways: by conformational changes within each network chain from $G(r)_o$ to $G(r)$ [conformation term] and by spatial redistributions of network junctions [dispersion term]. Initially, the network is in an isotropic state with $\langle r^2 \rangle = \langle r^2 \rangle_o$. The dispersion term is given by the expansion of the volume to V from V_o similar to that of gases, i.e., $\Delta G_{disp} = - N_x k_BT \log_e(V/V_o)$. The number N_x of network junctions (cross-links) with functionality f is $N_x = 2 N_c/f$ for a perfect network (no loose ends) with N_c network chains. Thus

(10-12) ΔG_{elast} $= G(r) - G(r)_o + \Delta G_{disp}$

$= N_c k_BT[3 \langle r^2 \rangle/(2 \langle r^2 \rangle_o)] - N_c k_BT[3/2] - N_x k_BT \log_e(V/V_o)$

$= (3 N_c/2) k_BT [((\langle r^2 \rangle/\langle r^2 \rangle_o) - 1] - (2 N_c/f) k_BT \log_e(V/V_o)$

End-to-end distances of network chains cannot be measured, however. They must be replaced by macroscopic parameters. This transformation requires models.

10.4.4.   Models

For modeling, one usually assumes that the observed macroscopic elongational ratios $\lambda_i = L_i/L_{i,o}$ in the three spatial directions $i = x, y, z$ reflect the microscopic shifts of either the segments or the junctions, i.e., that the deformation is **affine**. For junction-affine deformations, elongational ratios λ_x in x-direction are related to the shift of the mean-square end-to-end distance of the network chains, $\langle r^2 \rangle_o$, by $\lambda_x^2 = \langle x^2 \rangle / \langle x^2 \rangle_o = \langle x^2 \rangle / [\langle r^2 \rangle_o / 3]$. In all three spatial directions, we get:

$$(10\text{-}13)\quad \langle r^2 \rangle = \langle x^2 \rangle + \langle y^2 \rangle + \langle z^2 \rangle = (\lambda_x^2 + \lambda_y^2 + \lambda_z^2)\langle r^2 \rangle_o / 3$$

Note that "elongation" may mean $\lambda = L/L_o$ in theoretical treatments of rubber elasticity but $\varepsilon = \lambda - 1$ in mechanics (see Section 10.2.2.). In order to avoid confusion, λ will be called strain ratio for tensile drawing and elongational ratio for rubber theory. The term elongation will be reserved for $\varepsilon = \lambda - 1$.

The initial volume before the deformation is $V_o = L_o^3$ where $L_o = L_{x,o} = L_{y,o} = L_{z,o}$. Using $\lambda_i = L_i/L_{i,o}$ ($i = x, y, z$), the volume V after the deformation can be written as $V = L_x L_y L_z = \lambda_x L_{x,o} \lambda_y L_{y,o} \lambda_z L_{z,o} = L_o^3 (\lambda_x \lambda_y \lambda_z) = V_o (\lambda_x \lambda_y \lambda_z)$. Introducing this equation and Eqn.(10-13) into Eqn.(10-12) leads to:

$$(10\text{-}14)\quad \Delta G_{el} = N_c (k_B T/2)\{[\lambda_x^2 + \lambda_y^2 + \lambda_z^2 - 3] - [(4/f)\log_e(\lambda_x \lambda_y \lambda_z)]\}$$

The change of the Gibbs elastic energy on deformation is thus controlled only by the number N_c of network chains, the temperature T, the elongational ratios λ_i and the functionality f of the junctions but not by the chemical structure of junctions and network chains. The junctions may be furthermore chemical or physical; it is only required that their number does not change during deformation.

Junctions may fluctuate about their positions at rest. In unswollen networks with short network chains, these fluctuations are completely surpressed by adjacent chains. Long network chains, especially those in the swollen state, may have some freedem to fluctuate but can do this only by coupling their movements with those of neighboring chains. All three resulting models are based on **phantom networks**, i.e., on the absence of excluded volumes.

For historical reasons, only phantom networks with coupled fluctuations of junctions are called "phantom networks" whereas phantom networks with surpressed fluctuations of junctions are said to be "affine". Since all three models are based on affine deformations of phantom networks, we will avoid this confusing nomenclature and use the symbols S, A and P instead (Table 10-4).

Table 10-4   Models for the rubber elasticity of phantom networks.

Symbol	Conventional name	Affinity	Fluctuation	$(\alpha_s)_\parallel^2 =$	$(\alpha_s)_\perp^2 =$
S	simple affine	segments	surpressed	λ^2	λ^{-1}
A	affine	junctions	surpressed	$[(\lambda^2+1)/2]$	$[(\lambda+1)/(2\lambda)]$
P	phantom network	junctions	coupled	$\{[f+2+(f-2)\lambda^2]/[2f]\}$	$\{[f+2+(f-2)\lambda^{-1}]/[2f]\}$

All three models deliver the same function $\sigma_{11} = K_F(\lambda - \lambda^{-2})$ for the dependence of the tensile stress σ_{11} on the elongational ratio $\lambda = L/L_o$ (see below). They differ in the meaning of the front factor K_F. These factors contain some molecular parameters that cannot be obtained from mechanical measurements and are difficult to determine by independent methods.

However, the three models can be distinguished experimentally by the dependence of the expansion factor α_s of the radius of gyration on the elongational ratio λ (Table 10-4). The expansion factor $\alpha_s = [\langle s^2 \rangle/\langle s^2 \rangle_o]^{1/2}$ is the ratio of the radius of gyration, s, of the deformed network to the radius of gyration, s_o, of the undeformed network (both as the appropriate averages). It can be measured by inelastic coherent neutron scattering in the two directions $\parallel$ and $\perp$ to the draw direction.

According to experiments on cross-linked poly(dimethylsiloxane)s, both $(\alpha_s)_\parallel$ and $(\alpha_s)_\perp$ are not much dependent on the degree of swelling of networks. The transverse expansion factors $(\alpha_s)_\perp$ follow the P-model reasonably well, regardless of the molar masses of network chains. The longitudinal expansion factors $(\alpha_s)_\parallel$ are molar mass dependent, however: small network chains adhere to the A-model, medium-sized network chains to the P-model and large network chains to yet another, undetermined model. The S-model does not apply at all. However, most of the literature treats experimental data in terms of the A-model because it is simpler than the P-model.

10.4.5. Uniaxial Stretching

The uniaxial elongation of a network in x direction results in $L/L_o = \lambda_x \equiv \lambda$. If the volume does not change, $\lambda_y = \lambda_z = (1/\lambda)^{1/2}$ applies whereas a volume change to V from V_o leads to $\lambda_y = \lambda_z = [(V/V_o)(1/\lambda_x)]^{1/2}$. Eqn.(10-14) thus converts into $\Delta G_{el} = N_c(k_B T/2)\{[\lambda^2 + 2\,(V/V_o)\lambda^{-1} - 3] - (4/f)\,\log_e(V/V_o)\}$. Differentiation of this expression with respect to the length $L = L_o\lambda$ gives the force F:

$$(10\text{-}15) \quad F = (\partial \Delta G_{el}/\partial L)_{T,V} = (\partial \Delta G_{el}/\partial \lambda)_{T,V}/L_o = N_c k_B T[\lambda - (V/V_o)\lambda^{-2}]/L_o$$

$$= N_c k_B T (V/V_o)^{2/3}(\alpha - \alpha^{-2})/L_{i,V}$$

where $\alpha = L/L_{i,V} = L/[L_o(V/V_o)^{1/3}] = \lambda(V/V_o)^{-1/3}$, i.e., the elongational ratio per cubic root of the relative volume change on deformation of an isotropic body.

The tensile strength is defined as $\sigma_{11} = F/A_o$. Introducing Eqn.(10-16) and the initial volume $V_o = A_o L_o$, the Boltzmann constant $k_B = R/N_A$ and the molar concentration $[M_c] = N_c/(N_A V_o)$ of the network chains results in:

$$(10\text{-}16) \quad \sigma_{11} = F/A_o = RT[M_c]\{\lambda - (V/V_o)\lambda^{-2}\} = RT[M_c]\{\alpha - \alpha^{-2}\}(V/V_o)^{1/3}$$

If the volume does not change on deformation, Eqn.(10-16) reduces to:

$$(10\text{-}17) \quad \sigma_{11} = RT[M_c]\{\lambda - \lambda^{-2}\}$$

In a similar manner, one can correct for the effect of swelling by solvents. The expression $V_0/V = V_0/(V_0 + \Delta V)$ for the volume change of unswollen networks on deformation corrsponds to $V_2/V = \phi_2$, i.e., the volume fraction ϕ_2 of the polymer in the swollen network. Eqn.(10-17) becomes $\sigma_{11} = RT[M_c]\{\lambda - \lambda^{-2}\}\phi_2^{1/3}$.

Eqn.(10-17) was derived for the junction-affine deformation of networks with surpressed fluctuations of junctions (A-model). Coupled fluctuations must of course depend on the molar concentration $[M_x]$ of junctions. The P-model predicts that the front factor will then change to $RT([M_c] - [M_x])$ from $RT[M_c]$.

In all these equations, the molar concentration $[M_c]$ of network chains represents the *initial* concentration of *chemical* junctions (cross-links). Experiments deliver however the *actual* concentration of all *effective* junctions. Not all chemical cross-links are equally effective since the maximum elongation is controlled by the shortest network chains. A part of the total functionality of cross-linking sites is furthermore wasted by the formation of loops and loose ends. Physical junctions, on the other hand, increase the number of effective junctions. Examples are entanglements and crystalline regions.

Exprimental data are reproduced well by Eqn.(10-17) for compression ($\lambda < 1$) and reasonably well for not too high elongational ratios $1 < \lambda < 5$ (Fig. 10-8). The rubber crystallizes at high extensions; the crystallites act as physical cross-links and increase the tensile strength.

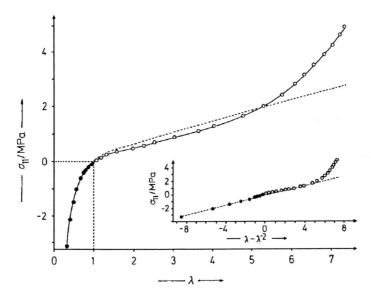

Fig. 10-8 Tensile stress σ_{11} of a vulcanized natural rubber as function of the strain ratio $\lambda = L/L_0$. O Tension, ● Compression. - - - - Calculated with Eqn.(10-17) and $RT[M_c] = 0.364$; the value of $RT[M_c]$ is obtained from a plot of $\sigma_{11} = f(\lambda - \lambda^{-2})$ which was made to fit the compression data (insert). The molar concentration of network chains is obtained as $[M_c] = 1.47 \cdot 10^{-4}$ mol/cm³ at 25°C. The density of natural rubber is $\rho = 0.91$ g/cm³; it is little changed by vulcanization. The average molar mass of network chains between junctions is thus calculated as $M_c = \rho/[M_c] = 6190$ g/mol. Data of [3].

10.4.6. Real Stress–Strain Behavior

The two classical models, affine model (A) and phantom network (P), often do not reproduce real stress–elongation curves, especially not for swollen elastomers. Entanglements suppress fluctuations of junctions at small deformations and the deformation is almost affine. At high deformations, entanglements de-entangle at small deformation rates and deformations approach those predicted for phantom networks. Chains then deform less than is indicated by the macroscopic elongational ratio.

The transition from the affine model to the phantom network is accompanied by a decrease of the apparent tensile modulus $E_{app} = RT[M_c] = \sigma_{11}/(\lambda - \lambda^{-2})$ (Eqn.(10-17)) with increasing deformation. This decrease is often described by the semi-empirical **Mooney–Rivlin equation** which results from symmetry considerations. It gives for the elastic Gibbs energy of volume-invariant networks:

$$(10\text{-}18) \quad \Delta G_{el} = C_1'[\lambda_x^2 + \lambda_y^2 + \lambda_z^2 - 3] + C_2'[\lambda_x^{-2} + \lambda_y^{-2} + \lambda_z^{-2} - 3]$$

The uniaxial stretching ($\lambda_x = \lambda$; $\lambda_y = 1/\lambda_x^{1/2}$; $\lambda_z = 1/\lambda_x^{1/2}$) of a volume-invariant elastomer ($\lambda_x\lambda_y\lambda_z \equiv 1$) results in $\Delta G_{el} = C_1'[\lambda^2 + (2/\lambda) - 3] + C_2'[\lambda^{-2} + 2\lambda - 3]$. Differentiation of this equation with respect to $L_o\lambda$ delivers the force $F = (\partial \Delta G_{el}/\partial L)_{T,V} = C_1'[2\lambda - 2\lambda^{-2}]/L_o + C_2'[-2\lambda^{-3} + 2]/L_o$. Introduction of $\sigma_{11} = F/A_o = FL_o/V_o$, $C_1 = C_1'/V_o$, $C_2 = C_2'/V_o$ and rearranging gives:

$$(10\text{-}19) \quad E_{app} = \sigma_{11}/(\lambda - \lambda^{-2}) = 2C_1 + 2C_2\lambda^{-1}$$

where C_1 and C_2 are constants. The reduced tensile stress $\sigma_{11}/(\lambda - \lambda^{-2})$ often becomes independent of the degree of swelling, $1/\phi_2$, of the polymer at infinitely large elongational ratios $\lambda^{-1} \to 0$ (Fig. 10-9), where ϕ_2 = volume fraction of polymer in the elastomer swollen by solvent. The value of $2C_1$ is therefore often identified with the statistical expressions $RT[M_c]$ (Eqn.(10-17)) or $RT\{[M_c] - [M_x]\}$ (P-model).

The term with C_2 is always the smaller part of the observed stress. The slope constant $2C_2$ decreases with decreasing volume fraction ϕ_2 of the polymer in the swollen rubber; its largest contribution is at small strains. It is affected by the cross-linking conditions, for example, the presence of liquids, chain orientation or degree of cross-linking. It does not depend, however, on the chemical structure of elastomers if cross-linking conditions are kept constant (Fig. 10-9, insert). Especially high C_2 values are observed for high entanglement densities.

10.4.7. Shearing

Shearing involves deformations in the 2-1 (y-x) direction (Fig. 7-4); dimensions in the 3-direction (z) remain constant and $\lambda_z = 1$. Volume-invariant shearing leads to $\lambda_x\lambda_y\lambda_z = 1$. The specimen is extended in x-direction (1) ($\lambda_x = \lambda$) and shortened in y-direction ($\lambda_y = 1/\lambda_x = 1/\lambda$). In Eqn.(10-14), the term $(4/f) \log_e(\lambda_x\lambda_y\lambda_z)$ becomes zero

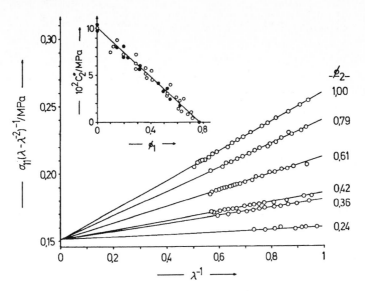

Fig. 10-9 Reduced stress of cross-linked natural rubber as function of the inverse elongational ratio at 45°C in the non-swollen state ($\phi_2 = 1$) and at different degrees of swelling, $1/\phi_2$, in decane (data of [4]). Insert: slopes $C_2^* = 2\,C_2$ as function of the volume fraction ϕ_1 of the solvent in swollen natural rubber (O), butadiene-styrene elastomers (●), and butadiene-acrylonitrile elastomers (⊙). Data of [5].

and the term $(\lambda_x^2 + \lambda_y^2 + \lambda_z^2 - 3)$ for stretching converts into $(\lambda^2 + \lambda^{-2} - 2) = (\lambda - \lambda^{-1})^2$ for shearing. Since shear deformation is defined as $\gamma = \lambda_x - \lambda_y = \lambda - \lambda^{-1}$, Eqn.(10-14) converts into:

$$(10\text{-}20)\quad \Delta G_{\mathrm{el}} = N_{\mathrm{c}}(k_{\mathrm{B}}T/2)[\lambda^2 - \lambda^{-2} - 2] = N_{\mathrm{c}}(k_{\mathrm{B}}T/2)\gamma^2$$

Differentiation of Eqn.(10-20) with respect to the *dimensionless* shear deformation $\gamma \equiv [\lambda - \lambda^{-1}]$ delivers the shearing energy $E_{\mathrm{s}} = (\partial \Delta G_{\mathrm{elast}}/\partial \gamma)_{T,V} = N_{\mathrm{c}}k_{\mathrm{B}}T[\lambda - \lambda^{-1}] = k_{\mathrm{B}}TN_{\mathrm{c}}\gamma$. Shearing involves the total volume; the shearing stress is thus $\sigma_{21} = E_{\mathrm{s}}/V_{\mathrm{o}}$. This expression can be compared to that for the tensile stress, $\sigma_{11} = F/A_{\mathrm{o}}$, where the drawing force F acts on the area A_{o}.

Insertion of the shearing energy $E_{\mathrm{s}} = k_{\mathrm{B}}TN_{\mathrm{c}}\gamma$, $k_{\mathrm{B}} = R/N_{\mathrm{A}}$ and the molar concentration $[M_{\mathrm{c}}] = N_{\mathrm{c}}/(N_{\mathrm{A}}V_{\mathrm{o}})$ of the network chains into $\sigma_{21} = E_{\mathrm{s}}/V_{\mathrm{o}}$ leads to

$$(10\text{-}21)\quad \sigma_{21} = RT[M_{\mathrm{c}}]\gamma$$

Comparison with Eqn.(7-14) shows that $RT[M_{\mathrm{c}}] = G$, the shear modulus.

The term $RT[M_{\mathrm{c}}]$ also appears in Eqn.(10-15) as $RT[M_{\mathrm{c}}] = \sigma_{11}/(\lambda - \lambda^{-2})$ which is experimentally true for $\lambda \to 1$ and $\varepsilon \to 0$ (Fig. 10-8). Setting $\lambda = \varepsilon + 1$, one obtains $\lambda - \lambda^{-2} = 1 + \varepsilon - (1 + \varepsilon)^{-2} \approx 1 + \varepsilon - (1 - 2\,\varepsilon) = 3\,\varepsilon$. In the limit of small deformations, one gets with Hooke's law, $\sigma_{11} = E\varepsilon$:

$$(10\text{-}22)\quad RT[M_{\mathrm{c}}] = \lim_{\lambda \to 1} \{\sigma_{11}/(\lambda - \lambda^{-2})\} = \sigma_{11}/(3\,\varepsilon) = E/3$$

Comparison of $RT[M_c] = G$ (Eqn.(10-21)) with $RT[M_c] = E/3$ (Eqn.(10-22)) delivers $E = 3G$ as indicated by Eqn.(10-4) for volume-invariant deformations (Poisson ratio $\mu = 1/2$). On shearing, elastomers thus behave as Hookean bodies. They are not Hookean on stretching, however, since the tensile stress is not directly proportional to the elongation (see Eqn.(10-15)).

10.5. Viscoelasticity

10.5.1. Fundamentals

Previous sections of this chapter discussed energy-elastic and entropy-elastic bodies. Such bodies return immediately and completely to their original positions if loads are removed. In most polymers, this takes time, however. Some polymers are also irreversibly deformed.

Bodies are called **viscoelastic** if they exhibit simultaneously time-independent elastic properties and time-dependent viscous properties. A viscoelastic behavior can be described by a combination of Hookean behavior (elasticity; Section 10.2.2.) and Newtonian behavior (viscosity; Section 7.4.3.). The elastic behavior of **Hookean bodies** can be modeled by a spring which contracts on application of a load but expands spontaneously to its original position after the load is removed (Fig. 10-10). The model of a **Newtonian liquid** is a dashpot with a viscous liquid through which a perforated plug is moved.

The viscoelastic behavior can be modeled by two simple combinations of springs and dashpots. The connection of spring (Hookean body) and dashpot (Newtonian liquid) in series leads to the **Maxwell element** and in parallel to the **Voigt–Kelvin element** (Fig. 10-10). The Maxwell element is a model for relaxations and the Voigt–Kelvin element a model for retardations (see below). Both models describe **linear viscoelasticity** since they combine stresses, deformations and deformation rates linearly. Additional combinations of springs and dashpots lead to more complicated elements, for example, the **Burgers element** (4-parameter element) as a Maxwell element and a Voigt–Kelvin element in series (see below).

10.5.2. Models

The **Maxwell element** contains a spring and a dashpot in series. Spring and dashpot are subjected to the same stress ($\sigma = \sigma_S = \sigma_D$); the deformations of spring and dashpot are additive ($\varepsilon = \varepsilon_S + \varepsilon_D$).

The dashpot (viscosity) causes the deformation to be time-dependent. Differentiation of the tensile deformation $\varepsilon = \varepsilon_S + \varepsilon_D$ with respect to time delivers the rate equa-

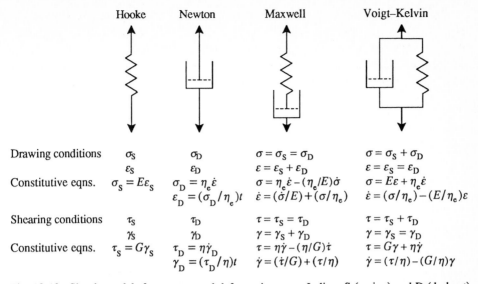

Hooke	Newton	Maxwell	Voigt–Kelvin

Drawing conditions	σ_S	σ_D	$\sigma = \sigma_S = \sigma_D$	$\sigma = \sigma_S + \sigma_D$
	ε_S	ε_D	$\varepsilon = \varepsilon_S + \varepsilon_D$	$\varepsilon = \varepsilon_S = \varepsilon_D$
Constitutive eqns.	$\sigma_S = E\varepsilon_S$	$\sigma_D = \eta_e \dot\varepsilon$	$\sigma = \eta_e \dot\varepsilon - (\eta_e/E)\dot\sigma$	$\sigma = E\varepsilon + \eta_e \dot\varepsilon$
		$\varepsilon_D = (\sigma_D/\eta_e)t$	$\dot\varepsilon = (\dot\sigma/E) + (\sigma/\eta_e)$	$\dot\varepsilon = (\sigma/\eta_e) - (E/\eta_e)\varepsilon$

Shearing conditions	τ_S	τ_D	$\tau = \tau_S = \tau_D$	$\tau = \tau_S + \tau_D$
	γ_S	γ_D	$\gamma = \gamma_S + \gamma_D$	$\gamma = \gamma_S = \gamma_D$
Constitutive eqns.	$\tau_S = G\gamma_S$	$\tau_D = \eta\dot\gamma_D$	$\tau = \eta\dot\gamma - (\eta/G)\dot\tau$	$\tau = G\gamma + \eta\dot\gamma$
		$\gamma_D = (\tau_D/\eta)t$	$\dot\gamma = (\dot\tau/G) + (\tau/\eta)$	$\dot\gamma = (\tau/\eta) - (G/\eta)\gamma$

Fig. 10-10 Simple models for stresses and deformation rates. Indices S (spring) and D (dashpot) have been omitted in constitutive equations and the subsequent ones since one cannot distinguish between the various contributions to stress and deformation. Tensile stresses $\sigma_S = \sigma_{11}$ are symbolized by σ. In order to avoid mix-ups with shear stresses $\sigma_D = \sigma_{21}$, the latter were given the traditional symbol τ instead of the recommended symbol σ_{21}.

Drawing (elasticity see Eqn.(10-1), viscosity see Section 7.4.6.): σ = tensile stress, $\dot\sigma = d\sigma/dt$ = rate of drawing, E = tensile modulus, ε = elongation, $\dot\varepsilon = d\varepsilon/dt$ = rate of extension, η_e = extensional viscosity.

Shearing (elasticity see Eqn.(7-13), viscosity see Eqn.(7-14)): η = viscosity, τ = shear rate, $\dot\tau$ = $d\tau/dt$ = shear deformation rate, G = shear modulus, γ = shear deformation, $\dot\gamma$ = shear rate.

tion $\dot\varepsilon = d\varepsilon/dt = d\varepsilon_S/dt + d\varepsilon_D/dt = \dot\varepsilon_S + \dot\varepsilon_D$. The two deformation rates are expressed by the constitutive equations: $\sigma_S = E\varepsilon_S \rightarrow d\varepsilon_S/dt = d(\sigma_S/E)/dt \rightarrow \dot\varepsilon_S = \dot\sigma_S/E$ for the drawing of a Hookean body and $(\sigma_D/\eta_e) = \dot\varepsilon_D$ for the drawing (not shearing!) of a Newtonian liquid. The resulting equation $\sigma = \eta_e\dot\varepsilon - (\eta_e/E)\dot\sigma$ is transformed into the **constitutive equation** for the deformation of a Maxwell element: $\sigma = \eta_e\dot\varepsilon - (\eta_e/E)\dot\sigma$. The constitutive equation for the shearing is obtained by analogy as $\tau = \eta\dot\gamma - (\eta/G)\dot\tau$.

Spring and dashpot are parallel in the **Voigt–Kelvin element**. Deformations are identical for spring and dashpot ($\varepsilon = \varepsilon_S = \varepsilon_D$); the stresses are additive ($\sigma = \sigma_S + \sigma_D$). Insertion of the expressions for σ_S and σ_D (Fig. 10-10) delivers for the extensional rate $\dot\varepsilon = (\sigma/\eta_e) - (E/\eta_e)\varepsilon$ and for the total stress $\sigma = E\varepsilon + \eta_e\dot\varepsilon$ (indices S and D omitted). Constitutive equations for shear stress τ and shear rate $\dot\gamma$ are obtained in a similar manner (see Fig. 10-10).

In a similar way, constitutive equations can be derived for the four possible **3-parameter models**: spring in series with a Voigt–Kelvin element (S-VK), spring parallel with a Maxwell element (S‖M), dashpot in series with a Voigt–Kelvin element (D-VK), and dashpot parallel with a Maxwell element (D‖M). These 3-parameter models are said to represent **standard linear solids**. An even more realistic model is the **Burgers element**, a 4-parameter model with a spring in series with a Voigt–Kelvin element and a dashpot (S-VK-D).

The constitutive equation of the Burgers element is difficult to solve, however, since it is a linear second order differential equation. The standard linear solids all have (various) linear first order differential equations. For example, the constitutive equation for the S-VK model is $(G_S+G_{VK})\sigma/\eta_{VK} + d\sigma/dt = (G_SG_{VK})\varepsilon/\eta_{VK} + G_S(d\varepsilon/dt)$ where G_S = modulus of the spring, G_{VK} = modulus of the VK–element, and η_{VK} = viscosity of the VK–element.

In experiments, one cannot distinguish between the contributions of the various elements to the observed behavior. Subscripts are therefore dropped, all stresses are simply denoted by σ, all deformations by γ (or ε), and all moduli by G (or E). The constitutive equations are therefore written as:

Maxwell element $\qquad\qquad\qquad\qquad \sigma = \eta(d\gamma/dt) \qquad\qquad - (\eta/G)(d\sigma/dt)$

Voigt–Kelvin element $\qquad\qquad\quad \sigma = \eta(d\gamma/dt) + G\gamma$

Standard linear solid S-VK $\qquad\quad \sigma = \eta'(d\gamma/dt) + G\gamma - (\eta/G')(d\sigma/dt)$

where $G_S\eta_{VK}/(G_S + G_{VK}) = \eta'$, $(G_SG_{VK})/(G_S + G_{VK}) = G$, and $\eta_{VK}/(G_S + G_{VK}) = \eta/G'$.

These equations have 3 variables (σ, γ, t) and can therefore be solved only for special cases: (I) constant deformation, (II) constant stress, (III) constant strain rate, and (IV) constant stress rate. Experiments with constant deformation are called **creep experiments** whereas those with constant stress are known as **stress relaxation experiments**. Results of these two experiments (and also dynamic experiments, see Section 10.5.7.) allow one to predict the polymer behavior for constant strain rate and constant stress rate. The reverse is not true.

10.5.3. Stress Relaxation

In a stress relaxation experiment, a body is suddenly deformed at time $t = 0$ by a value of ε (or γ). An example is the compression of a plastic seal between the bottle mouth and the cap. The deformation is held constant ($\varepsilon = \varepsilon_0$) during the total time of the experiment between $t = 0$ and $t = t_E$; the deformation rate $\dot\varepsilon$ is zero.

The body responds to this compression (negative drawing) by a change in stress. Since $\dot\varepsilon = 0$, the differential equation becomes $\dot\varepsilon = (\dot\sigma/E) + (\sigma/\eta_e) = 0$ for the deformation of a Maxwell element (Fig. 10-10). Integration of this equation delivers:

(10-23) $\quad \sigma = \sigma_0\exp[- (E/\eta_e)t] = \sigma_0\exp[- (t/t_{rlx})] = \sigma_0\exp[- (1/De)]$

where $t_{rlx} \equiv \eta_e/E$ is the **relaxation time**. The body thus reacts to the instantaneous deformation ε_0 by a sudden increase of the stress to a value of $\sigma = \sigma_0$. On continued constant deformation ($\varepsilon = \varepsilon_0$), stresses then decline exponentially. They approach zero at infinite time, i.e., the body relaxes completely (Fig. 10-11). This relaxation is the reason why plastic seals have to be retightened from time to time.

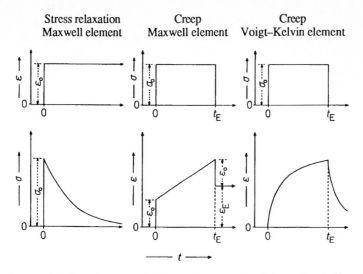

Fig. 10-11 Stress relaxation of a Maxwell element after a sudden deformation (left) and change of deformation by creep for a Maxwell element (center) and a Voigt–Kelvin element (right). Top: excitation; bottom: response.

Deformations ε (relaxation experiments) and stresses σ (creep experiments) are applied at time $t = 0$ and removed at time $t = t_E$. The relaxation experiment shows only the relaxation but not the recovery of stresses after removal of the deformation at time t_E. Note that the stress relaxation of a VK element is not shown since it is physically unrealistic (see below).

The relaxation time t_{rlx} indicates the time after which the stress is reduced to $(1/e)$th $\therefore$ 36.8 % of its original value. It is the intercept $t_{rlx} = \eta_e/E$ on the time axis if the tangent to the curve at σ_0 is extended to $\sigma = 0$. The viscosity is an extensional viscosity η_e and not a shear viscosity η. The modulus E in Eqn.(10-23) represents a relaxation modulus which is not identical with the modulus E of Eqn.(10-1) because of the differences in models.

The ratio of relaxation time t_{rlx} to the time scale of the experiment is called the **Deborah number** De after the prophetess of the Old Testament [Judges 5.5: "The mountains flowed before the Lord ... "]. By definition, the Deborah number is zero for Newtonian liquids and infinite for Hookean bodies. It adopts values of ca. 10^{-12} for water, 10^{-6} for mineral oil, 0.1-10 for polymer melts, and 10^9-10^{11} for glasses. The Deborah number is approximately unity at the glass temperature.

Polymers do not have *one* relaxation time but a whole **relaxation time spectrum**. In perfect elastomers, network chains have the same length. Stresses produced by short deformation times can relax within $t_{rlx} \approx 10^{-5}$ s because of the "free" rotation around chain bonds. Longer deformation times cause network junctions to shift. These displacements of junctions need long times to relax. Between these two very different relaxation times is a region where the modulus stays constant. A similar reasoning applies to entanglements which act as physical cross-links (see below).

In real polymers, distances between chemical cross-links (elastomers, thermosets) and physical cross-links (thermoplastics) are not uniform. The distribution of these distances generates distributions of the relaxation times of the two relaxation processes. The resulting relaxation time spectrum can be modeled by parallel Maxwell elements.

Relaxations cannot be described by the Voigt–Kelvin element. Here, the spring can respond immediately to a sudden deformation. The dashpot cannot do this because it has to overcome an infinitely large resistance which requires an infinitely large stress and that is not realistic.

10.5.4. Creep Experiment

In creep experiments, a body is subjected to a constant tensile stress σ_0 at time $t = 0$ and allowed to deform during the time $t = t_E$ of the experiment (Fig. 10-11, II and III). In a *Maxwell element*, the spring reacts immediately to this excitation because it is in series with the dashpot; the Maxwell element adopts at $t = 0$ a deformation $\varepsilon_0 = \sigma_0/E$ (Fig. 10-11, center). If the loading remains uniform ($\sigma =$ constant, i.e., $\dot{\sigma} = d\sigma/dt = 0$), Maxwell elements will elongate with constant speed σ_0/η_e because the dashpot is in series with the spring. The Maxwell element shows a **creep** or **cold flow**.

The cold flow is a **retardation process**. Integration of the differential equation $\dot{\varepsilon} = (\dot{\sigma}/E) + (\sigma/\eta_e)$ (see equation in Fig. 10-10) with the condition $\sigma = \sigma_0 = const.$ delivers the total deformation ε at time t:

$$(10\text{-}24) \quad \int \dot{\varepsilon}\, dt = \int (d\varepsilon/dt)dt = \varepsilon = (\sigma_0/E) + (\sigma_0/\eta_e)t = \varepsilon_0 + (\sigma_0/\eta_e)t$$

After removal of the load at the time $t = t_E$, a Maxwell element will contract by the elastic contribution $\varepsilon_0 = \sigma_0/E$ but will remain deformed permanently by a residual contribution of $\varepsilon_E = (\sigma_0/\eta_e)t_E$ (**permanent set**). A Maxwell element thus behaves like an elastic solid at the start of the creep experiment but like a viscous liquid during the experiment: it represents a **viscoelastic liquid**.

The linear increase of deformation with time at constant stress is not found for polymers: the Maxwell element is unrealistic for creep experiments. The time dependence of deformation in creep experiments on polymers is much better described by the *Voigt–Kelvin element* (Fig. 10-11, right). According to Fig. 10-10, the constitutive equation for tensile experiments is $\dot{\varepsilon} = (\sigma/\eta_e) - (E/\eta_e)\varepsilon$. Integration leads to:

$$(10\text{-}25) \quad \varepsilon = (\sigma_0/E)[1 - \exp(- Et/\eta_e)] = (\sigma_0/E)[1 - \exp(- t/t_{rtd})]$$

Immediately after the application of stress at $t = 0$, the deformation ε is zero (Fig. 10-11); however, this is not found for polymers either. One does observe a rapid increase of deformation with time at small times, tough, similar to that of liquids; the initial slope of the $\varepsilon = f(t)$ curve is σ_0/η_e. Deformations later increase more slowly.

The retardation time $t_{rtd} = \eta_e/E$ has about the same magnitude as the relaxation time t_{rlx}. The two times are not identical, however, because they apply to different models. The retardation time indicates the time at which the retarding proportion of the deformation has attained a value of $(1 - 1/e) = 0.632$ of the final deformation ε_E at time t_E if the load is kept constant. The Voigt–Kelvin element thus behaves as a **viscoelastic solid**: at time zero as a liquid but later, at $0 \le t \le t_E$, as a solid.

Removal of the load at time t_E causes the deformation to decline according to $\varepsilon = (\sigma_o/E)[1 - \exp(- t_E/t_{rtd})][\exp\{- (t - t_E)/t_{rtd}\}]$. The original state with $\varepsilon = 0$ is reached at infinitely large time. If the experiment is not terminated at time t_E but continued *ad infinitum*, a value of $\varepsilon_o = \sigma_o/E$ is obtained at $t \to \infty$.

The shear deformation is calculated similarly, starting with the differential equation $\dot{\gamma}= (\tau/\eta) - (G/\eta)\gamma$ for the shearing of the Voigt–Kelvin element (Fig. 10-10). The result is $\gamma = (\eta/G)[1 - \exp(- t/t_{rlx})]$. The retardation modulus G in this equation is neither identical with the retardation modulus E of Eqn.(10-25) nor with the shear modulus G of Eqn.(7-14) since the true viscoelastic behavior of polymers is only approximately described by the Voigt–Kelvin element or by the Maxwell element.

The four-parameter model of **Burgers** is more realistic. In this model, a Maxwell element and a Voigt–Kelvin element are connected in series. The total deformation $\varepsilon = \varepsilon_H + \varepsilon_N + \varepsilon_{VK}$ in tension is given by the contributions of the Hookean body (H), the Newtonian liquid (N), and the Voigt–Kelvin element (VK). Introduction of $\varepsilon_H = \sigma/E$ for the Hookean body, $\varepsilon_N = (\sigma/\eta_e)t$ for the Newtonian liquid (from Fig. 10-10), and the expression for ε_{VK} from Eqn.(10-25) delivers (subscripts H, N, VK omitted):

$$(10\text{-}26) \quad \frac{\varepsilon}{\sigma} = \frac{1}{E} + \frac{t}{\eta_e} + \frac{1}{E_{rtd}}\left[1 - \exp(-E_{rtd}t/\eta_e)\right] = \frac{1}{E_k}$$

The time-dependent ratio $\sigma/\varepsilon = E_k$ represents a creep modulus.

Creep curves are usually not analyzed in detail since the three contributions ε_H (elastic), ε_N (viscous) and ε_{VK} (viscoelastic) are not known. The time-dependent viscous and viscoelastic parts are however often combined in a new parameter ε_k. The time dependence of this parameter is modeled by **Findlay's law** $\varepsilon_k = \varepsilon_{k,o} \cdot t^n$. The creep curve can be thus described by:

$$(10\text{-}27) \quad \varepsilon = \varepsilon_H + \varepsilon_N + \varepsilon_{VK} = \varepsilon_H + \varepsilon_k = \varepsilon_H + \varepsilon_{k,o}t^n$$

Eqn.(10-27) allows one to extrapolate short-time deformations to long times. Similar extrapolations can be used to determine the time–dependence of strengths. Static tests of creep rupture strength subject specimens to various loads and measure the time to fracture. Periodic loading can involve changes of loads for the determination of flexural fatigue strengths or twisting of specimens for torsional fatigue strengths. These and many other testing methods consume much time and labor.

10.5.5. Boltzmann Superposition Principle

Specimens are stretched with constant rates $\dot{\varepsilon} = \dot{\varepsilon}_o$ in tensile tests (Section 10.2.2.). The tensile stress is zero at the beginning of the experiment ($\sigma_o = 0$). Integration of the differential equation $\dot{\sigma} = E \dot{\varepsilon}_o - (\sigma E/\eta_e)$ for the drawing rate of a Maxwell element (Fig. 10-10) results in:

$$(10\text{-}28) \quad \sigma = \eta_e \dot{\varepsilon}_0 \left[1 - \exp\left(\frac{-E}{\eta_e \dot{\varepsilon}_0} \cdot \varepsilon \right) \right]$$

Tensile stresses thus increase with increasing elongation ε first strongly and later not as strongly if $\eta_e \dot{\varepsilon}_0 = const. < \infty$; a limiting value of $\sigma_\infty = \eta_e \dot{\varepsilon}_0$ is approached for $\varepsilon \to \infty$. The two limits of this behavior are obtained if one sets $E\varepsilon/(\eta_e \dot{\varepsilon}_0) \equiv x$ and develops the exponential $\exp(- x)$ into an infinite series $1 - x + x^2/2! -\dots$ For infinitely large extensional rates (i.e., $\eta_e \dot{\varepsilon}_0 \to \infty$), Hooke's law $\sigma = E\varepsilon$ is recovered; the initial slope of the function $\sigma = f(\varepsilon)$ delivers the tensile modulus. For diminishing extensional rates (i.e., $\eta_e \dot{\varepsilon}_0 \to 0$), one obtains $\sigma = 0$. If the extensional viscosity is independent of the extensional rate ($\eta_e = const.$), moduli will decrease with decreasing extensional rates $\dot{\varepsilon}_0$ (or $\eta_e \dot{\varepsilon}_0$) (Fig. 10-12, top): Specimens behave as rigid materials (high E) for large drawing rates and as rubbers (low E) for small ones.

Calculated functions $\sigma = f(\varepsilon)$, and thus also the modulus E, behave at constant temperature and various extensional rates (Fig. 10-12, top) similar to functions that are measured at constant extensional rates and various temperatures (Fig. 10-12, bottom) if one neglects yield points which occur only in a certain temperature range. Functions $E = f(t)$ can thus be transformed into functions $E = f(T)$ and vice versa. Similar transformations can be performed for time- and temperature-dependent shear moduli G, tensile compliances $D = 1/E$, and shear compliances $J = 1/G$.

The transformation $E = f(t) \rightleftarrows E = f(T)$ makes use of the **Boltzmann superposition principle**. This principle states that a deformation which is caused by an additional load (or a recovery due to load removal) is independent of previous loading or unloading.

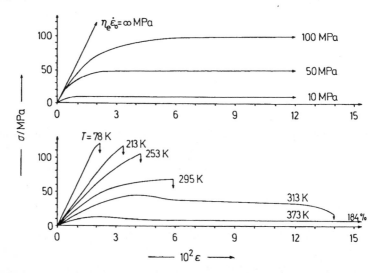

Fig. 10-12   Stress–strain behavior in tensile tests. $\downarrow$ Fracture.
   Top: Calculated for a Maxwell element with $E = 6000$ MPa and various values of $\eta_e \dot{\varepsilon}_0$.
   Bottom: Experimental for a poly(methyl methacrylate) at various temperatures [6].

Application of a constant shear stress $(\tau_o)_a$ at time $t_o = 0$ causes an increase of the shear deformation γ of a Voigt–Kelvin element. The shear deformation approaches a value of $\gamma_1 = (\tau_o)_a/G(t_1) = (\tau_o)_a J(t_1)$ at time t_1 (Fig. 10-13). Removal of the shear stress at t_1 causes a decrease of γ at time t_2; a shear stress $\gamma = \gamma_1 - \Delta\gamma_1$ is obtained. If the shear stress had not been removed at t_1, the specimen would have attained a value of shear deformation of γ_2. The actual value is however $\gamma = \gamma_2 - \Delta\gamma_2$. One thus recovers a deformation of $\Delta\gamma_2 = \gamma_2 - \gamma_1 + \Delta\gamma_1 = \gamma_b + \Delta\gamma_1$. This value is identical with the one that would have been obtained if an additional shear stress $(\tau_o)_b$ had been applied at time t_1. A deformation by an additional load (or stress) is therefore independent of all previous loads (or stresses).

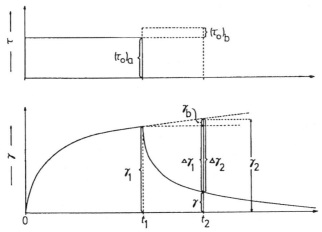

Fig. 10-13 Representation of the Boltzmann superposition principle for a creep experiment.

Superposition can be used to prepare a master curve for the dependence of shear moduli G on the time span of experiments. Moduli $G(t)$ are obtained as function of time at various temperatures (Fig. 10-14, left). These values can be superposed with the help of the shift factor a_T of the WLF Eqn.(9-14) (Fig. 10-14, right) provided that the relaxation time spectrum is independent of the temperature. The resulting master curve covers 16 decades of time $[(10^{-11}\text{-}10^5)\,\text{h};\ (40\ \text{nanoseconds to } 11.4\ \text{years})]$. Several dynamic methods have to be used to establish such a master curve since such a large time span cannot be covered by a single method.

10.5.6. Dynamic Measurements

In dynamic-mechanical measurements, specimens are subjected to periodic stresses. Two main types of experiments are employed: forced and free oscillations. These dynamic-mechanical methods are supplemented for higher frequency ranges by nuclear magnetic resonance, ultrasound and dielectric measurements. The combination of all these measurements allows one to cover very wide ranges of frequencies and temperatures.

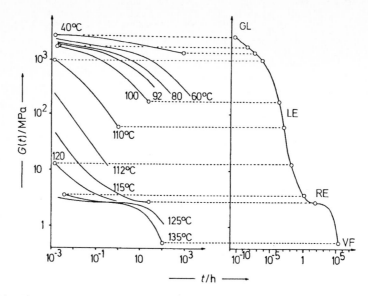

Fig. 10-14   Time dependence of shear modulus from stress relaxation experiments for a poly-(methyl methacrylate) of $\overline{M}_\eta$ = 3 600 000 g/mol at various temperatures (left). Data were shifted by the shift factor a_T of the WLF equation for a reference temperature of T_o = 388 K (T_G = 378 K); some of these shifted data points are indicated by o. The resulting time–temperature superposition (right) shows the ranges of viscous flow (VF), rubbery behavior (RE) and leathery state (LE); the glassy state is not obtained for this time scale. Compare the logarithmic time scale of Fig. 10-14 with the simple temperature scale of Fig. 10-5. With permission by Academic Press [7].

For free oscillations (**torsion pendulum** experiments), specimens are twisted once (see Fig. 10-1). The stress is then removed whereupon the specimen oscillates freely around its position at rest (Fig. 10-15). The frequency of the oscillations is constant since it depends on the damping characteristics of the polymer. The amplitudes θ_1, θ_2 ... of these oscillations decrease steadily but the ratio of two subsequent amplitudes is constant for ideal-viscoelastic polymers. The natural logarithm of this ratio is the **logarithmic decrement**, $\Lambda = \log_e(\theta_n/\theta_{n+1})$. The resulting torsional moduli are shear moduli.

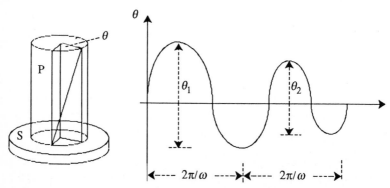

Fig. 10-15   Torsion pendulum experiment. A cylindrical polymer specimen P is mounted on a rigid non-polymeric support S and its upper part twisted by an angle θ. ω = Circular frequency.

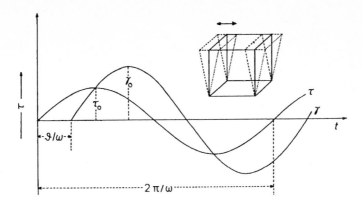

Fig. 10-16   Applied shear stress τ and resulting out-of-phase deformation γ as function of time for a sinusoidal uniaxial shearing. τ_o, γ_o = amplitudes of stress and deformation, respectively; ϑ = phase angle, ω = circular frequency.

In **forced oscillation experiments**, frequencies can be chosen at will. The most frequently used instrument for this type of measurements is the **Rheovibron**® which allows one to observe stress and deformations independently of each other. The applied stresses are tensile stresses; the resulting flexural moduli are thus tensile moduli.

The simplest types of experiments subject the specimen to sinusoidal stress, for example, by uniaxial shearing (Fig. 10-16). The shear stress $\tau(t) = \tau_o \sin \omega t$ varies with time t at constant circular frequency ω; τ_o is the amplitude. For Hookean bodies, an applied stress leads instantaneously to a shear deformation γ so that $\gamma(t) = \gamma_o \sin \omega t$. For viscoelastic bodies, shear deformations trail the applied stresses. In the ideal case, the phase angle ϑ remains constant in vector diagrams and one observes:

(10-29) $\gamma(t) = \gamma_o \sin (\omega t - \vartheta)$

Stress vectors can be treated as the sum of two components. One component is in phase with the deformation ($\tau' = \tau_o \cos \vartheta$) and the other one is not ($\tau'' = \tau_o \sin \vartheta$). The in-phase component τ' delivers the **shear storage modulus** G' (**in-phase modulus, real modulus,** "elastic modulus") which measures the rigidity of the specimen:

(10-30) $G' = \tau'/\gamma_o = (\tau_o/\gamma_o)\cos \vartheta = G^* \cos \vartheta$

The **shear loss modulus** G'' (**90° out-of phase modulus, imaginary modulus,** "viscous modulus") gives the loss of usable mechanical energy by dissipation into heat:

(10-31) $G'' = \tau''/\gamma_o = (\tau_o/\gamma_o)\sin \vartheta = G^* \sin \vartheta$

The two moduli G' and G'' are called "real" and "imaginary", respectively, because Eqns.(10-30) and (10-31) can also be derived with complex variables. The stress $\tau(t)$ is replaced by the complex stress $\tau^* = \tau_o \exp(i\omega t)$ and the shear deformation $\gamma(t)$ by the complex shear deformation $\gamma^* = \gamma_o \exp\{i(\omega t - \vartheta)\}$. The complex shear modulus becomes $G^* = G' + iG''$ or, with Eqns.(10-30) and (10-31), $G^* = [(G')^2 + (G'')^2]^{1/2}$.

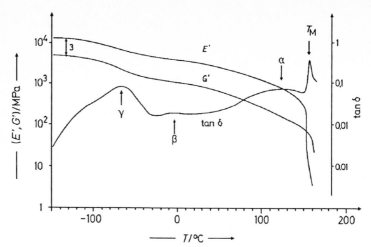

Fig. 10-17 Temperature dependence of the tensile storage modulus E', shear storage modulus G' and loss tangent tan δ of an acetal copolymer at 0.3 Hz $< v <$ 15 Hz [18]. At low temperatures, E' $\approx 3\,G'$ as predicted for energy-elastic bodies. The polymer melts at T_M. The α relaxation at ca. 120°C is caused by segmental movements in crystalline regions. The β relaxation at ca. 0°C represents the dynamic glass temperature of water-plasticized amorphous regions; it disappears on drying. The γ relaxation indicates the relaxation of chain segments in amorphous domains.

The ratio of imaginary modulus to real modulus is the **mechanical loss factor** Δ. In forced oscillations, it is given by the tangent of the loss angle ϑ, i.e., $\Delta = \tan \vartheta$, where tan ϑ is called the **loss tangent**. In free oscillations, the mechanical loss factor depends on the geometry of the test specimen; for cylindrical specimens, $\Lambda \approx \pi \tan \vartheta$. The loss factor has the same value for tensile moduli and shear moduli:

(10-32) $\Delta = G''/G' = E''/E'$

No shearing occurs on pressurizing. The ratio of the two bulk moduli thus does not equal the ratio of the corresponding shear moduli; one rather has $K''/K' < G''/G'$.

Because of the heat effects involved, mechanical loss factors go through maxima at transition and relaxation temperatures (Fig. 10-17). Since glass temperatures indicate relaxation phenomena, they are not thermodynamic transitions and thus strongly frequency dependent. Dynamic mechanical measurements are therefore the preferred methods for the determination of relaxation phenomena (Section 9.6.2.). Besides glass temperatures, a number of other, secondary relaxations are usually observed.

10.5.7. Shear Moduli of Melts

Shear storage moduli G' of *melts* and *concentrated solutions* of high molar mass polymers vary as follows with increasing frequency (Fig. 10-16). The logarithm of G' first increases linearly with the logarithm of ω (or $a_T\omega$) in an **end zone**. It then enters a **plateau** until at high frequencies it finally increases again linearly with the logarithm of the frequency in a **transition zone**.

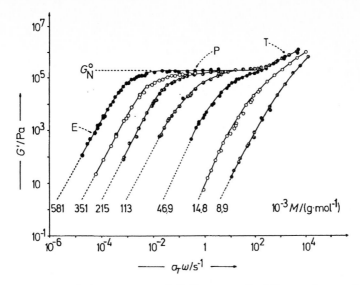

Fig. 10-18 Dependence of shear storage moduli G' on normalized frequencies $a_T\omega$ for melts of poly(styrene)s with narrow molar mass distributions. Data at different temperatures and circular frequencies are adjusted to 160°C by shift factors α_T (cf. Eqn.(9-13)). G_N^0 = Plateau modulus, E = end zone, P = plateau, T = transition zone. With permission by the American Chemical Society [8].

In the transition zone, all G' values approach asymptotically a common function $G' = f(a_T\omega)$ that is independent of molar masses; this zone must reflect the movement of chain segments. The function $G' = f(a_T\omega)$ does depend on molar masses in the end zone. It must here be effected by long-range conformational changes.

In the plateau region, shear storage moduli G_N^0 are independent of frequency and molar mass. Since the transition zone represents the viscous behavior and the end zone the viscoelastic one, plateaus must be representative of the rubbery behavior (see Fig. 10-14 for the time dependence of moduli). The rubber-like behavior must come from temporary cross-links that are generated by entanglements (Section 7.5.).

The rubbery behavior in *dynamic* experiments can be described by *equilibrium* theories of rubber elasticity. The amount-concentration $[M_c]$ of network chains is replaced by the molar concentration $[M_e] = \rho\phi/M_e$ of segments between entanglements. Similar to Eqn.(10-21), a molar mass M_e is calculated from the plateau modulus G_N^0, the density of the liquid (in melts: density of melt), the volume fraction ϕ_2 of the polymer (in melts: $\phi_2 \equiv 1$), and the molar energy RT:

(10-33) $M_e = (4/5)\, RT\rho\phi_2/G_N^0$

The numerical factor 4/5 results from a theoretical derivation of this equation for dynamic entanglements; sometimes, M_e is calculated from G_N^0 without the factor 4/5. Critical molar masses M_e from plateau moduli are lower than entanglement molar masses M_c from the molar mass dependence of rest viscosities of melts (Table 7-1) since M_e and M_c depend differently on entanglements (see also Section 7.5.2.). Their ratio is $M_c/M_e \approx 2.5$ (Table 7-1).

10.6. Failure

The term "failure" denotes an unwanted mechanical behavior. For load bearing applications, it may be the brittle fracture of rigid polymers or the yielding of ductile ones. For packaging materials, on the other hand, yielding may be a very desirable property since it allows one to stretch plastic films over edges without tearing.

10.6.1. Fracture Phenomena

Polymers fracture differently depending on chemical and physical structure, environment, and type, duration and/or frequency of deformation. Fractures can be smooth or splintery and happen "instantaneously" or after years. Elongations at break can vary between a few tenths of one percent and more than a thousand percent.

Two borderline cases exist. On **brittle fracture**, thermoplastics break perpendicular to the stress direction without flow; the fracture elongations are small and by definition smaller than a few percent (Section 10.2.3.). A **ductile fracture** ("tough fracture") occurs in the direction of shear stress by flow; here, fracture elongations may be several hundred percent. Similar large elongations at break are found for elastomers whereas fracture elongations of fibers are small.

Fracture strengths of thermoplastics, fibers and elastomers are almost zero below a critical molar mass M_c (Fig. 10-19). Measurable tensile strengths are only observed above certain critical molar masses that depend on the molar mass distributions of polymers and the type of preparation of test specimens. For example, the critical

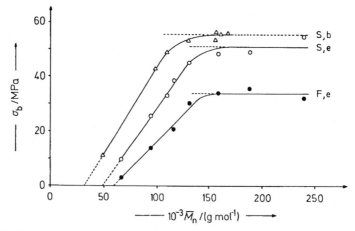

Fig. 10-19   Dependence of tensile strengths σ_b on number-average molar masses of atactic poly(styrene) at 23°C. Injection-molded test specimens (S; O, Δ) have higher tensile strengths than compression-molded ones (F; ●) since segments become oriented by injection molding. Polymers with broad molar mass distributions (Index b; Δ; $\overline{M}_w/\overline{M}_n \approx 1.77$-$2.85$) have high molar-mass tails and thus higher strengths than specimens with narrower molar mass distributions (Index e; O,●; $\overline{M}_w/\overline{M}_n \approx 1.04$-$1.18$). Data of [9].

molar mass is ca. 32 000 g/mol for injection-molded poly(styrene)s with broad molar mass distributions. This molar mass is in the same range as the critical molar mass M_c ≈ 35 000 g/mol for entanglements from melt viscosities at rest (Table 7-1). Entanglement densities increase with increasing molar masses and so do tensile strengths until finally entanglement densities and thus strengths become independent of molar masses at $\overline{M}_n$ > 150 000 g/mol.

10.6.2. Theoretical Fracture Strengths

The ultimate fracture strengths $\sigma_\parallel^\circ = F_c/A_c$ of fully aligned polymer chains can be estimated from the force F_c needed to break a chain and the cross-sectional area of the chain. In order to break a chain with the force $F_c = E_c/b_c$, covalently bonded atoms must move away from each other by a length b_c and then overcome the bond energy $E_c = E_m/N_A$ where E_m = bond strength (E_m = 348 kJ/mol for C-C bonds). The length b_c must be at least half of the bond length b (b = 0.154 nm for C-C bonds). The cross-sectional area $A_c = ac/N_u$ is given by the lattice constants a and c perpendicular to the chain axis and the number N_u of chains per unit cell (N_u = 2 for poly-(ethylene)). The ultimate fracture strength of poly(ethylene) is thus estimated as $\sigma_\parallel^\circ$ = $(2\,N_u E_m)/(abcN_A)$ = 40.8 GPa.

Another estimate is based on the theory of rate processes. It assumes that covalent bonds (and finally the specimen) are broken if the thermal fluctuations around a bond exceed a certain critical value. According to this theory, the fracture strength at T = 0 K is given by $\sigma_\parallel^\circ = [(4/3)b\Delta E_o^\ddagger E_\parallel]^{1/2}/A_c$ where b = bond length, $\Delta E_o^\ddagger$ = thermal activation energy for the fracture of bonds, $E_\parallel$ = modulus of elasticity in the chain direction (at 0 K), and A_c = cross-sectional area of the chain. This theory furnished a theoretical fracture strength of 32.5 GPa for poly(ethylene) (Table 10-5). For 13 polymers, the ratio $\sigma_\parallel^\circ/E_\parallel$ averaged 0.099 whereas a plot of $\log_{10}\sigma_\parallel^\circ$ = $f(\log_{10} E_\parallel)$ delivered $\sigma_\parallel^\circ/E_\parallel$ = 0.093.

Table 10-5   Cross-sectional areas A_c, Young's moduli $E_\parallel$, thermal activation energies $\Delta E_o^\ddagger$ for bond scission, theoretical fracture strengths $\sigma_\parallel^\circ$ at 0 K, and ratios $\sigma_\parallel^\circ/E_\parallel$ [13].

Polymer	$\dfrac{A_c}{nm^2}$	$\dfrac{E_\parallel}{GPa}$	$\dfrac{10^{20}\Delta E_o^\ddagger}{J}$	$\dfrac{\sigma_\parallel^\circ}{GPa}$	$\dfrac{\sigma_\parallel^\circ}{E_\parallel}$
Poly(ethylene)	0.182	340	50.0	32.5	0.096
Polyamide 66	0.203	200	28.5	16.9	0.085
Poly(vinyl chloride), atactic	0.286	200	65.9	18.2	0.091
Poly(tetrafluoroethylene)	0.277	156	56.3	15.3	0.098
Poly(oxymethylene)	0.172	150	18.9	14.0	0.093
Poly(propylene), isotactic	0.343	42	20.7	3.9	0.093
Poly(styrene), isotactic	0.698	12	38.2	1.4	0.117
Poly(vinyl t-butyl ether), isotactic	0.887	4.1	20.1	0.46	0.112

The data of Table 10-5 imply that the fracture strength is directly proportional to the the modulus of elasticity ($\sigma_{||}^o = K_{||}E_{||}$) and that the proportionality constant $K_{||}$ is universal for all polymers. This is not too surprising since both the modulus and the fracture strength are controlled by the bond energy. The **Frenkel theory** assumes that the dependence of forces per area (i.e., stresses σ) on bond distances L can be approximated by a sinusoidal function with the wavelength λ where $L_o =$ equilibrium distance, $\varepsilon = (L - L_o)/L_o$ and $\sigma_o =$ theoretical strength:

(10-34) $\sigma = \sigma_o \sin[2\,\pi\,(L - L_o)/\lambda] = \sigma_o \sin[2\,\pi\,L_o\varepsilon/\lambda]$

Differentiation for $L_o\varepsilon = 0$ delivers $(d\sigma/dL) = \sigma_o(2\,\pi/\lambda)[\cos(2\,\pi\,L_o\varepsilon/\lambda] = 2\,\pi\,\sigma_o/\lambda$. Another expression for the change of stress with distance is obtained from Hooke's law $\sigma = E_{||}\varepsilon = E_{||}[(L/L_o)-1]$ which leads to $d\sigma/dL = E_{||}/L_o$ so that $E_{||}/L_o = 2\,\pi\,\sigma_o/\lambda$. Chains are broken if atoms have moved by a distance $\lambda/2 = L_b - L_o$ from the equilibrium distance L_o. The theoretical tensile strength of parallel chains should thus be directly proportional to the tensile modulus:

(10-35) $\sigma_o = (E_{||}\lambda)/(2\,\pi\,L_o) = E_{||}(L_b - L_o)/(\pi\,L_o) = KE_{||}$; $K = (L_b - L_o)/(\pi L_o)$

Since experiments indicate $K = \sigma_{||}^o/E_{||} \approx 0.095$ for totally oriented chains (Table 10-5), the fracture distance $L_b = (1 + 0.095\,\pi)L_o$ must be ca. 30 % larger than the bond distance ($L_b \approx 1.3\,L_o$).

Fracture strengths of conventionally processed polymers are, however, much lower. Semicrystalline high-density poly(ethylene)s have typical fracture strengths of only ca. 30 MPa and amorphous at-poly(styrene)s ca. 36 MPa; these strengths are about 1000 times lower than the theoretical fracture strengths of covalent bonds.

Inspection of the stress–strain curves of *conventionally* processed polymers shows, however, that fracture strengths of polymers are not very different from yield strengths (Fig. 10-4). At the yield point, very little (if any) radicals are formed by chain scission according to electron spin resonance measurements. Even at fracture, only (0.01-1) % of all chains are broken. Since yielding involves flow phenomena, it must be the severance of intermolecular van der Waals and/or dipole interactions that is responsible for the failure at the yield point.

10.6.3. Failure Strengths

Polymers can be divided into two groups with respect to the dependence of the yield strength σ_y on the modulus E (Fig. 10-20). Polymers with glass temperatures T_G higher than the testing temperature T show $\sigma_y = K_I E$ (type I) whereas those with $T < T_G$ follow $\sigma_y = K_{II}E^{1/2}$ (type II). For example, plasticization of polyamides 66 by water (through conditioning at 65 % relative humidity) lowers the glass temperature of the dry polymer ($T_G = 50°C$) to well below the testing temperature ($T = 25°C$) and the functionality from type I to type II.

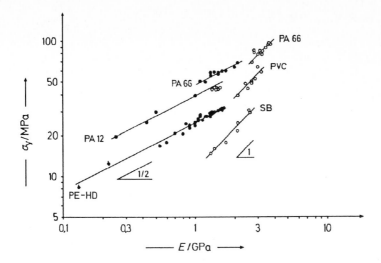

Fig. 10-20 Dependence of yield strengths σ_y on tensile moduli E at 23°C [14].
O Group I polymers($T < T_G$): PA 66 = dry polyamides 66 ($T_G = 50$°C); PVC = poly(vinyl chloride)s ($T_G = 81$°C); SB = thermoplastic styrene–butadiene copolymers ($T_G \approx 86$°C);
● Group II polymers ($T > T_G$): PE-HD = high-density poly(ethylene)s ($T_G = -80$°C) and Surlyn® ionomers (with pip); PA 12 = conditioned poly(laurolactam) ($T_G \ll 42$°C); conditioned PA 66 ($T_G \ll 50$°C).

The K_I of group I polymers varies between 0.030 (amorphous copolyamide with aromatic groups) and 0.011 (thermoplastic styrene-butadiene copolymers). The yield strengths of these unaligned chains are thus between ca. $E/30$ and $E/90$ whereas the fracture strengths of fully aligned chains are ca. $E/10$ (Section 10.6.2.). The differences in these σ_f/E values reflect the energies for the separation of bonds: covalent bonds with high bond energies *vs.* dipole and van der Waals bonds with low interaction energies. These differences also show up in the length ratios L_b/L_o (Eqn.(10-35)). Higher bond energies lead to higher separation lengths: ca. 1.30 for covalent bonds, ca. 1.09 for hydrogen bonds (polyamides), ca. 1.06 for dipole–dipole interactions (PVC), and ca. 1.04 for π–π interactions and dispersion forces (SB copolymers).

Values of $\sigma_y \approx E/30$ have also been calculated theoretically for metals. $E/30$ thus seems to reflect the maximum theoretical yield strength σ_y^o of materials. The data show that polymers almost attain their maximum yield strengths regardless of whether polymers are amorphous or crystalline. Metals are fundamentally different: their yield strengths are only a few percent of the maximum values (Table 10-6).

10.6.4. Fracture Strengths of Brittle Materials

Both yield strengths and fracture strengths of unaligned polymer chains are much lower than the theoretical fracture strengths of aligned chains (Sections 10.6.2. and 10.6.3.). Many researchers have therefore concluded that this behavior must be due

Table 10-6 Tensile moduli E, theoretical yield strengths $\sigma_y{}^\circ = E/30$, yield strengths σ_y and ratios $\sigma_y{}^\circ/E$ and $\sigma_y/\sigma_y{}^\circ$ of metals [15] and some commercial amorphous (a) or crystalline (c) polymers [16]. The values of $\sigma_y/\sigma_y{}^\circ$ of all metals lie between those of antimony and vanadium. The table contains physical property data of actual industrial polymer grades instead of the usual "typical" data for polymers since various grades of the same polymer can have widely different properties.

Material		$\dfrac{E}{MPa}$	$\dfrac{\sigma_y{}^\circ}{MPa}$	$\dfrac{\sigma_y}{MPa}$	$\dfrac{\sigma_y}{E}$	$\dfrac{\sigma_y}{\sigma_y{}^\circ}$
Antimony		80 000	2 670	11	0.000 138	0.0041
Iron		204 000	6 800	250	0.001 23	0.037
Aluminum		63 000	2 100	110	0.001 75	0.052
Vanadium		134 000	4 470	840	0.006 27	0.19
Poly(styrene)	Vestyron 114	3 300	110	60	0.018	0.55 a
Poly(vinyl chloride)	Hostalit Z 2060 C	2 700	90	50	0.019	0.56 a
Poly(ethylene)	Lupolen 6031 M	1 650	55	32	0.019	0.58 c
Poly(styrene-co-acrylonitrile)	Tyril 602	3 900	130	82	0.021	0.63 a
Poly(oxymethylene)	Delrin 500 NC-10	3 100	103	72	0.023	0.70 c
Polycarbonate A	Macrolon 2400	2 400	80	63	0.033	0.79 a
Poly(propylene), it-	Hostalen PP 41050	1 190	40	32	0.027	0.80 c
Poly(ε-caprolactam)	Ultramid B 35	3 200	107	90	0.028	0.84 c

to natural flaws which reduce the inherent strength of polymers. Such natural flaws (voids, cracks, etc.) have been found or suspected in glasses, ceramics, and metals.

This line of reasoning can be traced to the observation of **A.A.Griffith** that small glass objects have higher strengths than bigger ones. Griffith concluded that glass has natural defects, that these defects have a size distribution, that large bodies have more of the bigger defects than smaller ones, and that the fracture starts at the largest defects. Ergo: larger bodies have lower strengths than smaller ones.

The **Griffith theory** assumes that natural flaws grow in size, unite and cause macroscopic fracture. Flaws (cracks, voids, etc.) can only grow if the energy of the system decreases and the stress at fracture equals or exceeds the theoretical strength. Since the growth of natural defects changes the elastic energy of the system by U and the formation of new surfaces requires work W, the energy difference $W - U$ is available for the formation of new surfaces. The surface was assumed to be that of an elliptical hole in a plate, similar to the (much later discovered) crazes. The work is given by $W = S_h\gamma = Cd\gamma = 4\,aEd\gamma$, where γ = surface tension, S_h = surface of the hole, C = circumference of the ellipse, d = thickness of plate, a = length of the major semi-axis, and E = elliptical integral. The elliptical integral E becomes unity if the major semi-axis is much larger than the minor semi-axis. The work is thus $W = 4\,ad\gamma$.

The change U of the elastic energy of the plate was calculated earlier by **C.E.Inglis** as $U = -\pi a^2 d\sigma^2/E^\#$, where σ = stress and $E^\#$ = reduced tensile modulus. If the hole grows in the direction of the major axis, the change of total energy is:

$$(10\text{-}36)\qquad \frac{d(W-U)}{da} = \frac{d}{da}\left(4\,ad\gamma - \frac{\pi a^2 d\sigma^2}{E^\#}\right) \leq 0$$

Differentiation of Eqn.(1036) gives $[4\ d\gamma - (\pi a d\sigma^2/E^{\#})] \leq 0$. This expression equals zero if the stress becomes the critical stress, i.e., the fracture strength σ_b. Eqn.(10-36) converts into:

$$(10\text{-}37) \quad \sigma_b = \left(\frac{2\,E^{\#}\gamma}{\pi a}\right)^{1/2}$$

where the reduced modulus $E^{\#}$ equals the tensile modulus E in the case of thin sheets and plane stress (non-zero principal stresses in the sheet plane) and $E^{\#} = E(1-\mu)^2$ for thick plates and plane strain (all strains in the yield zone occur in the xy plane).

Eqn.(10-37) demands that tensile strengths decrease with the square root of the major semi-axis of elliptical holes. Notches can be approximated as half ellipses and a becomes the length L of the notch (sideways crack). The predicted square root dependence is indeed found for preformed *long* cracks of $L > 1$ mm in brittle polymers such as poly(styrene) (Fig. 10-21). However, the tensile strength becomes constant for *small* preformed cracks of $L < 1$ mm where the Griffith theory does not seem to apply. Indeed, the surface tension is only $\gamma = 0.83$ mN m^{-1} as calculated from the fracture strength $\sigma_b \approx 30$ MPa at $a = 2$ mm (Fig. 10-19) and $E^{\#} = E = 3.4$ GPa (Table 10-7); this value is about a factor 50 lower than the experimental value (Table 10-7). Conversely, very small crack lengths a of (10-60) nm result from Eqn.(10-37) if critical surface tensions γ_{crit} are used for γ (Table 10-7).

Since microscopy also does not reveal large *natural* cracks in these polymers, one must conclude such defects (voids, cracks, etc.) are formed by the drawing process. Electron spin resonance shows that unstressed it-poly(propylene) contains radical

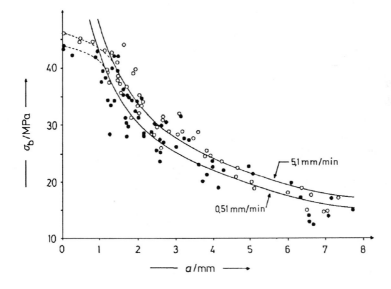

Fig. 10-21 Tensile strength σ_b of test rods from poly(styrene) as a function of the length L of artificial cracks at two different strain rates. The solid lines correspond to the functionality predicted by the Griffith theory (proportionality factor adjusted to fit the data). Data of [10].

Table 10-7   Tensile moduli E, fracture strengths σ_b and fracture elongations ε_b of some commercial polymers (CAMPUS® data), critical surface tensions γ_{crit} (generic data) and apparent crack lengths a calculated via Eqn.(10-37) with $E^\# = E$.

Polymer		$\dfrac{E}{\text{MPa}}$	$\dfrac{\gamma_{crit}}{\text{mN m}^{-1}}$	$\dfrac{\sigma_b}{\text{MPa}}$	$\dfrac{\varepsilon_b}{\%}$	$\dfrac{a}{\text{nm}}$
PA 66	Grilon T 300 FC	3800	46	90	5	14
PS	Styron 648	3400	33	53	2	25
PMMA	Degalan 6	3200	39	72	4	15
PC	Calibre 200-10	2200	45	58	168	19
PE-LLD	Clearflex MV 40	460	33	13.1	14.3	56

concentrations of less than 10^{14} radicals per cm^3. On application of stress, this concentration increases to ca. 10^{16} radicals per cm^3. At the same time, disk-like nanocracks with approximately the same concentration appear according to X-ray small angle measurements. These cracks are ca. 20 nm wide and ca. 10 nm long and are thus more or less elliptical.

The subsequent events depend on the alignment of chains and the rigidity of the polymers. In fibers, the numbers of cracks increase with time but their dimensions remain constant. The fiber fractures if the distance between cracks exceeds a critical value which equals approximately three times the diameter of the cracks. In ductile polymers, the stress is relieved by extending the plastic zones at the tips of elliptical cracks where the stress concentration is highest. Macroscopically, this leads to stress softening.

10.6.5. Fracture Strengths of Ductile Polymers

At present, tensile strengths, elongations at break and tensile moduli cannot be calculated for polymers of various morphologies since these mechanical properties may be affected by too many factors: voids, cracks, crazes, entanglements, segment orientation and shear flow in amorphous polymers and, in addition, orientation of crystalline lamellae and recrystallizations in semi-crystalline polymers. It seems that many of these effects act similarly on tensile strengths and elongations at break.

The **Hooke number** $He \equiv \sigma_b/E\varepsilon_b$ is a dimensionless quantity which scales with the elongation at break, ε_b, according to:

$$(10\text{-}38) \quad He \equiv \frac{\sigma_b}{E\varepsilon_b} = \frac{1}{[1+(\varepsilon_b/\varepsilon_{crit})^{ab}]^{1/b}}$$

In this empirical equation, ε_{crit}, a and b are parameters which adopt identical values for all thermoplastics (Fig. 10-22).

At small elongations, He assumes a value of unity (Hooke's law). The parameter ε_{crit} describes the fairly sharp (critical) transition from brittle polymers ($He = 1$) to

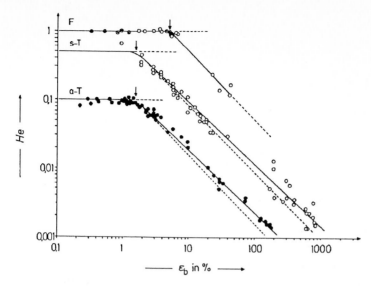

Fig. 10-22 Dependence of Hooke numbers *He* of thermoplastics T and some fibers F on fracture elongations ε_b (data of [11, 17]. For clarity, y-axes moved downward by a factor of 2 for semi-crystalline T and by a factor of 10 for amorphous T. Solid lines are empirical; dotted lines for a-T and s-T apply for a = 1. (●) a-T: PS, SAN, ASA, PMMA, PC; (○) s-T: PA 6, PA 66, PBT, PE-LLD; (○) F: natural fibers: hemp, cotton, jute, flax, ramie, silk, wool; (⊙) F: inorganic fibers: SiC, quartz, ceramics, glass, carbon.

those with *He* < 1 whose fracture behavior is dominated by flow processes. The critical elongation ε_{crit} increases with the orientation of chain segments *before* tensile drawing: it is 0.010 for non-oriented molecular composites, 0.0176 for thermoplastics slightly oriented by injection molding (Fig. 10-22), and 0.057 for highly oriented natural fibers.

The exponent *a* adopts a value of unity if test specimens are not additionally oriented *during* tensile testing, for example, natural fibers. Thermoplastics undergo additional orientation during tensile drawing which causes σ_b to increase gradually with increasing ε_b and lowers the exponent to *a* ≈ 0.90 from *a* = 1. The meaning of the exponent *b* (between 3 and 4) is unknown.

10.6.6. Impact Strengths

Conventional tensile testing deforms polymers with rates of ca. 0.1 m/s. In daily life, deformation rates are much higher though: people bang doors at ca. 3 m/s and a car at 75 mph (120 km/h) crashes into a standing obstacle at ca. 33 m/s. High-speed tensile tests up to 250 m/s can be performed with special instruments but these are very expensive.

Much less expensive are tests of **impact strengths** which employ impact speeds up to 4 m/s and extensional rates of up to 60 s^{-1}. The tips of notches may even experience extensional rates of up to 5000 s^{-1} in testing for **notched impact strengths**.

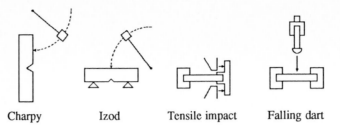

Charpy   Izod   Tensile impact   Falling dart

Fig. 10-23 Tests of notched impact strengths (Izod and Charpy) and impact strengths (tensile impact and falling dart).

In **Izod tests**, test specimens are clamped in vertical positions and struck by hammers (Fig. 10-23). Specimens are mainly subjected to bending stresses but also to shear stresses. In **Charpy tests**, specimens are placed horizontally and hit from above by a pendulum. The specimen is compressed on the impact side, subjected to bending stresses at the center, and exposed to strong tensile stresses on the opposite side. In tests of the **tensile impact strength**, specimens are clamped on both sides and hit on both sides of the clamp by a pendulum. **Falling dart tests** let a dart (or sphere) fall on a horizontally clamped specimen.

Impact strengths are measured as energies per area = forces per length that are required to cause fracture. For engineering purposes, it is useful to rearrange Eqn.(10-37) and multiply both sides of the new equation by a factor Y^2 that depends on the geometry of the test specimen:

(10-39) $Y^2\sigma_b{}^2 = Y^2(2\,E^\#\gamma/\pi)a^{-1} = K_{IC}{}^2a^{-1}$

The quantity $K_{IC} = Y(2\,E^\#\gamma/\pi)^{1/2}$ is the **critical stress intensity factor**. K_{IC} can be used to calculate the **fracture toughness** G_{IC} (**critical strain release rate**) for planar stresses:

(10-40) $G_{IC} = K_{IC}{}^2/(E^\#)$

Typical values of K_{IC} and G_{IC} are shown in Table 10-8.

For infinitely thin specimens, notched impact strengths measure only the energy that is required to initiate a break. The adequate quantity is the energy per width of notch (in the USA usually measured in ft·lbf/in ≈ 0.0535 kJ m^{-1}). For infinitely thick specimens, the energy for initiation is negligible compared with the energy for the propagation of a crack. The adequate quantity is here the energy per width of notch *and* thickness of specimen (in Europe mostly measured in kJ m^{-2}).

Notched impact strengths are very temperature dependent. At very low temperatures, all materials are brittle. Chain segments become more mobile with increasing temperature which allows stresses in high molar mass (i.e., entangled) polymers to be relieved by formation of crazes or shear bands. Short polymer chains cannot do that because they are not entangled. Segment mobilities increase especially near the glass temperature and so do impact strengths.

Table 10-8 K_{IC} and G_{IC} of thin plates at 23°C. T_G = Glass temperature in °C.

Material	$\dfrac{T_G}{°C}$	$\dfrac{E}{GPa}$	$\dfrac{K_{IC}}{MN\ m^{-3/2}}$	$\dfrac{G_{IC}}{kJ\ m^{-2}}$
Steel alloy	–	210	150	107
Poly(ethylene), medium density	– 80	0.89	5.0	28
Natural rubber, vulcanized	– 73	0.003	0.2	13
Poly(propylene), it-	– 15	1.4	4.6	15
Polyamide 6	50	3.0	2.8	2.6
Poly(vinyl chloride), at-	82	3.6	2.5	1.7
Polycarbonate, bisphenol A-	150	3.2	2.2	1.5
Poly(styrene), at-	100	3.3	1.0	0.30
Poly(styrene), at-. rubber-modified	?	2.0	2.0	2.0
Epoxy resin, hardened	?	2.8	0.5	0.089
Epoxy resin, rubber-modified	?	2.4	2.2	2.0
Wood	?	2.1	0.5	0.12
Glass	?	70	0.7 (?)	0.007

The mobility of chain segments in amorphous regions is also the reason why semi-crystalline polymers have high impact strengths if the glass temperatures of their amorphous regions are far below the testing temperature. The same effect increases the impact strength if thermoplastics are modified with rubber (see Section 14.6.3.).

10.6.7. Stress Crazing

Crazes (Fig. 10-3) develop fairly rapidly on surfaces of many polymers if the polymers are wetted by liquids, especially, if they are under stress. Crazes then travel into the interior of the polymer and develop into microcracks so that the polymer finally fractures. The effect is sometimes dramatic: polycarbonates resist tensile strengths of 10 MPa for hours at elongations of less than 5 % but disintegrate in minutes at the same stress if they are dipped into a mixture of toluene and octane.

This **stress crazing (stress cracking)** is also observed if no stresses are applied because most polymers have internal stresses generated by processing, for example, by segment orientation in injection-molded parts. It is a purely physical process; the often used term "stress corrosion" implies a chemical reaction which is not present.

The effect is small if liquids do not wet the polymer. In such systems, small defects first grow to visible hair-like cracks which then deepen until a limiting depth is reached. The material is however self-healing and susequently strengthens itself.

Wetting liquids swell the polymer surface which expands. The expansion causes stresses to develop between surface and interior which are relieved first by the generation of crazes and then by the formation of cracks. The degree of swelling depends on the thermodynamic quality of the liquid. The largest stress crazing is observed if solubility parameters of polymer and liquid match because this causes the largest swelling (Fig. 6-1).

Stress crazing decreases with increasing mobility of chain segments as well as with increasing cross-linking which seems to be contradictory. In chemically cross-linked polymers, stresses can relax elastically; such polymers are only slightly susceptible to stress crazing. The same is true for physical cross-linking by entanglements; stress crazing of linear polymers thus decreases with increasing molar mass. Polymeric plasticizers increase segment mobilities and decrease stress crazing; for the same reason, no stress crazing is observed at temperatures above the glass temperature.

10.7. Surface Properties

Average chemical compositions and morphologies of surfaces of solid polymers are usually different from those in the interior. The surface of aromatic polyimides is, for example, rich in imide-carbonyl groups whereas the aromatic residues face the interior. Crystallinities of semi-crystalline polymers can be larger or smaller at the surface than in the interior, depending on the processing conditions.

With air as contact partner, surfaces in equilibrium will be enriched by the polymer segments with the lowest Gibbs surface energy. Since surface energies depend on the partner (air, water, metals, etc.) and are influenced by kinetics (e.g., thermal history), surfaces of the same polymer can have various compositions and different properties.

10.7.1. Surface Tension

An interfacial tension γ_{sl} exists between a solid and a liquid. If the liquid resides as a drop on the surface of the solid, additional surface tensions liquid–vapor (γ_{lv}) and solid–vapor (γ_{sv}) exist. The vector diagram of these three interfacial tensions indicates that liquids do not spread totally on solid, smooth surfaces but rather form a somewhat flattened droplet with a contact angle ϑ between the surface of the solid and the edge of the droplet according to the **Young equation** $\gamma_{sv} = \gamma_{sl} + \gamma_{lv}\cos \vartheta$. The droplet spreads completely if $\vartheta = 0°$ and not at all if $\vartheta = 180°$. The cosine of the contact angle determines the **wettability** of the solid by the liquid.

Various liquids form different contact angles on the same polymer. If $\cos \vartheta$ is plotted against the surface tension γ_{lv} of the liquid (l) against air (v), straight lines are obtained that deliver a so-called **critical surface tension** at $\cos \vartheta \rightarrow 0$ (Table 10-9).

Most polymers have critical surface tensions γ_{crit} that are lower than the surface tension of water (72 mN m^{-1}) but higher than the surface tensions of oils and fats (20-30 mN m^{-1}); they are not wetted by water but by oils and fats. Frying pans and cooking pots are therefore coated with poly(tetrafluoroethylene) (Teflon) which has a very low critical surface tension of $\gamma_{crit} = 18.5$ mN m^{-1} that prevents the adherence of food. On the other hand, fillers for polymers such as titanium dioxide are not wetted by molten and solid polymers; they need adhesion promoters.

Table 10-9 Critical surface tensions γ_{crit} of solid polymers (20°C), surface tensions γ_{lv} of liquid polymers against air (150°C), and interfacial tensions γ_{ll} between liquid polymers (150°C).

Polymer 1		γ_{crit} in mN m^{-1}	γ_{lv} in mN m^{-1}	γ_{ll}/(mN m^{-1}) between polymer 1 and			
				PDMS	PE	PS	PEOX
Poly(tetrafluoroethylene)	PTFE	18.5	-	-	-	-	-
Poly(dimethylsiloxane)	PDMS	23	13.6	0	5.4	6.0	9.8
Poly(ethylene)	PE	33	28.1	5.4	0	5.7	9.5
Poly(styrene)	PS	34	30.8	6.0	5.7	0	-
Poly(oxyethylene)	PEOX	-	33.0	9.8	9.5	-	0
Titanium dioxide (anatase)	TiO$_2$	91	-	-	-	-	-

10.7.2. Hardness

The **hardness** of a body is defined as its mechanical resistance against penetration by another body. Hardness is always a property of the surface regions and not of the interior. It depends on moduli of elasticity, yield points and strain hardening. There is thus neither a general definition of "hardness" that is applicable to all materials, nor a universally applicable testing method.

Mohs hardness is determined by scratching test specimens with materials of arbitrarily set degrees of hardness (for example, talcum 1, quartz 7, diamond 10). It is not used for polymers but scratching with pencils of different Mohs hardnesses is sometimes practiced. **Brinell hardnesses** are also not used for polymers; this method measures the penetration of a sphere into a material (mostly metals) that remains *after* the load has been removed. Plastics are however characterized by their **Rockwell hardnesses** that determine the penetration depth of a sphere *under load* which measures both plastic *and* elastic deformations. Since polymers creep under load (Fig. 10-11, right), their Rockwell hardnesses are large compared with metals.

Elastomers and soft plastics are tested *statically* by a **durometer** for their **Shore hardnesses** as resistance against the penetration by a cone; this measures the compression of a calibrated spring. The Shore hardness of metals is determined *dynamically* with a scleroscope as the rebound of a small steel sphere. The latter method delivers the **resilience (falling ball rebound)** of elastomers.

10.7.3. Friction

A body contacting a support is held in place by the **normal force F_N** perpendicular to the surface of the support (Fig. 10-24). The body remains stationary as long as an applied tangential force F does not exceed a critical value F_S. The ratio $\mu_S = F_S/F_N$ is the **static coefficient of friction**. It depends on the chemical surface structure of the body and the support as well as the roughness of the surface (microscopic surface) but not on the geometry of the surface (the macroscopic surface).

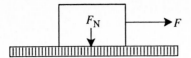

Fig. 10-24   A body presses on a plane plate with a normal force F_N. A tangential force moves the body in the horizontal direction only if it is greater than a critical value $F = F_S$.

The body moves on the support if F exceeds a critical value F_S. In order to move with constant velocity across a horizontal plane, a constant force $F_R > F_S$ is required. This force is essentially independent of the velocity. The ratio $\mu_k = F_R/F_N$ is the **kinetic coefficient of friction**. It is usually smaller than the static coefficient and also depends on the material and the surface structure.

Coefficients of friction are controlled by the surface roughness and/or deformation of bodies. The *rolling* of hard bodies on soft supports produces a friction which is almost completely due to energy losses by the deformation (but not the roughness) of the soft (viscoelastic) surface. Elastomers have coefficients of friction between 0.5 and 3.0, depending on the contacting body and the type of movement (rolling, gliding).

The *gliding* of hard bodies on other hard bodies affects only the tops of the hills and not the valleys of rough (microscopic) surfaces. The applied load acts only on the very small effective hilltop areas, not on the much larger geometric (macroscopic) surface. The contour of the microscopic surface is somewhat smoothened and the resulting contact area A_w is larger than the original surface area of the tops but still much smaller than the geometric surface.

In order to overcome adhesion, bonds between groups on contact areas have to be broken. This can be achieved by shearing the material near the surface which requires a shearing stress τ_b. The adhesion frictional force is thus $F_A = A_w \tau_b$. Since soft materials (large A_w) can be easily sheared (small τ_b) and hard materials (small A_w) only with difficulty (large τ_b), coefficients of friction are often very similar for plastics, ceramics and metals. Examples are poly(methyl methacrylate) PMMA on PMMA ($\mu_k = 0.8$), PMMA on steel ($\mu_k = 0.5$), and steel on PMMA ($\mu_k = 0.45$). Poly(tetrafluoroethylene) has especially low coefficients of friction: 0.04 for PTFE on PTFE and PTFE on steel and 0.10 for steel on PTFE.

10.7.4. Abrasion

Abrasion is the loss of material caused by surface friction. It depends on the hardness of the surface. Abrasion is usually measured as loss of volume $\Delta V = K_a F v t$ where F = applied force, v = linear velocity, and t = time. The coefficient of abrasion, K_a, has the physical unit of an inverse pressure.

Coefficients of abrasion depend not only on the two contacting materials but also on which one of the materials moves and which is at rest. The coefficient of abrasion of polyamide 66 against polycarbonate A is $K_a = 11 \cdot 10^{-13}$ Pa^{-1} for the system moving PA 66 + resting PC but only $0.25 \cdot 10^{-13}$ Pa^{-1} for resting PA 66 + moving PC. For polycarbonate one finds however $K_a = 9.8 \cdot 10^{-13}$ Pa^{-1} for resting PA 66 + moving PC and $200 \cdot 10^{-13}$ Pa^{-1} for moving PA 66 + resting PC.

10.8. Electrical Properties

10.8.1. Introduction

Most polymers resist the transport of electricity; they are **insulators** (or **dielectrics**) with electrical conductivities $\sigma < 10^{-9}$ S cm^{-1}. On doping with other chemical compounds, some polymers convert into **semiconductors** with $10^{-9} < \sigma/(\text{S cm}^{-1}) < 10^{-2}$ and some even into **conductors** with $\sigma \geq 10^2$ S cm^{-1}.

The scientific and technical literature uses both older terms and newer IUPAC names. The present IUPAC terms are as follows:

Matter generates an **electrical resistance** $R = U/I$ in ohms $\Omega = $ V A^{-1} if an **electric potential** U in volts V is applied for an electric current I in amperes A. The power $P = UI$ is reported in watts W = VA = J/s. The electrical resistance of a material depends on the distance between electrodes. It is thus a resistance of the measuring system and not that of the material itself, an **electrical surface resistance**. This quantity was formerly called "specific surface resistivity".

The **resistivity** ρ is defined as the electrical resistance R between opposite sides of a cube; it is therefore also called **volume resistivity**. It is proportional to the cross-sectional area and inversely proportional to the flowed-through length L. The resistivity therefore equals $\rho = RA^2/L$; it is usually reported in Ω cm. Since the volume resistivity is "specific" for a material, it is usually called "specific resistivity" in the older literature. It is thus not "specific" in the modern meaning which denotes a physical quantity divided by the mass (a true specific resistivity would have the unit Ω g^{-1} and not Ω cm).

The inverse of the resistivity is the **electrical conductivity** $\sigma = 1/\rho$; it is now measured in S cm^{-1}. This quantity has also been called "specific conductance" or "specific conductivity". The **electrical conductance** $G = 1/R = I/U$ is measured in siemens S = $1/\Omega$ = A/V. Since siemens is the inverse of ohm, one often finds S replaced by $\Omega^{-1} = \text{ohm}^{-1} \equiv$ mho in the older literature.

The **electric charge (quantity of electricity)** Q is reported in coulombs C = A s. It determines the **capacitance** $C = Q/U$ in farads F = C/V.

10.8.2. Relative Permittivity

Application of an electric potential difference to metals causes atoms of the outer shells to flow. Since these electrons are shared by all atomic nuclei, electrical conductivities are relatively little effected by the chemical nature of metals: silver $5.9 \cdot 10^5$ S/cm, aluminum $3.3 \cdot 10^5$ S/cm, iron $1 \cdot 10^5$ S/cm.

In uncharged polymers, electrons are relatively tightly bound to atomic nuclei. In most covalent bonds, binding electron pairs are somewhat nearer to one nucleus than to the other. One nucleus becomes more negative and the other one more positive: a permanent dipole is generated. Application of an electric field causes dipoles to orient in the direction of the field. In some other groups, electric fields induce negatively charged electrons and positively charged nuclei to move in opposite directions to produce induced dipoles.

The resulting polarizability is difficult to measure. It is therefore usually characterized by the **relative permittivity** ε_r as the ratio of the capacities of a condenser in vacuum and in the medium (formerly: dielectric constant). The relative permittivity of a vacuum is defined as unity. In air at 20°C and 50 Hz, it is $\varepsilon_r \approx 1.000\ 58$.

Table 10-10   Dielectric properties of plastics and elastomers. ε_r = Relative permittivity ("dielectric constant") at 1 MHz; ρ = volume resistivity; R = surface resistivity; tan δ = dissipation factor (loss tangent) at 1 MHz; S = electric strength ("dielectric strength"); U = tracking resistance (obtained by the standard testing method KC).

Polymer		ε_r	$\dfrac{\rho}{\Omega \text{ cm}}$	$\dfrac{R}{\Omega}$	tan δ	$\dfrac{S}{\text{kV mm}^{-1}}$	$\dfrac{U}{\text{V}}$
Poly(tetrafluoroethylene)	PTFE	2.15	10^{18}	-	0.0001	40	> 600
Poly(ethylene)	PE	2.3	10^{17}	10^{13}	0.0007	70	600
Poly(styrene), at-	PS	2.5	10^{17}	10^{15}	0.0002	140	500
Polycarbonate, bisphenol A	PC	2.9	10^{16}	-	0.01	-	200
Poly(methyl methacrylate), at-	PMMA	3.7	10^{15}	-	0.02	30	600
Poly(vinyl chloride), at-	PVC	<3.7	10^{15}	10^{13}	0.015	<50	< 600
Polyamide 6, dry	PA 6	3.7	10^{15}	-	0.03	<150	600
Polyamide 6, conditioned in air	PA 6	7	10^{12}	-	0.3	80	600
Polyester, unsaturated	UP	4.5	10^{13}	10^{12}	0.01	50	500
Phenolic resin, cross-linked	PF	8	10^{12}	-	0.05	10	-
-, -, mineral filled	PF	-	10^{10}	-	0.5	14	< 150
Poly(isoprene), 1,4-cis-	IR	2.6	10^{14}	-	0.0002	23	-
-, -, vulcanized	-	3	10^{14}	-	0.002	-	-
-, -, -, carbon black filled	-	>15	10	-	0.1	-	-

Relative permittivities of *apolar polymers* (PTFE, PE) are ca. 2-2.5 (Table 10-10). They decrease slightly with increasing temperature since intermolecular atomic distances become larger and effects of dispersion forces correspondingly smaller.

Polar polymers possess larger relative permittivities of ca. 3-8. In contrast to apolar polymers, relative permittivities of polar polymers increase with increasing temperature because dipoles become more mobile, especially near the glass temperature (see also Fig. 10-26). Relative permittivities of polymers rise if polar additives are present, for example, water (ε_r = 81) in conditioned polyamide 6, carbon black in cross-linked poly(isoprene), or metal powders as fillers (up to $\varepsilon_r \approx 170$ in thermoplastics and $\varepsilon_r \approx 18\,000$ in elastomers).

10.8.3. Resistivities

Application of small electric fields to a dielectric causes dipoles to orient. Stronger fields remove some electrons from atoms. It is the resulting ions, and not electrons as in metals, that make polymers electrically conducting.

The same factors that increase relative permittivities decrease *volume resistivities ρ* (Fig. 10-25): polar groups in polymers, high mobilities of chain segments, polar additives, and higher temperatures.

Surface resistivities are affected by far more parameters than volume resistivities. Humidity is especially adverse since it generates considerable ionic conductivities if polar surface impurities are present. For this reason, surface resistivities are usually two to three decades lower than volume resistivities (Table 10-11).

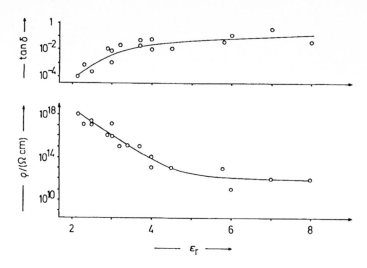

Fig. 10-25 Volume resistivities ρ and loss tangents $\tan\delta$ as functions of relative permittivities ε_r of plastics.

Surface impurities can lead to **tracking**. If a voltage is imposed between two electrodes, the current may find a conducting track across ionic impurities on the surface. The resulting voltaic arc burns a path into the insulating polymer which finally leads to a flashover and a current discharge.

Tracking is absent if the voltaic arc generates volatile degradation products, for example, monomer from poly(methyl methacrylate) or low molar mass products from poly(ethylene) and aliphatic polyamides. Poly(*N*-vinylcarbazole) is an excellent insulator but has a bad tracking resistance since it does not degrade to volatile products.

Supports for printed circuits require especially high surface resistances. They should neither possess polar groups nor attract humidity (see also Section 10.8.5.).

10.8.4. Dissipation Factors

Dipoles of a dielectric try to orient themselves in an alternating electric field. The faster the change of direction of the field (the higher the frequency), the more difficult becomes the orientation (see the mechanical analog in Section 10.5.7.) and the more energy is converted into heat (**power loss**) and lost as **available power**. The relative proportions of power loss and available power are determined by the dielectric phase angle $\phi = 90° - \delta$ between current and voltage where δ = **loss angle**. Power losses P_1 are zero at a phase angle of $\phi = 90°$. The available power P_a is zero if current and voltage are in phase; all electric energy is then converted into heat.

The ratio of power loss to available power is the **dissipation factor** or **loss tangent**:

$$(10\text{-}41) \quad \frac{P_1}{P_a} = \frac{UI\cos(90°-\delta)}{UI\sin(90°-\delta)} = \frac{\sin\delta}{\cos\delta} \equiv \tan\delta$$

The power loss measures the loss of energy per unit time. It is calculated from the frequency v of the alternating field, the electrical potential U, the capacity C_0 in the absence of the dielectric, the relative permittivity ε_r, and the loss tangent $\tan \delta$:

$$(10\text{-}42) \quad P_v = 2\,\pi v U^2 C_0 \varepsilon_r \cdot \tan \delta$$

The product $\varepsilon_r \cdot \tan \delta$ is the (**dielectric**) **loss index**. Polymers with high dielectric loss indices can be heated by high frequency fields; they soften and can be welded. Poly(vinyl chloride) is such a polymer; it is thus useless as an insulator for high-frequency conductors. Such insulators are apolar polymers with small dissipation factors and low relative permittivities (Table 10-7).

Dissipation factors can be treated as the ratio of imaginary and real permittivity:

$$(10\text{-}43) \quad \tan \delta = \varepsilon''/\varepsilon' = \varepsilon_r''/\varepsilon_r' \quad ; \qquad \varepsilon = \varepsilon' - i\varepsilon'' = \varepsilon' - (-1)^{1/2}\varepsilon''$$

The frequency dependence of relative real permittivities, $\varepsilon_r' = f(v)$, corresponds to a dispersion and the frequency dependence of relative imaginary permittivities, $\varepsilon_r'' = f(v)$, to an absorption. Maxima in $\varepsilon_r'' = f(v)$ thus indicate energy absorbing transitions or relaxations, for example, glass temperatures (Fig. 10-26). The maximum moves to larger frequencies at higher temperatures; the dynamic glass temperature increases with increasing frequency (see Section 9.6.2.).

Glass temperatures and other relaxations or transitions of *polar polymers* can thus be investigated via the behavior of their dipoles in alternating electric fields. Dipoles behave differently when they are in a side group or in the main chain. The main chain dipoles of poly(oxyethylene) $\{O{-}CH_2{-}CH_2\}_{\overline{n}}$ can orient themselves only

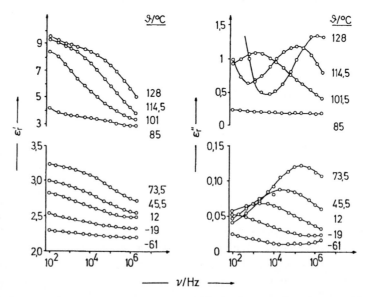

Fig. 10-26 Real (ε_r') and imaginary relative permittivities (ε_r'') of a poly(vinyl chloride) at different temperatures and frequencies. Top: β dispersion; bottom: γ dispersion. Data of [12].

above the glass temperature T_G since it is only then that the chain segments are sufficiently mobile. The side group dipoles of poly(vinyl ether)s $+CH_2-CH(OR)+_{\overline{n}}$, on the other hand, can align by either segmental movements of the main chain or by orientations of side groups. Poly(vinyl ether)s thus show two dispersion ranges at $T > T_G$: at low frequencies (long times) due to segmental movements and at high frequencies (short times) due to orientation of side groups. At $T < T_G$, only the effects of side groups are observed.

10.8.5. Electrostatic Charging

Static electricity is caused by either an excess or a shortage of electrons on insulated or non-grounded surfaces. This **electrostatic charging** is obtained if the electrical conductivity is smaller than ca. 10^{-8} S cm^{-1} and the relative humidity is less than 70 %. It is produced by the rubbing of two surfaces against each other (**triboelectric charging**) or by a contact of the surface with ionized air (e.g., cloth dryers).

Electric charging attracts dust to plastics, jams electric and electronic devices, makes plastic films sticky, causes dust to explode in closed spaces, etc. It is utilized for the electrostatic spraying of paints and for the flock-finishing of velvety surfaces.

Electrostatic charges depend on the counter material and the contact time. For example, an electric field of +120 V/cm was produced by an antistatically treated ABS polymer against poly(acrylonitrile) but a field of –1700 V/cm against polyamide 6. Charge densities vary widely, for example, from –14 C/g for phenolic resins to + 8 C/g for poly(trifluorochloroethylene). These charge densities are average values since charges are not distributed evenly; "islands" with positive charges may even occur in a "sea" of negative charges and vice versa.

Discharge times may vary between hours and fractions of seconds. In the first approximation, they depend on capacities $C = Q/U$. Both charges Q and electric potentials U are difficult to control for polymers, however. A discharge can practically only be reduced via the discharge resistance $R = U/I$ which in turn is controlled by the surface and volume resistivities of the polymers and the electric resistance of air. In the textile industry, electrostatic charges produced by fiber spinning are therefore removed externally by ionized air. Elastomers are internally protected against electrostatic charging by addition of carbon black.

10.8.6. Electrically Conducting Polymers

Conventional polymers can be made electrically conducting by addition of electrically conducting materials (carbon black, metal powders). Such formulated polymers prevent electrostatic charging or electromagnetic interference.

Since graphite exhibits an electrical conductivity of ca. 10^4 S/cm in its two-dimensionally conjugated layer planes, it was long suspected that one-dimensionally conjugated chains may also have considerable conductivities. Examples are emeraldine EM (a half-oxidized poly(aniline)), poly(pyrrole) PPY, trans-poly(acetylene) PAC, and poly(1,4-phenylene) PPP. These polymers are however insulators.

EM ... $\sigma = 10^{-9}$ S/cm

PPY ... $\sigma = 10^{-8}$ S/cm

PAC ... $\sigma = 10^{-9}$ S/cm

PPP ... $\sigma = 10^{-15}$ S/cm

Large electrical conductivities require high concentrations of charge carriers with great mobilities along and between chains. Considerable activation energies are however needed to generate charge carriers in EM, PPY, PAC and PPP because these polymers are compounds with half-filled valence bands. The concentration of charge carriers is thus small. The transfer of charges from chain to chain is also difficult as one can see for graphite with an electrical conductivity of only 1 S/cm perpendicular to the layer planes.

Electrical conductivities of such conjugated polymers can be considerably enhanced by addition of **dopants**. Treatment of emeraldine with HCl results in the protonation of –N= groups to $(-NH=)^+Cl^-$ groups; the electrical conductivity rises to ca. 5 S/cm from 10^{-9} S/cm. The repeating unit of emeraldine carries two $(-NH=)^+Cl^-$ groups. Such a dication should be diamagnetic. Protonated emeraldine is however paramagnetic; it must therefore exist as a di(radical cation):

EM$^+$ ... $\sigma = 5$ S/cm

The stretching of protonated emeraldine films increases the electrical conductivity to $\sigma_\| = 91$ S/cm in the draw direction and to $\sigma_\perp = 17$ S/cm perpendicular to it. Cross-linking enlarges conductivities further to $\sigma_\| = 217$ S/cm and $\sigma_\perp = 48$ S/cm. Orientation of chain segments and cross-linking of chains obviously ease the intermolecular transfer of charges. The same effect can be obtained by crystallization: films of EM$^+$ are amorphous ($\sigma = 0.2$ S/cm) if cast from chloroform solutions but crystalline ($\sigma = 178$ S/cm) if cast from *m*-cresol. The latter films contain however ca. 12 % *m*-cresol. The firmly bound cresol expands distances between chains which allows chains to become untwisted. This, in turn, probably removes defects in the π-conjugation. Similar effects are produced by other phenols, either alone or as additives to acids.

Other polymers can also be doped. AsF_5 as dopant allows the electrical conductivity of poly(pyrrole) PPY to increase to 100 S/cm from 10^{-8} S/cm, of trans-poly(acetylene) PAC to 1200 S/cm from 10^{-9} S/cm, and of poly(p-phenylene) PPP to 500 S/cm from 10^{-15} S/cm. The electrical conductivity of the non-conjugated poly(p-phenylene sulfide) even climbs to 10 S/cm from 10^{-16} S/cm upon doping with AsF_5.

Conductivities increase with increasing concentrations of dopants AsF_5, I_2, BF_3, etc.; they attain their maximum values often only at a 1:1 ratio of dopant to polymer. This behavior indicates that the doping of organic polymers does not produce physical defects as the doping of inorganic semiconductors does. Dopants such as AsF_5, I_2, or BF_3 rather oxidize polymers (p-doping), a chemical transformation. The reaction of vinylene groups with AsF_5 leads to carbonium ions:

(10-44) $2 \leftarrow CH=CH \rightarrow_{\overline{n}} + 3\ AsF_5\ \rightarrow\ 2 \leftarrow (CH=CH)^+ \rightarrow_{\overline{n}} + 2\ (AsF_6)^- + AsF_3$

There are also reducing dopants (n-doping), for example, lithium naphthalide.

Electrically conducting polymers are used in many small batteries. Poly(2-vinyl pyridine) complexed by iodine has an electrical conductivity of ca. 10^{-3} S/cm; it has been used for years as a cathode in Li/I_2 batteries for pacemakers. This solid state battery has a higher energy density than lead accumulators and a life-time of ca. 10 years. Since 1987, button batteries from poly(aniline) as cathode and alkali metal as anode serve as reserve batteries for personal computers and pocket calculators. Rechargeable batteries from poly(pyrrole), poly(acetylene) or poly(p-phenylene) are under development. They have approximately twice the energy density of nickel–cadmium batteries. Possible applications range from electrochromic windows, redox condensators and light diodes in non-linear optics.

10.9. Optical Properties

10.9.1. Refractive Index

If a ray of light is incident on a transparent body at an angle β with respect to the normal to its surface, it passes inside the body and continues to travel at a different angle β' with respect to the normal: the light is "refracted" (L: *refringere* (past participle *refractus*) $\rightarrow$ *re*: away, *frangere*: to break). The ratio of the sine of the incident beam to the sine of the refracted beam is defined as the **refractive index**. It equals the ratio of light velocities c_0 in vacuum and c in the body according to **Snellius' law**:

(10-45) $n \equiv \sin \beta / \sin \beta' = c_0/c$

The **Maxwell equation** $n^2 = \varepsilon_r$ relates refractive indices n and relative permittivities ε_r which in turn determine polarizabilities according to the **Clausius-Mosotti** equation

$\alpha = [3/(4 \pi N/V)][(\varepsilon_r-1)/(\varepsilon_r+2)]$ (see Section 5.2.1.). Insertion of the number concentration $N/V = N_A\rho/M$ delivers the **Lorentz–Lorenz equation**

$$(10\text{-}46) \quad \frac{n^2-1}{n^2+2} = \frac{4\pi}{3}\frac{N}{V}\alpha = \frac{4\pi}{3}\frac{N_A\rho}{M}\alpha$$

Polarizabilities $\alpha = \mu/E$ depend on dipole moments μ and applied electric field strengths E. Refractive indices thus increase with increasing number and mobility of electrons per molecule.

Carbon atoms have therefore larger polarizabilities than hydrogen atoms. Since the contribution of the latter can be neglected, most organic polymers have refractive indices of ca. 1.5 (Table 10-11). Exceptions to this rule exist for polymers with especially strong polarizabilities (e.g., fluorine atoms) or with very large side groups (e.g., N-carbazole). Even such exceptional molecules still have very similar molecular structures with respect to polarizabilities; the refractive indices of all organic polymers are between ca. 1.33 and ca. 1.73. They vary with the wavelength of incident light (optical dispersion).

Table 10-11 Densities ρ and refractive indices n of completely amorphous (index a) and totally crystalline (index c) polymers. $\parallel$ = Chain direction, $\perp$ = normal to chain direction.

Polymer	$\dfrac{\rho_a}{\text{g cm}^{-3}}$	$\dfrac{\rho_c}{\text{g cm}^{-3}}$	n_a	$(n_c)_{\parallel}$	$(n_c)_{\perp}$
Poly(tetrafluoroethylene)	(2.00)	2.302	1.345	1.345	1.345
Poly(dimethylsiloxane)	0.96		1.404		
Poly(propylene), it-	0.852	0.943	1.49	1.53	1.496
Poly(ethylene)	0.855	1.00	1.49	1.582	1.52
Poly(methyl methacrylate), at-	1.188		1.491		
Poly(vinyl alcohol), at-	1.265	1.340	1.50	1.55	1.505
Cellulose (flax)		1.50		1.595	1.531
Poly(vinylidene chloride)	1.67	1.875		1.603	1.611
Poly(N-vinylcarbazole), at-	1.2		1.675		

10.9.2. Light Transmission and Gloss

Some of the light incident on a homogeneous, transparent body is reflected from the surface (external reflection) and some passes inside, where it is reflected at an interior boundary of the body (internal reflection). According to **Fresnel**, the ratio of the intensity I_r of reflected light to the intensity I_0 of the incident light depends on both the angle of incidence, β, and the angle of reflection, β':

$$(10\text{-}47) \quad R = \frac{I_r}{I_0} = \frac{1}{2}\left[\frac{\sin^2(\beta-\beta')}{\sin^2(\beta+\beta')} + \frac{\tan^2(\beta-\beta')}{\tan^2(\beta+\beta')}\right]$$

The **reflection** R is small for low angles of incidence and begins to increase sharply at high β. If the light enters an optically homogeneous, plane-parallel body perpendicular to its surface, both β and β' become zero and the reflexion reduces to $R_0 = (n-1)^2/(n+1)^2$. The **light transmission (transmittance)** is therefore:

$$(10\text{-}48) \quad \tau_i = 1 - R_0 = 1 - (n-1)^2/(n+1)^2$$

Since refractive indices vary only between 1.33 and 1.73, maximum transmittances can only be between 98.0 % ($n = 1.33$) and 92.8 % ($n = 1.73$). These ideal transmittances are rarely obtained since some light is always absorbed and/or scattered. The most transparent polymer is poly(methyl methacrylate) ($n = 1.492$); but even PMMA has a transparency of only 93 % instead of 96.1 % for wavelengths between 430 nm and 1110 nm.

Industry defines **gloss** as ratio of reflection R of the specimen to the reflection R_{st} of a standard which is a specimen with $n_D = 1.567$ in the paint industry. Gloss R/R_{st} thus increases with increasing refractive index (see Eqn.(10-47)). Theoretical glosses are rarely achieved, however, since surfaces are always somewhat rough and scatter light. Optical inhomogeneities below the surface also scatter considerably.

Literature

10.2-10.6. MECHANICAL PROPERTIES
L.R.G.Treloar, The Physics of Rubber Elasticity, Clarendon Press, Oxford, 3rd ed. 1975
T.Murayama, Dynamic Mechanical Analysis of Polymeric Materials, Elsevier, Amsterdam 1978
J.D.Ferry, Viscoelastic Properties of Polymers, Wiley, New York, 3rd ed. 1980
J.F.Vincent, J.D.Currey, The Mechanical Properties of Biological Materials, Cambridge
    University Press, Cambridge 1980
R.A.Pethrick, R.W.Richards, eds., Static and Dynamic Properties of the Polymeric Solid State,
    Reidel, New York 1982
I.M.Ward, Mechanical Properties of Solid Polymers, Wiley, New York, 2nd ed. 1983
A.E.Zachariades, R.S.Porter, ed., The Strength and Stiffness of Polymers, Dekker, New York
    1983
J.J.Aklonis, W.J.MacKnight, Introduction to Polymer Viscoelasticity, Wiley-Interscience,
    New York, 2nd ed. 1983
R.P.Brown, ed., Handbook of Plastics Test Methods, Godwin Ltd., London 1981
V.Shah, Handbook of Plastics Testing Technology, Wiley, New York 1984
W.Brostow, R.D.Corneliussen, eds., Failure of Plastics, Hanser, Munich 1986
A.G.Atkins, Y.W.Mai, Elastic and Plastic Fracture, Halsted Press, New York 1986
H.H.Kausch, Polymer Fracture, Springer, Heidelberg, 2nd ed. 1986
A.E.Zachariades, R.S.Porter, eds., High Modulus Polymers – Approaches to Design and
    Development, Dekker, New York 1987
J.E.Mark, B.Erman, Rubberlike Elasticity: A Molecular Primer, Wiley, New York 1988
B.W.Rossiter, R.C.Baetzold, Determination of Elastic and Mechanical Properties, Wiley, New
    York 1991
S.Matsuoka, Relaxation Phenomena in Polymers, Hanser, Munich 1992
I.M.Ward, D.W.Hadley, An Introduction to the Mechanical Properties of Solid Polymers, Wiley,
    Chichester 1993
L.E.Nielsen, R.F.Landel, Mechanical Properties of Polymers and Composites, Dekker,
    New York, 2nd ed. 1993

10.7. SURFACE PROPERTIES
B.W.Cherry, Polymer Surfaces, Cambridge University Press, Cambridge 1981
G.M.Bartenev, V.V.Lavrentev, Friction and Wear of Polymers, Elsevier, Amsterdam 1981
W.J.Feast, H.S.Munro, Polymer Surfaces and Interfaces, Wiley, New York 1987
F.Garbassi, M.Morra, E.Ochiello, Polymer Surfaces – From Physics to Technology, Wiley,
      Chichester 1994
C.-M.Chan, Polymer Surface Modification and Characterization, Hanser, Munich 1994

10.8. ELECTRICAL PROPERTIES
M.E.Baird, Electrical Properties of Polymeric Materials, Plastics Institute, London 1973
P.Hedvig, Dielectric Spectroscopy of Polymers, Halsted, New York 1977
A.R.Blythe, Electrical Properties of Polymers, Cambridge University Press, Cambridge 1979
D.A.Seanor, ed., Electrical Properties of Polymers, Academic Press, New York 1982
J.Mort, G.Pfister, eds., Electronic Properties of Polymers, Wiley, Chichester 1982
G.Heinicke, Tribochemistry, Hanser, Munich 1984
H.Kuzmany, M.Mehring, S.Roth, Electronic Properties of Polymers and Related Compounds,
      Springer, Berlin 1985
C.C.Ku, R.Liepins, Electrical Properties of Polymers, Hanser, Munich 1987
T.A.Skotheim, ed., Handbook of Conducting Polymers, Dekker, New York 1986 (2 vols.)
H.Kiess, ed., Conjugated Conducting Polymers, Springer, Berlin 1992
C.P.Wong, ed., Polymers for Electronic and Photonic Applications, Academic Press, San Diego
      (CA) 1992
W.R.Salaneck, I.Lundström, B.Roanby, eds., Conjugated Polymers and Related Materials, Oxford
      University Press, New York 1993

10.9. OPTICAL PROPERTIES
G.H.Meeten, ed., Optical Properties of Polymers, Elsevier, London 1986
R.Ross, A.W.Birley, Optical Properties of Polymeric Materials and their Measurement,
      J.Phys. **D** [Appl.Phys.] **6** (1973) 795
R.A.Hann, D.Bloor, eds., Organic Materials for Non-Linear Optics, CRC Press, Boca Raton (FL)
      1989
J.Zyss, ed., Molecular Nonlinear Optics, Academic Press, Orlando (FL) 1993

References

[1]   H.-G.Elias, An Introduction to Plastics, VCH, Weinheim 1993, Fig. 7-1
[2]   H.-G.Elias, Makromoleküle, Hüthig and Wepf, Zug, 6th ed., vol. I (in preparation)
[3]   L.R.G.Treloar, Trans.Faraday Soc. **40** (1944) 59, data taken from Fig. 5
[4]   G.Allen, M.J.Kirkham, J.Padget, C.Price, Trans.Faraday Soc. **67** (1971) 1278, Fig. 3a
[5]   S.M.Gumbrell, L.Mullins, R.S.Rivlin, Trans.Faraday Soc. **49** (1953) 1495, Figs. 7 and 8
[6]   S.Rabinowitz, P.Beardmore, Crit.Rev.Macromol.Sci. **1** (1972) 1, Fig. 1
[7]   J.R.McLoughlin, A.V.Tobolsky, J.Coll.Sci. **7** (1952) 555, Fig. 5
[8]   S.Onogi, T.Masuda, K.Kitagawa, Macromolecules **3** (1972) 109, Fig. 2
[9]   H.W.McCormick, F.M.Brower, L.Kin, J.Polym.Sci. **39** (1959) 87, Table 1-3
[10] J.P.Berry, J.Polym.Sci. **50** (1961) 313, data selected from Fig. 3 for consistency
[11] H.-G.Elias, Macromol.Chem.Phys. **195** (1994) 3117, (a) Fig. 3, (b) Fig. 8
[12] Y.Ishida, M.Matsuo, K.Yamafuji, Kolloid-Z. **180** (1962) 108, Fig. 1-6
[13] T.He, Polymer **27** (1986) 253, data selected from Table 1
[14] H.-G.Elias, submitted to J.Polym.Sci.-Phys.
[15] E.Rabinowicz, Friction and Wear, Wiley, New York, 2nd ed. 1995
[16] CAMPUS® data from various companies
[17] H.-G.Elias, J.Polym.Sci., Part B, Polym.Phys. 33 (1995) 955, Fig. 5
[18] H.Domininghaus, Plastics for Engineers, Hanser, Munich 1993, Fig. 211 (modified)

11. Polymer Auxiliaries

11.1. Polymer Systems

Names of materials in polymer technology and biochemistry have in common that they relate to the action of materials and not to chemical structures *per se*. "Hormones" are a group of biochemically active materials that differ vastly in their chemical structures but all of them stimulate biological receptors. "High-impact plastics" may also be very different materials. A "high-impact poly(styrene)" is not homopoly-(styrene) $\text{-CH}_2\text{-CH(C}_6\text{H}_5)\text{-}_n$ but may be a styrene copolymer or a blend of a styrene polymer with another polymer, etc.

According to their main constituents, industrial plastics can be divided into several **polymer families**. The "styrenics" family comprises all polymers that contain styrene units as main monomer units. "Acetal plastics" are polymers with $\text{-O-CH}_2\text{-}$ units that are derived from formaldehyde HCHO or its cyclic trimer, trioxane. Polymers derived from ethylene as monomer are subdivided into several families: high-density poly-(ethylene), low-density poly(ethylene), linear low-density poly(ethylene), and so on, because each of these families has characteristic material properties. On the other hand, "acrylics" consist only of plastics from poly(methyl methacrylate) and "vinyls" denotes exclusively plastics based on vinyl chloride as main monomer.

Each polymer family usually consists of several **grades** which differ in molar mass, molar mass distribution, tacticity (if any), configurational statistics, type and extent of branching, constitution of comonomeric units (if any), type and proportion of additives, and so on. With very few exceptions, polymers are only **raw materials** which require additives for their industrial and economic use. Raw polymers for plastics and fibers are usually called **resins**. The resulting plastics, elastomers, fibers, adhesives, etc., are thus not polymers *per se* but **polymer systems** comprised of polymers and other materials. The various grades of a polymer family are usually not traded according to their chemical make-up but according to their use as easy-flow, heat-resistant, high-impact, extrusion-type, and so on.

Plastics are subdivided into thermoplastics and thermosets (Chapter 14). **Thermoplastics** contain linear or slightly branched polymers that soften at the glass temperature T_G (amorphous polymers) or melt at the melting temperature T_M (semi-crystalline polymers). They are usually processed at temperatures above these temperatures. The transformation solid $\rightleftarrows$ melt is reversible. Cooling below T_G (amorphous) or T_M (semi-crystalline) delivers slabs, tubes, films and other formed articles. Raw materials for thermoplastics are usually polymers, sometimes oligomers, and rarely monomers.

Thermosets result from the chemical cross-linking of monomers or oligomers to highly cross-linked polymers. These monomers and oligomers are sometimes called **reactive resins** or **thermosetting resins** if they are solid or semi-solid; oligomers are also known as **prepolymers**. Thermosetting resins are simultaneously reacted and shaped. They cannot be remelted and reshaped without chemical degradation after the thermosetting operation; they are "set" by heat, hence the name "thermoset". The transformation liquid $\rightleftarrows$ solid is irreversible.

Rubbers are cross-linkable polymers with glass temperatures below use temperatures. The irreversible chemical cross-linking of rubbers to polymers with wide mesh sizes results in **elastomers** (Chapter 12). **Thermoplastic elastomers** are elastomeric materials from polymers with reversible (physical) cross-links.

Fibers are "one-dimensional" entities with large aspect ratios = length/diameter (Chapter 13). **Natural fibers** are found in nature; they may be organic polymers formed by biosyntheses (e.g., silk, wool, cotton) or inorganic polymers (e.g., asbestos). Industrial syntheses lead to **man-made fibers** which are subdivided into **semi-synthetic fibers** based on natural fibers (e.g., rayon) and **synthetic fibers** (e.g., polyester, nylon, and acrylic fibers).

Polymers are usually processed as melts or concentrated solutions and only occasionally as solids. The viscoelasticity of polymers leads to a processing behavior that differs fundamentally from that of the classic working materials wood, metals and stone. This different behavior required new techniques which were first developed empirically for elastomers and semi-synthetic fibers and were later adopted for plastics. Other important processing techniques evolved after the peculiar rheological and mechanical properties of polymers became known and understood in greater detail.

11.2. Formulation

Polymers cannot be processed directly after polymerization. Batch-type polymerizations deliver products that differ slightly in structure and properties. Batches are thus blended in order to obtain polymers with constant specifications (**microhomogenization**). **Macrohomogenization** is the blending of powders, pellets, beads, etc.

Microhomogenization leads to polymers with broader molar mass distributions. The molar mass ratio $\overline{M}_w/\overline{M}_n$ of a blend of polymers A and B is easily derived from the definitions of number-average and mass-average molar masses of its constituents. Setting $w_A = 1 - w_B$ = mass fraction of polymer A, $Q_i = (\overline{M}_w/\overline{M}_n)_i$ = molar mass ratio of one of the polymers (i = A, B), and $R_n = (\overline{M}_n)_B/(\overline{M}_n)_A$ = ratio of number-average molar masses of both polymers, one obtains (see also Section 2.2.1.):

$$(11\text{-}1) \quad \overline{M}_w/\overline{M}_n = w_A Q_A[(w_A R_n + w_B)/R_n] + w_B Q_B[w_A R_n + w_B]$$

Polymers are often combined with other polymers. Rubbers are **blended** with other rubbers and fibers are **mixed** with other fibers. Thermoplastics are **blended** with thermoplastics, **reinforced** with fibers, and **toughened** (**impact-modified**) with rubbers.

The combination of polymers with other ingredients also employs different terminologies in the various industries. Plastics are **compounded**, textiles are **finished**, and rubbers, inks and adhesives are **formulated**. A **compound** of the plastics industry is therefore not a chemical compound but a mixture of a polymer with other (usually low molar mass) ingredients (fillers, colorants, antioxidants, etc.).

Low molar mass ingredients comprise organic substances (plasticizers, organic colorants, antioxidants, etc.) and inorganic ones (inorganic fillers and pigments, etc.). These **auxiliaries** are subdivided into additives and modifiers. **Additives** are used in small proportions (usually less than 5 %). Except for nucleation agents, they practically do not affect mechanical properties. **Modifiers** are added in greater proportions; they modify mechanical properties. Both additives and modifiers are combined with polymers *after* the polymerization where they either aid processing (**process additives**) or improve use properties (**functional additives** and **modifiers**). They are to be distinguished from **polymerization auxiliaries** which assist polymerizations (initiators, catalysts, inhibitors, dispersants, emulsifiers, etc.).

Additives and modifiers are usually present in small proportions in fibers (exception: silk). Their mass fractions are, on average, 5 % in papers, 23 % in plastics, and 60 % in elastomers but actual proportions may vary widely, in plastics for example from zero to 95 wt.-%. On average, costs of ingredients often match those of polymers themselves since the price per weight of many auxiliaries is often hundred times higher than that of polymers. Without auxiliaries, polymers would acquire neither many of their useful properties nor long life expectancies.

11.3. Stabilizers

Stabilizers protect polymers against unwanted chemical reactions. They comprise antioxidants, light stabilizers, heat stabilizers, and flame retardants. Biocides hinder the growth of microorganisms.

11.3.1. Antioxidants

Antioxidants prevent degradations of polymers by oxygen. More than 90 % of all antioxidants are used for polymeric hydrocarbons such as poly(ethylene)s, poly(propylene)s, styrene polymers, and ABS polymers.

The direct production of radicals by oxygen according to $RH + O_2 \rightarrow R^{\bullet} + {}^{\bullet}OOH$ is a very slow reaction. Much faster is the formation of hydroperoxides ROOH which subsequently decompose into oxy radicals and hydroxy radicals:

(11-2) $RH + O_2 \rightarrow ROOH \rightarrow RO^{\bullet} + {}^{\bullet}OH$

The hydroperoxides are decomposed by transition metal ions Mt^{n+}. Depending on the redox potential of the metal ion, either oxidations or reductions occur:

(11-3) $RO^{\bullet} + OH^{-} + Mt^{(n+1)+} \leftarrow ROOH + Mt^{n+} \rightarrow ROO^{\bullet} + H^{+} + Mt^{(n-1)+}$

Peroxide radicals $ROO^{\bullet}$ can decompose into $R^{\bullet}$ and O_2. They can also attack polymer molecules and generate hydrocarbon radicals $R^{\bullet}$ by the reaction $ROO^{\bullet} + RH \rightarrow ROOH + R^{\bullet}$. Initiation reactions thus produce peroxy radicals $ROO^{\bullet}$, oxy radicals $RO^{\bullet}$, hydroxy radicals $HO^{\bullet}$ and alkyl radicals $R^{\bullet}$. Additional polymer radicals are yielded by thermal or mechanical chain cleavages or originate from catalyst residues. The overall process is a chain reaction which is initiated according to Eqn.(11-2).

Deinitiators prevent the formation of $RO^{\bullet}$ and $^{\bullet}OH$ or direct the decomposition of ROOH in such a way that less radicals are formed. They are also called **secondary antioxidants** because they remove initiating radicals rather than terminate kinetic chains. Phosphites are often used deinitiators; they are oxidized by hydroperoxides according to $(R'O)_3P + ROOH \rightarrow (R'O)_3PO + ROH$.

Kinetic chains grow by reactions $R^{\bullet} + O_2 \rightarrow ROO^{\bullet}$ and $ROO^{\bullet} + RH \rightarrow ROOH + R^{\bullet}$. They branch by reactions $ROOH \rightarrow RO^{\bullet} + {}^{\bullet}OH$, $RO^{\bullet} + RH \rightarrow ROH + R^{\bullet}$, $HO^{\bullet} + RH \rightarrow H_2O + R^{\bullet}$, and $ROOH + RH \rightarrow RO^{\bullet} + R^{\bullet} + H_2O$. The kinetic chain is terminated by **chain terminators (primary antioxidants)**. Such terminators are generally sterically hindered aromatic molecules. An example is 2,5-di(*t*-butyl)hydroxytoluene BHT [in Eqn.(11-4) with R" = C(CH₃)₃); R' = CH₃] which is usually used for poly-(propylene) in combination with phosphites. One molecule BHT eliminates two radicals $ROO^{\bullet}$:

(11-4)

Hydroxyphenylpropionates (with R' = CH₂-CH₂COO-C₁₈H₃₇; R" = C(CH₃)₃) or similar compounds are used for poly(ethylene)s, styrene polymers, polyamides, poly-urethanes, polycarbonates, polyacetals, and poly(ethylene terephthalate). Aromatic di-amines are employed for elastomers and also for polyamides, flexible polyurethane foams, and carbon-black-filled poly(styrene)s.

The combination of a deinitiator and a chain terminator often generates larger effects with respect to the inhibition period and the rate of oxygen consumption. These synergistic effects arise because deinitiator and chain terminator act successively and not simultaneously. Antagonistic effects are also known, for example, if carbon black is combined with certain antioxidants. Carbon blacks have large internal surface areas; they absorb the antioxidants and render them ineffective.

11.3.2. Light Stabilizers

Light stabilizers protect polymers against degradation by ultraviolet light. Since UV absorption is usually associated with multiple bonds, few polymers are expected to be attacked by UV light. Chains of poly(isoprene) $+CH_2-C(CH_3)=CH-CH_2+_n$ are split between the methylene groups $-CH_2-CH_2-$. On degradation of poly(methyl methacrylate) $+CH_2-C(CH_3)(COOCH_3)+_n$, radicals $\sim CH_2-^\bullet C(CH_3)\sim$ and $^\bullet COOCH_3$ are formed followed by a disproportionation of the chain.

Ultraviolet light is however absorbed by several polymers which do not contain multiple bonds. These polymers contain either UV-absorbing catalyst residues or structural mistakes from either the polymerization reaction or the processing. Poly-(ethylene)s and poly(propylene)s contain $HOO-$, $-O-O-$, $>C=O$ and $>CH=CH<$ groups; poly(styrene)s also embody $-CH_2-CO-C_6H_5$ and $-C_6H_4-CH=CH-C_6H_4-$ units. For example, hydroperoxide groups $-CR(OOH)-$ release hydroxy radicals $HO^\bullet$. The resulting $\sim CR(O^\bullet)\sim$ groups dissociate into $-CO-$ and $R^\bullet$. Due to these reactions, more than 80 % of all light stabilizers are used for poly(olefin)s.

Polymers can also be protected against UV degradation if their light absorption is reduced by pigments or by reflective coatings. Carbon black as a filler is not only an excellent UV absorber but also an efficient radical trap. However, the rutile modification of the white pigment TiO_2 is a sensitizer.

Transparent polymers must always be protected by **UV absorbers**. *o*-Hydroxy-benzophenone I and 2-(2'-hydroxyphenyl)benztriazole II absorb radiation energy via the hydrogen bond and convert it into infrared radiation (heat). Dialkyldithio-carbamates III absorb not only UV light but also transform hydroperoxides into non-radical compounds. UV absorbers must be compatible with the substrate, fast to light, stable during processing, and color stable (for fibers).

Hindered amine light stabilizers (HALS, e.g., IV) form nitroxyl radicals $>N-O^\bullet$ by photo-oxidation. These radicals add radicals $R^\bullet$ and the resulting hydroxylamine ethers $>N-O-R$ react rapidly with peroxy radicals $R'-O-O^\bullet$ whereby nitroxyl radicals are regenerated to $R'-O-O-R$.

11.3.3. Heat Stabilizers

Heat stabilizers prevent the dehydrochlorination of homopolymers and copolymers of vinyl chloride. Poly(vinyl chloride) PVC contains carbon-carbon double bonds: at chain ends from chain transfer to monomer (Section 3.7.7.) and in the chain from chain transfer to polymer. It is at these double bonds that dehydrochlorinations start according to ~CH=CH-CHCl-CH$_2$~ $\rightarrow$ ~CH=CH-CH=CH~ + HCl. Successive dehydrochlorinations result in sequences of conjugated double bonds: PVC discolors from white to yellow, brown and black. Mechanical properties also deteriorate.

The double bonds form complexes with organic compounds of barium, zinc, and tin as well as organic and inorganic lead derivatives. These primary heat stabilizers do not prevent the dehydrochlorination but rather diminish the effects caused by the HCl evolution. Zinc carboxylate reacts according to Zn(OOCR)$_2$ + 2 HCl $\rightarrow$ 2 RCOOH + ZnCl$_2$. RCOOH and ZnCl$_2$ form a coordinatively unsaturated zinc complex. This complex reacts with the dehydrochlorinated PVC chains whereby ZnCl$_2$ is released. The formation of ~CH=CH-CH(OOCR)-CH=CH~ interrupts the polyene sequence which in turn reduces the discoloration.

Primary heat stabilizers are often combined with secondary stabilizers, for example, organophosphites, dicyanodiamides, or epoxidized vegetable oils.

PVC also discolors in light by photochemical oxidation. HCl is split off and structures such as ~CH$_2$-CO-CH=CH~ are formed.

11.3.4.   Flame Retardants

Flame retardants protect polymers by either denying oxygen access to the burning polymer or by "poisoning" flames. Polycarbonate A is "self-extinguishing" because it releases carbon dioxide on burning. Aluminum hydroxide Al(OH)$_3$ sets free water which douses the flame. Phosphorus-containing flame retardants oxidize in fires to non-volatile phosphorus oxides which form a glass-like coating on the surface of the substrate; oxygen can no longer access the polymer. Phosphorus oxides also react with water to give phosphoric acids which catalyze the splitting-off of water. Bromine or chlorine-containing organic compounds generate halogen radicals which combine with radicals produced by the burning polymer; this action interrupts the kinetic chain and poisons the flame.

The behavior of a polymer in a fire is often characterized by the **limiting oxygen index (LOI)** of the polymer. This index is defined as that volume fraction of oxygen in a mixture of O$_2$ and N$_2$ which just allows a polymer to burn. Materials with LOI < 22.5 % are usually called flame-retardant; those with LOI > 27 %, self-extinguishing. For example, limiting oxygen indices are 14 % for poly(oxymethylene), 17 % for poly(ethylene), 20 % for cellulose, 25 % for wool, 32 % for poly(vinyl chloride), and 95 % for poly(tetrafluoroethylene).

11.4. Colorants

Polymers are colored for various reasons: fashion, protection, warning, etc. Colorants are either dyes or pigments. **Dyes** consist of dye molecules which dissolve molecularly in plastics or are bound to fibers, either by adsorption to the fiber surface or by chemical reaction (reactive dyes). **Pigments** are dispersed aggregates of dye molecules or minerals that are insoluble in polymers.

Textiles are practically exclusively colored by dyes; the bulk coloration of fibers and fabrics is relatively rare. Plastics are however predominantly colored in bulk by pigments (98 %) and not by dyes (2 %) because pigments are more light-fast and do not migrate to the surface. Most commonly used is titanium dioxide as a white pigment for plastics and carbon black as black pigment for elastomers; carbon black is also a reinforcing agent for elastomers.

Contrary to many dyes, pigments do not require special affinities to polymers. They need to be wetted by polymer melts, however. Wetting can often be achieved by surface treatment of pigments. Dispersions of pigments are furthermore only stable in polymers if densities of pigments and polymers do not differ very much. The densities in question are not the true densities of pigments but rather the effective densities because aggregates of pigments always trap some air. For example, chromium yellow $PbCrO_4$ has a far higher density ($\rho = 5.8$ g/cm^3) than most plastics ($0.9 < \rho/(g\ cm^{-3}) < 2.1$); the effective density of $PbCrO_4$ is however only $\rho = 0.23$ g/cm^3. A part of the trapped air is removed by vacuum which not only adjusts the effective density of pigments to the density of the polymer but also prevents lumps of aggregates.

11.5. Fillers

Fillers are solid materials that are dispersed in plastics, elastomers, fibers, coatings, adhesives, papers, etc. **Inactive fillers** (**inert fillers**, **extenders**) extend expensive polymers and make plastics less costly. **Active fillers** (**reinforcing agents**) improve certain mechanical properties (see Section 14.5.4.).

Added liquid or elastomeric materials are not considered fillers. Examples are liquids acting as **plasticizers** (Section 9.6.4), rubbers used as **toughening agents** to increase the resistance to impact (Section 14.6.3.), and **compatibilizers** that bond the phases in blends of immiscible polymers (Section 8.5.).

Very many different fillers are used for plastics (Table 11-1). Short glass fibers are the most popular reinforcing fillers, especially for unsaturated polyester resins but also for poly(propylene)s, aliphatic polyamides, and saturated aromatic polyesters. Carbon black is the dominant filler for elastomers.

Table 11-1 Fillers for thermoplastics TP, thermosets TS, and elastomers EL. For abbreviations and acronyms of names of polymers see Chapter 15.

Filler	Application	wt-%	Improved properties
Inorganic fillers			
Chalk	PE, PVC, PPS, PB, UP	< 33	Cost, gloss
Heavy spar	PVC, PUR	< 25	Density
Talcum	TP, TS, papers	< 50	White pigment, impact strength
Mica	PUR, UP, PP	< 25	Hardness, rigidity
Kaolin	UP, vinyl polymers, papers	< 60	Demolding
Glass spheres	TP, TS	< 40	Modulus, shrinkage,
Glass fibers	TP, TS	< 40	Fracture and impact strength
SiO_2, pyrogenic	TP, TS, EL	< 30	Fracture strength, viscosity
Quartz	PE, PMMA, EP	< 45	Heat stability, fracture
Sand	EP, UP, PF	< 60	Shrinkage
Al, Zn, Cu, Ni	PA, POM, PP	< 100	Thermal + electrical conductivity
MgO	UP	< 70	Rigidity, hardness
ZnO	PP, PUR, UP, EP	< 70	UV stability, heat conductivity
Organic fillers			
Carbon black	PVC, HDPE, PUR, PI, EL	< 60	UV stability, coloration
Graphite	EP, MF, PB, PI, PPS, UP, PTFE	< 50	Rigidity, creep
Wood flour	PF, MF, UF, UP, PP	< 50	Shrinkage, impact strength
Starch	PVAL, PE, papers	< 7	Biological degradation

Many fillers are not applied as such but only after a surface treatment. Grafting polymers onto filler surfaces makes fillers more compatible with polymers. Glass fibers are treated with **sizes** (**finishes, lubricants**) which increase the slip and thus reduce the breakage of glass fibers. These sizes do not need to bed down fiber ends (as required for the spinning of textile fibers) since glass fibers possess smooth surfaces.

Some fillers act in more than one way. Carbon black is not only a reinforcing filler for elastomers but also a black colorant and an electrical conductor that prevents static electricity. Zinc oxide is an active filler, a white pigment and a vulcanization promoter for elastomers. Air acts as filler and pigment in coatings and as foaming agent and plasticizer in expanded plastics and elastomers. Water is not only a plasticizer for hydrophilic polymers but also a filler for unsaturated polyester resins and other polymers where it is dispersed in droplets.

11.6. Emigration of Auxiliaries

Auxiliaries in polymers are often not in thermodynamic equilibrium with the polymeric matrix. They try to demix and emigrate. These processes have different names. A transport of liquid auxiliaries (e.g., plasticizers) to the surface of the polymer is called **exudation** or **bleed-through** and that of solid auxiliaries **efflorescence** or **blooming**. In the latter case, one often differentiates between **chalking** (white auxiliaries), **bleeding** (colorants) and **flooding** (pigments). If a solid auxiliary deposits on the surface of a mold, it is called **plate-out**.

The transport of an auxiliary from a polymer into a surrounding liquid is a **bleeding** (solid auxiliaries; often for colorants only) or an **extraction** (plasticizers). **Migration** is the transport of a plasticizer from the polymer matrix into another contacting solid; "migration" is however often also used in the sense of "emigration".

The divisiveness of the various industries is documented by the fact that there is no generally accepted term that covers *all* processes. This book will use "emigration" as this general term in accordance with the usual meaning "to leave one region in order to settle in another one".

No transport of the auxiliary from the polymer to the surface or the surrounding air (exudation, efflorescence) occurs if auxiliary and polymer matrix are thermodynamically miscible. This is not true if the auxiliary is transported from a thermodynamically stable polymer–auxiliary system into a liquid or solid contact material (bleeding, extraction, migration) since the auxiliary has different chemical potentials in the polymer matrix and in the contact material. The rate of the transport is controlled by the diffusion of the auxiliary in the polymer. Kinetic effects (tortuosity, etc.) delay the transport but do not prevent it. This is the reason why polymeric plasticizers do migrate despite their high viscosities, protecting layers are only partially effective against migrating plasticizers, etc.

An auxiliary X will be completely extracted from the plastic P if the contacting liquid L is a good solvent for X. However, not only will X move into L but also L into P. This diffusion of L into P is faster than the relaxation of the polymer chains; the transport is directly proportional to time according to Case II (see Section 9.7.2.). If, however, L is a precipitant for P and a bad solvent for X, then X will be transported into L but L will not move into P. The transported amount will then be proportional to the square root of time (Fickian behavior; Case I). Finally, a thermodynamically controlled distribution equilibrium will be established for X between P and L: the auxiliary will not be completely extracted.

Emigration of an auxiliary is toxicologically unobjectionable if the transported amount does not exceed the admissible amount. It is however often aesthetically unpleasant and technologically undesirable.

Emigration is sometimes desirable, however. Lubricants and antistatic agents are supposed to move to the surface; they do not act in the interior of plastics and fibers. Another example are the so-called **low-profile resins** (Section 14.2.3.).

Literature

O.Lauer, K.Engels, Aufbereiten von Kunststoffen, Hanser, Munich 1971

J.Edenbaum, ed., Plastics Additives and Modifiers Handbook, Van Nostrand Reinhold, New York 1992

R.Gächter, H.Müller, eds., Plastics Additives, Hanser, Munich, 4th ed. 1993

G.Scott, ed., Atmospheric Oxidation and Antioxidants, Elsevier, Amsterdam 1993 (2 vols.)

12. Elastomers

12.1. Introduction

The name "rubber" originated from one of the first European uses of natural rubber as an eraser. It has almost completely eliminated the older name caoutchouc. **Natural rubber** NR was harvested by the Mayas by inserting grooves into wild growing rubber trees and collecting the rubber sap (**latex**) (Quechua: *caa* = wood, *o-chu* = weeping; Old Spanish: *cauchuc*; F: *caoutchouc*; E: *caoutch(o)uc*). L: *latex* = liquid). Today, almost all natural rubber is obtained from rubber plantations.

A **latex** (pl. *latices*) is a dispersion of polymer droplets ($T > T_G$) or particles ($T < T_G$) in a liquid, usually water. The dispersion may be an emulsion (rubber; $T > T_G$) or a suspension (poly-(styrene); $T < T_G$). Droplet and particle sizes are usually between 0.05 μm and 5 μm.

Synthetic rubbers SR are produced by industry. The first synthetic rubber was methyl rubber from 2,3-dimethylbutadiene $CH_2=C(CH_3)-C(CH_3)=CH_2$; it was produced in Germany during World War I. The first successful commercial rubber was poly(chloroprene) $+CH_2-CCl=CH-CH_2\}_n$ (DuPont 1929). The most important synthetic rubber is SBR, a copolymer of styrene and butadiene.

Annual world production of all rubbers is ca. $15 \cdot 10^6$ t (1/3 NR, 2/3 SR). Rubbers consume annually ca. $5 \cdot 10^6$ t carbon black (95 % of the total carbon black production!), ca. $2 \cdot 10^6$ t mineral oils as plasticizers, and ca. $0.7 \cdot 10^6$ t rubber chemicals.

Rubbers are mostly amorphous materials based on linear polymers with glass temperatures below the use temperature. The chemical cross-linking of rubbers to **elastomers** with relatively large mesh sizes is called **vulcanization** (sometimes **curing**). Elastomers show entropy elasticity (Section 10.4.) whereas rubbers are predominantly viscoelastic. The term "rubber elasticity" is thus rarely justified. **Hard rubbers** are rubbers that have been severely cross-linked; they are thermosets and do not show rubber elasticity.

12.2. Processing

Raw rubbers are highly viscous materials; the working-in of rubber chemicals, carbon black, mineral oil, etc., requires heavy machinery such as kneaders or roll mills (Fig. 12-1). **Kneaders** are built in many versions, for example, with blades, moving screws and stationary teeth. **Roll mills** have two heated mixing rolls that rotate towards each other with different velocities which causes the rubber to stick to one roll only. The rubber is continuously torn apart in the roll gap by the combined action of kneading and shearing. Heat and shearing causes rubber molecules to degrade to

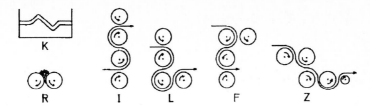

Fig. 12-1   Kneader K (here a sigma kneader with crankshaft-like motion), roll mills R, and various calenders I, L, F, and Z [1].

lower molar masses by this **mastication** (L: *masticare* = to chew). The rubber becomes less viscous and rubber chemicals, fillers, etc., can be worked in. **Calenders** (G: *kulindein* = to roll; L: *cylindrus* = cylinder) have large rolls that revolve with differential speed against each other; they are used to prepare rubber films and sheetings.

Proportions of rubber chemicals and other auxiliaries are not reported in percent of the final compound but in parts of added material per 100 parts of rubber (phr = *p*arts per *h*undred parts of *r*ubber). Formulations vary with the intended use of the elastomer. A typical recipe calls for 100 parts rubber, 20-100 phr carbon black (filler), 5-20 phr mineral oil (plasticizer), 1-3 phr sulfur (cross-linking agent), 1-2 phr vulcanization accelerator, 5 phr zinc oxide (activator), and 1 phr stearic acid (lubricant). The working-in of auxiliaries is promoted by **peptizers**, for example pentachlorothiophenol. These chemical compounds generate radicals which help degrade the rubber to lower molar masses.

12.3. Rubber Types

Synthetic rubbers are produced in ca. 30 families with more than 1800 grades. Names of families are usually symbolized by combinations of capital letters. Examples are: BR = poly(butadiene) rubber, CR = poly(chloroprene rubber), CSM = chlorosulfonated poly(ethylene), EPDM = rubber from ethylene + propylene + non-conjugated diene, EPM = saturated ethylene–propylene rubber, EVM = ethylene–vinyl acetate rubber, IIR = isobutylene–isoprene rubber (butyl rubber with ca. 4 % isoprene units), IR = synthetic isoprene rubber, MVQ = silicone rubber with methyl and vinyl side groups (e.g., -(CH$_2$CH$_2$CH=CH$_2$)), NBR = nitrile rubber from acrylonitrile and butadiene, SBR = styrene–butadiene rubber, T = thioplast rubber (sulfide rubber). BR with high 1,4-cis contents are called **stereo rubbers.**

Rubbers are usually subdivided into **all-purpose rubbers** (NR, BR, EPDM, EPM, IR, SBR), **special diene rubbers** (CR, IIR, NBR), and **exotics** (acryl, fluoro, silicone, thio, etc.). All-purpose rubbers comprise ca. 82 % of all synthetic rubbers. They are dominated by diene polymers because these polymers have chains with low rotational barriers (Section 2.4.1.), easy to cross-link units, and weak intermolecular dispersion forces (low glass temperatures). Diene rubbers differ in processing, vulcanization behavior, resistance against aging, green strength (strength before curing), tackiness, and so on.

12.4. Vulcanization

Cross-linking of natural rubber by sulfur under the action of heat was discovered accidentally by Charles Goodyear in 1839 as he was looking for a "drying agent" for the sticky material. Since sulfur and heat are the attributes of the Roman god Vulcanus, the process was called vulcanization by Thomas Hancock.

The vulcanization (**curing**) of *diene rubbers* by sulfur at (120-160)°C is today called **hot vulcanization**. Diene rubbers can also be cured by **cold vulcanization** with disulfurdichloride or magnesium oxide. *Saturated* hydrocarbon rubbers (EPM) can be cured by free radicals from peroxides. Vinyl groups in side chains (ca. 0.2 %) of *silicone rubbers* $+O\text{-}Si(CH_3)_2\frac{1}{n}$ can also be cross-linked at higher temperatures by peroxides or by addition of SH-groups of corresponding multifunctional compounds. Silicone rubbers with silanol side groups are cured in the cold by methyltriacetoxysilane, tetrabutyltitanate, etc., those with acetoxy end groups with moisture (RTV = room temperature vulcanizing).

Hot vulcanizations are performed in presses or autoclaves. The mechanism of cross-linking by sulfur is not known in detail. Cross-linking is not accelerated by radicals but by organic acids and bases; it is thus probably an ionic process. Added sulfur is present as cyclooctasulfur S_8 which seems to dissociate into $S_m^{\oplus}$ and $S_n^{\ominus}$ at vulcanization conditions ($m + n = 8$). The sulfur cation $S_m^{\oplus}$ adds to the diene rubber, e.g., cis-1,4-poly(isoprene) $+CH_2\text{-}C(CH_3)=CH\text{-}CH_2\frac{1}{n}$ (IR) to give I. This cation reacts with IR to give II and III. The allylic carbenium cation III adds S_8. The resulting IV forms a sulfur bridge by reaction with IR to V. The cation undergoes a transfer reaction to IR which regenerates III, etc.:

One sulfur bridge is formed from ca. 50 initially present sulfur atoms. Other sulfur structures must thus be present besides the polysulfide bridges of structure V, for example, shorter bridges and sulfur-containing rings. Only relatively long, sufficiently spaced sulfur bridges provide the polymer with elastic properties, however.

Hot vulcanizations are always performed in the presence of accelerators and activators. Such accelerators are tetraalkylthiurame disulfide VI, zinc dialkyldithiocarbamate VII and 2,2'-dithiobisbenzothiazole VIII:

The chemical structures VI, VII, and VIII are shown.

For VI:
$$R_2N-C(S)-S-S-C(S)-NR_2$$

For VII:
$$Zn[S-C(S)-NR_2]_2$$

For VIII:
benzothiazole—S—S—benzothiazole

VI and VII insert S_8 and form polysulfides, e.g., $R_2N-C(S)-S-S_x-C(S)-NR_2$ from VI. Accelerators increase the addition of S_8 to the allyl positions of III. They probably also generate shorter sulfide bridges; at the same sulfur content, a greater concentration of cross-links is produced. Typical activators are combinations of zinc oxide and salts of fatty acids. ZnO is a chelating agent that yields -S-Zn-S- bridges.

Different applications demand different cross-link densities which are defined as amounts of network chains per volume. Surgical gloves must be very soft, flexible and extensible; junctions are here separated by 100-150 monomeric units. Rubber gloves for household purposes are much more robust and rigid; they have 50-80 monomeric units between cross-links. Even more strongly cross-linked are inner tubes of tires (20-30), tire treads (10-20), and hard rubber (5-10).

Surgical gloves and other thin-walled articles are manufactured by dipping a positive mold into rubber latex, followed by *cold vulcanization* at room temperature. Disulfur dichloride as cross-linking agent produces monosulfide cross-links:

$$(12\text{-}1) \quad 2 \text{~~~}CH_2-CH=CH-CH_2\text{~~~} \xrightarrow{S_2Cl_2} \begin{array}{c} \text{~~~}CH_2-CH-CHCl-CH_2\text{~~~} \\ | \\ S \\ | \\ \text{~~~}CH_2-CH-CHCl-CH_2\text{~~~} \end{array}$$

Vulcanizations by *free radicals* are straightforward; they consist in radical forming reactions on polymer chains with subsequent chain combinations (see Section 3.7.7.). A related procedure is **dynamic vulcanization** which is usually performed with EPDM rubber that is blended with poly(ethylene), poly(propylene), aliphatic polyamides or other polymers. On mastication of these incompatible blends, usually in the presence of cross-linking agents, polymer chains are cleaved. The resulting polymer radicals cross-link polymer chains in the rubbery domains.

12.5. Properties

If two pieces of freshly cut raw natural rubber (and also some synthetic rubbers) are pressed together, they will form an irreversible joint within seconds. The full pull strength is obtained after ca. one hour. This **tack** is caused by a self-diffusion of rubber chains (Section 7.5.1.); it is very useful for the manufacture of tires.

About 50 % of all rubbers go into tire production. Another 25 % are used for other vehicle applications and in the construction industry. Tires demand rubbers with very special properties.

Only very few elastomers, usually exotics, exhibit stress-hardening (Fig. 10-2) and show correspondingly good tensile strengths, tear strengths, and abrasion resistances. All other elastomers must be compounded with a reinforcing filler (Section 11.5.). For diene rubbers, this is usually carbon black and for silicone rubbers, mainly highly disperse, pyrogenic SiO_2 (Aerosil®). Reinforcing fillers lead to greater hystereses and larger permanent sets. **Hysteresis** (G: *hysteron* = to be behind) is present if the stress–strain curve on unloading stays behind the stress–strain curve on loading (at equal stress, larger elongation on unloading than on loading, producing a hysteresis loop). Large hystereses produce great heat.

SBR surpasses natural rubber with respect to processability, vulcanization behavior, aging resistance, and especially abrasion resistance. It is mainly used for tires of passenger cars. Natural rubber, on the other hand, tops SBR in some mechanical properties; it also shows smaller hystereses. Natural rubber thus dominates the market for trucks and airplanes. Stereorubbers exhibit good skid resistance but have a high heat build-up. They are therefore blended with natural rubber for use in tire treads.

Inner tubes of tires should have low gas permeabilities; the rubbers of choice are butyl rubber and synthetic poly(isoprene). EPDM rubbers contain only a few carbon–carbon double bonds; they are thus weather resistant and used in the car, cable and construction industries. The high solubility parameters of polar rubbers (NBR, CR) prevent their swelling in mineral oils and gasoline (very low solubility parameters); they are the choice for fuel lines, gaskets and sealants. Polysiloxane rubbers have low glass temperatures and thus good flexibilities at low temperatures. Since their dimethyl siloxane units $-O-Si(CH_3)_2-$ are not easily attacked by oxygen and radicals, they are also highly heat-resistant. In addition, they repel water, insulate against electrical currents, and are biologically compatible. Some exotics are used in the arctic and in space because of their high resistance to cold and heat.

Thermoplastic elastomers (TPEs) are usually triblock copolymers with rubbery center blocks and "hard" (high T_G) end blocks; some are also graft copolymers. The blocks are incompatible and demix; end blocks cluster in domains and form physical cross-links (Section 8.5.). TPEs can be processed like thermoplastics at temperatures above the glass temperatures of their end blocks.

12.6. Reclaiming

Retreading is the most economical use of old tires; about 60 % of old truck tires and 10 % of old passenger car tires are retreaded in the United States. Worn retreaded tires are usually discarded legally in landfills and illegally in woods and creeks because scrap rubber is difficult to reclaim.

Practically all reclamation methods require the rubber to be shredded to small particles. Shredding requires considerable energy, however, because elastomers are entropy-elastic materials. Some shredded scrap rubber is added to asphalt. Pyrolysis of rubber delivers carbon black and fuel oils but there is an overcapacity of carbon black and a surplus of mineral oil. Scrap rubber can serve as an energy source but its burning is forbidden in some countries because of environmental concerns.

Only about 4-5 % of all scrap rubber is reclaimed world-wide. Reclamation methods have to separate fibers from tires, i.e., rayon, polyamide, polyester, glass, or steel. Glass fibers have to be removed mechanically. Cellulose fibers are hydrolyzed by cooking tire scraps with alkali. Other methods use steam or ditolyldisulfide plus tall oil. These processes cleave not only sulfur bridges but also polymer chains; there is also radical formation and subsequent transfer and recombination reactions. The properties of the resulting reclaimed rubber therefore differ from those of virgin rubber. Reclaimed rubber can however be vulcanized. It is used in tire formulations and in adhesives.

Literature

A.K.Bhowmick, H.L.Stephens, ed., Handbook of Elastomers, Dekker, New York 1988
N.P.Cheremisinoff, ed., Elastomer Technology Handbook, CRC Press, Boca Raton 1993
J.E.Mark, B.Erman, F.R.Eirich, ed., Science and Technology of Rubber, Academic Press, San Diego, 2nd ed. 1994
L.White, Rubber Processing, Hanser, Munich 1995

References

[1]   H.-G.Elias, An Introduction to Plastics, VCH, Weinheim 1993, Fig. 11-9

13. Fibers

13.1. Introduction

Natural fibers are subdivided into animal, vegetable and mineral fibers. Mineral fibers are mostly silicates; an example is asbestos. All presently used animal fibers consist of proteins (silk, wool, hair), all vegetable fibers are cellulosic. Animal fibers are either **secretion fibers** (silk) or **hair fibers** (wool, hair). Vegetable fibers are subdivided into bast fibers, leaf fibers, seed-hair fibers, and palm fibers. **Bast fibers** are obtained from bark or stems of certain plants (hemp, jute, ramie, linen (from flax), etc.); for textile use, they are called **soft fibers**. **Leaf fibers** are also called **hard fibers**; they are mainly used for cordage (sisal, manila hemp, etc.). **Seed-hair fibers** comprise cotton and kapok. **Palm fibers (brush fibers)** originate from various other parts of plants, for example, coir from coconut husks, broom corn from flower heads of *sorghum vulgare*, palmira palm from palm leaf stems, etc.

Man-made fibers were formerly called artificial fibers; they are subdivided into semi-synthetic fibers and synthetic fibers. **Semi-synthetic fibers** are obtained from biological materials that are converted into fibers by chemical processes and subsequent spinning. In most cases, this involves a reconstitution of originally present fiber structures; these fibers are therefore also called **regenerated fibers** (L: *regeneratus* = born again). Rayon from wood is the most important semi-synthetic fiber. Acetate fibers (cellulose acetates) are generally not considered regenerated fibers although they are usually based on regenerated cellulose. Another regenerated fiber is based on the protein casein.

Synthetic fibers result from the polymerization of monomers; synthetic mineral fibers such as glass fibers and rock wool are not considered synthetic fibers. The most important synthetic fibers are polyester, polyamide, acrylic, and olefin fibers.

For textile processes, the length of fibers is important. Fibers are therefore subdivided into **filament fibers** of "infinite" length and **staple fibers** of short length. The last term is only applied to fibers from chemical-industrial processes.

The use of fibers leads to another subdivision into textile and industrial fibers. **Textile fibers** are used for clothing (dresses, underwear, etc.) and for home textiles (carpets, draperies, upholstery, etc.). **Industrial fibers** serve as reinforcing materials for plastics and elastomers and for technical textiles such as ropes and filter cloths.

Despite the name, **dietary fibers** are, in general, not fibers at all. They are rather defined as those food polysaccharides and lignins that cannot be digested by enzymes in the human gastrointestinal tract. Dietary fibers are complex mixtures of cellulose, lignin and polysaccharides such as hemicelluloses, pectic substances, poly(fructosan)s, etc.; they may be insoluble or soluble

Fibers are traded under protected trade names. Examples are Dacron® (PES), Terylene® (PES), Trevira® (PES), Perlon® (PA 6), and Orlon® (PAN); nylon and spandex are no longer protected trade names. Names of fiber classes carry abbreviations and acronyms which however neither agree with those of corresponding plastics nor from country to country or organization to organization. For example, the acronym for poly(ethylene terephthalate) plastics is either PET (ASTM, DIN, ISO) or PETP (IUPAC), except for recycling, where it is PETE (or "2"). Fibers from PET carry the symbol PES (DIN) or PL (European Textile Characterization Law). Acrylic fibers from poly(acrylonitrile) generally have the acronym PAN, except in Germany where it is PAC because PAN is a registered trademark. Polyamide fibers PA are called just that in most countries except in English speaking ones where they are known as nylons, except if they are aromatic polyamides or are based on α-amino acids. Fibers from poly(ethylene)s and it-poly(propylene)s are classified as "olefin fibers" by the U.S. Federal Trade Commission (FTC). "Vinyl fibers" (ISO) refers to fibers from homopolymers and copolymers of poly(vinyl chloride) whereas the FTC calls the copolymers "vinyon". The "chlorofibers" of ISO comprise not only homopolymers and copolymers of vinyl chloride but also copolymers of vinylidene chloride; the latter fibers are called "saran" by the FTC.

13.2. Production

Wool of wild animals is probably the oldest fiber used by man; imprints of strings have been found on ca. 35 000-year-old ceramics. Cotton was known in Egypt about 12 000 B.C. Flax and linen were used by lake-dwellers in ca. 8000 B.C., hemp in Southeast Asia in ca. 6000 B.C., and silk in China in ca. 2700 B.C. The first semi-synthetic fibers were produced by denitration of cellulose nitrate, in 1879 for filaments of incandescent bulbs and in 1884 for textile fibers. The first synthetic fibers were poly(vinyl chloride) (1931), polyamide 66 (1935) and polyamide 6 (1937).

The world production of natural *textile fibers* is dominated by cotton (1990: ca. $19 \cdot 10^6$ t/a); wool (ca. $2 \cdot 10^6$ t/a) and silk (ca. $0.06 \cdot 10^6$ t/a) are produced in considerably smaller amounts. The world production of regenerated fibers has remained at ca. $3.5 \cdot 10^6$ t/a for years due to the high cost of to work-up of process water. The global synthetic fiber market ($16 \cdot 10^6$ t/a) is ruled by three polymers: polyesters (54 %), polyamides (24 %) and acrylics (15 %). The remaining 7 % are mainly olefin fibers; their total production of ca. $1.5 \cdot 10^6$ t/a includes textile and industrial fibers. All important textile fibers are based on semi-crystalline polymers; it is the crystallization and orientation of polymer chains and crystalline domains which provide these fibers with their textile properties.

Industrial fibers comprise natural hard fibers ($4 \cdot 10^6$ t/a) and glass fibers ($1.4 \cdot 10^6$ t/a), the latter mainly for construction and for reinforcing plastics. The annual world production of *high-performance fibers* is small: ca. 4000 t for carbon fibers and ca. 17 000 t for aramide fibers (aromatic polyamides).

The model for the first man-made fibers was silk, an "endless" fiber of ca. 4000 m length. Its creation by secretion was the archetype for the production of semi-synthetic and synthetic fibers by fiber spinning. The resulting "endless" fibers could however not be processed to yarns, etc., by the usual textile machines since these machines were designed for the relatively short wool and cotton fibers. The filaments were thus cut to short fibers, i.e., staple fibers.

Filaments and staple fibers have quite different textile properties even if they are from the same polymer. Filaments are silk-like whereas staple fibers are either more wool-like or more cotton-like, depending on the polymer and the fiber processing. Acrylics are almost exclusively produced as wool-like staple fibers whereas silk-like filaments dominate fibers from polyamides 6 and 66.

13.3. Fiber Spinning

Practically all man-made fibers are obtained from polymers by melt, solution or dispersion spinning. **Reaction spinning** of monomers or prepolymers is only suitable for very reactive starting materials, for example, for the manufacture of spandex fibers from isocyanates (Section 4.1.8.).

In **melt spinning**, melted polymers are pressed under nitrogen through a spinneret with many narrow holes of ca. (50-400) μm diameter. For *filaments,* the resulting many fibrils are pulled off at high speed whereby they cool and solidify while coming together to form a yarn. The undrawn yarn is then drawn with much higher rates up to 4000 m/min whereupon polymer chains and crystalline domains are oriented. For *staple fibers*, the tow emerging from the spinneret is drawn and crimped. After a heat setting to stabilize the physical structure, drawn tow is cut to staple fibers.

Melt spinning is performed at high temperatures; it requires meltable and thermally stable polymers, for example, poly(olefin)s, aliphatic polyamides, aromatic polyesters, and glass. Some polymers degrade slightly during melt spinning, delivering monomers, oligomers and other low molar mass degradation products which are deposited on fibers and spinning equipment.

Blowing hot air at high speed below the spinneret perpendicular to the emerging fibrils tears apart fibrils to **microfibers** of ca. (0.1-0.3) dtex. Microfibers are bundled like spaghetti and united into a kind of loose fleece. The resulting textiles have small pores which allow the permeation of water vapor but not of water droplets. The textiles "breathe" but repel water.

Dry spinning is suitable for solutions of thermally unstable polymers in volatile solvents; these solutions are called **dopes**. Examples are ca. 30 % poly(acrylonitrile) PAN in *N,N*-dimethylformamide DMF, 20 % cellulose triacetate in dichloromethylene, and (15-20) % aromatic polyamides in DMF + 5 % LiCl. Warm air or nitrogen is blown against the emerging fibrils whereupon solvents evaporate and fibers solidify. The pull-off speed is (300-400) m/min. Higher speeds of ca. 500 m/min and greater form stabilities of fibers are obtained by **gel spinning** of concentrated solutions, for example, 50 % PAN in DMF or (1-2) % ultrahigh molar mass ($M > 3 \cdot 10^6$ g/mol) poly(ethylene)s in xylene for high-modulus fibers.

Wet spinning is employed for thermally labile polymers for which no volatile solvent exists. The precipitation of fibers does not take place in nitrogen or air but in a precipitating solvent. Examples are the spinning of rayon fibers from (7-10) % solutions of sodium cellulose xanthate into sulfuric acid solutions of sodium or zinc acetate, (10-18) % aqueous poly(vinyl alcohol) solutions into aqueous solutions of sodium sulfate, or (15-20) % acetone solutions of modacrylics into aqueous acetone.

Insoluble and unmeltable polymers are subjected to **dispersion spinning.** The fiber-forming polymer is dispersed in a solution of another polymer, for example, poly(tetrafluoroethylene) in aqueous poly(vinyl alcohol) solutions. After the extrusion of the dispersion through spinnerets, solvents are evaporated by heating. The auxiliary polymer is removed by burning which causes the fibers to sinter.

Spinning requires polymers with sufficiently high molar masses. Low molar masses do not provide the emerging (sometimes wet!) fibers with sufficient strength (Fig. 10-17); they also prevent the formation of long threads. The maximum thread length obtainable is called the **spinnability**. High spinnabilities are obtained if the extensional viscosity increases faster than the reduction of the cross-sectional area, i.e., if the polymer undergoes stress-hardening (Section 10.2.2.). This requires an extrusion rate that is much greater than the rate of disappearance of entanglements. Melts and dopes must thus have high Deborah numbers (Section 10.5.3.). Since relaxation times are a function of viscosities and experimental times are proportional to spinning velocities, spinnabilities depend on the product of viscosity η and spinning velocity v.

With increasing product ηv, spinnabilities pass through maxima (Fig. 13-1). At low rates and low viscosities, they are controlled by **capillary fracture** (a ductile fracture, see Section 10.6.1.): the liquid "fractures" into droplets because of the effects of surface tension (similar to a slow jet of water). At high rates and viscosities, liquids can be elongated. If a certain amount of stored energy is exceeded, threads break by **cohesion fracture**, a brittle fracture perpendicular to the draw direction.

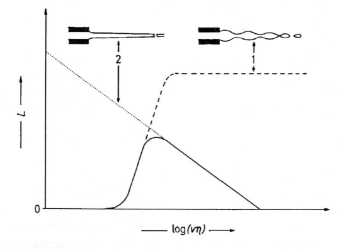

Fig. 13-1 Spinnability L (maximum thread length) as function of the logarithm of the product of spinning velocity v and viscosity η. - - - Capillary fracture 1, ····· Cohesion fracture 2.

Fig. 13-2   Parison swelling on extrusion.

The spinning of fibers is an extrusion of the melt or dope (see also Section 14.2.4.). Polymer coils are deformed on entry into the narrow holes of the spinneret (Fig. 13-2). Because of the short residence time in the holes, molecule segments can no longer de-entangle and a normal stress is produced. This stress disappears at the exit of the hole and the coils (and thus the fiber) expand perpendicular to the draw direction. This effect is known by various names, for example, **parison swell** (extrusion), **swelling** (hollow body forming), **Barus effect** or **memory effect** (melt viscosity), and **Weissenberg effect** or **rock climbing** (solutions).

This effect causes round filaments to have larger diameters than the circular holes through which they were melt spun. Non-cylindrical holes deliver fibers whose cross-sections deviate considerably from those of the holes because normal stresses vary in different directions (Fig. 13-3). Such **profile fibers** have properties which are highly desirable for certain textile purposes, for example, a higher gloss of triangular cross-sections, a greater stiffness if shapes are trilobal or cross-like, a better water-uptake of hollow fibers, etc. Fibers from solution spinning through circular holes always have non-circular cross-sections because of differences in diffusion rates of solvents out of the fiber and precipants into the fiber surface on solidifying.

Wet, dry and dispersion spinning (and, to some degree, melt spinning) can also be used to prepare **bicomponent fibers**. Two dopes of different composition (e.g., two different polyamides) are fed separately into the spinneret and joined shortly before the exit. Depending on the design of the spinneret, both fiber components are present in the fiber either side-by-side (S-S) or as core–mantle. If both components take up different proportions of water, S-S fibers will crimp. Bicomponent fibers are also used to create various color effects.

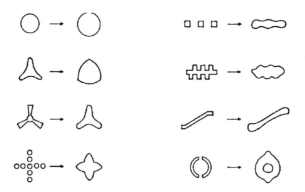

Fig. 13-3   Some cross-sections of holes in spinnerets and resulting cross-sections of fibers.

13.4. Textile Fibers

Fibers have high aspect ratios L/d of length L to diameter d (Table 13-1). The spinning of fibers into yarns requires, for example, aspect ratios of at least 800. Yarns can thus not be obtained from bagasse, the fibrous materials that remains after sugar cane is pressed (Table 13-1). Short fibers are converted into papers, felts, and fleeces.

Table 13-1 Density ρ, length L, diameter d, aspect ratio, and fineness of dry fibers.

Fiber	ρ/(g cm^{-3}) min.	L/cm min.	max.	d/μm min.	max.	Aspect ratio	Fineness in dtex
Silk	1.34	< 120 000		8	15	6·10^9	1
Wool (sheep)	1.30	< 13			20	6500	4
Ramie	1.50	2.5	25	16	120	830	24
Cotton	1.54	2.7	7.5		20	1400	4
Flax	1.50	0.4	7.0	8	30	1200	6
Rayon (staple fiber)	1.52	3	8	10	50	1800	11
Synthetic fibers (staple)	various	3	8	10	50	1800	various
Hemp	1.48	0.5	2.0	15	28	1400	7
Jute	1.50	0.07	0.6	15	25	180	2
Bagasse		0.1	0.4	40	100	35	60

Further processing of filaments and staple fibers to yarns, knits, fabrics, etc., is dictated by their **fineness (titer)**. Titers are length-related masses that are now measured in 1 tex ≡ 1 g/km and formerly in denier = g/(9000 m). They are sometimes (and erroneously) also called "linear densities".

Diameters of natural fibers vary from fiber to fiber and even within a fiber. It is therefore expedient for textile fibers to report moduli of elasticity and tensile strengths in force per titer = energy per mass and not as force per cross-sectional area = energy per volume as for plastics and elastomers. Because of the physical unit energy/mass, elasticities and strengths of fibers are often said to be **specific elasticities** and **specific tensile strengths**. These quantities can be converted into elasticities and strengths by multiplication by the density: (property in N/tex)×(density in g/cm^3) = property in GPa. Industrial fibers are much more uniform with respect to diameters; their moduli and strengths are given in GPa or MPa.

Fine fibers are generally used for textiles and coarser ones for industrial purposes. Textile microfibers have titers of (0.1-0.3) dtex and fibers for clothing (1-5) dtex. Upholstery requires fibers of (5-15) dtex, carpets (12-50) dtex, and filter cloths (25-100) dtex. Industrial microfibers have however titers of only ca. 0.08 dtex (L = (0.15-0.3) mm, d = (3-5) μm, $\rho \approx$ 2 g/cm^3); they are used as reinforcing agents.

The classical natural fibers cotton, wool and silk are still the models for synthetic textile fibers. Silk was always a luxury fiber because of its price, look and feel. Wool has been traditionally used for overwear in cold and moderate climates. Cotton is the fiber of choice for underwear and, in tropical climates, also for overwear. Perception and use properties of textiles are only in part determined by the chemical structure of polymers. Much more important are physical structures, shapes of fibers, the way fibers are assembled to yarns and yarns to wovens, knits, etc., and the finishing. Appropriate textile engineering can make a given fiber-forming polymer more silk-like, wool-like or cotton-like.

Chemical structures determine *dyeing* and water uptake. Poly(olefin)s contain no reactive and polar groups. They are therefore dyed in the melt by pigments and then spun into fibers; since spun-dyed fibers are manufactured by the fiber producer, they cannot easily adapt to fashion changes. The monomeric units of polyesters, polyamides and acrylics also show little affinity to dyestuffs. Fiber grades of these polymers thus usually incorporate small proportions of basic or acidic comonomers. Since dyestuffs cannot enter crystalline domains, such polymers may also contain bulky comonomer units that reduce crystallization.

The wear comfort of textiles is predominantly determined by the *moisture pick-up*. The excellent feel of cotton in hot, humid weather is not only due to the many hydroxyl groups of its hydrophilic cellulose molecules but also due to its hollow fiber structure (Fig. 13-4) which allows cotton to retain significant amounts of water even at 100 % relative humidity. This capillary effect can also be obtained if one makes hydrophobic fibers microporous. Paper from such fibers can be written on by ink and roller pens.

The glass temperature of dry cellulose is ca. 220°C; it is lowered by water to ca. 20°C. Ironing of damp cotton fabrics causes water to evaporate and cellulose molecules to freeze-in in the desired way. For the same reason, cotton fabrics also crease easily under their own weight. They are thus converted into non-iron fabrics by treatment with UF resins or so-called reactant resins. UF resins are prepolymers based on

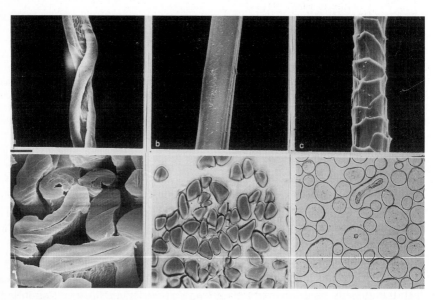

Fig. 13-4 Lengthwise (top) and cross-sectional (bottom) shape of natural fibers. Left: cotton with a hollow fiber structure; center: silk with triangular cross-sections; right: wool with a scaly surface and a bicomponent structure [2].

urea and formaldehyde. They enter intermicellar voids of cellulose fibers where they react and cross-link. The resulting crease-resistance is thought to be due to the stiffening action of the cured resin and less to the intermolecular cross-linking of cellulose chains. Reactant resins are mostly methylol compounds of cyclic urea resins that predominantly cross-link cellulose chains.

The *wear comfort* of fibers is mainly determined by their mechanical behavior on deformation. Silk is the most rigid of all natural textile fibers (highest Young's modulus); it also has the highest fracture strength (Fig. 13-5). Wool is the softest fiber (lowest Young's modulus); it can also be stretched easily but is not very strong. Rayon as staple fiber behaves like wool. Acrylics are also wool-like. Polyamides 6 and 66 resemble natural silk in their force–elongation diagrams but are considerably more "plastic" (greater elongations at break). The first polyester fibers were silk-like due to their smooth surface. By using various cross-sectional shapes, crimping and texturizing, they can now be made more wool-like or more cotton-like.

The *perception* of fibers is mainly due to the fiber shape. Silk fibers have triangular cross-sections and rather smooth surfaces (Fig. 13-4). Such cross-sections provide silk fabrics not only with the silky hand and rustle but also with a high gloss (increased light refraction similar to the cut of diamonds). Similar fiber geometries can be imposed on synthetic fibers by suitable spinneret geometries (Fig. 13-3).

Wool has a scaly surface which creates the typical wooly perception (and the itching!). Crimping and bulky volume of wool are caused by their bicomponent structure (Fig. 4-6). Paracortex and orthocortex have different chemical compositions; the two fiber halves thus take up different proportions of moisture. The same effect can be obtained from synthetic fibers with a side-by-side bicomponent structure.

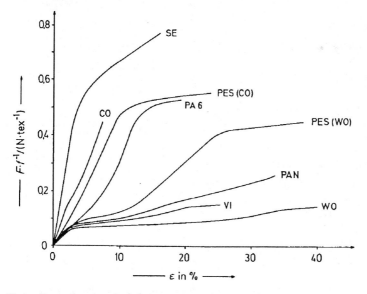

Fig. 13-5 Force F per titer t as function of elongation ε for cotton [CO], polyamide 6 [PA 6], acrylics [PAN], polyester fibers as cotton-type [PES(CO)] and wool-type [PES(WO)], rayon [VI], silk [SE], and sheep's wool [WO]. CO, SE, VI, and WO are international acronyms of fibers.

Elastan fibers are fibrous elastomeric materials. Their elasticity can be obtained by various means. It may be due to a chemical cross-linking similar to that of rubber-based elastomers. It may also be caused by a physical cross-linking via domains as in block copolymers with "hard" urethane blocks and elastic polyether segments; they are called **spandex fibers** if they are comprised of at least 85 % of a segmented poly-urethane. Elasticity may also be imposed by crimping of fibers.

13.5. Industrial Fibers

Industrial fibers are fibers that are used for non-textile purposes. They are usually very rigid and hydrophobic. Their applications range from cords and ropes to filter cloth and reinforcing fillers for plastics (Section 14.5.).

Industrial natural fibers are always based on cellulosic plant fibers. Some of these fibers are globally harvested in considerable amounts: 2 000 000 t/a jute, 700 000 t/a flax, 630 000 t/a sisal, 250 000 t/a hemp, 200 000 t/a coir, 150 000 t/a henequen, etc.

Olefin fibers are dominated by poly(ethylene) fibers (ca. 1 700 000 t/a); poly(pro-pylene) fibers are produced in smaller amounts (ca. 200 000 t/a). They are increas-ingly used for carpets. "Olefin fibers" sometimes include *chlorofibers* (20 000 t/a).

Aramid fibers comprise those based on poly(m-phenylene isophthalamide) (No-mex®), Poly(p-phenylene terephthalamide) (Kevlar®) and the copolymer Techno-ra®. Their global production is ca. 17 000 t/a which includes polybenzimidazole PBI.

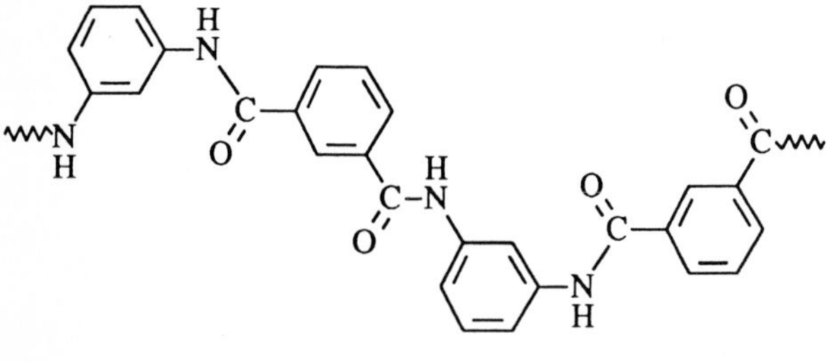

Nomex™

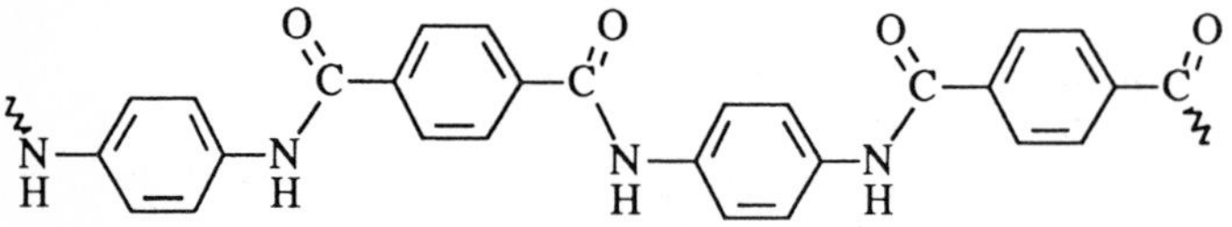

Kevlar™

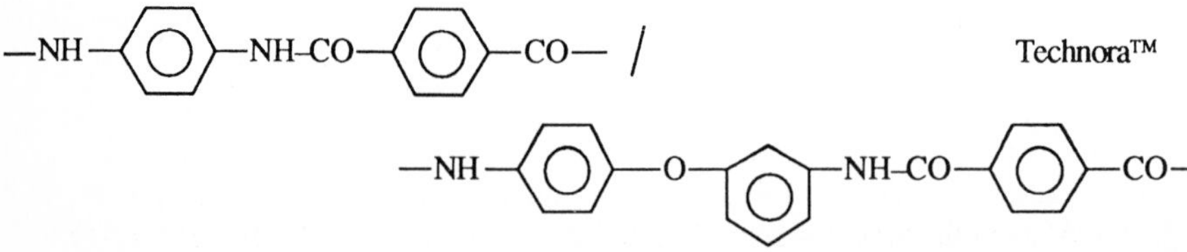

Technora™

PBI

Poly(vinyl alcohol) fibers are manufactured in East Asian countries in amounts of ca. 200 000 t/a; the fibers are cross-linked by aldehydes via intermolecular acetal bridges. They are used for fishing nets, conveyor belts, and, in smaller amounts, also for textiles.

Carbon fibers are produced from pitch or by pyrolysis of poly(acrylonitrile) fibers at (1000-1500)°C. Their carbon content is (80-95) %. *Graphite fibers* have far higher carbon contents of ca. 99 %; they are formed pyrolytically at temperatures of ca. 2500 °C. All these fibers (ca. 4 000 t/a) are very rigid; they have Young's moduli of up to 690 GPa. Their main use is as reinforcing fibers for thermosets.

The global demand for *glass fibers* is ca. 1 000 000 t/a. Glass fibers are produced as long fibers for insulation and short fibers for reinforcement of plastics. A small proportion of glass fibers is used for home textiles such as draperies.

13.6. Recycling

Old woolen textiles and yarns are mechanically shredded. Reclaimed wool is called **shoddy** if it was obtained from woven or knitted goods. Shoddy is spun again to yarns, usually in combination with virgin wool from sheep.

Old cellulosic textiles (cotton, rayon) are shredded and used as fillers for cushions, etc. **Rags** were the exclusive raw material for the manufacture of paper in the medieval times.

Literature

B. von Falkai, ed., Synthesefasern, Verlag Chemie, Weinheim 1981
B.P.Corbman, Textiles. Fiber to Fabric, McGraw-Hill, New York, 6th ed. 1983
W.Bobeth, ed., Textile Faserstoffe - Beschaffenheit und Eigenschaften, Springer, Heidelberg 1993
H.Brody, ed., Synthetic Fibre Materials, Longman Higher Education, Harlow (Essex) 1994

References

[1]   By courtesy of the former Institute for Applied Microscopy of the Fraunhofer Society, Karlsruhe, Germany

14. Plastics

14.1. Introduction

Plastics are working materials that are based on polymers as raw materials (G: *plastein* = to form, to shape). Their name is based on their ability to pass through "plastic states" during processing (cf. Section 7.4.4.). Plastics comprise ca. 500 constitutionally and/or configurationally different polymers which are available in ca. 13 000 grades with ca. 25 000 trade names.

Only about 30-40 of the constituent polymers are industrially important. Plastics derived from these polymers are subdivided according to their performance into 4 or 5 groups which are not sharply separated, however.

Commodity plastics (bulk plastics, volume plastics) comprise thermoplastics that are produced in large amounts: poly(vinyl chloride) PVC, poly(ethylene) PE (different densities), isotactic poly(propylene) PP, styrene polymers (PS, SB, HIPS, SAN, ABS), and poly(methyl methacrylate) PMMA.

Technical plastics comprise engineering plastics and high-performance plastics. **Engineering plastics** have improved mechanical properties. They are load-bearing and can often replace metals and ceramics in engineering applications. Engineering plastics include poly(ethylene terephthalate) PET, poly(butylene terephthalate) PBT, polyamides PA, polycarbonates PC, poly(oxymethylene)s POM, and blends of polymers with PPE [poly(oxy-2,5-dimethyl-1,4-phenylene), "polyphenylene ether"].

High-performance plastics (specialty plastics) are technical plastics with even more improved mechanical properties. This group includes aromatic polyamides (aramides, PAR), various polysulfones (PSU, PPSU, PAS, etc.), polyetherketones (PEK, PEEK, PEEKK), polyimides (PI, PAI, PEI), poly(phenylene sulfides) PPS, liquid-crystalline polymers (LCP), and fluoropolymers (see Chapter 15 for abbreviations).

Thermosets form a special group of plastics. Some thermosets are however often considered commodity plastics and others engineering plastics. The commodity-type group comprises alkyd resins, phenol–formaldehyde resins PF, amino resins (melamine–formaldehyde resins MF and urea–formaldehyde resins UF), and unsaturated polyester resins UP. The engineering plastics group consists of epoxy resins EP, polyurethanes PUR, and various allyl polymers (e.g., diallyl phthalate PDAP). The curing of thermosets requires chemical reactions. The resins are therefore sometimes considered a subclass of **reaction polymers** which also include thermoplastics prepared by the RIM process (see below).

Commodity, engineering, and high-performance plastics can be used for very different applications, **functional polymers** however only for one purpose. Functional polymers are not **functionalized polymers (function polymers)**, i.e., they are not polymers that have been retroactively fitted with reactive chemical groups "functional groups"). An example of a functional polymer is poly(ethylene-*co*-vinyl alcohol) which is exclusively used as a barrier polymer for packaging purposes. Another example is poly(p-phenylene sulfide), used for coatings. **Fluoroplastics** may also be considered functional polymers since they are usually only applied because of their excellent surface properties.

The global production of plastics was ca. $100 \cdot 10^6$ t/a in 1993. This includes ca. $2.6 \cdot 10^6$ t/a engineering plastics, ca. $0.12 \cdot 10^6$ t/a high-performance plastics, and ca. $20 \cdot 10^6$ t/a thermosets. These annual production data include ca. $4 \cdot 10^6$ t/a rubber-modified plastics and ca. $0.34 \cdot 10^6$ t/a polymer blends.

Raw materials for thermoplastics are polymers (exception: RIM); for thermosets, monomers or oligomers (prepolymers). All oligomeric and polymeric raw materials for plastics are often called **resins** (L: *resine*; G: *rhetine*) because the first prepolymers for thermosets resembled natural resins. These natural resins are organic solids that break with conchoidal fracture instead of the planar surfaces created upon fracture of crystalline materials or the drawn-out zones formed upon the breaking of gums and waxes. "Natural resin" refers mainly to oleoresins from tree saps but is also used for shellac, insect exudations, and mineral hydrocarbons.

14.2. Processing

14.2.1. Overview

Processing methods for plastics can be subdivided into four groups according to the state of matter of raw materials and the chemical/physical processes involved:

    A. Monomers to final products with simultaneous polymerization and shaping.

    B. Monomers to oligomers, followed by polymerization and shaping.

    C. Monomers to solid polymers, which are then melted, shaped and solidified.

    D. Machining and other methods for solid polymers.

Group A methods need only one step for the conversion of a monomer into the shaped product; they should thus be the most economical. They are used for the continuous coating of wires and fabrics and for the batchwise casting of shaped products in small numbers. The mass production of shaped articles requires however the very rapid conversion of a low-viscosity starting material into a very viscoelastic end product. Only very few monomers and types of polymerization satisfy this condition; an example is the anionic polymerization of ε-caprolactam by the RIM process.

Group B methods are used for thermosets (Section 14.4.). Polymerizations are stopped before the gel point is reached (so-called **B-stage**) since viscosities are low here and auxiliaries can be worked-in easily. Shaped articles usually need to be cured in molds because of low viscosities of B-stage materials.

Group C processes are the method of choice for thermoplastics (Section 14.3.). The polymerization to uncross-linked polymers does not proceed through a gel point which causes a "sudden" solidification of the reacting material. Thermoplastics can

thus be prepared, stored as pellets, beads, etc., for long times before they are compounded and processed. Some thermoplastics need to be dried before processing
since moisture in polymers will be converted to water vapor at high processing temperatures which in turn generates internal voids. For testing, shaped articles of these
thermoplastics must be **conditioned**, in Europe usually at 65 % relative humidity (RH)
and in the United States commonly at 50 % RH. For example, polyamide 66 takes up
ca. 3.5 % moisture but poly(propylene) only 0.004 %.

Group D processes are only used for difficult-to-process polymers or for complicated shapes. Methods comprise cutting, sawing, stamping, drilling, welding, etc.

Shaped articles produced by all four groups have frequently to be after-treated by
degrating, polishing, surface treatments, etc.

14.2.2. Processing Steps

Processing usually proceeds in three steps: softening by heating, shaping, and
stabilization of the shape.

Heating can only be avoided for some fast polymerizing, liquid polymers that are
processed by the RIM technique. Most monomers and prepolymers for thermosets
and all polymers for thermoplastics are however either solids or highly viscous liquids. They have to be heated in order to obtain the desired viscosity.

Heat can be provided externally or generated internally. *External heat* is usually
taken up by contact with hot surfaces and less frequently by convection (hot air) or
radiation (infrared); it is then distributed within the plastic by heat conduction. *Internal heat* is generally produced by friction on shearing, far less by atomic vibrations
on treatment with ultrasound or high frequences. A softening by contact heat alone is
called **plastifying**. It is used less often then softening by both contact heat *and* friction which is called **plastification.** Plastifying and plastification should not be confused with **plasticization**, the softening of polymers by copolymerization or added
plasticizers (see Section 9.6.4.).

Temperatures increase by $\Delta T = \Delta p/(\rho c_p)$ on adiabatic shearing where $\Delta p = \eta \, \dot{\gamma} =$ pressure difference, $\eta =$ dynamic viscosity, $\dot{\gamma} =$ shear rate, $\rho =$ density, and $c_p =$ specific heat capacity. A poly-
(ethylene) with $\rho = 0.96$ g/cm^3, $c_p = 2.1$ J/(g K) (Table 9-1) and $\eta = 10^4$ Pa s (Fig. 7-16) will be
heated 52 K by adiabatic friction alone if it is injection molded with shear rates of $\dot{\gamma} = 10^4$ s^{-1}.

Processing is facilitated by **processing aids. Lubricants** reduce friction. For example, easy-flow poly(styrene)s contain (3-4) % mineral oil as external lubricant
which reduces the friction between the thermoplastic and the metal walls of the processing unit. Internal lubricants break up polymer aggregates; they reduce Barus effects and melt fractures. Examples are long-chain fatty acid esters.

Shaping is controlled by melt strengths F. For viscoelastic melts, melt strengths are
given by the cross-sectional area A_0 of the parison, the extensional viscosity η_e, the
rate of extension, $\dot{\varepsilon}$, and the time t:

(14-1) $F/A_0 = \eta_e \, \dot{\varepsilon}/[\exp(\dot{\varepsilon} t)]$

Monomers of Group A and oligomers of Group B processes have fairly low melt strengths. The processing of these low-viscosity materials requires supports or molds. Typical processing methods are casting, coating, molding, and compression molding (Table 14-1).

In thermoplastics, polymer chains are usually entangled because polymers have molar masses that exceed the critical molar mass for entanglements. Because of entanglements, thermoplastics can be processed at higher extensional or shear rates. Most common is injection molding. Extrusion and blow forming do not even need external supports for shaping.

Shape stabilization takes place by cooling below the glass temperature (amorphous thermoplastics) or melting temperature (semi-crystalline thermoplastics) or by cross-linking (thermosets). Semi-crystalline polymers have to be cooled to temperatures far below the melting temperature because of the latent heat of crystallization. Cooling can only proceed by heat conduction (Section 9.2.4.). This fairly slow process controls the **cycle time** of discontinuous processes.

Table 14-1 Processing methods for plastics. ++ Most often used, + usually used, (+) as monomers, {+} as plasticized polymer. Other processing methods: Calendering (6 %), coating (5 %), powder methods (2 %), and other methods (9 %) such as coating, dipping, laminating, winding, thermoforming. $\dot\gamma$ = Shear rate. For abbreviations, see Chapter 15 [3].

Plastic	With external stabilization			Without external stabilization			
	Casting, dipping	Laminating, winding	Press molding	Injection molding	Extrusion	Blow forming	Thermo-forming
$\dot\gamma$ in s^{-1}	$10\text{-}10^2$	$10\text{-}10^2$	$1\text{-}10$	$10^3\text{-}10^4$	$10^2\text{-}10^3$	$10^2\text{-}10^3$	$10^2\text{-}10^3$
Thermosets from monomers							
DAP	+						
PUR	++						
EP	+	+	+				
Thermosets from prepolymers							
UP	+	++	++	+			
PF			++	+			
MF			+	+			
Semi-crystalline thermoplastics							
POM				++	+		
PA 6, 66				++	+		
PET				+	+	++	
PE, PP			+	++	++	++	+
Amorphous thermoplastics							
PC				+	+	+	+
PS				++	+	+	++
PMMA	(+)			+	++		+
PVC	{+}			+	++	++	+
Mass fraction (%):			3	31	35	9	

14.2.3. Processing of Viscous Materials

Casting requires liquid monomers (methyl methacrylate, styrene, ε-caprolactam, N-vinyl carbazole, diallyl phthalate) or prepolymers (phenolic resins, epoxy resins) which are polymerized after being poured into open forms (Fig. 14-1). Exothermic polymerizations are difficult to control. During casting, plastisols (suspensions of finely divided polymers in plasticizer) and organosols (ditto, in organic solvents) gelatinize; this process is called **fusion** (not to be confused with the melting of crystalline polymers, Section 9.4.2.). Hollow bodies are produced by **rotational casting**.

Another casting method is **hand lay-up molding (contact molding, impression molding)**. Glass fiber mats are impregnated with unsaturated polyester resins, a mixture of unsaturated polyester molecules and monomers (usually styrene or methyl methacrylate). Impregnated mats are put on a positive mold (e.g., a boat hull) and cured by free-radical polymerization whereby polyester molecules act as cross-linking agents. Pre-impregnated mats are known as **prepregs**. **Sheet molding compounds (SMCs)** are pre-impregnated pieces of glass-fiber mats which are placed in negative molds before curing. Pre-mix molding compounds with glass fibers and mineral fillers are called **bulk molding compounds (BMCs)**. Increasingly used are thermoplastics reinforced by glass-fiber mats **(GMTs)**.

Unsaturated polyester resins UP shrink on curing by (6-8) % and BMTs still by (0.2-0.4) %. Since fiber bundles are immovable in BMTs, **shrinkage** is uneven and the resulting profile will not be smooth. Shrinkage is avoided if solutions of thermoplastics are added to the resin, e.g., 25 % solutions of poly(styrene) or poly(ε-caprolactone) in styrene. A part of the styrene monomer evaporates on curing and "foams" the resin which compensates for the shrinkage caused by polymerization (see Section 3.1.4.). The added polymer is incompatible with the cured polyester and migrates in part to the surface. Foaming and demixing generate smooth surfaces; these **low-profile resins** also have improved dimensional stability.

Resin-impregnated glass fiber rovings are wound around a rotating cylinder in **filament winding** (Fig. 14-2). The cylinder is removed after curing. **Rovings** are collections of parallel strands or filaments that are assembled without intentional twist. Uniaxial semi-finished goods are also prepared by **pultrusion** from resin-impregnated glass fiber rovings. Despite its name, this process is not an extrusion since the material is pulled out and not pushed out (see below).

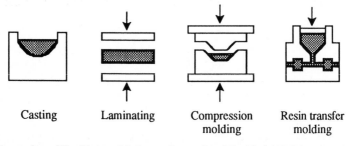

Casting   Laminating   Compression   Resin transfer
                                     molding   molding

Fig. 14-1   Processing of liquid materials by casting and molding (schematic).

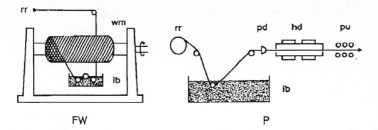

Fig. 14-2   Impregnation of rovings by filament winding FW and pultrusion P [2].
  Filament winding: Rovings rr are impregnated with the resin in a bath ib and led by a recipro-
cating fiber guide onto a winding mandrel wm. The collapsible mandrel is removed after curing
(cross-linking of the resin).
  Pultrusion: Rovings rr are drawn by tension rolls through an impregnating bath ib into a pre-
forming die pd. The resin is shaped and cured in a long heated die hd. A gap between the die hd and
the puller unit pu allows the cured laminate to cool and develop strength.

Molding processes are known in many variations and by different names. They all
have in common that powders or pelletized plastics are put into a negative mold and
heated under pressure (Fig. 14-1). **Compression molding** employs cold raw materials
and a heated mold. In **resin transfer molding**, thermosetting materials are preheated
in a transfer chamber and then forced by high pressure through sprues, etc., into the
already closed mold where they are cured. **Press molding** uses impregnated mats or
fabrics.

14.2.4. Processing of Viscoelastic Materials

Extrusion is the most important method for the processing of elastomers and pelletized thermoplastics (Table 14-1). Extruders have single or twin screws which transport, compact, melt (or soften), mix, shape, and finally extrude the melt (Fig. 14-3).
The extruded **parison** exudes unsupported. It must therefore be able to bear its own
weight until it finally solidifies. The required melt strength is provided by chain
entanglements; extrusion grade polymers must thus have high molar masses.
  Parisons swell on exiting the die block due to a positive Barus effect if melts of
polymer coils are extruded to amorphous thermoplastics (Fig. 13-2). Negative Barus
effects are observed for highly crystallizing or liquid-crystalline polymers.

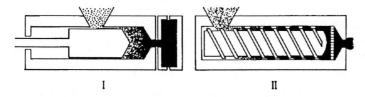

Fig. 14-3   Processing by (I) injection molding (here with a plunger (torpedo)) and (II) extrusion
(here with a single screw). Modern injection molding machines almost exclusively use screws.

Die blocks should be relatively short. A force $F_f = \pi R^2 p$ acts on a liquid if it flows through a nozzle with radius R and length L. This force is counteracted by a friction force $F_\tau = 2 \pi R L \tau$ where $\tau = \eta \dot{\gamma}$ for Newtonian liquids (Eqn.(7-14)). In the steady state, $F_f = F_\tau$ and the **Bagley equation** results:

(14-2) $p = 2 \tau (L/R) = 2 \eta \dot{\gamma}(L/R) = K'(L/R)$

For non-Newtonian liquids, one finds (Fig. 14-4)

(14-3) $p = p_o + K(L/R)$

A plot of pressure p against die geometry L/R delivers an intercept p_o at $L/R \rightarrow 0$ (Fig. 14-4). The pressure p_o is controlled by two effects: the pressure loss by the elastically stored energy of the flowing liquid and the formation of stationary flow profiles on both sides of the capillary. The higher L/R, the more pressure needs to be applied for processing.

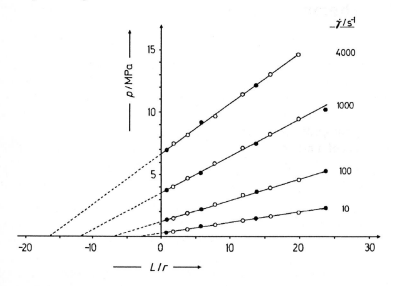

Fig. 14-4   Bagley diagram of a high-impact poly(styrene) at 189°C, using capillaries with diameters of 0.6 mm (O) or 1 mm (●) and shear rates of 10 s^{-1}, 100 s^{-1}, 1000 s^{-1}, and 4000 s^{-1} [3]. The dotted lines have no physical meaning.

In **injection molding**, pellets are melted by the combined action of external heating and internal friction caused by the transport of the material by single or twin screws. Torpedos, as shown in Fig. 14-3, are still used for special purposes.

The melted material is injected through a sprue into the cavity of a closed mold that is often slightly heated. On injection, a ca. 0.05-mm-thick skin of polymer is formed at the colder wall of the mold. The melt continues to flow in the center of the mold which generates a radial orientation of polymer segments and filler particles (if any). Crystallizable polymers must crystallize fast since otherwise shrinkage and warpage would occur. Such polymers thus require nucleating agents. The mold is then opened and the shaped article is ejected.

Several processes utilize the existence of yield values (Section 10.2.2.), especially for the fabrication of semifinished goods. Preheated films are pressed by compressed air against a mold in **blow forming (blow molding)**. In **deep drawing (drawing)**, a film is pressed by a plug against a cavity whereas in **vacuum forming** films are sucked into cavities by applied vacuum. Three-dimensional bodies are obtained analogously by **thermoforming**.

In **extrusion blow molding**, an endless tube is blown downwards through a tube extrusion die into a mold. Closing the mold delivers a hollow body with a line joint at the bottom. **Injection blow molding** works similarly, the parison is however prepared by injection molding and not by an extruder.

14.3. Thermoplastics

In discussing polymers, one should keep in mind that there is no such thing as *the* poly(styrene) or *the* poly(ethylene). Rather, there are many different grades of the same polymer which have very different properties. Examples are some grades of poly(styrene)s from the same company (Table 14-2). Some companies have more than 100 different grades of the same polymer!

The selection of a material for an application is controlled by the costs of material and processing as well as the desired use properties. Materials are rarely used because of their mass (example: fly-wheels). In most cases, one tries to use as little material as

Table 14-2 Some of the 34 grades of poly(styrene) from a single company ("Polystyrol" grades of BASF). VEF = Very easy flow, EF = easy flow, HM = high molar mass, HR = heat resistant.

Physical property	Test condition	Unit	Grade VEF	EF	HM	HR	HM-HR
Viscosity number	?	mL/g	74	96	119	96	119
Heat distortion temperature B	0.45 MPa	°C	80	82	84	98	98
Heat distortion temperature A	1.8 MPa	°C	70	72	76	86	86
Vicat temperature A	10 N	°C	88	88	92	106	106
Vicat temperature B	50 N	°C	84	84	89	101	101
Young's modulus	1 mm/min	MPa	3150	3200	3150	3200	3250
Creep modulus	1000 h	MPa	?	2300	2830	2700	2850
Tensile strength	5 mm/min	MPa	46	50	56	50	63
Fracture elongation	5 mm/min	%	1.5	2	2	2	3
Impact strength	-30 to +23°C	kJ/m^2	6	9	11	10	13
Notched impact strength	-30 to +23°C	kJ/m^2	2	2	2	2	2

possible, for example, because of the energy saved in transportation. The important property is the cost per volume, not the cost per mass as used in trading. Plastics are more expensive than metals, wood, concrete and glass on a mass basis (Table 14-3). They do compete with many metals on a volume basis.

Properties *per se* are also not the decisive criteria for the application of materials: Bodies can be made more rigid by either materials with higher Young's moduli *or* by thicker walls *or* by reinforcement of walls with ribs. The decisive factor is rather the *performance* of the material, i.e., its cost per volume *and* property (Table 14.3). On this basis, unfilled plastics cannot compete with metals with respect to rigidity but fiber-reinforced ones can (Section 14.5.). Plastics are successful competitors with metals and concrete with respect to yield stresses.

Commodity thermoplastics often have only very weak intermolecular forces between non-oriented chain segments. Their fracture strengths are low but can be in-

Table 14-3   Some properties and prices of important, unfilled materials. ρ = Density, E = tensile modulus, σ_y = yield stress, HDT = heat distortion temperature (method A), Pr = price = cost per mass, Q = cost per volume (data for 1987). * Flexural modulus; ** injection molding; *** E-glass. For abbreviations and acronyms see Chapter 15.

Materials	$\dfrac{\rho}{\text{g cm}^{-3}}$	$\dfrac{E}{\text{GPa}}$	$\dfrac{\sigma_y}{\text{MPa}}$	$\dfrac{\text{HDT}}{^\circ\text{C}}$	$\dfrac{Pr}{\$\,\text{kg}^{-1}}$	$\dfrac{Q}{\$\,\text{L}^{-1}}$	$\dfrac{10^3 Q/E}{\$/(\text{MPa L})}$	$\dfrac{10^3 Q/\sigma_y}{\$/(\text{MPa L})}$
Commodity thermoplastics								
PE-HD	0.96	1.1	32	49	1.15	1.10	1000	35
PP	0.90	1.2	36	56	1.10	0.99	830	28
PS	1.05	3.2	46	73	1.10	1.16	360	25
PVC	1.35	2.9	60	68	0.88	1.19	410	20
Technical thermoplastics								
ABS	1.07	2.9	56	93	1.39	1.49	510	27
PA 66	1.10	2.9	65	75				
POM	1.42	3.7	72	124				
PC	1.20	2.9	79	135	3.90	4.68	1600	59
PSU	1.24	26	73	203				
Thermosets								
UF	1.56	10.0	43					
PF	1.36	8.6	50	121				
EP	1.20	3.6	72	110	2.65	3.18	880	44
PI	1.40	5.0	72	243				
Metals								
Steel	7.8	210*	430			0.113	0.0006	
Al **	2.8	69*	93		2.42	6.46	88	69
Cu	8.77	122	72		2.26	19.8	160	280
Zn **	6.6	41*	157					
Miscellaneous								
Wood	0.5	11	70		0.13	0.065	6	0.9
Concrete	2.3	30	5.5		0.056	0.13	4	24
Glass ***	2.54	72	125		0.26*	0.66	9	5.3

creased by orientation of the plastics as shown in Table 14-4 for a low-density poly-(ethylene) PE-LD after uniaxial drawing to films. Semi-crystalline polymers such as PE and PP are physically cross-linked via their crystalline regions whereas their amorphous regions are rubber-like because of their low glass temperatures. Low-density poly(ethylene)s PE-LD are thus used mainly for packaging purposes (films, bags, etc.) which demand stretchability and tear resistance. The more rigid high-density poly(ethylene)s PE-HD are mainly blow-molded to hollow bodies. Poly(styrene) serves in Europe mainly for packaging purposes (50 %) and technical parts (20 %), in the United States however predominantly as expanded plastics for thermal insulation (buildings, cups, etc.).

Table 14-4 Characteristic values of glass temperatures T_G, melting temperatures T_M, heat distortion temperatures HDT (at 1.8 MPa), Young's moduli E, yield stresses σ_y, fracture strenghts σ_b, fracture elongations ε_b, and notched impact strengths I_N of some thermoplastics. NB = no break; [a] not crystalline if conventionally processed; [b] conditioned at 65 % relative humidity; ‖ in draw direction. PHBA = poly(p-hydroxybenzoate). For abbreviations and acronyms see Chapter 15.

Polymers	T_G °C	T_M °C	HDT °C	E GPa	σ_y MPa	σ_b MPa	ε_b %	I_N J m^{-2}
Commodity thermoplastics								
PS	100	–	98	3.2	–	55	2	2
PVC	82	–	68	3.3	46	48	40	16
PE-LD	– 80	115	40	0.15	8	10	600	NB
PE-HD	– 80	135	82	1.4	30	30	1200	5.2
PP	– 15	176	55	1.7	37	35	700	2.5
Engineering thermoplastics								
PC	150	[a]	132	2.2	60	65	130	NB
ABS		–	99	3.0	60		15	5
POM	– 82	181	136	3.1	69	70	40	130
PMMA	110	–	95	3.2	84	60	4	2
PA 6, dry	50	215	73	3.8	95		15	4
PA 6, conditioned[b]	≪ 50		73	1.6	60		50	12
High-performance thermoplastics								
PSU (Table 4-1)	185	–	210	2.4	85	83	60	
PPS (Table 4-1)	185	–	138	3.8		70	1.6	
PEEK (Table 4-1)	144	–	156	4.0	91	100	150	
PHBA		> 550	295	7.3		100	8	85
P(HBA-co-HNA)			180	11.0 ‖				
P(BZO), uniaxial				270 ‖		2000 ‖	0.88	
P(BZO), biaxial				34		550	2.5	

Mer HBA of
4-hydroxybenzoic acid

Mer HNA of
2,6-hydroxynaphthoic acid

Mer BZO of
cis-1,4-phenylenebenzbisoxazole

Commodity thermoplastics usually have low impact strengths. The brittle poly-(styrene) can be made more resistant against impact by copolymerization of styrene with acrylonitrile to SAN, by blending with poly(oxy-2,6-dimethyl-1,4-phenylene), or by copolymerization of styrene and various monomers in the presence of rubbers. The last group comprises:

- ABS   Styrene + acrylonitrile in presence of SBR;
- ACS   Styrene + acrylonitrile in presence of chlorinated PE, + SAN;
- AES   Styrene + acrylonitrile in presence of EPDM;
- ASA   Styrene + acrylonitrile in presence of acrylate rubber;
- MBS   Styrene + methyl methacrylate in presence of SBR;
- SMA   Styrene + maleic anhydride in presence of rubbers.

Due to polar groups and/or more rigid chain segments, *engineering thermoplastics* and *high-performance thermoplastics* have higher strengths and heat distortion temperatures than commodity plastics. They also command considerably higher prices because raw materials for their manufacture are more complex and produced in smaller amounts.

14.4. Thermosets

Thermosets are plastics that are based on chemically cross-linked polymers with high glass temperatures of relatively short network chains. They are made from monomers or oligomers (**prepolymers**) that are also designated as **reaction polymers** and ought to be called **thermosetting materials** (but not thermosetting "polymers"). Unfortunately, raw materials for thermosets are often also called thermosets.

Thermosetting materials can be liquid or solid. Powdery raw materials must have glass temperatures of at least (40-50)°C in order to avoid agglomeration on storage.

Two processes happen simultaneously on **curing**: gelation (gelling) by chemical cross-linking and vitrification by freezing-in of segmental movements. Cross-linking can occur by free-radical chain polymerization, by polycondensation and by polyaddition of multifunctional monomers and oligomers. Examples include:

- Free-radical polymerization of diallyl phthalate DAP and copolymerization of unsaturated polyesters (industrially, a *mixture* of unsaturated polyester molecules and styrene or other monomers);

- Polycondensation of formaldehyde with phenols (PF), urea (UF), or melamine (MF) or formation of polyimide thermosets (Eqn.(4-21));

- Polyaddition of epoxides with dicarboxylic acid anhydrides or multifunctional amines (EP) or isocyanates with multifunctional alcohols (PUR).

Gelling (Section 4.1.9.) is controlled by functionalities and molar proportions of all participating reactants. It starts at a certain (critical) extent of reaction and depends only slightly on reaction temperatures (slight change of the ratio of intramolecular to intermolecular bond formation with temperature). The transformation to cross-linked polymers is avalanche-like because functional groups *per polymer molecule multiply* with each step of the reaction. This gelling by chemical cross-linking of multifunctional molecules is not to be confused with the gel effect, a viscosity effect in the chain polymerization of bifunctional monomers (Section 3.7.8.).

Curing is controlled by three types of glass temperatures:

(I)   the glass temperature $T_{G,o}$ of the amorphous thermosetting material (crystalline monomers are always processed at temperatures above their melting temperatures; oligomers are amorphous at these temperatures);

(II)   the glass temperature $T_{G,gel}$ of the thermoset at the gel point;

(III)   the glass temperature $T_{G,\infty}$ of the completely cured thermoset.

Four curing regions can be distinguished according to the position of the hardening temperature T_h relative to the positions of these three glass temperatures as shown in Fig. 14-5 for thermosets by polyaddition:

(1) $T_h < T_{G,o}$: The thermosetting material remains in the glassy state during the whole experimental time (sol glass); there is no cross-linking and curing.

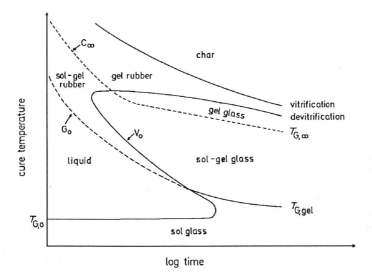

Fig. 14-5   Schematic representation of the time–temperature–transformation diagram (TTT diagram) for the time dependence of *athermal* gelling, vitrification, deglassing, and char formation of thermosetting materials. At cure temperatures below the glass temperature $T_{G,o}$ of the prepolymer, the glassy material is slowly transformed into a sol glass. The line - - - G_o indicates the onset of gelation at temperatures above $T_{G,o}$. Vitrification starts above the line ----- V_o. Full cure is obtained at temperatures above the line - - - C_{∞}..

The prepolymer exists as a sol glass at low temperatures. On increased heating at *short heating times*, the sol glass of the prepolymer first liquifies. It then reacts and finally cross-links (gels) at a cure temperature above G_o where it becomes a sol-gel-rubber. The material is completely reacted at temperatures above C_{∞} where it becomes a gel rubber. At still higher temperatures, it chars.

Increasing heating temperatures at *long heating times* first convert the sol glass into a sol-gel glass at temperatures above the glass temperature $T_{G,gel}$. Residual prepolymer (sol) is completely reacted at temperatures above $T_{G,\infty}$. The material becomes a gel glass. At still higher temperatures, movements of chain segments become so strong that the completely cured glass first devitrifies, then degrades, and finally chars. With permission by Wiley Publ. [4].

(2) $T_{G,o} < T_h < T_{g,gel}$: Thermosetting materials convert into prepolymers with increasing time. Prepolymers are liquid at low curing temperatures but vitrify with increasing extent of reaction (increasing polymer concentration *and* higher molar mass).

(3) $T_{G,gel} < T_h < T_{G,\infty}$: The polymerization progresses to the gel point where for the first time a cross-linked gel is formed in a sol (branched oligomers and unreacted monomers). The lower the curing temperature, the earlier the gel-sol mixture solidifies to a glass. At high curing temperatures, polymer segments remain mobile. The gel-sol mixture continues to react. Finally, a gel glass is formed that consists of the completely cross-linked polymer.

(4) $T_{G,\infty} < T_h$: The gel point is rapidly surpassed. The gel can however only exist as an elastomer since it is at a temperature above the glass temperature $T_{G,\infty}$ of the fully cured thermoset. A sol-gel elastomer is formed first and later, after complete chemical reaction, a gel elastomer. Because of the high temperatures, the gel elastomer is heterolytically cleaved. The resulting radicals cause transfer and cross-linking reactions with the network chains. The gel elastomer is increasingly cross-linked. It finally becomes a glass that starts to char.

Thermosets by polycondensation (PF, MF, UF, PI) generate low molar mass leaving molecules during curing (in most cases, water). These molecules are volatile at high curing temperatures but cannot escape because of the high viscosity of the polymerizing material; they form voids. In order to reduce void formation, a precondensation is carried out to a so-called **B-stage** just before gelation. B-stage oligomers have less functional groups; the subsequent curing to the C-stage delivers less water vapor and less voids. Bubble formation is also prevented by addition of water-absorbing materials to the reacting substance, for example, wood flour (wood meal). No glasses are formed by thermosets that are based on formaldehyde (PF, MF, UF) since they start to decompose at temperatures below their glass temperatures $T_{G,\infty}$.

No leaving molecules are generated by polyaddition (EP, PUR); the viscosities of the thermosetting materials are also already high. A B-stage is therefore unnecessary and one can react thermosetting EP and PUR directly to thermosets.

For thermosets by *chain polymerization*, B-stages are unnecessary for the highly viscous unsaturated polyester resins (UP). They are required, however, for the low-viscosity allyl monomers (DAP, etc.).

The cross-linking of thermosetting materials proceeds at random; it dictates the structure of networks and reduces the mobility of chain segments. Diffusion becomes increasingly difficult and some functional groups do not find reactive partners during the normal course of reaction. These residual groups can undergo after-reactions, however. Even if the materials have been cooled to room temperature, **green strengths** of thermosets still increase albeit very slowly.

"Green strength" of thermosets refers to the strength immediately *after* curing but "green strength" of rubbers to strengths *before* vulcanization to elastomers.

Unreacted functional groups of thermosets can be attacked by atmospheric agents (water, oxygen, etc.) which reduces the weatherability. Some special purpose thermosetting materials can thus be cross-linked only via end groups. An example are so-called **vinyl ester resins** which are not vinyl esters but bifunctional macromonomers with long residues between (meth)acryl end groups, e.g.,

$$CH_2=C-CO-OCH_2CH_2O-\underset{}{\bigcirc}-\underset{\underset{CH_3}{|}}{\overset{\overset{CH_3}{|}}{C}}-\underset{}{\bigcirc}-OCH_2CH_2O-CO-\underset{\underset{CH_3}{|}}{C}=CH_2$$

The many short network chains between junctions (high cross-linking density) lead to good creep resistances and improved heat distortion temperatures of all thermosets, especially filled and reinforced ones (Table 14-5). Disadvantages of thermosets are the often long processing times and the irreversible cross-linking.

Table 14-5 Characteristic values of unfilled thermosetting materials cured by chain polymerization (CPM), polycondensation (PCD), or polyaddition (PAD). ρ = Density, Δw = moisture uptake after 24 hours, ΔV = shrinkage, HDT = heat distortion temperature at 1.8 MPa, E = Young's modul, σ_B = fracture strength, ε_B = elongation at break, I_N = notched impact strength.

Polymer		ρ g cm^{-3}	Δw %	ΔV %	HDT °C	E GPa	σ_B MPa	ε_B %	I_N J m^{-1}
DAP	CPM	1.27	0.2	1.0	155	2.2	28		17
UP	CPM	1.3	0.4	0.6	130	3.4	70	2	16
PF	PCD	1.25	0.15	1.1	121	2.8	65	1.8	16
MF	PCD	1.48		0.7	148				
UF	PCD			1.3					
EP	PAD	1.2	0.13	0.5	170	2.5	70	6	35
PUR	PAD	1.05	0.2	1.0	91				21

14.5. Filled and Reinforced Plastics

Polymer composites are three-dimensional working materials. They consist of continuous polymeric matrices and materials of higher modulus (fillers) imbedded therein. **Fillers** may be minerals, glass, metals or polymers; they may be spherical, spheroidal, platelets, fibers, or fabrics.

Some fillers are added to improve economics (**inactive fillers, extenders**) and others to improve certain mechanical properties (**active fillers, reinforcing agents**). Inactive fillers are usually corpuscular particles with axial ratios of ca. 1 (chalk, glass spheres, metal powders, etc.). Active fillers have axial ratios considerably greater than unity; they may be short fibers (e.g., glass fibers), long fibers (glass fibers, polymer fibers, carbon fibers, etc.), or platelets (kaolin, talcum, mica, etc.). No sharp dividing line exists between inactive and active fillers. The term "**reinforcement**" is also not defined; it may be increased fracture strength, improved impact strength, etc.

14.5.1. Mixing Rules

Properties of composites change with the proportion of the filler F. The appropriate measure of concentration is the volume fraction ϕ_F and not the mass fraction w_F since all physical properties depend on volume requirements of components and not on their mass. The volume fraction $\phi_F \equiv 1 - \phi_M$ of the filler can be calculated from the mass fraction w_F and the densities ρ_F (filler) and ρ (composite) if volumes are not additive [$\phi_F = w_F(\rho/\rho_F)$]. The density of the composite need not be known for additive volumes; densities ρ_M (matrix) and ρ_F (filler) suffice:

$$(14\text{-}4) \quad \phi_F = \frac{w_F}{w_F + w_M(\rho_F/\rho_M)}$$

In the simplest cases, properties vary with composition according to one of the three mixing rules, for example, the modulus E:

$$(14\text{-}5) \quad E^n = (E_M)^n \phi_M + (E_F)^n \phi_F$$

where n may adopt values of $+1$ or -1. These rules carry various names in different academic disciplines (Table 14-6).

Table 14-6   Names of mixing rules. * After some mathematical manipulation.

Discipline	Exponent in Eqn.(14-5)		
	n = 1	(n → 0)*	n = – 1
Mathematics	Arithmetic mean	Geometric mean	Harmonic mean
Chemical engineering	Rule of mixtures	Logarithmic rule	Inverse mixing rule
Mechanical engineering	Voigt model	–	Reuss model
Electrical engineering	Parallel	–	Series
Materials science	Upper bound	–	Lower bound

14.5.2. Moduli

In a simple case, stress is taken up by a matrix and transferred to *long fibers*. Matrix and fibers are assumed to have the same Poisson ratios; they are also supposed to deform only elastically and not plastically. Matrix and fibers should be in good physical contact though they need not interact via chemical or physical bonds. Slippage should be absent. These conditions are fulfilled for elastic moduli since moduli are determined at vanishingly small stresses below the proportionality limit, i.e., in the Hookean range with $\sigma_{\parallel} = E\varepsilon$ (Section 10.2.2.).

Infinitely long fibers with Young's moduli E_F are assumed to lie parallel in a matrix with the modulus E_M. If the composite is elongated in the longitudinal fiber di rection by $\varepsilon = (L - L_o)/L_o$, the force $F = \sigma_{\parallel}A = E_{\parallel}\varepsilon A$ at the cross-sectional area A of the composite is distributed among the cross-sectional areas A_F and A_M of the two components according to $F = F_M + F_F = \sigma_M A_M + \sigma_F A_F = \varepsilon E_M A_M + \varepsilon E_F A_F$. The cross-sections of fibers are perpendicular to the fiber axes; the volume fractions of the components are therefore $A_M/A = \phi_M$ and $A_F/A = \phi_F$. Insertion of $\phi_M + \phi_F \equiv 1$, $A_M + A_F = A$ and $F/A = E_{\parallel}\varepsilon$ delivers the rule of mixtures, i.e., a linear increase of the longitudinal modulus $E_{\parallel}$ of the composite with increasing fiber fraction ϕ_F (Fig. 14-6):

(14-6) $E_{\parallel} = E_M \phi_M + E_F \phi_F = E_M + (E_F - E_M)\phi_F$

Fiber and matrix are subjected to the same stress $\sigma = \sigma_F = \sigma_M$ if the stress is perpendicular to the fiber direction. Neither stresses σ nor moduli E are additive but the proportions of elongations ε and stress compliances $D = 1/\sigma$ are. With $\varepsilon = \varepsilon_M \phi_M + \varepsilon_F \phi_F = (\sigma_M/E_M)\phi_M + (\sigma_F/E_F)\phi_F = \sigma/E_{\perp}$, one obtains the inverse rule of mixtures:

(14-7) $1/E_{\perp} = (\phi_M/E_M) + (\phi_F/E_F) = D_M \phi_M + D_F \phi_F = D_{\perp}$

Long fibers are difficult and expensive to work into plastics (see hand lay-up, filament winding, pultrusion, etc.). Composites with short fibers can however be processed with conventional machines; they are thus preferred fillers for many plastics. The efficiency of short fibers depends not only on modulus and orientation, like that of long fibers, but also on their aspect ratio (ratio of length to diameter).

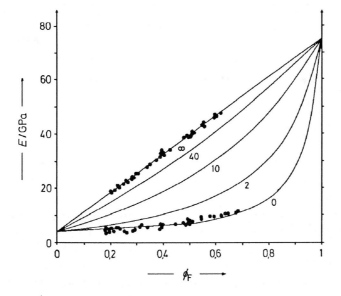

Fig. 14-6 Dependence of Young's moduli E on the volume fraction ϕ_F of fillers for various values of the parameter K_1 of the Halpin–Tsai Eqn.(14-8). $K_1 = \infty$ corresponds to the simple rule of mixtures, $K_1 = 0$ to the inverse rule of mixtures. $K_1 = 2$ applies to the modulus of spherical fillers. ● Experimental longitudinal moduli $E_{\parallel}$ ($K_1 = \infty$) and transverse moduli $E_{\perp}$ ($K_1 = 0$) of a composite of unidirectional long glass fibers in a cured unsaturated polyester resin [5].

The **Kerner model** tries to assess the effect of these quantities on the modulus. It models the composite as fibers F in a cylinder of the matrix M; the cylinder is imbedded in matter with the properties of the composite itself. The matrix calculation for stresses and elongations acting in various directions leads, after some simplifications (e.g., equal Poisson numbers of F and M), to the **Halpin–Tsai equation**, which is usually written as:

$$(14\text{-}8) \qquad \frac{E}{E_M} = \frac{1 + K_1 K_2 \phi_F}{1 - K_2 \phi_F} \quad ; \quad K_2 = \frac{(E_F/E_M) - 1}{(E_F/E_M) + K_1}$$

and can be easily transformed into a more transparent expression:

$$(14\text{-}9) \qquad E = \frac{E_F + K_1(E_F \phi_F + E_M \phi_M)}{K_1 + E_F \left(\dfrac{\phi_F}{E_F} + \dfrac{\phi_M}{E_M} \right)}$$

K_1 is an empirical constant that is controlled by the shape and geometry of packing of fillers and the direction of the load relative to the orientation of anisotropic fillers, among other factors. $K_1 = 2\,L_F/d_F$ for longitudinal and transverse Young's moduli if plastics are unidirectionally reinforced by short fibers or platelets of length L_F and thickness d_F whereas $K_1 = 3^{1/2} \log_{10}(L_F/d_F)$ for shear moduli.

The Halpin–Tsai equation delivers the upper limit $E_\parallel$ of the rule of mixtures for $K_1 = \infty$ and the lower limit $E_\perp$ for $K_1 = 0$. It can not describe synergistic or antagonistic effects. **Synergism** is defined differently: either as $E > E_\parallel$ (Young's moduli always greater than the upper limit) or $E > E_F$ (Young's moduli always greater than the modulus of the filler; the function $E = f(\phi_F)$ has a maximum of E). **Antagonism** is present if either $E < E_\perp$ or $E < E_M$, depending on definition.

14.5.3. Interlayers

Phases M and F are usually not in direct contact; they are rather connected by **interlayers** Z which have a different physical (and sometimes also chemical) structure. Contrary to the Kerner model, filler particles F are surrounded by Z (and not by M) and imbedded in M (and not in the composite V). Interlayers may have thicknesses d between a few atomic diameters (ca. 0.5 nm) and some micrometers. Only sufficiently thick interlayers fulfill thermodynamic requirements for phases ($d > 5$ nm) and can be truly called **interphases**.

For example, many fillers absorb water which may form interlayers that are up to three molecular diameters thick (ca. 0.5 nm). This water can react with prepolymers or the polymeric matrix. In carbon-fiber filled epoxy resins, epoxy groups react with water molecules absorbed on fibers to give hydroxyl groups. The OH groups are unable to react with amine curing agents in the cold-curing of epoxy resins. Only plasticizing, low molar mass compounds are produced near the fiber surface. The loss of cross-linkable epoxy groups also generates only wide-mesh networks near the fiber surface. The resulting interlayer is about 500 nm thick and much more ductile than the rigidly cross-linked epoxy matrix. The effect disappears if carbon fibers are coated with cured phenolic resins.

Fillers can also act as nucleating agents. Poly(*p*-phenylene terephthalamide) fibers in polyamide 66 as matrix nucleate trans-crystallizations of PA 66 on the fiber surface. The thickness of these epitaxially grown PA 66 layers varies between 5 μm and 23 μm. The layers are true interphases because their physical structure is quite different from that of the partially crystallized matrix.

Composites of thermoplastics or rubbers with corpuscular fillers often have **bound polymers**, i.e., polymers that cannot be extracted by solvents and remain "bound" to the filler. In carbon-black-filled styrene–butadiene rubbers, these interlayers are ca. 3.5 nm thick; according to dynamic-mechanical measurements, they appear glassy. The interlayer is ca. 2 nm for pyrogenic SiO_2 in poly(ethylene); it probably consists of absorbed poly(ethylene) chains in brush-like conformations.

The surfaces of fillers and/or interfaces are modified industrially by **coupling agents**. Glass fibers are treated with vinyl triethoxysilane $CH_2=CHSi(OC_2H_5)_3$ which presumbly first hydrolyzes to $CH_2=CHSi(OH)_3$ in the presence of primary or secondary amino groups. This compound then forms covalent bonds with silanol groups on the surface of the glass fibers, resulting in $>Si\text{-}O\text{-}Si(OH)_2CH=CH_2$ which can then react with the matrix (e.g., unsaturated polyesters) to form covalent bonds.

14.5.4. Strengths and Elongations

Composites of *long fibers* in polymeric matrices should exhibit a maximum fracture strength (upper limit) of:

(14-10) $\sigma_{b,\parallel} = (E_M\phi_M + E_F\phi_F)\varepsilon_{b,C}$; $\varepsilon_{b,C}$ = fracture elongation of the composite C

according to Hooke's law and the simple rule of mixtures if stresses are in the longitudinal fiber direction and both phases are deformed with equal elongations until the more brittle phase fractures first (usually the fiber). Experimental tensile strengths are however only one half to one third of those predicted by Eqn.(14-10).

The lower limit is obtained if stresses are perpendicular to fiber axes, elongations are additive, and fracture occurs if stresses approach the strength of the weakest component, usually the matrix or the interlayer. As a safety margin, only properties of the matrix are often considered in engineering. Even this safety margin may be too high, however, if the fiber does not adhere to the matrix, for example, because of voids.

Ends of *short fibers* cannot transfer stresses from the matrix to the fiber; fiber ends act as stress concentrators. Tensile stresses that are parallel to the long axes of fibers do not jump at the fiber end from zero to the maximum stress σ_{max}, however, but increase gradually over a distance $L_c/2$ at each fiber end (Fig. 14-7). Since the matrix

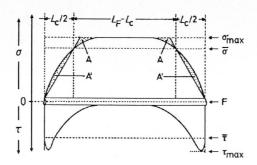

Fig. 14-7   Distribution of tensile stresses σ and shear stresses τ along a fiber F with length L_F and critical length L_c. See text for definitions of mean and maximum tensile and shear stresses.

(lower E_M) deforms more strongly than the fiber (higher E_F), shear stresses appear at the interface of matrix and fiber. These shear stresses lead to tensile stresses that are different from those at the matrix and the fiber themselves.

It is usually assumed that the average tensile stress $\overline{\sigma}_{M,C}$ at the matrix in the composite C is identical with the tensile stress σ_M of the pure matrix itself. The average tensile stress $\overline{\sigma}_{F,C}$ of the fiber in the composite is however smaller than the tensile stress σ_F of the fiber.

The distribution of tensile stresses along the fiber of length L_F can be approximated by an isosceles trapezoid with height σ_F and widths L_F and $L_F - L_c$. Tensile stresses are zero at each fiber end. They increase linearly from 0 to σ_F between fiber ends and $L_c/2$, attain a constant value of $\sigma_{max} = \sigma_F$ at $L_F - L_c$, and finally decrease linearly to zero along $L_F - (L_c/2)$ at the other fiber end. Since the two areas A and A' under the curves must be equal, average tensile stresses at the fiber become $\overline{\sigma}_{F,C} = \overline{\sigma} = \sigma_F[L_F - (L_c/2)]$. Insertion of the simple rule of mixtures and the average tensile strength $\overline{\sigma}_{M,C} = \sigma_M$ at the matrix delivers the tensile strength $\sigma_{b,C}$ of the composite:

$$(14\text{-}11) \quad \sigma_{b,C} = \overline{\sigma}_{M,C}\phi_M + \overline{\sigma}_{F,C}\phi_F = \sigma_M\phi_M + \sigma_F\{1 - [L_c/(2\,L_F)]\}\,\phi_F$$

The **critical fiber length** L_c is the length of fibers which can be pulled out of the composite without breaking the fiber. It can be calculated from the equilibrium between forces. Tensile forces act across the cross-sectional area $A_F = \pi r_F^2$ of the fiber whereas shear forces are distributed across the cylindrical surfaces $A_E = 2\,\pi r_F(L_c/2)$ of fiber ends. The tensile force $F_\sigma = \sigma_F A_F$ along the fiber must equal the shear force $F_\tau = \tau_{F,M}A_E$ along the fiber–matrix interface. The equilibrium condition $F_\sigma = F_\tau$ thus leads to the critical fiber length $L_c = \sigma_F r_F/\overline{\tau}_{F,C}$.

The same approach can be used to calculate the equilibrium of forces for fiber lengths that are smaller than the critical ones. The average tensile strength of fibers is $\overline{\sigma}_{F,C} = \overline{\tau}_{F,C}L_F/r_F$. The tensile strength of composites thus increases with increasing fiber aspect ratio L_F/d_F (Fig. 14-8; fiber diameter $d_F = 2\,r_F$):

$$(14\text{-}12) \quad \sigma_b = \sigma_M\phi_M + \overline{\sigma}_{F,C}\phi_F = \sigma_M\phi_M + [(\overline{\tau}_{F,C}L_F)/r_F)]\phi_F$$
$$= \sigma_M\phi_M + 2\,\overline{\sigma}_{F,C}(L_F/d_F)\phi_F$$

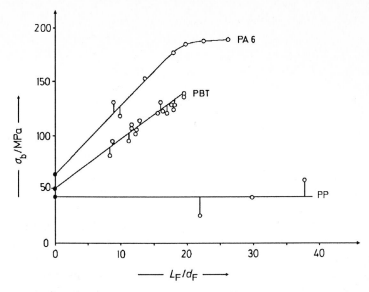

Fig. 14-8 Tensile strength σ_b as function of aspect ratio L_F/d_F of fibers in glass-fiber-reinforced poly(ε-caprolactam) PA 6, poly(butylene terephthalate) PBT and it-poly(propylene) PP (after data of [6]. Lines are empirical; ● tensile strengths of unfilled matrices. It is unclear whether the tensile strength of poly(propylene) is indeed independent of the aspect ratio (as shown) or passes through a minimum before it increases linearly with increasing L_F/d_F.

Strengths of composites with *particulate fillers* depend on very many parameters. Tensile strengths of composites with the same matrix and equal weight fractions of fillers increase generally from spheroids (three-dimensional) to platelets ("two-dimensional") and fibers ("one-dimensional") (Table 14-7); unfortunately, no systematic experiments exist for the more appropriate *volume* fractions. Tensile strengths are effected only slightly by spheroids and platelets but considerably by fibers. All fillers decrease tensile strengths of composites of poly(vinyl chloride) PVC-S from suspension polymerization, probably because polymer aggregates formed during polymerization break up during the working-in of fibers.

Fillers can be elongated far less than polymers, however. The *extensibility* of a composite is therefore far smaller than that of the matrix. Filler surfaces also constitute barriers for the coil molecules of the matrix. Coils are forced to adopt less probable macroconformations near the surfaces (see Section 5.5.3.); they become more rigid and the extensibility drops even more. At similar surface structures, extensibilities are decreased the more, the larger the specific surfaces (cf. Table 14-7: glass spheres vs. glass fibers and the series chalk–talcum–mica–asbestos.

14.5.5. Thermal Properties

Thermal expansion coefficients of composites are lower than those of matrices for the very same reasons as extensibilities. Fillers with negative thermal expansion coefficients may even reduce thermal expansion coefficients of composites to zero if they are applied in appropriate concentrations. Examples of fillers with such expansion coefficients are carbon fibers and the mineral eucryptite ($LiAl[SiO_4]$). Smaller expansion coefficients reduce shrinkage after processing.

Table 14-7 Ratio $\sigma_b/\sigma_{b,M}$ of tensile strengths of composites with $w_F = 30$ wt.-% (* at 40 wt.-%, ** at 20 wt.-%) of fillers to those of the pure matrix [7]. $w_F = 0.3$ corresponds to $\phi_F \approx 0.15$ for ρ_M = 1 g/cm³ and $\rho_F = 2.5$ g/cm³. Asbestos is chrysotile. Axial ratios refer to fillers *before* compounding. For example, short glass fibers are broken during processing; their axial ratios after processing are only 10-20.

Property of filler	Spheroids		Platelets		Fibers	
	Glass	Chalk	Talcum	Mica	Asbestos	Glass
Young's modulus E_F in GPa	72	26	20	30	145	73
Density ρ_F in g/cm³	2.48	2.71	2.8	2.82	2.5	2.49
Axial ratio L_F/d_F	1	1-5	4-15	30-100	17	100
Specific surface in m²/g	0.3	1-15	6-17	30	40	1

Polymer matrix				$\sigma_b/\sigma_{b,M}$ of polymers filled with 30 wt.-%					
Type	$\dfrac{E_M}{MPa}$	$\dfrac{\sigma_{b,M}}{MPa}$	$\dfrac{\varepsilon_{b,M}}{\%}$	Glass	Chalk	Talcum	Mica	Asbestos	Glass
PE-LD	210	10	500	1.00	1.60*	1.60	1.30	2.00	2.40
PA 6	1200	64	220	1.01	0.78	-	-	1.92	2.31
PA 12	1200	60	270	0.75	-	-	-	-	1.38
PE-HD	1400	27	> 550	-	0.74	-	-	1.11	2.22
PA 66	1500	63	180	1.29	-	-	0.62	2.11	2.43
PP	1600	31	620	-	0.81*	0.97*	0.97	1.45*	1.58
POM	2700	63	45	0.67**	0.83**	0.97	-	-	1.65
PVC-S	2700	60	8	-	0.77	0.57**	-	0.55	0.42*
PBT	2800	60	120	0.83**	1.03	-	-	-	2.17
PS	3800	55	4	-	0.27	0.71*	-	1.29**	1.73

Polymer matrix				$\varepsilon_b/\varepsilon_{b,M}$ of polymers filled with 30 wt.-%					
Type	$\dfrac{E_M}{MPa}$	$\dfrac{\sigma_{b,M}}{MPa}$	$\dfrac{\varepsilon_{b,M}}{\%}$	Glass	Chalk	Talcum	Mica	Asbestos	Glass
PE-LD	210	10	500	0.15	0.44*	0.08	0.092	0.040	0.089
PA 6	1200	64	220	0.091	0.14	-	-	0.014	0.016
PA 12	1200	60	270	0.093	-	-	-	-	0.022
PE-HD	1400	27	> 550	-	< 0.016	-	-	< 0.018	< 0.004
PA 66	1500	63	(180)	0.028	-	-	-	-	0.017
PP	1600	31	620	-	0.29*	0.017*	-	0.011*	0.008
POM	2700	63	45	0.016**	0.11**	0.067	-	-	0.089
PVC-S	2700	60	(8)	-	1.0	0.75**	-	0.15	0.38*
PBT	2800	60	120	0.83**	1.03	-	-	-	0.46
PS	3800	55	4	-	0.5	0.4*	-	-	0.5

Static glass temperatures of amorphous polymers and *melting temperatures* of semi-crystalline polymers do not vary appreciably on addition of fillers since only a small proportion of macroconformations is changed (only those near filler surfaces).

The situation is quite different for *heat distortion temperatures* HDT (Section 9.2.1.) since they are determined by a kind of creep experiment (Section 10.5.4.). HDTs of composites from amorphous polymers depend only slightly on loads; at 30 wt.-% of fillers and 1.82 MPa (method A) they are somewhat larger than static glass temperatures (polyethersulfone 174°C *vs.* PSU–kaolin 177°C; poly(styrene) 86°C *vs.* PS–glass fibers 93°C). Small (positive or negative) effects are also observed for spheroidal fillers in semi-crystalline matrices.

Much stronger effects of type and proportion of fillers and of applied loads are found for platelets and fibers as fillers. The heat distortion temperature of it-poly-(propylene) (62°C) in presence the of 40 wt.-% filler increases to 73°C (kaolin), 95°C (talcum), and 120°C (glass fibers). An addition of 30 wt.-% glass fibers to polyamide 6 increases the HDT to 208°C from 80°C, probably because of the presence of trans-crystalline interphases.

14.6. Polymer Blends

Polymer blends are homogeneous or heterogeneous mixtures of two chemically different polymers. Some blends are prepared for economical reasons, most however because they usually have higher impact strengths than their components. Since the price per property (see Section 14.3.) of blends is often lower than that of their components, some polymer blends can even command higher prices per mass than their components (Table 14-8). About 10 % of all thermoplastics and 75 % of all elastomers are polymer blends.

Table 14-8 Properties of commercial polymer blends and their components. E = Young's modulus, σ_b = tensile strength, a_n = impact strength, HDT = heat distortion temperature.

Components	$\dfrac{E}{\text{GPa}}$	$\dfrac{\sigma_b}{\text{MPa}}$	$\dfrac{a_n}{\text{J·m}^{-1}}$	$\dfrac{\text{HDT}}{\text{°C}}$	Rel. cost
ABS: Poly(acrylonitrile-*co*-butadiene-*co*-styrene)	2.06	35	320	86	1.00
PC: Polycarbonate of bisphenol A	2.41	66	800	132	1.53
ABS + PC	2.55	43	530	105	1.19
PVC: Poly(vinyl chloride)	2.76	48	< 530	66	1.00
PMMA: Poly(methyl methacrylate)	3.10	69	25	85	2.08
PVC + PMMA	2.34	45	800	75	2.44
PS: Poly(styrene)	2.06	33	64	88	1.00
PPE: Poly(2,6-dimethyl-1,4-oxyphenylene)	2.55	72	85	192	4.10
PS + PPE	2.41	66	160	129	1.31

Polymers are usually not miscible in the industrially important concentration ranges (Section 6.2.3.). Such thermodynamically immiscible polymer blends may still be compatible, however; ie., their properties may behave as if the blends were miscible. Compatibilities are promoted by **compatibilizers** (Section 8.5.) which may be diblock copolymers or graft copolymers. Compatibilizers are often added to polymer blends; they may however also be generated *in situ* by grafting during melt mixing of polymers.

14.6.1. Mechanical Mixing

Polymer blends are prepared by mechanical mixing of melts, latices or solutions of different polymers or by polymerization of a monomer in the presence of another polymer. Highly viscous *melts* can be mixed thoroughly only at high temperatures and by strong shear fields. These conditions cause polymer chains to cleave homolytically. The resulting polymer radicals undergo transfer and graft reactions which generate graft copolymers that act as compatibilizers.

Viscosities and volume fractions of components determine which of the two components of a polymer blend enters the continuous phase. In the first approximation, one obtains:

$$(14\text{-}13) \quad Q = \frac{\eta_1}{\eta_2} \cdot \frac{\phi_2}{\phi_1}$$

The continuous phase is formed from component 1 if $Q < 1$ and from component 2 if $Q > 1$. The component with higher viscosity and/or lower concentration thus enters the dispersed phase if all other conditions are equal.

No polymer chains are split and no compatibilizers are formed in *solution* or *latex mixing*. Latices have very low viscosities (Fig. 3-13) and are thus easy to mix. The subsequent coagulation leaves the intimate mixture intact but the material has then to be melted which presents similar problems as in melt mixing.

Thickening of solutions of polymers 1 and 2 in a common solvent results in single-phase or multi-phase blends, depending on interaction parameters $\chi_{i,j}$. Single-phase blends are obtained if on evaporating (thickening) no two-phase regions are encountered in the total concentration and temperature range. Such two-phase regions are present at sufficiently large differences $\chi_{31} - \chi_{32}$. The initially single-phase solution begins to demix at a certain polymer concentration and forms two phases. This 2-phase region is abandoned on further thickening but the microphases are now too large and the diffusion coefficients too small for a dissolution of the phases: the thermodynamically miscible blend remains 2-phase for kinetic reasons.

Only a few commercial blends of two thermoplastics are single-phase blends:
– poly(styrene) PS plus poly(2,6-dimethyl-1,4-phenylenoxide) PPE (PPO®);
– poly(vinyl chloride) PVC plus poly(methyl styrene-*co*-acrylonitrile) MeSAN;
– poly(vinyl chloride) PVC plus chlorinated poly(ethylene) C-PE.

All single-phase blends possess negative or slightly positive interaction parameters χ (Section 6.2.3.). They are amorphous; their glass temperatures vary monotonically with composition. The amorphous blend of poly(vinyl chloride) and poly(methyl methacrylate) is compatible but not thermodynamically miscible (see Table 14-8).

Many industrial blends are composed of an amorphous and a semi-crystalline polymer. Most of these blends are compatible. Only the blend from poly(vinylidene fluoride) and poly(methyl methacrylate) is thermodynamically miscible. Films from this blend are optically clear because regions of different refractive indices are absent.

Blends of two semi-crystalline polymers are rarely used. Components of these blends have usually very similar chemical structures. Examples are it-poly(propylene) + it-poly(1-butene) and poly(ethylene terephthalate) + poly(butylene terephthalate) (see Section 8.3.3.).

14.6.2. In-situ Polymerizations

In-situ polymerizations generate polymer blends by polymerization of a monomer A in the presence of a smaller proportion of polymer B. In the most important industrial blends, polymer B is a graftable and cross-linkable diene rubber and monomer A leads to a thermoplastic. The blends are thermodynamically immiscible.

Initially present is a solution of polymer B in excess monomer A. An example is the free-radical polymerization of solutions of 8 % poly(butadiene) in styrene. In the first stage, poly(styrene) is produced in the presence of unchanged poly(butadiene). Since poly(styrene) and poly(butadiene) are not miscible, poly(styrene) domains are formed in poly(butadiene) solutions at fairly small poly(styrene) concentrations. The further fate of the system depends on whether it is stirred or not.

Non-stirred systems lead to a grafting of styrene chains on poly(butadiene) chains which are finally cross-linked. Poly(styrene) domains grow and are embedded in the network of grafted and cross-linked poly(styrene). It results in a dispersion of poly-(styrene) particles in a continuous network of cross-linked poly(butadiene).

On stirring, phase separation occurs if styrene conversions exceed ca. (9-12) %, i.e., if the concentration of newly formed poly(styrene) becomes comparable to the concentration of the initially present poly(butadiene). The system converts to a dispersion of grafted and cross-linked poly(butadiene) domains in a continuous poly-(styrene) matrix. Smaller initial rubber concentrations lead to lower monomer conversions at which the phase reversal occurs.

Such blends are known as **rubber-modified plastics**. They are produced by in-situ polymerizations or by melt mixing. In-situ polymerizations are employed for

- (5-20) % solutions of styrene–butadiene rubber in styrene (delivers impact-resistant poly(styrene) IPS);
- solutions of cis-poly(butadiene) in styrene (generates high-impact poly-(styrene) HIPS);
- acrylonitrile–butadiene rubber in styrene + acrylonitrile (produces ABS);
- acrylic rubber in styrene +acrylonitrile (leads to ACS).

All other important industrial rubber-modified plastics are made by melt mixing. These blends always require added compatibilizers whereas in-situ polymerizations generate their own compatibilizers by grafting. Examples for melt mixing are
- NBR-modified PVC: the Q and e values (Section 3.8.5.) of vinyl chloride are unfavorable for a free-radical grafting of VC on nitrile rubber NBR.
- Saturated rubbers are difficult to graft. Examples are butyl rubber-modified poly(ethylene) and acrylic rubber-modified poly(methyl methacrylate).
- EPDM-modified it-poly(propylene); propylene can only be polymerized by Ziegler–Natta catalysts which react with EPDM.
- ABS-modified polycarbonate A; the thermoplastic component can only be produced by polycondensation.

14.6.3. Rubber-Modified Thermoplastics

Rubber-modified plastics consist of (5-20) % dispersions of rubber particles in plastics, mainly thermoplastics and sometimes thermosets. Very different morphologies and mechanical properties are obtained (Fig. 14-9), depending on blend formation and proportions of matrix and rubber. The addition of rubbers increases tensile strengths and notched impact strengths of plastics; it decreases Young's moduli and fracture elongations. Plastics become tougher although the glass temperature of the matrix does not change. Added rubbers are therefore also called **impact modifiers** and the rubber-modified plastics also **impact-modified plastics**.

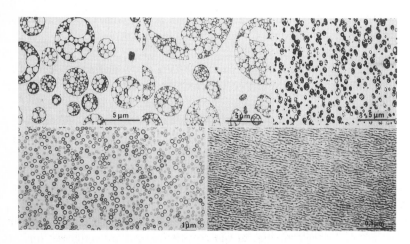

Fig. 14-9 Morphologies of different rubber-modified poly(styrene)s.
Top left: conventional impact-resistant poly(styrene).
Top center: high impact poly(styrene) with increased resistance against stress crazing.
Top right: high impact poly(styrene) with high surface gloss.
Bottom left: capsule morphology of high impact poly(styrene) with increased transparency.
Bottom right: finely divided rubber phase in glass-clear high-impact poly(styrene).
By permission of Hanser Publishers [8].

Homopoly(styrene) can only be elongated by ca. 2 % (Table 14-4) but rubber-modified poly(styrene) by ca. 15 %. The higher deformation and the improved resistance against impact are not due to the rubber particles *per se* since soft phases can absorb considerable energy only near their glass temperatures (Section 10.5.7.) and the T_G of the rubber is far lower. This energy absorption is furthermore small; it can rarely effect the response to stresses.

Impact strengths vary with the size of rubber particles as shown in Fig. 14-10 for rubber-modified polyamide 66. All specimens with notched impact strengths $a_n \leq$ 200 J/m were brittle and all with notched impact strengths $a_n \geq 500$ J/m were ductile. Intermediate values $200 \leq a_n/(\text{J m}^{-1}) \leq 500$ were not observed. The transition brittle $\rightarrow$ ductile must therefore occur at a certain critical value. Since the efficiency of impact modifiers depends on both the volume fraction ϕ_F and the size of the rubber particles (surface!), the critical parameter must be the critical distance L_{crit} between the surfaces of rubber particles.

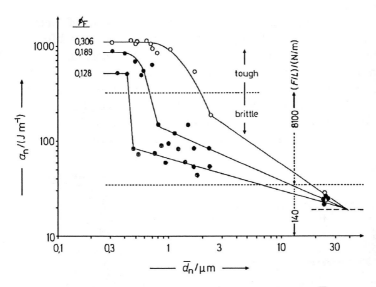

Fig. 14-10   Notched impact strength a_n as function of the number-average $\bar{d}_n$ of the diameter of the rubber particles in a rubber-modified polyamide 66 [9]. ------- Empirical functions; — — — impact strength of matrix (a_n = 18.6 J/m); ········ border between tough and brittle impact fracture and high and low adhesion.
Rubbers were two hydrocarbon rubbers with the same chemical structure, except that the high adhesion rubber (adhesion F/L = 8100 N/m) contained 1 % reactive groups and the low-adhesion rubber (F/L = 140 N/m) contained none. Volume fractions of rubber particles were 30.8 %, 18.9 % or 12.8 %. The tough materials deform 25 % by crazing and 75 % by shear flow. The critical distance between rubber particles is ca. 300 nm (see Eqn.(14-14)).

The critical distance can be calculated by various models. If the particles are assumed to be in a cubic lattice (6 neighbors), then the critical distance is:

(14-14) $L_{crit} = d_F \{ [\pi/(6 \ \phi_F)^{1/3}] - 1 \}$

Rubber particles do not keep the matrix together like rubber bands since the deformable dispersed rubber phases are enclosed by the rigid matrix. Rubber and matrix are deformed jointly but differently. The greatest deformation is at the circumference perpendicular to the direction of deformation of rubber particles. Since the rubber can be deformed but the matrix cannot, stress concentrates here. The stress is relieved by shear deformation or by crazing. Shear flow comprises the whole specimen whereas crazing is localized. Shear flow thus absorbs more energy than crazing.

Shear flow is however only possible if the glass temperature T_G is far below the testing temperature (usually 25°C). An example is it-poly(propylene) with $T_G \approx$ –15°C. A rubber-modified poly(propylene) thus deforms to ca. 90 % by shear flow. Atactic poly(styrene) has a glass temperature of ca. $T_G \approx$ +100°C; a rubber-modified poly(styrene) deforms to ca. 96 % by crazing.

In order to reduce stress concentrations homogeneously, rubber particles must adhere well to the thermoplastic matrix (hence presence of compatibilizers) and must furthermore be either chemically cross-linked or highly entangled (hence high molar masses). Crazes grow equatorially (perpendicular to the stress direction) until they hit an obstacle, e.g., another rubber particle or a shear band, or until the stress concentration becomes too low at the tip of the craze.

A rubber modification is therefore the more effective, the higher the volume fraction ϕ_F of the rubber particles (Fig. 14-10). Large particles are especially good as initiators of craze formation; their overall efficiency is however not high since their number concentration is low. Notched impact strengths increase with decreasing particle diameter d_F and finally become constant at small particle sizes.

14.7. Disposal

Plastics were easy to discard in the good old days: they were put into a landfill like many other old materials. The increasing amount of plastics (although still small in percentage terms), the growing demand for more land for landfills and the groundwater contamination by landfills (though not by plastics!) stimulated the search for recycling and other disposal methods. The discussion about the "best" method is often emotionally charged. The following discussion is thus restricted to facts and present technical and economical possibilities; political decisions (which differ from state to state and country to country) and utopian daydreams are not considered.

It is a fact that synthetic polymers cannot be replaced in modern societies by natural products. The reasons are technical, economic, ecological, and humanitarian. Transportation requires tires from elastomers, many articles can be only produced from plastics (CD disks!), packaging materials from plastics reduce food spoilage, plastic parts serve in medicine, etc. Plastics have low densities; they need less energy for production and transport. The replacement of synthetic fibers by natural fibers is possible but would require the same land area as is now used for food production, etc.

14.7.1. Landfills

Citizens of highly industrialized countries consume large amounts of materials. The annual per capita consumption in the United States comprises 8500 kg non-metallic minerals (sand, stone, concrete, etc.), 1000 kg wood (construction, paper, etc., excluding fuel), 500 kg metals (iron, aluminum, etc.), 120 kg plastics, 25 kg fibers, 14 kg elastomers, 6 kg leather, etc. (see also Table 1-2).

Broken concrete, lumber discards, etc., are usually deposited in special landfills. A large proportion of metals is recycled as scrap but about 65 kg per year and capita are placed in landfills. Annual municipal waste per capita consists of 265 kg paper and cardboard, 65 kg metals, 70 kg plastics, fibers, and elastomers, and 270 kg other (wood, rubble, glass, plants, food leftovers, etc.). The proportion of plastics in municipal refuse is 6 wt.-% or 7.5 vol.-%.

The proportion of plastics in refuse is relatively small but very visible because it usually consists of discarded packaging with bold colors. All these plastic boxes, bottles, etc., do *not* require large volumes in landfills because plastic containers are easily compressed by the weight of materials placed upon them. Plastics scarcely add to groundwater contamination since polymers do not dissolve or degrade in landfills and are also not resorbed by organisms. Low molar mass additives which may be dissolved or resorbed have long been regulated for toxicological and/or ecological reasons; their diffusion (if any) is very slow (Sections 9.7.2. and 11.6.).

14.7.2. Biological Degradation

Most polymers cannot degrade naturally by light, oxygen, water, microorganisms, etc., because of their carbon chains and the lack of chromophores. Even cellulose papers degrade in landfills only very slowly: headlines of newspapers can still be read after 25 years in landfills. It is possible to produce biologically degradable polymers but these polymers are unsuitable for long-lived goods. Biodegradable polymers for short-lived goods must lead to degradation products that are ecologically safe. They are thus usually restricted to medical goods (e.g., surgical sutures) or agricultural materials (films, seed coatings, etc.). The energy content of plastics is of course wasted during biodegradation.

Such biodegradable synthetic polymers comprise:
– poly(glycolide) = poly(glycolic acid) PGL by chain polymerization of the cyclic dimer of glycolic acid $HO–CH_2–COOH$;
– poly(lactide) = poly(lactic acid) PLT by chain polymerization of the cyclic dimer of lactic acid $HO-CH(CH_3)-COOH$ or by microbiological synthesis of lactic acid followed by polycondensation and removal of the water by evaporation;
– poly(ε-caprolactone) PCL by chain polymerization of the lactone of 6-hydroxy-hexanoic acid;
– poly(3-hydroxybutyrate-*co*-hydroxyvalerate) PHB-HV from glucose by microbiological synthesis.

These polymers have the following monomeric units:

$$-O-CH_2-CO- \qquad -O-\underset{CH_3}{\underset{|}{CH}}-CO- \qquad -O-(CH_2)_5-CO-$$
$$\qquad \text{PGL} \qquad\qquad\qquad \text{PLT} \qquad\qquad\qquad \text{PCL}$$

$$-O-\underset{CH_3}{\underset{|}{CH}}-CH_2-CO- \;+\; -O-\underset{C_2H_5}{\underset{|}{CH}}-CH_2-CO-$$
$$\qquad\qquad\qquad\qquad\qquad\qquad\qquad\qquad \text{PHB-HV}$$

14.7.3. Recycling

Plastics can be recycled in three different ways:
- as materials by remelting or reshaping,
- as raw materials by degradation to monomers or petrochemicals,
- as energy providers by incineration.

Recycling as *materials* is especially suitable for polymers with the same grade, i.e., scrap from polymer production or processing. Such scrap can be macrohomogenized with virgin materials if it is not soiled. Clean thermoplastics of the same family can be microhomogenized albeit to new grades. The same applies to discarded consumer goods if they are of the same type, for example, PET soft drink or milk bottles.

The situation is quite different for plastics discarded by consumers which are mixtures of plastics with different grades and from different polymer families. These goods have to be collected, disassembled (if necessary), sorted, cleaned and then pelletized, compounded, etc. Such reclaimed plastics usually cost two or three times as much as virgin plastics. The mixing of many grades of the same family leads of course to a new grade which cannot be tailor-made for a certain application within narrow specifications. The problem is compounded if plastics from different polymer families are blended (Table 14-9). Such blends require appropriate compatibilizers.

Table 14-9   Compatibility of important thermoplastics. 1 = Very compatible, 6 = incompatible. Reprinted by permission of Hanser Publishers [10].

	PS	HIPS	SAN	ABS	PA	PC	PMMA	POM	PVC	PP	LDPE	HDPE
PS	1											
HIPS	6	1										
SAN	6	6	1									
ABS	6	6	1	1								
PA	5	4	6	6	1							
PC	6	5	2	2	6	1						
PMMA	4	4	1	1	6	1	1					
POM	6	6	6	5	6	6	5	1				
PVC	6	6	2	3	6	5	1	6	1			
PP	6	6	6	6	6	6	6	6	6	1		
LDPE	6	6	6	6	6	6	6	6	6	6	1	
HDPE	6	6	6	6	6	6	5	6	6	6	1	1

Another factor is that recycled plastics never have the same properties as virgin plastics, even if they are of the same grade. Plastics are slightly degraded by processing due to the combined action of heat and shearing. Degradation also occurs during use and recycling. Properties of recycled plastics decrease with the number of cycles. They eventually become so low that the material can only be converted into raw materials or incinerated. Blending of recycled plastics with virgin plastics slows the process but does not prevent it.

Sorting is also prone to error. Automatic sorting is preferred but requires either spectroscopic equipment or specially added markers or both. Flotation is often vitiated by the presence of different types and amounts of fillers; dissolution requires solvents which have to be worked up.

Hand sorting is eased by letter combinations or numbers that are stamped onto the goods in inconspicious places. Common codes are:

1 = poly(ethylene terephthalate)	or	PETE		(not PET)
2 = high density poly(ethylene)	or	HDPE	or	PE-HD
3 = poly(vinyl chloride)	or	V		(not PVC)
4 = low density poly(ethylene)	or	LDPE	or	PE-LD
5 = poly(propylene)	or	PP		
6 = poly(styrene)	or	PS		
7 = all others, including blends.				

Recycling as *raw materials* for polymers comprises several processes. Polycondensation polymers are usually converted by solvolysis into their monomers or monomer derivatives. An example is poly(ethylene terephthalate) which delivers dimethyl terephthalate and ethylene glycol by methanolysis or diethylene glycol terephthalate and ethylene glycol by glycolysis.

Plastics with different grades or soiled plastics can be degraded to low molar mass products which can serve as petrochemical raw materials after purification and fractionation. These processes can be subdivided into three groups:

– Thermal processes, for example, pyrolysis at (700-900)°C with exclusion of air. This process delivers relatively large amounts of gas and coke. PVC requires a dechlorination before processing.

– Reductive processes, for example, hydrogenation in the sump phase at ca. 470°C and 280 bar with addition of residual oil from crackers. The process leads mainly to liquid saturated hydrocarbons, called syncrude. Heteroatoms such as chlorine (from PVC), oxygen (from PET) and nitrogen (from PA or PUR) are converted into the corresponding hydrogen compounds.

– Oxidative processes, for example, in the presence of oxygen and water vapor at temperatures of (1350-1600)°C and pressures up to 150 bar. The process leads to a mixture of H_2 and CO that serves as synthesis gas for many chemicals.

Energy recycling is advantageous because of the high heat content of polymers (Section 9.2.3.). Combustion enthalpies vary between 18 MJ/kg for PVC to 87 MJ/kg for ABS. Plastic refuse in municipal waste reduces the amount of heating oil which otherwise has to be added to guarantee complete incineration of waste. HCl from

PVC is mainly bound to ash as chloride; it can also bind heavy metal ions. The remaining HCl is removed by scrubbers.

Incineration of plastics refuse is however politically incorrect since it is feared that dioxine emissions increase. However, much more dioxin is generated by the burning of wood. Furthermore, far less dioxines are produced at higher incineration temperatures, e.g., ca. 1200 °C instead of the conventional 800°C.

Literature

GENERAL
G.W.Becker, D.Braun, eds., Kunststoff-Handbuch, Hanser, Munich 1986 ff. (several volumes)
P.J.Corish, ed., Concise Encyclopedia of Polymer Processing and Applications, Pergamon, London 1991
A.W.Birley, B.Haworth, J.Batchelor, Physics of Plastics. Processing, Properties and Materials Engineering, Hanser, Munich 1992
H.-G. Elias, An Introduction to Plastics, VCH, Weinheim 1993
W.Michaeli, H.Greif, H.Kaufmann, F.-J.Vossebürger, Plastics Technology, Hanser, Munich 1994

HISTORY
R.Friedel, Pioneer Plastic. The Making and Selling of Celluloid, University of Wisconsin Press, Madison (WI) 1983
S.T.I.Mossman, P.J.T.Morris, The Development of Science, Royal Society of Chemistry, London 1994

14.2. PROCESSING
C.D.Han, Rheology in Polymer Processing, Academic Press, New York 1976
G.Astarita, L.Nicolais, eds., Polymer Processing and Properties, Plenum, New York 1984
D.H.Morton-Jones, Polymer Processing, Chapman and Hall, New York 1989
J.L.White, Principles of Polymer Engineering Rheology, Wiley, New York 1990
W.Michaeli, Plastics Processing, Hanser, Munich 1995

14.3. THERMOPLASTICS
C.A.Brighton, G.Pritchard, G.A.Skinner, Styrene Polymers: Technology and Environmental Aspects, Applied Science, Barking (Essex) 1979
J.M.Margolis, ed., Engineering Thermoplastics, Dekker, New York 1985
L.I.Nass, C.A.Heiberger, eds., Encyclopedia of PVC, Dekker, New York, 2nd ed. 1988 (4 vols.)
S. van der Ven, Polypropylene and Other Polyolefins, Elsevier, Amsterdam 1990
J.E.Mark, H.R.Allcock, R.West, Inorganic Polymers, Prentice Hall, Englewood Cliffs (NJ) 1993

14.4. THERMOSETS
S.H.Goodman, ed., Handbook of Thermoset Plastics, Noyes, Park Ridge (NJ) 1986
S.-C.Lin, E.M.Pearce, High Performance Thermosets: Chemistry, Properties, Applications, Hanser, Munich 1994

14.5. FILLED AND REINFORCED PLASTICS
E.Baer, A.Moet, eds., High Performance Polymers, Hanser, Munich 1991 (composites)
P.K.Mallick, Fiber-Reinforced Composites, Dekker, New York, 2nd ed. 1993
F.Jones, ed., Handbook of Polymer-Fibre Composites, Longman Higher Education, Harlow (Essex) 1994

14.6. POLYMER BLENDS

L.A.Utracki, A.P.Plochoko, Industrial Polymer Blends and Alloys, Hanser, Munich 1985
L.A.Utracki, Polymer Alloys and Blends, Hanser, Munich 1989
I.S.Miles, S.Rostami, eds., Multicomponent Polymer Systems, Wiley, New York 1993
A.A.Collyer, ed., Rubber-Toughened Engineering Plastics, Chapman and Hall, London 1994

14.7. DISPOSAL

D.P.Mobley, ed., Plastics from Microbes - Microbial Synthesis of Polymers, Hanser, Munich 1994
G.J.L.Griffin, ed., Chemistry and Technology of Biodegradable Polymers, Chapman and Hall, New York 1994
N.Mustafa, ed., Plastics Waste Management - Disposal, Recycling, and Reuse, Dekker, New York 1994
A.L.Bisio, M.Xanthos, eds., How to Manage Plastics Waste, Hanser, Munich 1995
A.-C.Albertsson, S.J.Huang, eds., Degradable Polymers, Recycling, and Plastics Waste Management, Dekker, New York 1995
G.Scott, D.Gilead, Degradable Polymers: Principles and Applications, Chapman and Hall, London 1995

References

[1]   H.-G.Elias, An Introduction to Plastics, VCH, Weinheim 1993, Table 11-1
[2]   H.-G.Elias, An Introduction to Plastics, VCH, Weinheim 1993, Fig. 11-8
[3]   J.Meissner, in R.Vieweg, G.Daumiller, eds., Kunststoff-Handbuch **5** (1969) 162, Fig. 10
[4]   J.B.Enns, J.K.Gillham, J.Appl.Polym.Sci. **28** (1983) 2567, Fig. 1
[5]   After data of H.Brintrup, PhD Thesis, RWTH Aachen
[6]   G.W.Ehrenstein, R.Wurmb, Angewandte Makromol.Chem. **60/61** (1977) 157
[7]   A.W.Bosshard und H.P.Schlumpf, various tables in R.Gächter, H.Müller, ed.,
         Plastics Additives Handbook, Hanser, Munich, 3rd ed. 1990
[8]   J.Jenne, Kunststoffe **44** (1984) 551, Fig. 2
[9]   S.Wu, Polymer **26** (1985) 1855, Tables 1-3
[10]   J.Brandrup, G.Menges, W.Michaeli, M.Bittner, Die Wiederverwertung der Kunststoffe, Hanser, Munich 1995

15. Appendix

15.1. Physical Quantities and Units

Many physical properties can be described in quantitative terms by *quantity calculus*. The value of a physical quantity (symbols in *italics*) equals the product of a numerical value (upright numbers) and a physical unit (symbols in upright letters):

$$\text{physical quantity} = \text{numerical value} \times \text{physical unit}$$

This equation can be manipulated by the ordinary rules of algebra. If, for example, a certain item has a length of 0.002 meters, then this may be written as:

$$L = 0.002 \text{ m} \quad \text{or} \quad L = 2 \cdot 10^{-3} \text{ m} \quad \text{or} \quad 10^3 \, L/\text{m} = 2 \quad \text{or} \quad L = 2 \text{ mm} \quad \text{or} \quad L/\text{mm} = 2$$

A column head $10^2 \, F/(\text{N m}^{-2})$ for a column entry of 7.35 thus indicates a force $F = 7.35 \cdot 10^{-2}$ N/m^2. Literature data often do not follow these SI rules (Systéme International) by the International Standardization Organization (ISO) which are adopted by IUPAP (International Union of Pure and Applied Physics), IUPAC (International Union of Pure and Applied Chemistry), etc. Instead one finds various non-rational notations such as F, N m^{-2} or F [N m^{-2}] for a column figure of $7.35 \cdot 10^{-2}$, often with wrong algebraic statements such as $10^{-2} \, F$, N m^{-2}.

American technical literature and sometimes also American scientific literature still uses mainly Anglo-Saxon physical units although by law all U.S. government units were supposed to convert to the SI system ("metric system") by the end of 1992. Only SI units can be lawfully used for commercial purposes in many other countries. In order to provide access to older and Anglo-Saxon technical literature, the following tables list names and symbols of SI units, SI prefixes, and conversions from SI units (in American notation) into Anglo-Saxon units. In many cases, IUPAC/IUPAP recommmendations for symbols for physical quantities are also given.

Table 15-1 Names and symbols for physical quantities and their SI units

| Physical quantity | | SI unit | |
Symbol	Name	Name	Symbol
Base units			
l	length	meter	m
m	mass	kilogram	kg
t	time	second	s
I	electric current	ampere	A
T	thermodynamic temperature	kelvin	K
n	amount of substance	mole	mol
I_v	luminous intensity	candela	cd
Supplementary units			
α, β, γ	plane angle	radian	$\text{rad} = \text{m m}^{-1}$
ω, Ω	solid angle	steradian	$\text{sr} = \text{m}^2 \text{m}^{-2}$

Table 15-2 Important SI derived units with special names and symbols

Physical quantity Symbol	Name	Physical unit Name	Symbol	Derived units		Base units
v	frequency	hertz	Hz [1]	$= \text{s}^{-1}$		
E	energy, work, heat	joule	J	$= \text{J}$	$= \text{N m}$	$= \text{m}^2 \text{ kg s}^{-2}$
F	force	newton	N	$= \text{J m}^{-1}$	$= \text{N}$	$= \text{m kg s}^{-2}$
γ	surface tension	–	–	$= \text{J m}^{-2}$	$= \text{N m}^{-1}$	$= \text{kg s}^{-2}$
p, σ	pressure, stress	pascal	Pa	$= \text{J m}^{-3}$	$= \text{N m}^{-2}$	$= \text{m}^{-1} \text{kg s}^{-2}$
Q	electric charge	coulomb	C	$= \text{A s}$		
U	electric potential, electromotive force	volt	V	$= \text{J C}^{-1}$		$= \text{m}^2 \text{ kg s}^{-3} \text{ A}^{-1}$
R	electric resistance	ohm	Ω	$= \text{V A}^{-1}$		$= \text{m}^2 \text{ kg s}^{-3} \text{ A}^{-2}$
G	electric conductance	siemens	S	$= \Omega^{-1}$		$= \text{m}^{-2} \text{ kg}^{-1} \text{ s}^3 \text{ A}^2$
C	electric capacitance	farad	F	$= \text{C V}^{-1}$		$= \text{m}^{-2} \text{ kg}^{-1} \text{ s}^4 \text{ A}^2$
P	power, radiant flux	watt	W = VA	$= \text{J s}^{-1}$		$= \text{m}^{-1} \text{ kg s}^{-2}$
B	magnetic flux density	tesla	T	$= \text{V s m}^{-2}$		$= \text{kg s}^{-2} \text{ A}^{-1}$
Φ	magnetic flux	weber	Wb	$= \text{V s}$		$= \text{m}^2 \text{ kg s}^{-2} \text{ A}^{-1}$
L	inductance	henry	H	$= \text{V A}^{-1} \text{ s}$		$= \text{m}^2 \text{ kg s}^{-2} \text{ A}^{-2}$
A	activity (radioactive)	becquerel	Bq	$= \text{s}^{-1}$		
D	absorbed dose (radiation)	gray	Gy	$= \text{J kg}^{-1}$		$= \text{m}^2 \text{ s}^{-2}$
–	dose equivalent	sievert	Sv	$= \text{J kg}^{-1}$		

[1] The unit hertz should be used only for frequency in the sense of cycles per second. Circular frequency and angular velocity have the unit rad s^{-1} which may be simpified to s^{-1} but *not* to Hz.

Table 15-3 SI Prefixes (origin: D = Danish, G = Greek, I = Italian, L = Latin, N = Norwegian)

Factor	SI System Prefix	Symbol	Trivial name England, Germany	USA, France	Origin
10^{18}	exa	E	trillion	quintillion	G: *ex* = six [$10^{18} = (10^3)^6$]
10^{15}	peta	P	billiard	quadrillion	G: *pente* = five [$10^{15} = (10^3)^5$]
10^{12}	tera	T	billion	trillion	G: *teras* = monster
10^9	giga	G	milliard	billion	G: *gigas* = giant
10^6	mega	M	million	million	G: *megas* = large
10^3	kilo	k	thousand	thousand	G: *khilioi* = thousand
10^2	hecto	h	hundred	hundred	G: *hekaton* = hundred
10^1	deca	da	ten	ten	G: *deka* = ten
10^{-1}	deci	d	one tenth	one tenth	L: *decima pars* = tenth
10^{-2}	centi	c	one hundredth	one hundredth	L: *pars centesima* = hundredth
10^{-3}	milli	m	one thousandth	one thousandth	L: *pars millesima* = thousandth
10^{-6}	micro	μ	one millionth	one millionth	G: *mikros* = little, small
10^{-9}	nano	n	one milliardth	one billionth	G: *nan(n)os* = dwarf
10^{-12}	pico	p	one billionth	one trillionth	I: *piccolo* = small
10^{-15}	femto	f	one billiardth	one quadrillionth	D, N: *femten* = fifteen
10^{-18}	atto	a	one trillionth	one quintillionth	D, N: *atten* = eighteen

In the computer industry, k or K is used as symbol for $2^{10} = 1024 \approx 1000$, M as symbol for $2^{20} = 1\,048\,576 \approx 10^6$, and G as symbol for $2^{30} = 1\,073\,741\,800 \approx 10^9$.

In finance and gas industries, the following symbols are used: M for 10^3 (L: *mille* = thousand), MM or $\overline{\text{M}}$ for 10^6 (*mille × mille*), B for 10^9 (USA: billion), T for $10^{12.}$ (USA: trillion).

Table 15-4   Older non-SI units. * Only these units may be used together with SI units and/or SI prefixes.

Physical quantity	Physical unit Name	Symbol	Value in SI units	Notes
time	minute	min	60 s	1)
time	hour	h	3600 s	1)
time	day	d	86 400 s	1)
length	ångstrøm	Å	10^{-10} m $\equiv$ 0.1 nm	2)
area	barn	b	10^{-28} m^2	
volume	liter	l, L	10^{-3} m$^3 \equiv$ 1 L	*
mass	ton(ne)	t	10^3 kg	3)
mass	unified atomic mass unit[b]	u = $m_a(^{12}C)/12$	$\approx 1.66054 \cdot 10^{-27}$ kg	4,5)
energy	electronvolt	eV	$\approx 1.60218 \cdot 10^{-19}$ J	6)
pressure	bar	bar	10^5 Pa	2) *
plane angle	degree	°	$(\pi/180)$ rad	
plane angle	minute	'	$(\pi/10\ 800)$ rad	
plane angle	second	"	$(\pi/648\ 000)$ rad	
temperature	Celsius temperature	°C	$\theta/°C \equiv (T/K) - 273.15$	7)

[1] IUPAC allows the use of the non-SI units "minute", "hour", and "day "in appropriate contexts although these three physical units are not part of the SI system. These units should not be used with SI prefixes.

"Month" and "year" are not scientific units. In commercial data, the symbol for "month" is often "mo" and the symbol for "year" either "yr" or, preferably, "a" (from L: *annus*).

[2] This unit is approved for "temporary use with SI units" in fields where it is presently used.

[3] IUPAC allows the use of the physical unit "ton(ne)" = 1000 kg (especially for technical and commercial data) which is, however, not a IUPAC unit. "Ton(ne)" (symbol: ton) is not to be confused with "long ton" ($\approx$ 1016.047 kg) and "short ton" ($\approx$ 907.185 kg); both are often used without the adjectives "long" and "short".

[4] The value of this unit depends on the experimentally determined value of the Avogadro constant N_A; the value of the corresponding SI unit is therefore not exact.

[5] The unified atomic mass (physical unit: kg) is sometimes called the dalton (symbol Da); in the biosciences, "dalton" has erroneously come to mean the relative molecular mass (physical unit: 1) or the molar mass (physical unit: g/mol)!

[6] The value of this unit depends on the experimentally determined value of the elementary charge e; the value of the corresponding SI unit is therefore not exact.

[7] The SI unit of the Celsius temperature *interval* is the degree Celsius (symbol of the unit: °C), which is equal to the kelvin (*not* "degree kelvin"). The symbol of the unit kelvin (small k!) is K, *not* °K. Celsius is always written with a capital C.

When quoting temperatures in kelvin, a space should be written between the numerical value and the symbol K as it is customary for all physical properties. However, if temperatures are given in degrees Celsius, no space should exist between numerical values and symbol. Thus: $T = 298$ K but $\theta = 25$°C.

IUPAC now recommends the symbol θ for the physical property "Celsius temperature". In polymer science, a capital theta (Θ) is the traditional symbol for the property "theta temperature". Experience of this author has shown that θ and Θ are easily mixed up, even by experienced polymer scientists. This book thus uses the symbol T for both thermodynamic temperatures and Celsius temperatures; the possibility of a mix-up is remote since physical units are always given.

Table 15-5 Some fundamental constants. Numbers were recommended in 1986 by CODATA. Numbers in parentheses indicate the standard deviation uncertainty in the least significant digits.

Physical quantity	Physical value = number · physical unit	
Speed of light in vacuum	c_o	= 299 792 458 m s^{-1} (exactly)
Elementary charge	e	= 1.602 177 33 (49) 10^{-19} C
Faraday constant	F	= 9.648 530 9 (29)·10^4 C mol^{-1}
Planck constant	h	= 6.625 075 5 (40)·10^{-34} J s
Boltzmann constant	k	= 1.380 658 (12)·10^{-23} J K^{-1} [1]
Avogadro constant	N_A	= 6.022 136 7 (36)·10^{23} mol^{-1} [2]
(Molar) gas constant	R	= 8.314 510 (70) J K^{-1} mol^{-1}
Zero of the Celsius scale		= 273.15 K (exactly)
Molar volume of an ideal gas (p = 1 bar; θ = 0°C)	V_m	= 22.711 08 (19) L mol^{-1}
Standard atmosphere (atm)	p	= 101 325 (exactly)
Permittivity of vacuum	ε_o	= $1/(\mu_o c_o^2)$ = 8.854 187 816...·10^{-12} F m^{-1}
Permeability of vacuum	μ_o	= $4\pi \cdot 10^{-7}$ H m^{-1} (exactly)
Atomic mass constant (unified atomic mass unit)	m_u	= 1 u = 1.660 540 2 (10)·10^{-27} kg

[1] In this book, k_B is used as the symbol in order to avoid confusion with the general symbol k for rate constants.

[2] This physical quantity was never determined by the Italian Amadeo Avogadro, Conte di Quaregna (1776-1856) but rather by the Viennese Joseph Loschmidt (1821-1895) (alternative ISO symbol L!). It was erroneously called the Loschmidt *number* in German speaking countries and should not be called the Avogadro *number* since it is not a number (it has the physical unit of an inverse amount-of-substance, i.e., mol^{-1}). Avogadro's law referred originally to the number of particles per volume, i.e., a number concentration.

Table 15-6 Conversion of out-dated and Anglo-Saxon units into SI units. Non-SI units marked with * may be used with SI prefixes and together with SI units.

Name	Old unit	= SI unit
Length		
light year	1 ly	= 9.4605·10^{15} m
nautical mile (sea mile)	1 n	= 1852 m
statute mile (land mile)	1 mile	= 1609.344 m
furlong	1 furlong	= 201.168 m
rod (perch, pole)	1 rod	= 5.029 2 m
fathom	1 fathom	= 1.828 8 m
yard	1 yd	= 0.914 4 m
foot	1 ft = 1'	= 0.304 8 m = 30.48 cm
inch	1 in = 1"	= 0.025 4 m = 2.54 cm
mil	1 mil	= 2.54·10^{-5} m = 25.4 μm
micron	1 μ	= 10^{-6} m = 1 μm
millimicron	1 mμ	= 10^{-9} m = 1 nm
ångstrøm*	1 Å	= 10^{-10} m = 0.1 nm

Table 15-6   Conversion of out-dated and Anglo-Saxon units into SI units (continued).

Name	Old unit	= SI unit
Area		
square mile	1 sq. mile	$= 2.589\ 988\ 110 \cdot 10^6\ \mathrm{m}^2$
hectare	1 ha	$= 10\ 000\ \mathrm{m}^2$
acre	1 acre	$= 4046.856\ \mathrm{m}^2$
square yard	1 sq. yd.	$= 0.836\ 127\ 36\ \mathrm{m}^2$
square foot	1 sq. ft.	$= 9.20\ 304 \cdot 10^{-2}\ \mathrm{m}^2$
square inch	1 sq. in.	$= 6.451\ 6 \cdot 10^{-4}\ \mathrm{m}^2$
Volume		
stere	1 st	$= 1\ \mathrm{m}^3$
cubic yard	1 cu. yd.	$= 0.764\ 554\ 857\ \mathrm{m}^3$
imperial barrel	1 barrel	$= 0.1636\ \mathrm{m}^3$
US barrel petroleum	1 bbl = 42 US gal	$= 0.158\ 987\ \mathrm{m}^3$
US barrel	1 barrel	$= 0.119\ \mathrm{m}^3 = 119\ \mathrm{L}$
bushel	1 bu	$= 3.524 \cdot 10^{-2}\ \mathrm{m}^3 = 35.24\ \mathrm{L}$
cubic foot	1 cu. ft.	$= 2.381\ 684\ 659\ 2 \cdot 10^{-2}\ \mathrm{m}^3$
peck (US)	1 peck	$= 8.810\ \mathrm{L}$
board foot	$12 \cdot 12 \cdot 1$ cu. in.	$= 2.3597 \cdot 10^{-3}\ \mathrm{m}^3$
gallon (British or Imperial)	1 gal	$= 4.546\ 09 \cdot 10^{-3}\ \mathrm{m}^3 = 4.545\ 96\ \mathrm{L}$
gallon (US dry)	1 gal	$= 4.405 \cdot 10^{-3}\ \mathrm{m}^3$
gallon (US liquid)	1 gal = 4 US qt.	$= 3.785\ 41 \cdot 10^{-3}\ \mathrm{m}^3 = 3.785\ 412\ \mathrm{L}$
liter (cgs)	1 L	$= 1.000\ 028 \cdot 10^{-3}\ \mathrm{m}^3$
liter*	1 L	$\equiv 1 \cdot 10^{-3}\ \mathrm{m}^{-3}$
quart (US dry)	1 qt.	$= 1.101\ \mathrm{L}$
quart (US liquid)	1 qt. = 2 US pints	$= 0.946\ 335\ \mathrm{L}$
pint (US liquid)	1 pt. = 2 US cups	$= 0.473\ 168\ \mathrm{L}$
pint (US dry)	1 pt	$= 0.550\ 6\ \mathrm{L}$
cup (US)	1 cup = 8 fluid oz.	$= 0.236\ 534\ \mathrm{L}$
British liquid (fluid) ounce	1 oz.	$= 0.028\ 413\ \mathrm{L}$
US fluid (liquid) ounce	1 oz. = 2 tbl. sp.	$= 0.029\ 574\ \mathrm{L}$
cubic inch	1 cu. in.	$= 0.016\ 387\ 064\ \mathrm{L}$
tablespoon	1 tbl.sp. = 3 tea sp.	$= 0.014\ 79\ \mathrm{L}$
teaspoon	1 tea sp.	$= 0.004\ 93\ \mathrm{L}$
dram (US fluid)	1 dram	$= 0.003\ 697\ \mathrm{L}$
Mass		
long ton (UK)	1 ton = 2240 lbs	$= 1016.046\ 909\ \mathrm{kg}$
short ton (US)	1 sh.ton = 2000 lbs	$= 907.184\ 74\ \mathrm{kg}$
quintal	1 quintal	$= 100\ \mathrm{kg}$
hundredweight (UK)	1 cwt.	$= 50.802\ 3\ \mathrm{kg}$
short hundredweight	1 sh.cwt.	$= 45.359\ 2\ \mathrm{kg}$
slug	1 slug	$= 14.59\ 39\ \mathrm{kg}$
stone	1 stone = 14 lbs.	$= 6.350\ 293\ 18\ \mathrm{kg}$
pound (avoirdupois, US)	1 lb = 16 oz.	$= 453.592\ 37\ \mathrm{g}$
pound (apothecaries' or troy, US)	1 lb = 8 drams	$= 373.242\ \mathrm{g}$
ounce (avoirdupois, US)	1 oz.	$\approx 28.349\ 5\ \mathrm{g}$

Table 15-6 Conversion of out-dated and Anglo-Saxon units into SI units (continued).

Name	Old unit	= SI unit
Mass (continued)		
ounce (troy)	1 oz.	$\approx$ 31.103 5 g
dram (apothecaries')	1 dram	= 3.888 g
dram (avoirdupois)	1 dram	= 1.772 g
pennyweight	1 pennyweight	= 1.555 g
carat	1 ct	= 0.2 g
grain	1 gr	= 64.798 91 mg
unified atomic mass unit*	1 mu	= 1.660 540 2 (10)$\cdot 10^{-27}$ kg
electron mass*	-	$\approx$ 9.109 39$\cdot 10^{-31}$ kg
Time		
year	1 a	$\approx$ 365 days (for statistical purposes only)
month	1 mo	$\approx$ 30 days (for statistical purposes only)
Temperature		
degree Fahrenheit	$(z°F - 32°F)(5/9)$	$= y°C$
degree Celsius	$(\theta/°C)$	$\equiv (T/K) - 273.15$

Density $(1 \text{ kg m}^{-3} = 1\cdot 10^{-3} \text{ g cm}^{-3})$

–	1 lb/cu.in.	= 27.679 904 71 g cm^{-3}
–	1 oz/cu.in.	= 1.729 993 853 g cm^{-3}
–	1 lb/cu.ft.	= 1.601 846 337$\cdot 10^{-2}$ g cm^{-3}
–	1 lb/gal US	= 7.489 150 454$\cdot 10^{-3}$ g cm^{-3}

Energy, work (1 J = 1 N m = 1 W s)

quadrillion BTU (US)	1 Quad = 10^{15} BTU	= 1.055 EJ
coal unit (German)	1 ton SKE	$\therefore$ 29.31 GJ (SKE = Steinkohleeinheit)
coal unit (US)	1 ton coal	$\therefore$ 27.92 GJ
coal unit (UK)	1 ton coal	$\therefore$ 24.61 GJ
short ton bituminous coal	1 T	$\therefore$ 26.58 GJ
kilowatt-hour	1 kWh	= 3.6 MJ
horse power hour	1 hph	= 2.685 MJ
cubic foot-atmosphere	1 cu.ft.atm.	= 2.869 205 kJ
British thermal unit	1 BTU$_{mean}$	= 1.055 79 kJ
British thermal unit	1 BTU$_{IT}$	= 1.055 06 kJ
–	1 cu.ft.lb(wt)/sq.in.	= 195.237 8 J
liter atmosphere (cgs)	1 L atm	= 101.325 J
–	1 m kgf	= 9.806 65 J
calorie, international	1 cal$_{IT}$	= 4.186 8 J
calorie, thermochemical	1 cal$_{th}$	= 4.184 J
–	1 ft-lbf	= 1.355 818 J
–	1 ft-pdl	= 4.215 384 J
–	1 erg	= $1\cdot 10^{-7}$ J = 0.1 μJ = 1 g cm^2 s^{-2}
electronvolt*	1 eV	$\approx$ 1.602 18$\cdot 10^{-19}$ J

Table 15-6   Conversion of out-dated and Anglo-Saxon units into SI units (continued).

Name	Old unit	= SI unit
Force		
–	1 ft-lbf/in.	= 53.378 64 N
kilogram-force	1 kgf	= 9.806 65 N
pound force	1 lbf	= 4.448 22 N
ounce-force	1 oz.f.	= 0.2780 N
poundal	1 pdl	= 0.138 255 N
pond	1 p	= $9.806\ 65 \cdot 10^{-3}$ N
dyne	1 dyn	= $1 \cdot 10^{-5}$ N
Length-related force		
–	1 kp/cm	= 980.665 N m^{-1}
–	1 lbf/ft	= 14.593 898 N m^{-1}
Pressure, stress (1 MPa = 1 N mm^{-2})		
physical atmosphere	1 atm = 760 torr	= 0.101 325 MPa
bar*	1 bar	≡ 0.1 MPa
technical atmosphere	1 at	= 0.098 065 MPa
–	1 kp/cm^2	= 0.098 065 MPa
–	1 kgf/cm^2	= 0.098 065 MPa
pound-force per square inch	1 psi = 1 lb/sq.in.	= $6.894\ 757 \cdot 10^{-3}$ MPa
inch mercury (32° F)	1 in.Hg	= $3.386\ 388 \cdot 10^{-3}$ MPa
inch water (39.2°F)	1 in.H$_2$O	= 249.1 Pa
torr	1 torr	= (101 325/760) Pa ≈ 133.322 Pa
millimeter mercury	1 mm Hg	= $13.5951 \cdot 9.806\ 65$ Pa ≈133.322 Pa
–	1 dyn/cm^2	= $1 \cdot 10^{-5}$ MPa
millimeter water	1 mm H$_2$O	= $9.806\ 65 \cdot 10^{-6}$ MPa
–	1 pdl/sq.ft.	= $1.488\ 649 \cdot 10^{-6}$ MPa
Power (1 W = 1 J·s^{-1})		
horsepower (UK)	1 hp	= 745.7 W
horsepower (boiler)	1 hp	= 9810 W
horsepower (electric)	1 hp	= 746 W
horsepower (metric)	1 PS	= 735.499 W
–	1 BTU/h	= 0.293 275 W
–	1 cal/h	= $1.162\ 222 \cdot 10^{-3}$ W
Heat conductivity		
–	1 cal/(cm s °C)	= 418.6 W m^{-1} K^{-1}
–	1 BTU/(ft h °F)	= 1.731 956 W m^{-1} K^{-1}
–	1 kcal/(m h °C)	= 1.162 78 W m^{-1} K^{-1}
Heat transfer coefficient		
	1 cal/(cm^2 s °C)	= $4.186\ 8 \cdot 10^4$ W m^{-2} K^{-1}
	1 BTU/(ft^2 h °F)	= 5.682 215 W m^{-2} K^{-1}
	1 kcal/(m^2 h °C)	= 1.163 W m^{-2} K^{-1}

Table 15-6   Conversion of out-dated and Anglo-Saxon units into SI units (continued).

Name	Old unit	= SI unit
Fineness = titer = linear density		
tex*	1 tex	$= 1 \cdot 10^{-6}$ kg m^{-1}
denier	1 den	$= 0.111 \cdot 10^{-6}$ kg m^{-1}
Tenacity		
–	1 gf/den	$= 0.082\ 599$ N tex$^{-1} = 0.082\ 599$ m^2 s^{-2}
		$= 98.06$ MPa $\times$ (density in g cm^{-3})
Dynamic viscosity		
poise	1 P	$= 0.1$ Pa s
centipoise	1 cP	$= 1$ mPa s
Kinematic viscosity		
stokes	1 St	$= 1 \cdot 10^{-4}$ m^2 s^{-1}
Heat capacity		
clausius	1 Cl	$= 1$ cal$_{th}$/K $= 4.184$ J K^{-1}
Molar heat capacity		
entropy unit	1 e.u.	$= 1$ cal$_{th}$ K^{-1} mol$^{-1} = 4.184$ J K^{-1} mol^{-1}
Electrical conductivity		
reciprocal ohm	1 mho	$= 1$ S^{-1}
Electrical resistance		
Average international ohm	-	$= 1.000\ 49$ Ω
US international ohm	-	$= 1.000\ 495$ Ω
Electrical field strength		
–	1 V/mil	$= 3.937\ 008 \cdot 10^4$ V m^{-1}
Electrical dipole moment		
Debye	1 D	$= 10^{-18}$ Fr cm $\approx 3.335\ 64 \cdot 10^{-30}$ C m
		(Fr = Franklin)
Radioactivity		
Curie	1 Ci	$= 37$ GBq
Röntgen	1 R	$= 2.58 \cdot 10^{-4}$ C kg^{-1}
	1 rem	$= 10^{-2}$ Sv
	1 rad	$= 0.01$ J kg$^{-1} = 0.01$ Gy

15.2.   Concentrations

Concentrations measure the abundance of a substance 1 in all substances $i = 1, 2, 3, ...$ present.

Mass fraction $= w_1 = m_1/\Sigma_i\ m_i = m_1/m = c_1/c$. Mass of substance 1 divided by the sum of masses m_i of all substances i. Since all masses reside in the same gravity field, a mass fraction can also be called a **weight fraction** ("weight" is not an accepted ISO term; it was formerly the name of a mass in a gravity field).

The value of 100 w_i is called weight percent (wt-%) and the value of $1000w_i$ is called weight promille (wt-‰). The English-language literature also uses part per million (1 ppm $= 10^{-4}$ %), part per (American) billion (1 ppb $= 10^{-7}$ %), and part per (American) trillion (1 ppt $= 10^{-10}$ %).

Volume fraction $= \phi_i = V_i/(\Sigma_i\ V_i = V_i/V$: Volume of substance i divided by the sum of volumes of all substances i. Volumes V_1, V_2 ... relate to the volumes of substances *before* the mixing process.

Mole fraction, amount fraction, number fraction $= x_1 = n_1/(\Sigma_i\ n_i = n_1/n$. Amount of substance divided by the sum of the amounts n_1 of all substances i. "Amount-of-substances" (short: amounts) are measured in moles, never in kilograms; they are not masses. The amount-of-substance was (and still is) erroneously called "mole number" in the literature.

Mass concentration = mass density. In polymer science, in general as $c_1 = m_1/V$), i.e., as mass of substance 1 per volume V of mixture *after* mixing. IUPAC recommends the symbols γ_1 or ρ_1 instead of c_1 but ρ_1 may be confused with the same symbol for the mass density of a *neat* substance. Mass concentrations are usually called "concentrations" in the literature.

Number concentration = number density of entities. $C_1 = N_1/V$). Number of entities 1 (molecules, atoms, ions, etc.) per volume V of mixture *after* mixing. IUPAC recommends the name "concentration" for this physical quantity but this may be confused with the much more common "concentration" = mass concentration.

Amount-of-substance concentration = amount concentration. In polymer science, it is usually as [1] $= n_1/V$), i.e., amount-of-substance 1 per volume of mixture after mixing. IUPAC recommends $c_1 = n_1/V$ which may be confused with the symbol c_1 for the mass concentration. The amount concentration is often called "mole concentration" or molarity and given the symbol M; the latter symbol is not recommended by IUPAC and should not be used with SI prefixes (i.e., not mM for an amount concentration of "millimole" per Liter).

Molality of a solute. $a_1 = n_1/m_2$, i.e., amount n_1 of substance 1 per mass m_2 of solvent 2. Molalities are often denoted by m which should not be used as symbol for the unit mol kg^{-1}.

15.3.   Ratios of physical quantities

The terms "normalized", "relative", "specific", and "reduced" are sometimes used with different meanings in the literature although they are clearly defined.

Normalized requires that the quantities in the numerator and the denominator are of the same kind. A normalized quantity is always a fraction (quantity of the subgroup divided by the quantity of the group); the sum of all normalized quantities is usually normalized to unity.

Relative also refers to quantities that are of the same kind in numerator and denominator but the quantity in the denominator may be in any defined state. Example 1: relative viscosity $\eta_r = \eta/\eta_2$ as the ratio of the viscosity η of a solution to the viscosity η_2 of the solvent 2. Example 2: relative humidity = ratio of moisture content of air to moisture content of air saturated with water (both at the same temperature and pressure).

Specific refers to a physical quantity divided by the mass. The symbol of a specific quantity is the lower case form of the symbol of the quantity itself. Example: $c_p = C_P/m$ = specific heat capacity = heat capacity (in heat per temperature) divided by mass. The so-called specific viscosity $(\eta - \eta_2)/\eta_2$ is not a *specific* quantity.

Reduced refers to a quantity that is divided by a specified other quantity. Example: reduced osmotic pressure Π/c = osmotic pressure Π divided by the mass concentration c.

Dimensionless quantity: a product or ratio of two or more different physical quantities which are combined in such a way that the resulting physical quantity has the physical unit of unity (i.e., is "dimensionless").

15.4. Abbreviations of Plastics, Fibers, Elastomers, etc.

Several organizations have issued recommendations for the use of abbreviations and acronyms of thermoplastics, thermosets, fillers, fibers, elastomers, etc., for example:

ASTM = American Society for the Testing of Materials;
ANSI = American National Standards Institute;
DIN   = German Industrial Standards;
ISO   = International Standardization Organization;
IUPAC = International Union of Pure and Applied Chemistry ;
(–)   = abbreviations used but not recommended by ASTM, DIN, ISO, or IUPAC.

For plastics names, the following standards apply: D 1600-86a [ASTM/ANSI], 7728 (1988) [DIN], 1043-1: 1987(E) [ISO], and the list in Pure Appl.Chem. **59** (1987) 691 [IUPAC].

The recommended abbreviations and acronyms for plastics are often neither identical with those of fibers of the same chemical structures nor with those of elastomers containing the same monomeric units; some of these also deviate from those recommended for recyclable plastics (see Chapter 14). Several abbreviations have different meanings in either two or more of the ASTM, DIN, ISO, and IUPAC systems and/or the technical literature.

See also Chapter 12 for abbreviations and acronyms of rubbers and Chapter 13 for fibers.

ABS   thermoplastic from acrylonitrile, butadiene and styrene [ASTM, DIN, ISO, IUPAC]
ACR   acrylic rubber from acrylic esters and butadiene
AES   rubber from acrylonitrile, EPDM and styrene
ASA   rubber from acrylonitrile, styrene and ACR (DIN, ISO)
BMC   bulk molding compound
BOPP   biaxially oriented poly(propylene)
CA   cellulose acetate [ASTM, DIN, ISO, IUPAC]
CF   thermoset from cresol and formaladehyde [ASTM, DIN, ISO, IUPAC]
CMC   carboxymethyl cellulose [ASTM, DIN, ISO, IUPAC]
CPE   chlorinated poly(ethylene) (ASTM); see PE-C
CR   chloroprene rubber [IUPAC]
CSM   chlorsulfonated poly(ethylene)
EC   ethyl cellulose [ASTM, DIN, ISO, IUPAC]
EEA   copolymer of ethyl acrylate and ethylene (ASTM); E/EA [DIN, ISO]
EMA   copolymer from ethylene and methacrylic acid [ASTM]
E/MA   copolymer from ethylene and methyl methacrylate [DIN, ISO]
EP   epoxy resin [ASTM, DIN, ISO, IUPAC]
EPDM   rubber from ethylene, propylene and a non-conjugierted diene [DIN, ISO]
EPS   expanded (foamed) poly(styrene)
ETFE   copolymer from ethylene and tetrafluoroethylene [ASTM]; E/TFE [DIN, ISO]
EVA   copolymer from ethylene and vinyl acetate [ASTM]; E/VA [DIN]
E/VAC   copolymer from ethylene and vinyl acetate [ISO]
FEP   copolymer from hexafluoropropylene and tetrafluoroethylene [ASTM, DIN, ISO]
GRS   old symbol for butadiene–styrene rubber (government rubber with styrene); now SBR
HDPE   thermoplastic poly(ethylene) with high density [ASTM]
HEMA   poly(hydroxyethyl methacrylate)
HIPS   high-impact poly(styrene); a rubber-modified poly(styrene)
IIR   rubber from isobutylene and some isoprene [IUPAC]
LDPE   thermoplastic poly(ethylene) with low density [ASTM]
LLDPE   thermoplastic linear poly(ethylene) with low density [ASTM]
MBS   thermoplastic terpolymer from methyl methacrylate, butadiene and styrene [DIN]
MF   thermoset from melamine and formaldehyde [ASTM, DIN, ISO, IUPAC]
NBR   nitrile rubber = rubber from acrylonitrile and butadiene [IUPAC]

NR   natural rubber
OPP   oriented poly(propylene)
PA   polyamide [ASTM, DIN, ISO, IUPAC]
PAI   polyamideimide [ASTM, DIN, ISO]
PAN   poly(acrylonitrile) [ASTM, ISO, IUPAC]
PB   thermoplastic poly(1-butene) [ASTM, DIN, ISO]
PBT   thermoplastic poly(butylene terephthalate) [ASTM, DIN, ISO]
PC   thermoplastic polycarbonate (usually from bisphenol A) [ASTM, DIN, ISO, IUPAC]
PCTFE   thermoplastic poly(chlorotrifluoroethylene) [ASTM, DIN, ISO, IUPAC]
PDAP   thermoset poly(diallyl phthalate) [ASTM, DIN, ISO, IUPAC]
PDMS   poly(dimethylsiloxane)
PE   poly(ethylene) [ASTM, DIN, ISO, IUPAC]
PEBA   polyether-*block*-polyamide [ASTM, DIN, ISO]
PE-C   chlorinated poly(ethylene) (DIN, ISO); see also CPE
PEEK   polyetheretherketone [ASTM, ISO]
PEEKK   polyetheretherketoneketone
PE-HD   thermoplastic poly(ethylene) with high density [DIN, ISO]
PEI   polyesterimide [ASTM, ISO]; polyetherimide [DIN]
PEK   polyetherketone
PE-LD   thermoplastic poly(ethylene) with low density [DIN, ISO]
PE-LLD   linear poly(ethylene) with low density [ISO]
PEO   poly(ethylene oxide) [ASTM, IUPAC]
PEOX   poly(ethylene oxide) [DIN, ISO]
PES   polysulfone from bisphenol A and dichlorodiphenyl sulfone [ASTM, DIN, ISO]
PET   poly(ethylene terephthalate) [ASTM, DIN, ISO]
PETP   poly(ethylene terephthalate) [IUPAC]
PF   thermoset from phenol and formaldehyde [ASTM, DIN, ISO, IUPAC]
PHB   poly(3-hydroxybutyrate) *or* poly(*p*-hydroxybenzoate)
PI   polyimide
PIB   poly(isobutylene) [ASTM, DIN, ISO, IUPAC]
PMMA   poly(methyl methacrylate) [ASTM, DIN, ISO, IUPAC]
POM   poly(oxymethylene) [ASTM, DIN, ISO, IUPAC]
PP   isotactic poly(propylene) [ASTM, DIN, ISO]
PPE   poly(oxy-(2,6-dimethyl)-1,4-phenylene) [ASTM, ISO]; this polymer is commonly (and
          erroneously) called polyphenylene oxide and abbreviated as PPO; PPO® is however a
          trade mark.
PPG   poly(propylene glycol)
PPOX   poly(propylene oxide) [ASTM, DIN, ISO]
PPS   poly(*p*-phenylenesulfide) [ASTM, DIN, ISO, IUPAC]
PS   poly(styrene) [ASTM, DIN, ISO, IUPAC]
PTFE   poly(tetrafluoroethylene) [ASTM, DIN, ISO, IUPAC]
PUR   poly(urethane) [ASTM, DIN, ISO, IUPAC]
PVAC   poly(vinyl acetate) [ASTM, DIN, ISO, IUPAC]
PVAL   poly(vinyl alcohol) [ASTM, DIN, ISO, IUPAC]
PVB   poly(vinyl butyral)
PVC   poly(vinyl chloride) [ASTM, DIN, ISO, IUPAC]
PVDC   poly(vinylidene chloride) [ASTM, DIN, ISO, IUPAC]
PVDF   poly(vinylidene fluoride) [ASTM, DIN, ISO, IUPAC]
PVK   poly(*N*-vinyl carbazole) [ASTM, DIN, ISO]
PVP   poly(*N*-vinyl pyrrolidone [ASTM, DIN, ISO]
SAN   polymer from styrene and acrylonitrile [ASTM, DIN, ISO, IUPAC]
S/B   thermoplastic from styrene and butadiene [DIN, ISO, IUPAC]
SI   silicone plastics [ASTM, DIN, ISO]
SMA   thermoplastic copolymer from styrene and maleic anhydride [ASTM, ISO]; S/MA [DIN]
SMC   sheet molding compound
SR   synthetic rubber

T	thioplast rubber (Thiokol®)
TEO	olefin-based thermoplastic elastomer
TES	styrene-based thermoplastic elastomer
TPE	thermoplastic elastomer
UF	thermoset from urea and formaldehyde [ASTM, DIN, ISO, IUPAC]
UP	with styrene or methyl methacrylate cured thermoset from unsaturated polyester resins synthesized from ethylene glycol and maleic anhydride [ASTM, DIN, ISO, IUPAC]
V	poly(vinyl chloride) (recycling)

15.5. Generic and Trade Names of Fibers and Plastics

There are ca. 25 000 different trade names of plastics and at least 10 000 trade names of fibers. Some of the more often encountered names are

Araldit®	epoxy resins (CIBA)
Bakelite®	phenol–formaldehyde resins
Buna	historical German name for synthetic rubbers. The first of these rubbers were synthesized from butadiene (Bu) with metallic sodium as initiator (Na), hence Buna; these Buna grades were distinguished by numbers. Later Bunas were manufactured by free radical copolymerization, e.g., Buna S = copolymer of butadiene and styrene, Buna N = copolymer of butadiene and acrylonitrile.
Cellophane	films of cellulose hydrate, sometimes laminated. Free name in the United States; registered trademark in other countries, e.g., as Cellophan® in Germany.
Dacron®	poly(ethylene terephthalate) fiber (DuPont)
Delrin®	poly(oxymethylene) from formaldehyde (DuPont)
Diolen®	poly(ethylene terephthalate) fiber (Glanzstoff)
Dralon®	poly(acrylonitrile) fiber (Bayer)
Epikote®	epoxy resin (Shell)
GR	*government rubber* = synthetic rubbers that were based on German patents (see Buna) and developed during World War II by a crash program under U.S. government auspices. Examples: GR-S, GR-N (see Buna S and Buna N).
Hostaform®	poly(oxymethylene) from trioxane (Hoechst)
Hostalen®	poly(ethylene)s (Hoechst); Hostalen PP® = poly(propylene)s
Hostalit®	poly(vinyl chloride) (Hoechst)
Hycar®	various synthetic elastomers (Goodrich)
Kevlar®	poly(*p*-phenylene terephthalamide) (DuPont)
Lexan®	bisphenol A polycarbonate (General Electric)
Lucite®	poly(methyl methacrylate) (DuPont)
Lupolen®	poly(ethylene) (BASF)
Luran®	copolymer from styrene and acrylonitrile (BASF)
Lycra®	spandex fiber (DuPont)
Makrolon®	bisphenol A polycarbonate (Bayer)
Marlex®	poly(ethylene) (Phillips Petroleum)
Moltopren®	foamed polyurethane (Bayer)
Neoprene®	poly(chloroprene) rubber (DuPont)
Nomex®	poly(*m*-phenylene isophthalamide) (DuPont)
nylon	generic name for polyamides except those from α-amino acids (a former trade name)
Orlon®	poly(acrylonitrile) fiber (DuPont)
Parylen N®	poly(*p*-xylylene) (Union Carbide)
Perlon®	poly(ε-caprolactam) fiber (Perlon Association, Germany)
Plexiglas®	poly(methyl methacrylate) (Röhm)
Pluronics®	segmented copolymers from ethylene oxide and propylene oxide (Wyandotte)
Polyox®	high molar mass poly(ethylene oxide) (Union Carbide)

Quina®   polymer fiber from *trans,trans*-diaminodicyclohexylmethane and dodecanedicarboxylic
               acid (DuPont)

saran   a manufactured fiber in which the fiber-forming substance is any long-chain synthe-
               tic polymer composed of at least 80 % by weight of vinylidene chloride units
               -CH_2CCl_2- (definition of the U.S. Federal Trade Commission)

spandex   a manufactured fiber in which the fiber-forming substance is a long-chain synthetic
               polymer comprised of a least 85 % of a segmental polyurethane (definition of the
               U.S. Federal Trade Commission)

Styrofoam®   foamed poly(styrene) (Dow Chemical)

Teflon®   a family of fluoropolymers (DuPont). Examples: poly(tetrafluoroethylene), copoly-
               mers of tetrafluoroethylene and hexafluoropropylene (Teflon FEP), copolymers of
               tetrafluoroethylene and perfluorinated vinyl ethers (Teflon PFA).

Terylene®   poly(ethylene terephthalate) fiber (ICI)

Thiokol®   rubbers based on various thioethers (DuPont)

Trevira®   poly(ethylene terephthalate) fibers (Hoechst)

Ultramid®   polyamides (PA 6, 66, 610) (BASF)

16. Index

Entries are listed in strict alphabetical order; they may consist of a single word, abbreviations, acronyms, or combinations thereof. Qualifying letters and numbers as well as hyphens, parentheses, brackets, and braces in names of chemical compounds such as 1-, 1,4-, α-, β-, o-, m-, p-, L-, D-, ( ), [ ], etc., have been disregarded for alphabetization.

The following abbreviations are used: def. = definition; eqn. = equation; ff. = and following; PM = polymerization.